Salinity Responses and Tolerance in Plants, Volume 1

Vinay Kumar • Shabir Hussain Wani
Penna Suprasanna • Lam-Son Phan Tran
Editors

Salinity Responses and Tolerance in Plants, Volume 1

Targeting Sensory, Transport and Signaling Mechanisms

Springer

Editors
Vinay Kumar
Department of Biotechnology
Modern College of Arts, Science
and Commerce, (Savitribai Phule
Pune University)
Pune, India

Department of Environmental Science
Savitribai Phule Pune University
Pune, India

Shabir Hussain Wani
Genetics and Plant Breeding
Mountain Research Centre for Field Crops
Khudwani, Anantnag
Sher-e-Kashmir University of Agricultural
Sciences and Technology of Kashmir
Jammu and Kashmir, India

Plant Soil and Microbial Sciences
Michigan State University
East Lansing, MI, USA

Penna Suprasanna
Nuclear Agriculture and Biotechnology
Bhabha Atomic Research Centre
Mumbai, Maharashtra, India

Lam-Son Phan Tran
Signaling Pathway Research Unit
RIKEN Center for Sustainable
Resource Science
Yokohama, Japan

ISBN 978-3-030-09294-8 ISBN 978-3-319-75671-4 (eBook)
https://doi.org/10.1007/978-3-319-75671-4

Softcover re-print of the Hardcover 1st edition 2018

Printed on acid-free paper

This Springer imprint is published by the registered company Springer International Publishing AG part of Springer Nature.
The registered company address is: Gewerbestrasse 11, 6330 Cham, Switzerland

Preface

Hyper soil salinity has emerged as a key abiotic stress factor, and poses serious threats to crop yields and quality of produce. Owing to the underlying complexity, conventional breeding programmes have met with limited success. The conventional genetic engineering approaches via transferring/overexpressing a single 'direct action gene' per event did not yield optimal results as expected. Nevertheless, the biotechnological advents in the last decade coupled with availability of genomic sequences of major crops and model plants have opened new vistas for understanding salinity responses and improving salinity tolerance in important glycophytic crops. Through this two-volume book series entitled 'Salinity Responses and Tolerance in Plants', we are presenting critical assessments of the potent avenues to be targeted for imparting salt-stress tolerance in major crops in the post-genomic era.

Specifically, this book series is an attempt to update the state of the art on intrinsic mechanisms underlying salinity responses and adaptive mechanisms in plants, as well as novel approaches to target them with assistance of sound biotechnological tools and platforms for developing salt-tolerant crops to feed the ever-increasing global population. Volume 1 deals with plant response to salinity stress and perspectives for improving crop salinity tolerance via targeting the sensory, ion transport and signalling mechanisms, whereas Volume 2 will be aimed at discussing the potencies of post-genomic-era tools like DNA helicases, RNAi, genomics intervention, genome editing and systems biology in addition to other approaches, which together may enable us to develop salt-tolerant crops in the era of climate change.

This book series is an excellent and comprehensive reference material for plant breeders, geneticists, biotechnologists, under/post-graduate students of agricultural biotechnology as well as for agricultural companies, providing novel and powerful options for understanding and targeting molecular mechanisms for producing salt-tolerant and high-yielding crops. The chapters are written by internationally reputed scientists, and all the chapters underwent review process for assuring scientific accuracy and high-quality presentation.

We express our sincere thanks and gratitude to our esteemed authors to come together on this important task and contribute their excellent work. We are grateful to Springer Nature for giving us the opportunity to complete this book project.

Pune, India	Vinay Kumar
Srinagar, India	Shabir Hussain Wani
Mumbai, India	Suprasanna Penna
Yokohama, Japan	Lam-Son Phan Tran

Contents

Contributors

Chedly Abdelly Laboratory of Extremophile Plants, Centre of Biotechnology of Borj-Cedria, Hammam-Lif, Tunisia

Monika Agacka-Mołdoch Department of Plant Breeding and Biotechnology, Institute of Soil Science and Plant Cultivation—State Research Institute, Puławy, Poland

Parinita K. Agarwal Plant Omics Division, CSIR-Central Salt and Marine Chemicals Research Institute (CSIR-CSMCRI), Council of Scientific & Industrial Research (CSIR), Bhavnagar, Gujarat, India

Pradeep Agarwal Plant Omics Division, CSIR-Central Salt and Marine Chemicals Research Institute (CSIR-CSMCRI), Council of Scientific & Industrial Research (CSIR), Bhavnagar, Gujarat, India

Mohammad Alnayef School of Land and Food, University of Tasmania, Hobart, TAS, Australia

Susana de Sousa Araújo Plant Cell Biotechnology Laboratory, Instituto de Tecnologia Química e Biológica António Xavier (ITQB NOVA), Oeiras, Portugal

Plant Biotechnology Laboratory, Department of Biology and Biotechnology 'L. Spallanzani', Università degli Studi di Pavia, Pavia, Italy

Abdallah Atia Department of Biology, College of Science, King Khalid University, Abha, Saudi Arabia

Research Unit, Nutrition and Nitrogen Metabolism and Stress Protein, Department of Biology, Faculty of Sciences of Tunis, Tunis, Tunisia

Laboratory of Extremophile Plants, Centre of Biotechnology of Borj-Cedria, Hammam-Lif, Tunisia

Zouhaier Barhoumi Department of Biology, College of Science, King Khalid University, Abha, Saudi Arabia

Laboratory of Extremophile Plants, Centre of Biotechnology of Borj-Cedria, Hammam-Lif, Tunisia

Melike Bor University of Ege, Department of Biology, İzmir, Turkey

Jayakumar Bose ARC Centre of Excellence in Plant Energy Biology, School of Agriculture, Food and Wine, Waite Research Institute, The University of Adelaide, Glen Osmond, SA, Australia

Marian Brestic Department of Plant Physiology, Slovak University of Agriculture, Nitra, Slovakia

David J. Burritt Department of Botany, University of Otago, Dunedin, New Zealand

Artemio Cerda Department of Geography, Universitat de Valencia, Valencia, Spain

Sara Francisco Costa Plant Cell Biotechnology Laboratory, Instituto de Tecnologia Química e Biológica António Xavier (ITQB NOVA), Oeiras, Portugal

Anna Czubacka Department of Plant Breeding and Biotechnology, Institute of Soil Science and Plant Cultivation—State Research Institute, Puławy, Poland

Sagar Satish Datir Department of Biotechnology, Savitribai Phule Pune University, Pune, India

Ahmed Debez Laboratory of Extremophile Plants, Centre of Biotechnology of Borj-Cedria, Hammam-Lif, Tunisia

Punita L. Devineni Department of Genetics, Osmania University, Hyderabad, India

Divya Gohil Plant Omics Division, CSIR-Central Salt and Marine Chemicals Research Institute (CSIR-CSMCRI), Council of Scientific & Industrial Research (CSIR), Bhavnagar, Gujarat, India

Houda Gouia Research Unit, Nutrition and Nitrogen Metabolism and Stress Protein, Department of Biology, Faculty of Sciences of Tunis, Tunis, Tunisia

Geetha Govind Department of Agricultural Chemistry, College of Bio-resources and Agriculture, National Taiwan University, Taipei, Taiwan

Department of Crop Physiology, College of Agriculture, Hassan, Karnataka, India

Chiraz Chaffei Haouari Research Unit, Nutrition and Nitrogen Metabolism and Stress Protein, Department of Biology, Faculty of Sciences of Tunis, Tunis, Tunisia

Vokkaliga T. Harshavardhan Department of Agricultural Chemistry, College of Bio-resources and Agriculture, National Taiwan University, Taipei, Taiwan

Xiaolan He Jiangsu Academy of Agricultural Sciences (JAAS), Nanjing, China

Safa Hkiri Research Unit, Nutrition and Nitrogen Metabolism and Stress Protein, Department of Biology, Faculty of Sciences of Tunis, Tunis, Tunisia

Tahsina Sharmin Hoque Department of Soil Science, Bangladesh Agricultural University, Mymensingh, Bangladesh

Chwan-Yang Hong Department of Agricultural Chemistry, College of Bio-resources and Agriculture, National Taiwan University, Taipei, Taiwan

Mohammad Anwar Hossain Department of Genetics and Plant Breeding, Bangladesh Agricultural University, Mymensingh, Bangladesh

Laboratory of Plant Nutrition and Fertilizers, Graduate School of Agricultural and Life Sciences, University of Tokyo, Tokyo, Japan

Gandra Jawahar Department of Genetics, Osmania University, Hyderabad, India

Rajesh Kalladan Institute of Plant and Microbial Biology, Academia Sinica, Taipei, Taiwan

Sunita Kataria School of Biochemistry, Devi Ahilya University, Indore, India

Tushar Khare Department of Biotechnology, Modern College of Arts, Science and Commerce (Savitribai Phule Pune University), Pune, India

Polavarapu B. Kavi Kishor Department of Genetics, Osmania University, Hyderabad, India

Kundan Kumar Department of Biological Sciences, Birla Institute of Technology & Science Pilani, Goa, India

Somanaboina Anil Kumar Department of Genetics, Osmania University, Hyderabad, India

Vinay Kumar Department of Biotechnology, Modern College of Arts, Science and Commerce (Savitribai Phule Pune University), Pune, India

Department of Environmental Science, Savitribai Phule Pune University, Pune, India

Davide Martins Genetics and Genomics of Plant Complex Traits Laboratory, Instituto de Tecnologia Química e Biológica António Xavier (ITQB NOVA), Oeiras, Portugal

Sonia Mbarki National Research Institute of Rural Engineering, Water and Forests (INRGREF), Ariana, Tunisia

Laboratory of plant Extremophile Plants, Center of Biotechnology of Borj-Cedria, Hammam-Lif, Tunisia

Filiz Özdemir University of Ege, Department of Biology, İzmir, Turkey

Maheshwari Parveda Department of Genetics, Osmania University, Hyderabad, India

Guddimalli Rajasheker Department of Genetics, Osmania University, Hyderabad, India

Anshu Rastogi Department of Meteorology, Poznan University of Life Sciences, Poznan, Poland

Ankush Ashok Saddhe Department of Biological Sciences, Birla Institute of Technology & Science Pilani, Goa, India

Sergey Shabala School of Land and Food, University of Tasmania, Hobart, TAS, Australia

Samrin Shaikh Department of Biotechnology, Modern College of Arts, Science and Commerce (Savitribai Phule Pune University), Pune, India

Abderrazak Smaoui Laboratory of Extremophile Plants, Centre of Biotechnology of Borj-Cedria, Hammam-Lif, Tunisia

Nese Sreenivasulu Grain Quality and Nutrition Center, Plant Breeding Division, International Rice Research Institute, Los Baños, Philippines

Ankanagari Srinivas Department of Genetics, Osmania University, Hyderabad, India

Amrita Srivastav Department of Biotechnology, Modern College of Arts, Science and Commerce (Savitribai Phule Pune University), Pune, India

Oksana Sytar Department of Plant Physiology, Slovak University of Agriculture, Nitra, Slovakia

Mohsin Tanveer School of Land and Food, University of Tasmania, Hobart, TAS, Australia

Satpal Turan National Research Centre on Plant Biotechnology, IARI PUSA, New Delhi, India

Sandeep Kumar Verma Department of Biotechnology, Innovate Mediscience India, Indore, India

Marek Zivcak Department of Plant Physiology, Slovak University of Agriculture, Nitra, Slovakia

Aziza Zoghlami Institut National de la Recherche Agronomique de Tunisie (INRAT), Ariana, Tunisia

About the Editors

Vinay Kumar is an Assistant Professor at the Post-Graduate Department of Biotechnology, Progressive Education Society's Modern College of Arts, Science and Commerce, Ganeshkhind, Pune, India, and a Visiting Faculty at the Department of Environmental Sciences, Savitribai Phule Pune University, Pune, India. He obtained his Ph.D. in Biotechnology from Savitribai Phule Pune University (Formerly University of Pune) in 2009. For his Ph.D., he worked on metabolic engineering of rice for improved salinity tolerance. He has published 30 peer-reviewed research/review articles and contributed ten book chapters in edited books published by Springer, CRC Press and Elsevier. He is a recipient of Government of India's Science and Engineering Board, Department of Science and Technology (SERB-DST) Young Scientist Award in 2011. His current research interests include elucidating molecular mechanisms underlying salinity stress responses and tolerance in plants. His research group is engaged in assessing the individual roles and relative importance of sodium and chloride ions under salinity stress in rice, and made significant contributions in elucidating the individual and additive effects of sodium and chloride ions.

Shabir Hussain Wani is an Assistant Professor cum Scientist at the Mountain Research Centre for Field Crops, Khudwani Sher-e-Kashmir University of Agricultural Sciences and Technology of Kashmir, Srinagar, Jammu and Kashmir, India. He has published more than 100 papers/chapters in peer-reviewed journals, and books of international and national repute. He is Review Editor of Frontiers in Plant Sciences, Switzerland. He is editor of SKUAST Journal of Research, and LS: An International journal of Life Sciences. He has also edited ten books on current topics in crop improvement published by CRC press, Taylor and Francis Group and Springer. His Ph.D research fetched the first prize in North zone at national level competition in India. He is the fellow of the Linnean Society of London and Society for Plant Research, India. He received various awards including Young Scientist Award (Agriculture) 2015, Young Scientist Award 2016 and Young Achiever Award 2016 by various prestigious scientific societies. He has also worked as visiting scientist in the Department of Plant Soil and Microbial Sciences, Michigan State University, USA, for the year 2016–2017 under the Raman Post-Doctoral Research Fellowship programme sponsored by University Grants Commission, Government of India, New Delhi. He is a member of the Crop Science Society of America.

Penna Suprasanna (Ph.D. Genetics, Osmania University, Hyderabad) is a Senior Scientist and Head of Plant Stress Physiology and Biotechnology Group in the Nuclear Agriculture and Biotechnology Division, Bhabha Atomic Research Centre, Mumbai, India. He made significant contributions to crop biotechnology research through radiation-induced mutagenesis, plant cell and tissue culture, genomics and abiotic stress tolerance. His research on radiation-induced mutagenesis and in vitro selection in sugarcane yielded several agronomically superior mutants for sugar, yield and stress tolerance. He has made intensive efforts to apply radiation mutagenesis techniques in vegetatively propagated plants through collaborative research projects. He is actively associated with several national and international bodies (IAEA, Vienna) in the areas of radiation mutagenesis, plant biotechnology and biosafety. He is the recipient of the "Award of Scientific and Technical Excellence" by the Department of Atomic Energy, Government of India. He is the Fellow, Maharashtra Academy of Sciences, Andhra Pradesh Academy of Sciences, Telangana Academy of Sciences; Fellow, Association of Biotechnology; and Faculty Professor, Homi Bhabha National Institute, DAE. Dr. Suprasanna has published more than 225 research papers/articles in national and international journals and books. His research is centred on molecular understanding of abiotic stress (salinity, drought and arsenic) tolerance and salt-stress adaptive mechanism in halophytic plants. The research group led by him on crop genomics has successfully identified novel microRNAs, early responsive genes besides validating the concept of redox regulation towards abiotic stress tolerance and plant productivity.

Lam-Son Phan Tran is Head of the Signaling Pathway Research Unit at RIKEN Center for Sustainable Resource Science, Japan. He obtained his M.Sc. in Biotechnology in 1994 and Ph.D. in Biological Sciences in 1997, from Szent Istvan University, Hungary. After doing his postdoctoral research at the National Food Research Institute (1999–2000) and the Nara Institute of Science and Technology of Japan (2001), in October 2001, he joined the Japan International Research Center for Agricultural Sciences to work on the functional analyses of transcription factors and osmosensors in *Arabidopsis* plants under environmental stresses. In August 2007, he moved to the University of Missouri, Columbia, USA, as a Senior Research Scientist to coordinate a research team working to discover soybean genes to be used for genetic engineering of drought-tolerant soybean plants. His current research interests are elucidation of the roles of phytohormones and their interactions in abiotic stress responses, as well as translational genomics of legume crops, with the aim to enhance crop productivity under adverse environmental conditions. He has published over 125 peer-reviewed papers with more than 90 research and 35 review articles, and authored 8 book chapters to various book editions published by Springer, Wiley-Blackwell and American Society of Agronomy, Crop Science Society of America and Soil Science Society of America. He has also edited 11 book volumes, including this one, for Springer and Elsevier.

Chapter 1
Salinity Stress Responses and Adaptive Mechanisms in Major Glycophytic Crops: The Story So Far

Sunita Kataria and Sandeep Kumar Verma

Abstract In many areas of the world, salinity is a major abiotic stress-limiting growth and productivity of plants due to increasing use of poor quality of water for irrigation and soil salinisation. Various physiological traits, metabolic pathways and molecular or gene networks are involved in plant adaptation or tolerance to salinity stress. This chapter deals with the adaptive mechanisms that plants can employ to cope with the challenge of salt stress and provide updated overview of salt-tolerant mechanisms in major glycophytic crops with a particular interest in rice (*Oryza sativa*), soybean (*Glycine max*), wheat (*Triticum aestivum*) and *Arabidopsis* plants. Salt stress usually inhibits seed germination, seedling growth and vigour, biomass accumulation, flowering and fruit set in major glycophytic crops. In addition, elevated Na + levels in agricultural lands are increasingly becoming a serious threat to the world agriculture. Plants suffer osmotic and ionic stress under high salinity due to the salts accumulated at the outside of roots and those accumulated at the inside of the plant cells, respectively. Salinity stress significantly reduces growth and productivity of glycophytes, which are the majority of agricultural products. Plants tolerant to NaCl implement a series of adaptations to acclimate to salinity, including morphological, physiological, biochemical and molecular changes regulating plant adaptation and tolerance to salinity stress. These changes affect plant growth and development at different levels of plant organisation, e.g. they may reduce photosynthetic carbon gain and leaf growth rate and increase in the root/canopy ratio and in the chlorophyll content in addition to changes in the leaf anatomy that ultimately lead to preventing leaf ion toxicity, thus maintaining the water status in order to limit water loss and protect the photosynthesis process. Finally, we also provide an

---

Sunita Kataria and Sandeep Kumar Verma are contributed equally to this work.

S. Kataria (✉)
School of Biochemistry, Devi Ahilya University, Indore, India

S. K. Verma (✉)
Department of Biotechnology, Innovate Mediscience India, Indore, India

V. Kumar et al. (eds.), *Salinity Responses and Tolerance in Plants, Volume 1*,
https://doi.org/10.1007/978-3-319-75671-4_1

updated discussion on salt-induced oxidative stress at the subcellular level and its effect on the antioxidant machinery in major glycophyte crops plants. In response to salinity stress, the productions of ROS, such as singlet oxygen, superoxide, hydroxyl radical and hydrogen peroxide, are enhanced, and overexpression of genes leading to increased amounts and activities of antioxidant enzymes like superoxide dismutase (SOD), catalase (CAT) and glutathione-S-transferase (GST)/glutathione peroxidase (GPX) increases the performance of plants under stress. The molecular mechanism of stress tolerance is complex and requires information at the miRNA/omics level to understand it effectively. During abiotic stress conditions, the advancement of "omics" is providing a detailed fingerprint of proteins, transcripts or all metabolites upregulated or downregulated in plant cells. However, the regulatory mechanisms of these protein-coding genes are largely unknown; in this regard, the microRNAs (miRNAs) may prove extremely important in deciphering these gene regulatory mechanisms and the stress responses. Some miRNAs are functionally conserved across plant species and are regulated by salt stress. In major crops through transgenic technologies, miRNAs represent themselves as potent targets to engineer abiotic stress tolerance, due to the critical roles in post-transcriptional regulation of gene expression in response to salinity and resultant growth attenuation.

Keywords Adaptive mechanisms · Antioxidant activity · Glycophytic plants · Stress responses · Salinity stress

Abbreviations

·OH	Hydroxyl radical
CAT	Catalase
Ci	Intercellular CO_2 concentration
ETR	Electron transport rate
GPX	Glutathione peroxidase
gs	Stomatal conductance
GST	Glutathione-S-transferase
H_2O_2	Hydrogen peroxide
ICDH	NADP-specific isocitrate dehydrogenase
MDA	Malondialdehyde
miRNA	microRNAs
N	Nitrogen
NRA	Nitrate reductase activity
O_2^-	Superoxide
ROS	Reactive oxygen species
SOD	Superoxide dismutase
SOS	Salt overly sensitive signal pathway
SPAD	Soil and plant analyser development

1.1 Introduction

Salinity is one of the most serious factors limiting the productivity of agricultural crops, with adverse effects on germination, plant vigour and crop yield (Munns and Tester 2008). A major challenge towards world agriculture involves production of 70% more food crop for an additional 2.3 billion people by 2050 worldwide (Srivastava et al. 2016). Salinity is a major stress limiting the increase in the demand for food crops. More than 20% of cultivated land worldwide (~about 45 hectares) is affected by salt stress, and the amount is increasing day by day. More than 40% of irrigated agricultural land worldwide has been predicted to be soon affected by salinity (Munns and Gilliham 2015). Thus crop productivity worldwide is hampered by salinity, which is one of the most brutal environmental stresses (Munns and Tester 2008). Crops with improved tolerance to salt stress will be required to ensure food security into the future.

Salinity stress is also considered as a hyperionic stress. One of the most detrimental effects of salinity stress is the accumulation of Na^+ and Cl^- ions in tissues of plants exposed to soils with high NaCl concentrations. Entry of both Na^+ and Cl^- into the cells causes severe ion imbalance and excess uptake might cause significant physiological disorder(s). High Na^+ concentration inhibits uptake of K^+ ions which is an essential element for growth and development that results in lower productivity and may even lead to death (James et al. 2011). Plants are unable to cope, tolerate and survive in saline conditions long enough to supply sufficient photosynthates to the reproductive organs and produce viable seeds, if the former process progresses faster than the latter. Based on this two-phase concept, the osmotic effect exerted by salts in the medium around the roots would cause the initial growth reduction in both salt-tolerant and salt-sensitive genotypes (i.e. osmotic phase). However, the salt-sensitive genotypes are much more affected by the ionic phase, because of their inability to prevent Na^+ build-up in transpiring leaves to toxic levels (Munns et al. 2006). Because of this development, crops have been classified into two categories: (i) salt-includers and (ii) salt-excluders. Salt-includers take up Na^+ and translocate it to the shoot, where it is sequestered and used as vacuolar osmoticum (tissue tolerance), whereas the salt-excluders adapt to saline stress by avoiding Na^+ uptake (Mian et al. 2011). The salt-sensitive genotypes can be differentiated from the salt-tolerant ones at ionic phase, and the effect of salinity on crops may also be a result of the combination of osmotic and ionic salt effect. The ionic phase has been associated with the reduction in the stomatal conductance, photosystem II efficiency, decrease in photosynthesis capacity, reduced biomass and poor yield in plants (Isla et al. 1998; Tester and Davenport 2003; Netondo et al. 2004; Tavakkoli et al. 2011). These changes affect plant growth and development at different levels of plant organisation (Munns 2002), e.g. they may reduce photosynthetic carbon gain and leaf growth rate (Munns 1993). Osmotic stress in the initial stage of salinity stress causes various physiological changes, such as interruption of membranes, nutrient imbalance, impairment of the ability to detoxify reactive oxygen species (ROS), differences in the antioxidant enzymes and decreased photosynthetic activity and decrease in stomatal aperture (Munns and Tester 2008; Rahnama et al. 2010). Unlike animals, higher plants, which are sessile, cannot escape from the surroundings but adapt themselves to the changing environments by forming a series of molecular responses to cope with these problems (Shao et al. 2007). Under saline conditions, plants have to activate different physiological and biochemical mechanisms in order to cope with the resulting stress. Such mechanisms include changes in morphology, anatomy, water relations, photosynthesis, the hormonal profile, toxic ion distribution and biochemical adaptation (such as the antioxidative metabolism response) (Hernández et al. 2001; Parida and Das 2005; Ashraf and Harris 2013; Acosta-Motos et al. 2015). One of the biochemical changes occurring in plants subjected to environmental stress conditions is the production of ROS such as superoxide radicle, H_2O_2, singlet oxygen and hydroxyl radicals (Cho and Park 2000; Mahajan and Tuteja 2005; Ahmad and Prasad 2011a, b). The ROS can damage essential membrane lipid proteins and nucleic acids (Inzé and Van Montagu 1995; Garratt et al. 2002). Several studies, however, indicate that levels of ROS in plant cells are normally protective by antioxidant activity. Association between saline environment and endogenous level of water-soluble antioxidant enzymes has been reported (Foyer 1993; Gueta-Dahan et al. 1997; Tsugane et al. 1999). Reports indicate that antioxidant could be used as a potential growth regulator to improve salinity stress resistance in several plant species (Shalata and Neumann 2001; Khan et al. 2006; Gunes et al. 2007). Salinity-induced ROS formation can lead to oxidative damages in various cellular components such as proteins, lipids and DNA, interrupting vital cellular functions of plants. Genetic variations in salt tolerance exist, and the degree of salt tolerance varies with plant species and varieties within a species. Among major crops, barley (*Hordeum vulgare*) shows a greater degree of salt tolerance than rice (*Oryza sativa*) and wheat (*Triticum aestivum*). This chapter deals with the adaptive mechanisms that glycophyte plants can implement to cope with the challenge of salt stress. Most of the salinity adaptive mechanisms in plants are accompanied by certain morphological, anatomical, physiological and biochemical changes (Fig. 1.1).

1.2 Glycophytes

Plants on the basis of adaptive evolution can be classified generally into two major types: the halophytes (that can withstand salinity) and the glycophytes (that cannot withstand salinity and eventually die). The majority of major crop species belong to this second category. Glycophytes or sweet plants are not salt tolerant; most of the cultivated crops by humans belong to this group, such as rice, wheat and maize. Most of the crop species are glycophytes and, generally, show limited growth and development due to salinity. Most of the grain crops and vegetables are glycophytes (salt-sensitive flora) and therefore are highly susceptible to soil salinity even when the soil ECe is $<4 dSm^{-1}$. Table 1.1 shows how most of our essential food crops are susceptible to salinity stress. Different threshold tolerance ECe and a different rate

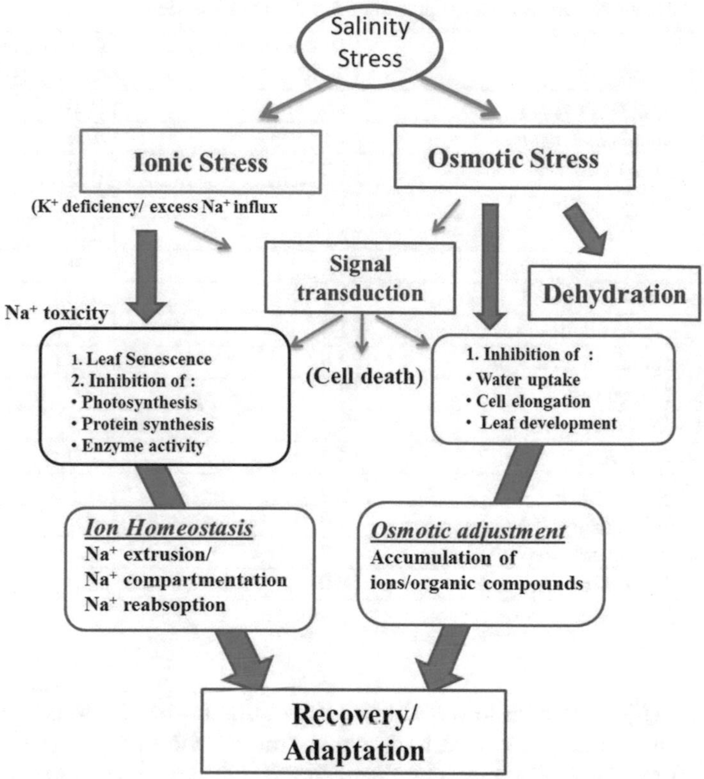

Fig. 1.1 Schematic summary of the stresses that plants suffer under high salinity growth condition and the corresponding responses that plants use in order to adapt the salt stress

of reduction in yield beyond threshold tolerance indicate variations in salt tolerance among those crops (Chinnusamy et al. 2005). However among glycophytes, there is considerable variation in salt tolerance; monocotyledonous crops, such as barley and wheat, are relatively tolerant, while most dicot crops, and also the model plant *Arabidopsis thaliana*, are at the sensitive end of the range (Munns and Tester 2008). With increasing amounts of arable land undergoing salinisation and increasing food demand from the growing human population, there is a need to ameliorate the harmful effect of salinity using various strategies (Szabolcs 1994). *Porteresia coarctata* or wild rice, an interesting example of the halophyte relative to the cultivated rice, is more efficient in the protection of its photosynthesis machinery against free radical produced by salinity stress when exposed to the salinity of 400 mM NaCl (Bose et al. 2013). Rice is the most salt sensitive among cereals (Munns and Tester 2008). In rice, it has been observed that the rate of Na^+ uptake into shoots mediated by the intrusive apoplastic ion transport is considerably high under salinity stress (Yeo et al. 1987; Yadav et al. 1996; Ochiai and Matoh 2002). The growth response of glycophytes to salinity (>40 mM NaCl) occurs in two phases: (i) a rapid response to increasing in external salt known as "osmotic

Table 1.1 Glycophytic crops are susceptible to soil salinity (Mass 1990)

Glycophytic crops	Threshold salinity (dSm^{-1})	Decrease in yield (Slope % per dSm^{-1})
Bean (*Phaseolus vulgaris* L.)	1.0	19.0
Eggplant (*Solanum melongena* L.)	1.1	6.9
Onion (*Allium cepa* L.)	1.2	16.0
Pepper (*Capsicum annuum* L.)	1.5	14.0
Corn (*Zea mays* L.)	1.7	12.0
Sugarcane (*Saccharum officinarum* L.)	1.7	5.9
Potato (*Solanum tuberosum* L.)	1.7	12.0
Cabbage (*Brassica oleracea* var. capitata L.)	1.8	9.7
Tomato (*Lycopersicon esculertum*)	2.5	9.9
Rice, paddy (*Oryza sativa* L.)	3.0	12.0
Peanut (*Arachis hypogaga* L.)	3.2	29.0
Soybean [*Glycine max* (L.) Merr.]	5.0	20.0
Wheat (*Triticum aestivum* L.)	6.0	7.1
Sugar beet (*Beta vulgaris* L.)	7.0	5.9
Cotton (*Gossypium hirsutum* L.)	7.7	5.2
Barley (*Hordeum vulgare* L.)	8.0	5.0

phase" and (ii) slower response with the accumulation of Na^+ ions in vacuoles referred to as "ionic phase". At both phases, the growth and yield of crops are significantly reduced (Munns and Tester 2008). The osmotic phase of growth reduction depends on the salt concentration outside the plant rather than the salt in the plant tissues, and growth inhibition is mostly due to a water deficit (drought stress) or osmotic stress, with little genotypic differences. However, the ionic phase of growth reduction takes time to develop (usually between 2 and 4 weeks) as results of an internal salt injury caused by excessive accumulation of toxic Na^+. At this phase, salinity would cause the plants to close its stomatal apertures and consequently reduced the photosynthetic rate due to the negative effect of toxic Na^+ that accumulated in the thylakoid membranes of the chloroplasts. This would increase ROS formation and oxidative stress that would result in leaf injury and loss of photosynthetic capacity of the plants. All these responses to salinity contribute to the deleterious effects on plants (Hernandez et al. 1993; Mittova et al. 2003).

1.3 Salinity Stress Responses in Major Glycophytic Crops

Soil salinity is a major factor that limits the yield of agricultural crops, jeopardising the capacity of agriculture to sustain the burgeoning human population increase (Flowers 2004; Parida and Das 2005; Munns and Tester 2008). At low salt concentrations, yields are slightly affected or not affected at all (Maggio et al. 2001). As the concentrations increase, the yields move towards zero, since most plants, glycophytes, including most crop plants, will not grow in high concentrations of salt and are severely inhibited or even killed by 100–200 mM NaCl. The reason is that they have evolved under conditions of low soil salinity and do not display salt tolerance (Munns and Termaat 1986). Measurements of ion contents in plants under salt stress revealed that halophytes accumulate salts, whereas glycophytes tend to exclude the salts (Zhu 2001a).

1.3.1 Arabidopsis

Although *Arabidopsis* is a typical glycophyte in being not particularly salt tolerant, various pieces of indirect evidence suggest that it might contain most, if not all, of the salt tolerance genes one might find in halophytes (Zhu 2000). It is hypothesised that halophytes generally use similar salt tolerance effectors and regulatory pathways that have been found in glycophytes but that subtle differences in regulation account for large variations in tolerance or sensitivity (Zhu 2000). To test this hypothesis directly, the salt tolerance mechanisms operating in halophytes must be discovered. Several halophytes have been used extensively in physiological and molecular biological investigations (Meyer et al. 1990; Niu et al. 1993). However, none of these plants is a suitable genetic model system. To be a genetic model system, a plant must have desirable life history traits (small size, short life cycle, the ability to self-pollinate and high seed number) and also certain genetic traits (small genome and easy transformation and mutagenesis). AtNHX1 is the Na^+/H^+ antiporter, localised to the tonoplast, predicted to be involved in the control of vacuolar osmotic potential in *Arabidopsis* (Apse et al. 1999). The first evidence showed that the overexpression of AtNHX1 in *Arabidopsis* plants promoted sustained growth and development in soil watered with up to 200 mM NaCl (Apse et al. 1999), although previously it has been reported that transgenic *Arabidopsis* do not show a significantly improved salt tolerance as compared to that of control plants (Yang et al. 2009). In *Arabidopsis*, ion homeostasis is mediated mainly by the salt overly sensitive (SOS) signal pathway. SOS proteins are a sensor for calcium signal that turns on the machinery for Na^+ export and K^+/Na^+ discrimination (Zhu 2001a, b). In particular, SOS1, encoding a plasma membrane Na^+/H^+ antiporter, plays a critical role in Na^+ extrusion and in controlling long-distance Na^+ transport from the root to shoot (Shi et al. 2000, 2002). This antiporter forms one component in a mechanism based on sensing of the salt stress that involves an increase of cytosolic [Ca^{2+}], protein interactions and reversible phosphorylation with SOS1 acting in concert with other two proteins known as SOS2 and SOS3 (Oh et al. 2010) (Fig. 1.2). Both the protein kinase SOS2 and its associated calcium-sensor subunit SOS3 are required for the post-translational activation of SOS1 Na^+/H^+ exchange activity in *Arabidopsis* (Qiu et al. 2002; Quintero et al. 2011) and in rice (Martínez-Atienza et al. 2007). The increased expression in tomato and rice of *Arabidopsis* arginine vasopressin 1 (AVP1), encoding a

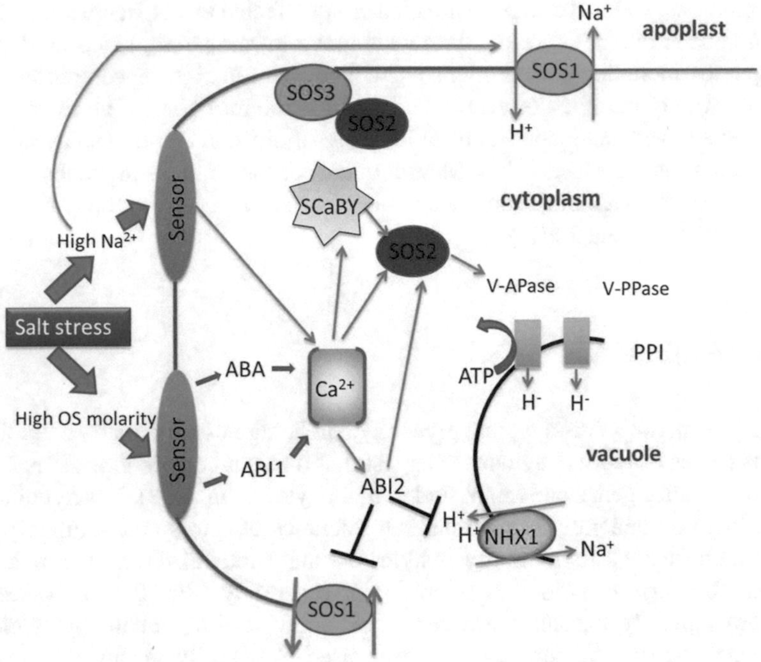

Fig. 1.2 Signalling pathways responsible for Na + extrusion in *Arabidopsis* under salt stress

vacuolar pyrophosphatase acting as a proton pump on the vacuolar membrane, enhanced sequestering of ions and sugars into the vacuole, reducing water potential and resulting in increased salt tolerance when compared to wild-type plants (Pasapula et al. 2011).

Excess Na^+ and high osmolarity are separately sensed by unknown sensors at the plasma membrane level, which then induce an increase in cytosolic [Ca^{2+}]. This increase is sensed by SOS3 which activates SOS2. The activated SOS3-SOS2 protein complex phosphorylates SOS1, the plasma membrane Na^+/H^+ antiporter, resulting in the efflux of Na^+ ions. SOS2 can regulate NHX1 antiporter activity and V-H^+-ATPase activity independently of SOS3, possibly by SOS3-like Ca^{2+}-binding proteins (SCaBP) that target it to the tonoplast. Salt stress can also induce the accumulation of ABA, which, by means of ABI1 and ABI2, can negatively regulate SOS2 or SOS1 and NHX1 (Silva and Gerós 2009) (Figs. 1.2 and 1.3).

1.3.2 Wheat

Wheat plants stressed at 100 to 175 mM NaCl showed a significant reduction in spikelets per spike, delayed spike emergence and reduced fertility, which results in poor grain yield. However, Na^+ and Cl^- concentrations in the shoot apex of these

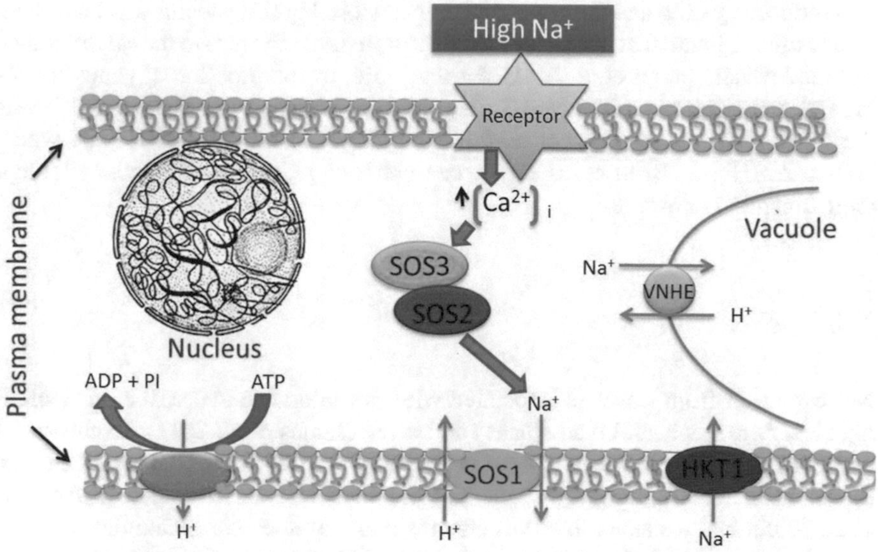

Fig. 1.3 The cellular machinery that contribute to salt tolerance in *Arabidopsis thaliana*, three *s*alt *o*verly *s*ensitive genes (*SOS1*, *SOS2* and *SOS3*) were found to function in a common pathway

wheat plants were below 50 and 30 mM, respectively, which is too low to limit metabolic reactions (Munns and Rawson 1999).

Durum wheat is a salt-sensitive species, and germination and seedling stages are the most critical phases of plant growth under salinity (Flagella et al. 2006). Its sensitivity to salt stress is higher than bread wheat, due to a poor ability to exclude Na^+ from the leaf blades and a lack of the K^+/Na^+ discrimination character displayed by bread wheat (Gorham et al. 1987; Läuchli et al. 2008). However, a novel source of Na^+ exclusion has been found in an unusual durum wheat genotype named Line 149. Genetic analysis has shown that Line 149 contains two major genes for Na^+ exclusion, named Nax1 and Nax2 (Munns et al. 2003). The proteins encoded by the Nax1 and Nax2 genes are shown to increase retrieval of Na^+ from the xylem in roots, thereby reducing shoot Na^+ accumulation. In particular, the Nax1 gene confers a reduced rate of transport of Na^+ from root to shoot and retention of Na^+ in the leaf sheath, thus giving a higher sheath-to-blade Na^+ concentration ratio. The second gene, Nax2, also confers a lower rate of transport of Na^+ from root to shoot and has a higher rate of K^+ transport, resulting in enhanced K^+ versus Na^+ discrimination in the leaf (James et al. 2006). The mechanism of Na^+ exclusion allows the plant to avoid or postpone the problem related to ion toxicity, but if Na^+ exclusion is not compensated for by the uptake of K^+, it determines a greater demand for organic solutes for osmotic adjustment. The synthesis of organic solutes jeopardises the energy balance of the plant. Thus, the plant must cope with ion toxicity on the one hand and turgor loss on the other (Munns and Tester 2008). By introgressing Nax genes from *Triticum monococcum* into hexaploid bread wheat (*Triticum aestivum*), the leaf blade Na^+ concentration was reduced by 60% and the proportion of Na^+ stored in leaf sheaths was increased. The results indicate that Nax genes have the potential to improve the salt tolerance of bread wheat (James et al. 2011). Furthermore, overexpression of genes encoding late embryogenesis abundant (LEA) proteins, which accumulate to high levels during seed development, such as the barley HVA1 (Xu et al. 1996) and wheat dehydrin DHN-5 (Brini et al. 2007), can enhance plant salt tolerance, although their function is obscure.

1.3.3 Rice

Na^+ exclusion from leaves is associated with salt tolerance in cereal crops including rice, durum wheat, bread wheat and barley (James et al. 2011). Exclusion of Na^+ from the leaves is due to low net Na^+ uptake by cells in the root cortex and the tight control of net loading of the xylem by parenchyma cells in the stele (Davenport et al. 2005). Na^+ exclusion by roots ensures that Na^+ does not accumulate to toxic concentrations within leaf blades. A failure in Na^+ exclusion manifests its toxic effect after days or weeks, depending on the species, and causes premature death of older leaves (Munns and Tester 2008). The effect of salinity on leaf ultrastructure can vary in relation to the tolerance of the plant to NaCl, as described in two rice (*Oryza sativa* L.) varieties (Flowers et al. 1985). The sensitive rice variety (Amber) showed differences in the integrity of the chloroplast compared with the tolerant variety (IR2153). Concretely, in response to high NaCl, most of the chloroplasts in the Amber variety showed signs of damage. In contrast, the chloroplasts in the tolerant variety IR2153 did not show any effect of salinity. This response correlated with a 30% decline in photosynthesis in Amber, whereas no significant change in photosynthesis was observed in IR2153. Many studies have focused on mapping QTLs for salt tolerance-related traits in rice because of its requirement for irrigation for maximum yield, its sensitivity to salinity and its relatively small genome (Flowers et al. 2000). Better results have been obtained at the seedling stage, while the expression and relationship of QTLs detected in different developmental stages are more difficult to study and fully understand at tillering and reproductive stages (Prasad et al. 2000; Alam et al. 2011). Transgenic tobacco plants overexpressing GhNHX1 from cotton and transgenic rice overexpressing the Na^+/H^+ antiporter gene clone from OsNHX1 exhibited higher salt tolerance (Fukuda et al. 2004; Wu et al. 2004). The increased expression in tomato and rice of *Arabidopsis* arginine vasopressin 1 (AVP1), encoding a vacuolar pyrophosphates acting as a proton pump on the vacuolar membrane, enhanced sequestering of ions and sugars into the vacuole, reducing water potential and resulting in increased salt tolerance when compared to wild-type plants (Shen et al. 2010). Feng et al. (2007) suggested that overexpression of sedoheptulose-1,7-bisphosphatase was an effective method for enhancing salt tolerance in rice.

1.3.4 Soybean

Soybean is classified as a moderately salt-tolerant crop, and the final yield of soybean is reduced when soil salinity exceeds 5 dS/m (Ashraf and Foolad 2007). Salt damage in soybean results from the accumulation of chloride in stems and leaves (Parker et al. 1983; Wang and Shannon 1999). Soybean is a strategic crop plant grown to obtain edible oil forage. High sensitivity to soil and water salinity is one of the biggest problems with soybean crop. Results have indicated that salinity affects growth and development of plants through osmotic and ionic stresses. Because of accumulated salts in soil under salt-stressed condition, plant wilts apparently, while soil salts such as Na and Cl disrupt normal growth and development of plant (Khajeh-Hosseini et al. 2003; Farhoudi et al. 2007). Chen and Yu (1995) reported salt stress led to decreased seedling growth of soybean cultivars. The reduction in plant growth under saline conditions may either be due to osmotic reduction in water availability which resulted in increasing stomatal conductance as reported by Günes et al. (1996) or to excessive ions, Na and Cl accumulation in the plant tissues (Yousif and Al-Saadawi 1997; Schuch and Kelly 2008). Salt damage in soybean results from the accumulation of chloride in stems and leaves (Parker et al. 1983; Wang and Shannon 1999). Salinity stress causes adverse effects on soybean growth and yield (Ghassemi-Golezani et al. 2009). Baghel et al. (2016) found that salt stress considerably suppressed shoot and root fresh and dry weights, leaf area as well as photosynthetic efficiency, nitrogen metabolism and soybean yield. Salinity affects plant growth due to changes in many physiological processes including photosynthesis (Kalaji and Pietkiewicz 1993; Kalaji and Guo 2008). A reduction in chlorophyll content, stomatal conductance and ribulose-1,5-bisphosphate carboxylase/oxygenase (RuBisCo) activity and an increase in the chlorophyll *a*/*b* ratio had been observed under salt stress (Kalaji and Nalborczyk 1991; Baghel et al. 2016).

Soybean assimilates significant amounts of nitrogen (N) in symbiosis with rhizobia through N_2 fixation. Successful establishment of legume-Rhizobium symbiosis is dependent on salinity (Rao et al. 2002). N metabolism is also affected by salinity, probably through the depression of the activity of enzymes involved in N metabolism (Mansour et al. 2002; Santos et al. 2002). The addition of genistein (a nod gene inducer) enhances soybean nodulation and growth under saline conditions (Miransari and Smith 2007, 2009). In chickpea, salt-tolerant genotypes have greater nodulation and symbiotic N_2 fixation capacity than sensitive genotypes (Rao et al. 2002). These studies indicate that salt-induced reduction in growth and yield in legume crops including soybean is partly due to depressed N_2 fixation activity (Van Hoorn et al. 2001). Yasuta and Kokubun (2014) concluded that the super-nodulating genotype En-b0–1 was more tolerant to salinity than its parental normal-nodulating cultivar, due to its higher capacity for nodulation and better ability to prevent excessive accumulation of Na and Cl in shoots while withholding these toxic elements in roots and nodules.

The analysis of variance for seed characteristics of the economic yield of soybean, i.e. number of seeds, the number of seeds per pod, 100 seed weight and seed yield per plant, indicated highly significant differences. The increasing salinity levels decrease all the seed characteristics in soybean (Baghel 2017). Similar results of reduction in yield under saline conditions have been reported in rapeseed (Rameeh et al. 2012; Valiollah 2013). Identification of the genes that improve the salt tolerance of crops is essential for the effective utilisation of saline soils for agriculture. Guan et al. (2014) used fine mapping in a soybean (*Glycine max*) population derived from the commercial cultivars Tiefeng 8 and 85–140 to identify GmSALT3 (salt tolerance-associated gene on chromosome 3), a dominant gene associated with limiting the accumulation of sodium ions (Na^+) in shoots, and found a substantial enhancement in salt tolerance in soybean. The relationship between increased salt accumulations in leaves with decreased soybean yield has been reported (Bustingorri and Lavado 2013). Liu et al. (2016) results indicated that GmSALT3 mediated regulation of both Na^+ and Cl^- accumulation in soybean, and it contributes to improved soybean yield through maintaining a higher seed weight under salt stress.

Pi et al. (2016) confirmed the functional relevance of chalcone synthase, chalcone isomerase and cytochrome P450 monooxygenase genes using soybean composites and *Arabidopsis thaliana* mutants and found that their salt tolerance was positively regulated by chalcone synthase but was negatively regulated by chalcone isomerase and cytochrome P450 monooxygenase. They have proposed a novel salt tolerance pathway involving chalcone metabolism, mostly mediated by phosphorylated MYB transcription factors. Finally, maybe these flavonoids appropriately reduced the ROS or play roles in other functions for enhancing the soybean's tolerance to salinity.

1.4 Salinity Stress and Adaptive Mechanisms of Glycophytic Crops

The salinity limits crop plant production which is contrary to the increased demand for food all over the world. Therefore, the study of salinity tolerance in plants is of special importance. High concentrations of salts cause ion imbalance and hyperosmotic stress in plants. As a consequence of these primary effects, secondary stresses such as oxidative damage often occurred. Many plants developed mechanisms either to exclude salt from their cells or to tolerate its presence within the cells. During the onset and development of salt stress within a plant, all the major processes such as photosynthesis, nitrogen fixation, protein synthesis and energy and lipid metabolism were affected severely (Parida and Das 2005; Baghel et al. 2016).

Under high salinity, plants have to cope with two major stresses, osmotic stress and ionic stress (Figs. 1.1, 1.2 and 1.3). The osmotic stress immediately comes over plants in accordance with a rise in salt levels outside the roots, which caused the inhibitions of water uptake, cell expansion and lateral bud development (Fig. 1.1)

(Munns and Tester 2008). The ionic stress phase develops later when toxic ions such as Na^+ accumulate in excess in plants particularly in leaves over the threshold, which leads to an increase in leaf mortality with chlorosis and necrosis and a decrease in the activity of essential cellular metabolisms including photosynthesis (Fig. 1.1) (Glenn et al. 1999; Munns 2002; Munns et al. 2006; Baghel et al. 2016), which in turn significantly reduces growth and productivity of crops. Consequently, to cope with salinity stress, the effective strategies for glycophytes are to keep cytosolic Na^+ levels low at the cellular level and to keep shoot Na^+ concentrations low at the whole plant level. In addition to these factors, acquisition and maintenance of K^+ were found to have a considerable impact on plant salt tolerance (Wu et al. 1996; Zhu et al. 1998). For salt tolerance of glycophyte plants, the maintenance of high cytosolic K^+/Na^+ ratios has been strongly suggested to be crucial especially in shoots (Blumwald 2000; Horie et al. 2005; Ren et al. 2005; Yamaguchi and Blumwald 2005; Hauser and Horie 2010).

Different adaptive mechanisms may be involved in gradual acclimation to salinity in contrast to adjustment to a sudden shock. The sensitivity to salinity of a given species may change during ontogeny. Salinity tolerances may increase or decrease depending on the plant species and/or environmental factors. Crop plants differ greatly in their salinity tolerance as reflected in their different growth responses at different growth stages (Foolad 1999; Takeda and Matsuoka 2008; Baghel et al. 2016; Kataria et al. 2017). For some species, salt sensitivity may be greatest at germination, whereas for other species, sensitivity may increase during reproduction (Marschner and Rimmington 1988; Howat 2000).

1.4.1 Seed Germination

Seed germination and seedling establishment are the most vulnerable stages affected by salinity, which later affects growth and development of crop plants and results into low agricultural production (Garg and Gupta 1997). If the effect of salinity stress can be alleviated at the initial crop establishment stage, some economic yield can be realised from salinity-affected areas. The harmful effects of salinity on plant growth are associated with (1) low osmotic potential of soil solution, (2) nutritional imbalance, (3) specific ion effect and (4) a combination of all these factors (Welbaum et al. 1990; Ashraf and Wu 1994; Huang and Redmann 1995). Physiological studies to distinguish between these effects are limited (Bliss et al. 1986), but evidence suggests that low water potential of the germination medium is a major limiting factor (Bradford 1995). Rapid seed germination or early vigour of crop can be an adaptive strategy for alleviation of salinity stress under field conditions as the seedling establishment can be ensured (Sharma and Sen 1989).

Salinity affects the germination process manyfold. Slow rate of imbibition may lead to a series of metabolic changes including upregulation or downregulation of enzyme activities (Ashraf 2002; Gomes-Filho et al. 2008), disturbance in the mobility of inorganic nutrients to developing tissues (Ashraf and Wahid 2000;

Khan and Weber 2006), perturbance in nitrogen metabolism (Yupsanis et al. 1994), imbalances in the levels of plant growth regulators (Khan and Rizvi 1994), reduction in hydrolysis and utilisation of food reserves (Ahmed and Bano 1992; Othman et al. 2006) and accumulation of compatible osmotica such as soluble sugars, free proline and soluble protein (Poljakoff-Mayber et al. 1994; Zidan and Elewa 1995; Ashraf et al. 2003). Kaveh et al. (2011) found a significant negative correlation between salinity and the rate and percentage of germination which resulted in delayed germination and reduced germination percentage in *Solanum lycopersicum*. Bybordi (2010) reported that the germination percentage in *Brassica napus* significantly reduced at 150 and 200 mM NaCl. In a recent study, germination percentage also decreased on increasing concentration of salinity levels (0 mM NaCl to 100 mM NaCl) in soybean and maize (Kataria et al. 2017). However, it has been established that salinity adversely affects the process of germination in various plants like *Oryza sativa* (Xu et al. 2011), *Triticum aestivum* (Akbarimoghaddam et al. 2011), *Zea mays* (Carpýcý et al. 2009; Khodarahmpour et al. 2012), *Brassica* spp. (Akram and Jamil 2007) and soybean (Kataria et al. 2017). In wheat, soybean and maize salinity has been shown to negatively affect the rate of starch mobilisation by causing a decrease in α amylase activity (Lin and Kao 1995; Kataria et al. 2017). All these processes may lead to a poor or complete lack of seed germination under saline conditions (Poljakoff-Mayber et al. 1994; Kataria et al. 2017).

1.4.2 Morphological Adaptations

The harmful effect of salinity can vary depending on climatic conditions, light intensity, plant species or soil conditions (Tang et al. 2015). The root system is the first which perceived the salt stress and impairs plant growth both in the short term, by inducing osmotic stress caused by reduced water availability and, in the long term, by salt-induced ion toxicity due to the nutrient imbalance in the cytosol (Munns and Gilliham 2015). Salinity caused a significant reduction in root length and shoot length. The previous results demonstrated that the response of root length to salt stress was more severe than shoot length of wheat (Akbarimoghaddam et al. 2011), sorghum (El Naim et al. 2012), soybean and maize (Kataria et al. 2017). Demir and Arif (2003) observed that root growth was more adversely affected as compared to shoot growth by salinity. Jeannette et al. (Bayuelo-Jimenez et al. 2002) reported that total fresh weight of root and shoot of radish was reduced with increased salt stress. These results were also in agreement of Jafarzadeh and Aliasgharzad (2007) who observed that root length was significantly affected by salt composition, cultivars and salinity levels. The reduction in root and shoot development may be due to toxic effects of the higher level of NaCl concentration as well as unbalanced nutrient uptake by the seedlings.

From the results of the previous studies, which looked at the effect of salt stress on growth, one can notice a connection between the decrease in plant length and the increase in the concentration of sodium chloride (Beltagi et al. 2006; Mustard and Renault 2006; Gama et al. 2007; Jamil et al. 2007; Houimli et al. 2008; Liu et al. 2009; Memon et al. 2010). Numerous studies showed the affection of leaf area negatively by using different concentrations of NaCl (Netondo et al. 2004; Mathur et al. 2006; Chen et al. 2007; Zhao et al. 2007; Yilmaz and Kina 2008; Liu et al. 2009). The harmful influence of salinity on leaf number also increases with the increase in concentration, according to the studies done by Raul et al. (2003), Jamil et al. (2005), Gama et al. (2007) and Ha et al. (2008). The fresh and dry weights of the shoot system are affected, either negatively or positively, by changes in salinity concentration, type of salt present or type of plant species (Jamil et al. 2005; Niazi et al. 2005; Saqib et al. 2006; Turan et al. 2007; Saffan 2008; Liu et al. 2009; Memon et al. 2010; Baghel et al. 2016).

Munns (1992) found that salt accumulation in the old leaves accelerates their death and thus decreases the supply of carbohydrates and/or growth hormones to the meristematic regions, thereby inhibiting growth. The fact that plant growth is limited by a reduction in the photosynthesis rate and by an excessive uptake of salts affects the production of specific metabolites that directly inhibit growth (Azza et al. 2007). The same negative effects have also been reported by Shereen et al. (2001) and Murat et al. (2008). Abd El-Samad et al. (2005) who reported that due to decrease in field water capacity, inhibition occurs in plant dry weight. This inhibition was associated with a decline in nitrate reductase (NR) and nitrogenase (NA) activities in both shoots and roots of wheat plants. Researchers showed that environmental stresses may hasten the seed filling rate and decrease grain filling duration (Abd EL-samad et al. 2005). This can influence the final yield of all grain crops such as soybean, rice, maize, etc. Brevedan and Egli (2003) found that seed filling period is under genetic control and it is sensitive to salt stress.

1.4.2.1 Roots and Aerial Part Morphology

The anatomy of the root system (length, root diameter, etc.) determines root performance, enabling plants to acquire water and nutrients and thereby increase the replacement rate of plant water loss (Passioura 1988). Since roots serve as an interface between plants and the soil, so optimum root systems can support shoot growth and improve plant yields (Vamerali et al. 2003). A proliferated root system would, therefore, appear to be better for plants, for it allows them to penetrate deeper layers of soil to acquire water and nutrients (Franco et al. 2011). The permeability of roots to water also determines the other root characteristics, such as the number and diameter of xylem vessels, width of the root cortex, number of root hairs and the suberin deposition in both the root exodermis and endodermis (Steudle 2000; Ranathunge et al. 2010). Furthermore, environmental factors in the soil (changes in temperature, lack of O_2, mechanical impedance, salinity) can also produce marked impacts on root anatomy. The cell walls of root cells of salinised plants are often unevenly thickened and convoluted (Shannon et al. 1994). In roots of woody trees, salts often promote the suberisation of the hypodermis and endodermis and resultant in the formation of a well-developed Casparian strip closer to the root apex, different to that found in non-salinised roots (Walker et al. 1984).

Furthermore, the morphology of some plants shows their sensitivity to salinity. These morphological features limit the distribution of crops to those areas where irrigation water is of good quality. Salinity reduces plant growth through osmotic and toxic effects, and high sodium uptake ratio values cause sodicity, which increases soil resistance, reduces root growth and reduces water movement through the root with a decrease in hydraulic conductivity (Rengasamy and Olsson 1993).

Roots seem to be more resistant to soil salinity than aerial biomass which could be linked to its low chloride levels. Bustingorri and Lavado (2011) found that leaf senescence and premature leaf abscission were clear symptoms of the effect of salinity on aerial biomass. These processes had been assumed to be part of a plant strategy to adapt to high soil salinity (Tardieu and Davies 1993; Larcher 2003; Maggio et al. 2007). Different nutritional and physiological responses were observed, such as ionic regulation, appropriate photosynthetic performance or maintenance of sink strength.

Another typical response to salt stress described in different papers is a reduction in total leaf area. Indeed, decreased leaf growth is the earliest response of glycophytes exposed to salt stress (Munns and Termaat 1986). The observed reduction in the canopy area may be considered as an avoidance mechanism, which minimises water loss by transpiration when the stomata are closed (Ruiz-Sánchez et al. 2000). This effect can favour the retention of toxic ions in roots, limiting the accumulation of these ions in the aerial part of the plant (Colmer et al. 2006). Under saline conditions, cell wall properties change and leaf turgor and photosynthesis rates (PN) decrease, leading to a reduction in total leaf area (Franco et al. 1997; Rodrıguez et al. 2005). Furthermore, high salt concentrations normally reduced the stem growth also. Decreases in leaf and stem provoke a reduction in all aerial part sizes and in the plant height.

The increase in root-to-shoot ratio or decrease in shoot-to-root ratio is a common response to salt stress, related to factors associated with water stress (osmotic effect) rather than a salt-specific effect (Hsiao and Xu 2000). A greater root proportion under salt stress can favour the retention of toxic ions in this organ, controlling their translocation to the aerial parts. This response can constitute a typical mechanism of plant resistance/survival under saline conditions (Cassaniti et al. 2009, 2012).

High salt concentrations in the irrigation water result in reduced plant growth (Munns and Tester 2008), limiting leaf expansion (Cramer 2002) and changing the relationship between the aerial and root parts (Tattini et al. 1995). Under high salinity, different plant species have shown a higher dry root mass than shoot dry mass, resulting in an increased root-to-shoot ratio which is thought to improve the source/sink ratio for water and nutrients under such conditions (Zekri and Parsons 1989).

Morphometric analysis performed at the ultrastructural level in maize roots demonstrated that the radial width of the Casparian band on a cross-section of a root, a morphological parameter that should be related to the effectiveness of the Casparian band as an apoplastic barrier, increased under salinity stress, suggesting that the function of the band is enhanced under salinity stress (Karahara et al. 2004). A novel protein family mediating Casparian band formation in the endodermis of *Arabidopsis* root (Roppolo et al. 2011) has shed a new light on the study of the Casparian band development, which might also be relevant to salt sensitivity of plants.

1.4.2.2 Leaf Anatomy and Ultrastructure Changes in Leaves Under Salinity

Salinity causes increases in epidermal thickness, mesophyll thickness, palisade cell length, palisade diameter and spongy cell diameter in leaves of bean, cotton and *Atriplex* (Longstreth and Nobel 1979). In contrast, Parida et al. (2004) found that both epidermal and mesophyll thickness and intercellular spaces decrease significantly in NaCl-treated leaves of the mangrove *Bruguiera parviflora*. Intercellular spaces were also reduced in leaves under salinity (Delfine et al. 1998). Mitsuya et al. (2000) found that salt stress causes (1) vacuolation development and partial swelling of endoplasmic reticulum, (2) decrease in mitochondrial cristae and swelling of mitochondria, (3) vesiculation and fragmentation of tonoplast and (4) degradation of cytoplasm by the mixture of cytoplasmic and vacuolar matrices in leaves of sweet potato. In leaves of potato, salt stress causes rounding of cells, smaller intercellular spaces and a reduction in chloroplast number (Bruns and Hecht-Buchholz 1990). Romero-Aranda et al. (2001) found that salinity causes reduction of plant leaf area and stomatal density in tomato plants.

Hernández et al. (1995) studied the effect of NaCl on leaf chloroplast ultrastructure in two different pea cultivars with different levels of sensitivity to NaCl, one salt sensitive (cv. Challis) and the other relatively salt tolerant (cv. Granada), up to 70 mM NaCl and found that NaCl produced a disorganised thylakoid structure in both pea cultivars. These authors described two different chloroplast populations that showed different densities in Percoll gradients. The authors noticed that the higher density band contained higher starch contents than the low-density band by electron microscopy. Salinity affected the two chloroplast bands differently, and even the high-density band disappeared in the salt-treated sensitive plants (Hernandez et al. 1995). In the NaCl-tolerant plants, salinity reduced the diameter of the chloroplasts from the low-density band. In both cases, an increase in the number and size of plastoglobuli was observed by the effect of NaCl, and the change was more evident in the tolerant plants (Hernandez et al. 1995). Plastoglobuli are omnipresent in chloroplasts and chromoplasts (Austin et al. 2006). Hernández et al. (1995) found that NaCl had different effects on the starch content in salt-sensitive and salt-tolerant pea plants. Whereas the percentage of starch decreased in both chloroplast populations in tolerant plants, no change occurred in sensitive plants. These data indicated that tolerant plants could use starch for different physiological processes to cope with the salt stress challenge (Munns and Gilliham 2015).

1.4.3 Physiological Adaptations of Plants to Salinity

1.4.3.1 Photosynthesis and Chlorophyll Fluorescence

Plant growth is the result of integrated and regulated physiological processes. Physiological processes are affected by a number of environmental factors, and they determine the response of plants to stress. Limitation of plant growth by environmental factors cannot be assigned to a single physiological process. The photosynthesis is a dominant physiological process. Plant growth as biomass production is a measure of net photosynthesis, and so environmental stresses affecting growth also affect photosynthesis. Salt stress affects photosynthesis both in the short and long term. In the short term, salinity can affect photosynthesis by stomatal limitations, leading to a decrease in carbon assimilation (Parida and Das 2005). This effect can produce rapid growth cessation, even after just a few hours of salt exposure (Hernández and Almansa 2002). In the long term, salt stress can also affect the photosynthetic process due to salt accumulation in young leaves (Munns and Tester 2008) and decreases in chlorophyll and carotenoid concentrations in glycophyte plants (Parida et al. 2002; Stepien and Johnson 2009; Duarte et al. 2013; Baghel et al. 2016). The decrease in chlorophyll content (Jamil et al. 2007) and activity of photosystem can be attributed to a reduction in photosynthesis under salinity (Ganieva et al. 1998; Baghel et al. 2016). Baghel et al. (2016) reported that photosynthetic activity decreases when plants are grown under saline conditions leading to reduced growth and productivity of the soybean crop.

The photosynthesis rate can drop due to stomatal closure and/or other non-stomatal limitations, like the disturbance of the photosynthetic electron chain and/or the inhibition of the Calvin cycle enzymes, such as RuBisCo, phosphoenolpyruvate carboxylase (PECP), ribulose-5-phosphate kinase, glyceraldehyde-3-phosphate dehydrogenase or fructose-1,6-bisphosphatase (Parida and Das 2005; Meloni et al. 2008). Under salinity conditions, it is known that salt-tolerant species show increased or unchanged chlorophyll content, whereas chlorophyll levels decrease in salt-sensitive species, suggesting that this parameter can be considered a biochemical marker of salt tolerance in plants (Stepien and Johnson 2009; Ashraf and Harris 2013).

In plant physiology, analysis of chlorophyll fluorescence is an easy and popular technique used studies that can provide interesting information about the state of the PSII (Maxwell and Johnson 2000). Under saline conditions, a general decrease in photochemical quenching parameters (Fv/Fm, Y(II), qP) and in the electron transport rate (ETR) takes place, but increases in non-photochemical quenching parameters [qN, NPQ, Y(NPQ)] have also been reported (Maxwell and Johnson 2000; Moradi and Ismail 2007; Lee et al. 2013; Shu et al. 2013; Ikbal et al. 2014; Acosta-Motos et al. 2015). The response to NaCl stress is correlated with decreases in PS II efficiency and increases in non-photochemical quenching parameters as a mechanism to safely dissipate excess energy. Salt stress was found to reduce ETR in a salt-sensitive rice cultivar, whereas only a slight reduction in ETR occurred in a salt-tolerant cultivar. However, the effect of NaCl stress on non-photochemical quenching parameters was somewhat different, since qN increased more substantially in the salt-tolerant cultivar than in the salt-sensitive cultivar (Moradi and Ismail 2007). Besides reducing photochemical quenching parameters and ETR, salt stress also reduces non-photochemical quenching parameters in salt-sensitive plants (Lee et al. 2013; Shu et al. 2013). Previous work on durum wheat and soybean showed that salinity stress caused a large decrease in stomatal conductance (gs) (James et al. 2002; Baghel et al. 2016).

Interestingly, James et al. (2002) found that the efficiency of PS II in the tolerant wheat accession was unaffected, while in the sensitive genotype, there was a decline in the quantum yield of PS II photochemistry, coinciding with leaf ageing, higher Na^+ and Cl^- concentrations in the leaf and chlorophyll degradation. Following stomatal closure, the internal reduction of CO_2 decreases the activity of several enzymes including RuBisCo (Chaves et al. 2009), thus limiting carboxylation and reducing the net photosynthetic rate under salt stress (Baghel et al. 2016). To estimate the effects of salinity on photosynthesis, the intercellular CO_2 concentration (Ci) is another parameter that has been used in several studies (Seemann and Critchley 1985; Redondo-Gómez et al. 2007; Stepien and Johnson 2009).

Under salinity, a salt-tolerant species *Eutrema salsugineum* was shown better maintained CO_2 assimilation rate (as a function of Ci) as compared with a sensitive species, *Arabidopsis* (Stepien and Johnson 2009). However, it is difficult to differentiate cause-effect relationships between photosynthesis (source) and growth reduction (sink); also, the effects of salinity on photosynthesis can be caused by alterations in the photosynthetic metabolism or else by secondary effects caused by oxidative stress (Chaves et al. 2009). However, under salinity stress, leaf expansion, associated with changes in leaf anatomy (smaller and thicker leaves), results in higher chloroplast density per unit leaf area, which can lead to a reduction in photosynthesis as measured on a unit chlorophyll basis (Munns and Tester 2008). Noninvasive methods that capture photosynthetic responses include measurements by the infrared gas analyser (IRGA) and pulse-amplitude-modulated (PAM) chlorophyll fluorometers. In addition, the use of soil and plant analyser development (SPAD) metres to determine the chlorophyll content can also provide an estimate of leaf damage under stress. The chlorophyll content can be estimated using the SPAD index, which is the ratio between leaf thickness (as determined by the transmission of light in the IR range) and leaf greenness (as determined by the transmission of light in the red light range). The SPAD index has been shown to decrease under salinity compared to control conditions. The extent of this decrease has been shown to vary between barley accessions, suggesting a genetic control of this effect of salinity on the SPAD index (Adem et al. 2014).

1.4.3.2 Nitrogen Fixation

Under salt stress, nitrate reductase activity (NRA) of leaves decreases in many plants (Baki et al. 2000; Flores et al. 2000). The primary cause of a reduction of NRA in the leaves is a specific effect associated with the presence of Cl^- salts in the external medium. This effect of Cl^- seems to be due to a reduction in NO_3^- uptake and consequently a lower NO_3^- concentration in the leaves, although a direct effect of Cl^- on the activity of the enzyme cannot be discarded (Deane-Drummond and Glass 1982; Flores et al. 2000). The nitrate content of leaves decreases, but it increases in roots under NaCl stress, and NRA of leaves also decreases under salinity in *Zea mays* (Baki et al. 2000). Salinity inhibits nitrogen fixation by reducing nodulation and nitrogenase activity in chickpea (*Cicer arietinum* L.) (Soussi et al. 1999).

Considerable inhibition of nodulation and N_2 fixation has also been reported by other workers (Bekki et al. 1987). Severe salt stress reduces the leghaemoglobin content and nitrogenase activity in soybean root nodules (Comba et al. 1997). The exposure of nodulated roots of legumes such as soybean, common bean and alfalfa to NaCl results in a rapid decrease in plant growth associated with a short-term inhibition of both nodule growth and nitrogenase activity (Serraj et al. 1998). Khan et al. (1998) have reported that salinity inhibits nitrate content and uptake and NRA in leaves of maize plants. Nitrogenase activity measured with regard to acetylene reduction (C_2H_2) decreases in common bean by short-term exposure to salinity (Serraj et al. 2001). Decrease in NRA activity and in total nitrogen and nitrate uptake has been reported in leaves of *Bruguiera parviflora* (Parida and Das 2004).

NADP-specific isocitrate dehydrogenase (ICDH) is a key cytosolic enzyme that links carbon and nitrogen metabolism by supplying carbon skeletons for primary nitrogen assimilation in plants. The characterisation of the transcript Mc-ICDH1 encoding an NADP-dependent isocitrate dehydrogenase (NADP-ICDH; EC 1.1.1.42) has been reported from the facultative halophyte *Mesembryanthemum crystallinum* L., focusing on salt-dependent regulation of the enzyme (Popova et al. 1995). By immune cytological analyses, NADP-ICDH proteins are localised to most cell types with the strongest expression in epidermal cells and in the vascular tissue. In leaves of salt-adapted plants, signal intensities increase in mesophyll cells. In contrast to Mc-ICDH1, the activity and transcript abundance of ferredoxin-independent glutamate synthase (EC 1.4.7.1), which is the key enzyme of N assimilation and biosynthesis of amino acids, decrease in leaves in response to salt stress (Popova et al. 2002).

1.4.4 Biochemical Adaptations

1.4.4.1 Antioxidative Metabolism

Salt stress is complex and imposes a water deficit because of osmotic effects on a wide variety of metabolic activities (Greenway and Munns 1980; Cheeseman 1988). This water deficit leads to the formation of ROS such as superoxide (O_2^-), hydrogen peroxide (H_2O_2), hydroxyl radical (·OH) (Halliwell and Gutteridge 1985) and singlet oxygen ($^1O_2$) (Elstner 1987). These cytotoxic activated oxygen species can seriously disrupt normal metabolism through oxidative damage to lipids (Fridovich 1986a; Wise and Naylor 1987) and to protein and nucleic acids (Fridovich 1986b; Imlay and Linn 1988; Birben et al. 2012; Krishnamurthy and Rathinasabapathi 2013). Since internal O_2 concentrations are high during photosynthesis (Steiger et al. 1977), chloroplasts are especially prone to generate activated oxygen species (Asada 1987).

Once (O_2^-) produced is rapidly dismutated either enzymatically or non-enzymatically, to yield H_2O_2 and O_2. In addition, H_2O_2 may interact in the presence of certain metal ions or metal chelates to produce highly reactive (·OH) (Imlay and Linn 1988). Plants possess a number of antioxidants that protect against the potentially cytotoxic species of activated oxygen. The metallo-enzyme superoxide dismutase (SOD; EC 1.15.1.1) converts O_2^- to H_2O_2. The breakdown of H_2O_2 is catalysed by catalase and a variety of peroxidases (Chang et al. 1984). Although catalase is apparently absent in the chloroplast, H_2O_2 can be detoxified in a reaction catalysed by an ascorbate-specific peroxidase often present in high levels in this organelle (Chen and Asada 1989) through the ascorbate-glutathione cycle (Asada 1992; Halliwell and Gutteridge 2015). Both ascorbate and glutathione have been reported in millimolar concentrations within the chloroplast (Halliwell 1982). Ascorbate can also be oxidised by direct reaction with O_2^- or by serving as a reductant of the α-chromoxyl radical of oxidised α-tocopherol (Foyer et al. 1991). α-Tocopherols are affluent in the thylakoid membranes, which disrupts lipid peroxidation reactions not only by reacting with O_2^- but also by scavenging hydroxyl, peroxyl and alkoxyl radicals (Halliwell 1987).

It is well known that at the subcellular level, salinity induces oxidative stress in plants (Acosta-Motos et al. 2017). Different works have reported that salt stress induces an accumulation of (O_2^-) and (H_2O_2) in different cell compartments, including chloroplasts, mitochondria and apoplastic space, which correlates with increases in some oxidative stress parameters, such as lipid peroxidation and protein oxidation (Acosta-Motos et al. 2017). In spite of the inhibition of catalase (an H_2O_2-scavenger enzyme) and an increase in the H_2O_2-producing enzyme glycollate oxidase in pea leaf peroxisomes, the concentration of H_2O_2 was statistically lower in the presence of NaCl than in control plants (Kangasjärvi et al. 2012).

Previous reports have been shown that the response of the antioxidative defences of the salt-tolerant species is somewhat different to that of salt-sensitive species. For instance, some authors have described that in the salt tolerance response, a coordinated upregulation of the antioxidative machinery as one of the mechanisms is involved (Acosta-Motos et al. 2017). The significant increases have been observed in the antioxidant defences in chloroplasts and mitochondria in salt-tolerant pea cultivars (cv. Granada and cv. Puget) (Acosta-Motos et al. 2017). Salt-induced oxidative stress was alleviated by a selective upregulation of a set of antioxidant enzymes at the subcellular level in the wild salt-tolerant-related tomato species *Lycopersicon pennellii*. Salt-treated (100 mM NaCl) *L. pennellii* plants showed increases in SOD and the ASC-GSH cycle components in the chloroplasts, mitochondria and peroxisomes, whereas a drop occurred in most of these antioxidant defences in the cultivated tomato (*Solanum lycopersicum*), which is sensitive to 100 mM NaCl, (Acosta-Motos et al. 2017). Especially in roots, it has been also described that NADPH regeneration systems contribute to maintaining the redox state and this seems to be essential in the biochemical mechanisms of plant defence against oxidative stress situations (Meneguzzo et al. 1998).

The salt tolerance capability of soybean host is also important for nodulation; Comba et al. (1997) indicated a correlation between N fixation and secondary oxidative stress induced by salt treatment. These authors found a dramatic decrease in leghaemoglobin content, and nitrogen fixation activity was accompanied with an increase in malondialdehyde (MDA) and a decrease in glutathione reductase (GR), ascorbate peroxidase (APX), catalase (CAT) and superoxide dismutase (SOD) activities when the salt-sensitive soybean cultivar "411" was grown on medium containing 150 mM NaCl. No significant changes were observed in nitrogen fixation activity or leghaemoglobin and MDA contents, but it exhibited an increase in GR, APX, CAT and SOD activities when the salt-tolerant cultivar "377" was subjected to the same treatment (Comba et al. 1997). Overexpression of genes leads to increased amounts and activities of mitochondrial Mn-SOD, Fe-SOD, chloroplastic Cu/Zn-SOD, bacterial catalase and glutathione-S-transferase (GST)/glutathione peroxidase (GPX); it can increase the performance of plants under stress (Roxas et al. 2000). In leaves of the rice plant, salt stress preferentially enhances the content of H_2O_2 and the activities of SOD, APX and GPX, whereas it decreases catalase activity (Lee et al. 2001). On the other hand, salt stress has a little effect on the activity levels of glutathione reductase (Lee et al. 2001).

In general, salt stress affects ascorbate and glutathione levels much more in sensitive plants than in tolerant plants. In a NaCl-sensitive pea genotype, NaCl stress produced a 50% reduction in the total ascorbate pool after 15 days of stress, which may be due to the loss of both the oxidised (DHA) and the reduced (ASC) forms. In a NaCl-tolerant genotype, salt stress also reduced ASC and DHA levels but less so than the sensitive plants (Hernández and Almansa 2002). NaCl stress (100 mM) produced a decline in the ASC content in leaves of the salt-sensitive rice cultivar IR-29, but no effect was recorded in the leaves or roots of a NaCl-tolerant rice cultivar (Pokkali) (Lee et al. 2013).

1.4.4.2 Proline and/or Glycine Betaine

In glycophytes, proline and/or glycine betaine concentrations are much lower, but if partitioned exclusively to the cytoplasm, they could generate a significant osmotic pressure and then balance the vacuolar osmotic potential. In durum wheat seedlings, proline can contribute for more than 39% of the osmotic adjustment in the cytoplasmic compartments of old leaves, while the contribution of glycine betaine can account for up to 16% of the osmotic balance in younger tissues, independently of nitrogen nutrition, unlike proline (Carillo et al. 2008). Among the best known compatible solutes, proline and glycine betaine have been reported to increase greatly under salt and drought stresses (Munns 2002) and constitute the major metabolites found in durum wheat under salt stress, as in other Poaceae (Carillo et al. 2005; Ashraf and Foolad 2007).

1.4.5 Molecular Mechanism of Salt Tolerance

Physiologic or metabolic adaptations to salt stress at the cellular level are the main responses amenable to molecular analysis and have led to the identification of a large number of genes induced by salt (Ingram and Bartels 1996; Bray 1997; Shinwari et al. 1998). The expression of many plant genes is regulated by the salinity at the transcriptional and post-translational levels. The molecular mechanism of plant salt tolerance is very complex (Zhu 2001b; Munns and Tester 2008). Several studies have been conducted in many plant models to investigate this mechanism. Published analyses showed the expression profiles of many genes and proteins involved in salt stress responses in several plants like *Arabidopsis thaliana*, rice, wheat, soybean, tobacco and other plant species (Long et al. 2016). The root is the primary tissue involved in salinity perception and is one of the first to be injured following exposure to several types of stresses. The productivity of the entire plant is often limited due to the sensitivity of the root to stress (Steppuhn et al. 2010). Thus, a comprehensive understanding of root molecular responses to salt stress is necessary for researchers to be able to increase crop tolerance to salt stress.

Salt tolerance is a multigenic trait, and a number of genes categorised into different functional groups are responsible for encoding salt stress proteins:

(i) Genes for photosynthetic enzymes
(ii) Genes for synthesis of compatible solutes
(iii) Genes for vacuolar-sequestering enzymes
(iv) Genes for radical-scavenging enzymes

Most of the genes in the functional groups have been identified as salt inducible under stress conditions. Other genes have been detected by a salt-hypersensitivity assay in *Arabidopsis*, which led to the identification of mutants in potassium uptake as being critical in salt sensitivity (Wu et al. 1996). However, other physiological systems may be equally limiting under stress conditions, and mutants in these physiological pathways could lead to increased salt toxicity and would affect survival in a negative manner. Transcript regulation in response to high salinity has been investigated for salt-tolerant rice (variety Pokkali) with microarrays including 1728 cDNAs from libraries of salt-stressed roots (Kawasaki et al. 2001). NaCl at 150 mM reduces photosynthesis to one-tenth of the prestress value within minutes. Hybridisations of RNA to microarray slides were probed for changes in transcripts from 15 min to 1 week after salt shock. Beginning 15 min after the shock, Pokkali shows upregulation of transcripts. Approximately 10% of the transcripts in Pokkali are significantly up- or downregulated within 1 hour of salt stress. The initial differences between control and stressed plants continue for hours but become less pronounced as the plants adapted over time.

The "omics" concept including molecular marker technology with the development of high-throughput profiling techniques has made it possible to analyse the metabolites/solutes/QTLs responsible for salinity tolerance. The identification of these components is fundamental for helping the selection efficiency in breeding programmes and mapping the major genes controlling salt tolerance in order to genetically manipulate using the real candidate genes rather than nonspecific abiotic-responsive genes (Sharma et al. 2016). The first evidence showed that in soil watered with up to 200 mM NaCl, the overexpression of AtNHX1 in *Arabidopsis* plants promoted sustained growth and development (Apse et al. 1999). Plant salt tolerance increases by the overexpression of regulatory genes in signalling pathways, such as transcription factors (DREB/CBF) and protein kinases (MAPK, CDPK) (Chen et al. 2009). The overexpression of the vacuolar Na+/H+ antiporter has shown to improve salinity tolerance in several plants (Silva and Gerós 2009). Xue et al. (2009) identified QTL controlling Na^+ contents and Na^+/K^+ ratio in mature barley plants grown on salt-treated soil. Similarly, the molecular markers *wmc170* (2A) and *cfd080* (6A) are expected to facilitate breeding for salinity tolerance in bread wheat, the latter being associated with seedling vigour (Genc et al. 2010). The results indicate that Nax genes have the potential to improve the salt tolerance of bread wheat (James et al. 2011). Bread wheat (*Triticum aestivum* L.) has a major salt tolerance locus, Kna1, (*TaHKT1;5-D*, a candidate gene underlying the *Kna1* locus), responsible for the maintenance of a high cytosolic K+/Na + ratio in the leaves of salt-stressed plants. Recent finding shows there is potential to increase the salinity tolerance of bread wheat by manipulation of HKT1;5 genes (Byrt et al. 2014). Babgohari et al. (2013) suggested that the D genome is more effective regarding Na + exclusion in wheat and wild wheat relatives. Furthermore, in silico promoter analysis showed that TaHKT1;5 genes harbour jasmonic acid response elements (Babgohari et al. 2013). It is possible that other levels of NaCl and ABA treatments cause a change in the GSTF1 gene in wheat (Niazi et al. 2014). Furthermore, overexpression of genes encoding late embryogenesis abundant (LEA) proteins, which accumulate to high levels during seed development, such as the barley HVA1 (Xu et al. 1996) and wheat dehydrin DHN-5 (Brini et al. 2007), can enhance plant salt tolerance, although their function is difficult to understand.

For improved stress tolerance on the genetic level, the information regarding RNA-mediated stress regulatory networks offers an innovative path. In maize, miR396, miR167, miR164 and miR156 were found to be downregulated, while upregulation in the expression of miR474, miR395, miR168 and miR162 families was observed after salt shock in roots (Ding et al. 2009). Recent reports of Zhang and Wang (Zhang et al. 2011) have revealed that manipulation of microRNA (miRNA)-arbitrated gene regulations can help to engineer plants for enhanced abiotic stress tolerance. miRNAs may prove potent targets for plant improvement, with superior tolerance to abiotic stresses due to their central role in complex gene regulatory network (Zhang and Wang 2015). Recent findings have established that in response to numerous abiotic stresses, plants assign miRNAs as critical post-transcriptional regulators of gene expression in a sequence-specific manner (Shriram et al. 2016). The miRNAs are small (20–24 nt)-sized, non-coding, single-stranded riboregulator RNAs abundant in higher organisms. Several miRNAs have been recognised as abiotic stress regulated in important crops under soil salinity (Gao et al. 2011), nutrient deficiency (Liang et al. 2010), UV-B radiation (Casadevall et al. 2013), heat (Goswami et al. 2014) and metal stress (Gupta et al. 2014). Liu et al. (2008) evidenced a 117 miRNAs expression profiles under three abiotic stress conditions in *Arabidopsis* and found that 14 of them were differentially regulated by one or more stress conditions. Amongst them 10 were NaCl-regulated, 4 were drought-regulated, and 10 were cold (4 °C)-regulated. The transgenic miRNA 402 lines of *Arabidopsis* showed better seed germination, and seedling growth under salt stress which shows miR402 regulates salt stress tolerance in a positive manner (Kim et al. 2010). Interestingly, in *Arabidopsis* overexpression of *osa-miR393* resulted into enhanced salt tolerance, suggesting their regulatory role in salinity tolerance (Gao et al. 2011). In radish (*Raphanus sativus*), Sun et al. (2015) identified 49 known and 22 novel salt stress-responsive miRNAs, and interestingly the target prediction analysis revealed the implication of the target genes in signalling, regulating ion homeostasis besides modulating the decreased plant growth under salt stress. A stress-related NAC1 from *Phyllostachys edulis* was characterised. Its ectopic expression in *Arabidopsis* indicated that PeSNAC1 together with ped-miR164b participated in tolerance to salinity and drought stresses through regulation of root development (Wang et al. 2016). In a recent study, miRNAs including sly-miR156e-5p, sly-miRn23b, sly-miRn50a, etc. were involved in the salt stress response of *Solanum pimpinellifolium* (Zhao et al. 2017). Taken together, these results demonstrate that miRNAs act as critical regulators of gene expression for maintaining normal growth and development in adaptation to abiotic stresses.

1.5 Conclusions and Future Perspectives

Salt stress causes huge losses of agriculture productivity worldwide. Therefore, plant biologists aimed at overcoming severe environmental stresses that needs to be quickly and fully implemented. Together with conventional plant physiology, genetics and biochemical approaches to studying plant responses to abiotic stresses have begun to bear fruit recently. In agricultural crops, relevant information on biochemical indicators at the cellular level may serve as selection criteria for tolerance of salts. Though there were many transgenic plants with high-stress tolerance generated, plant abiotic stress tolerance is a complex trait that involves multiple physiological and biochemical mechanisms and numerous genes. Transgenic plants with commercial value should at the same time retain relatively high productivity and other traits important for agriculture. Moreover, genetic modification should be combined with marker-assisted breeding programmes with stress-related genes and QTLs, and ultimately, the different strategies should be integrated, and genes representing distinctive approaches should be combined to substantially increase plant stress tolerance.

Salinity profoundly affects various aspects of plant cell structure and metabolism. Salinity with its osmotic and ionic aspects induces cellular adjustment associated with a profound reorganisation of cell structure and metabolism. Proteins play a major role in salt stress acclimation and plant cellular adjustment since proteins are involved in a wide array of cellular processes associated with salt acclimation including signalling, regulation of gene expression and protein metabolism, energy metabolism, redox metabolism, osmolyte metabolism, defence-related proteins, mechanical stress-related proteins, phytohormone, lipid and secondary metabolism. Comparative proteomic analysis has enabled the researchers to identify differentially abundant proteins in genetically related plant materials revealing differential stress tolerance. Besides changes in protein relative abundance, changes in protein post-translational modification (PTM) pattern, as well as protein activity, have been observed under salinity.

The differences between the sensitive (glycophytes) and the tolerant (halophytes) plant at protein level represent only a piece of a complex adaptation to salinity found in salt-tolerant plants. It is becoming evident that the differences in salinity response at the protein level are conferred by differences at genome and transcriptome level including a higher gene copy number and an altered structure of promoter sequences in salt-tolerant plants with respect to salt-sensitive ones. The differences at genome level are reflected at transcript and protein levels resulting in the differential transcript and protein expression levels between susceptible and tolerant plants. Corresponding to differences at genome level (changes in gene copy number and promoter structure and organisation), comparative transcriptomic and proteomic studies have revealed that differences in transcript and protein levels regarding several stress-induced genes between salt-tolerant and salt-sensitive plants can be detected even under control (non-stressed) conditions similarly to situation reported in other stresses (e.g. COR/LEA proteins under cold). The changes at protein level reveal also a crucial impact on plant cell structure and metabolism. Regarding the functional aspects of the changes in proteome composition, not only alterations in protein relative abundance but also changes in protein post-translational modification (PTM) and protein-protein interactions deserve to be studied. Taken together, the results from comparative proteomic studies represent an important step to understanding plant salt response processes as a whole. To develop the salt-tolerant cultivars which will better able to cope with the increasing soil salinity constraints, the use of both genetic manipulation and traditional breeding approaches will be required to unravel the mechanisms involved in salinity tolerance. The developments in the area of plant molecular biology, particularly the complete sequencing of model plant genomes and the availability of microarray analysis tools which offer advantages and solutions to the complex interesting questions of salt resistance and tolerance, will certainly pave a way towards improvement of crop plants for better sustainability in the changing environmental conditions.

Acknowledgement The financial support for this work was received from Innovate Mediscience to Dr. Verma S. and DST Women Scientists-A Scheme (SR/WOSA/ LS-17/2017(C)) to Dr. Kataria S. is thankfully acknowledged.

References

Abd EL-Samad HM, Komy HM, Shaddad MAK, Hetta AM (2005) Effect of molubdenum on nitrogenase and nitrate reductase activities of wheat inoculated with *Azospirillium brasilense* growth under drought stress. Gen Appl Plant Physiol 31(1–2):43–54

Acosta-Motos J-R, Diaz-Vivancos P, Álvarez S, Fernández-García N, Sanchez-Blanco MJ, Hernández JA (2015) Physiological and biochemical mechanisms of the ornamental *Eugenia myrtifolia* L. plants for coping with NaCl stress and recovery. Planta 242(4):829–846

Acosta-Motos JR, Ortuño MF, Bernal-Vicente A, Diaz-Vivancos P, Sanchez-Blanco MJ, Hernandez JA (2017) Plant responses to salt stress: adaptive mechanisms. Agronomy 7(1):18

Adem GD, Roy SJ, Zhou M, Bowman JP, Shabala S (2014) Evaluating contribution of ionic, osmotic and oxidative stress components towards salinity tolerance in barley. BMC Plant Biol 14(1):113

Ahmad P, Prasad MNV (2011a) Abiotic stress responses in plants: metabolism, productivity and sustainability. Springer, New York/Dordrecht/Heidelberg/London

Ahmad P, Prasad MNV (2011b) Environmental adaptations and stress tolerance of plants in the era of climate change. Springer, New York/Dordrecht/Heidelberg/London

Ahmed J, Bano M (1992) The effect of sodium-chloride on the physiology of cotyledons and mobilization of reserved food in cicer-arietinum. Pak J, Bot 24(1):40–48

Akbarimoghaddam H, Galavi M, Ghanbari A, Panjehkeh N (2011) Salinity effects on seed germination and seedling growth of bread wheat cultivars. Trakia J Sci 9(1):43–50

Akram NA, Jamil A (2007) Appraisal of physiological and biochemical selection criteria for evaluation of salt tolerance in canola (*Brassica napus* L.) Pak J Bot 39(5):1593–1608

Alam R, Sazzadur Rahman M, Seraj ZI, Thomson MJ, Ismail AM, Tumimbang-Raiz E, Gregorio GB (2011) Investigation of seedling-stage salinity tolerance QTLs using backcross lines derived from *Oryza sativa* L. Pokkali. Plant Breed 130(4):430–437

Apse MP, Aharon GS, Snedden WA, Blumwald E (1999) Salt tolerance conferred by overexpression of a vacuolar Na+/H+ antiport in Arabidopsis. Science 285(5431):1256–1258

Asada K (1987) Production and scavenging of active oxygen in photosynthesis. Photoinhibition

Asada K (1992) Ascorbate peroxidase–a hydrogen peroxide-scavenging enzyme in plants. Physiol Plant 85(2):235–241

Ashraf M (2002) Exploitation of genetic variation for improvement of salt tolerance in spring wheat. In: Prospects for saline agriculture. Springer, New York/Dordrecht/Heidelberg/London, pp 113–121

Ashraf M, Foolad M (2007) Roles of glycine betaine and proline in improving plant abiotic stress resistance. Environ Exp Bot 59(2):206–216

Ashraf M, Harris P (2013) Photosynthesis under stressful environments: an overview. Photosynthetica 51(2):163–190

Ashraf M, Wahid S (2000) Time-course changes in organic metabolites and mineral nutrients in germinating maize seeds under salt (NaCl) stress. Seed Sci Technol 28(3):641–656

Ashraf M, Wu L (1994) Breeding for salinity tolerance in plants. Crit Rev Plant Sci 13(1):17–42

Ashraf M, Zafar R, Ashraf MY (2003) Time-course changes in the inorganic and organic components of germinating sunflower achenes under salt (NaCl) stress. Flora-Morphol Distribution Funct Ecol Plants 198(1):26–36

Austin JR, Frost E, Vidi P-A, Kessler F, Staehelin LA (2006) Plastoglobules are lipoprotein subcompartments of the chloroplast that are permanently coupled to thylakoid membranes and contain biosynthetic enzymes. Plant Cell 18(7):1693–1703

Azza A, Fatma E, Favahat M (2007) Responses of ornamental plants woody trees to salinity world. J Agric Sci 3:386–395
Babgohari MZ, Niazi A, Moghadam AA, Deihimi T, Ebrahimie E (2013) Genome-wide analysis of key salinity-tolerance transporter (HKT1; 5) in wheat and wild wheat relatives (A and D genomes). In Vitro Cell Dev Biol Plant 49(2):97–106
Baghel L (2017) Magnetopriming of soybean seeds – advantages, inheritance and stress tolerance. Ph.D Thesis, School of Life Sciences, DAVV, Indore, India
Baghel L, Kataria S, Guruprasad KN (2016) Static magnetic field treatment of seeds improves carbon and nitrogen metabolism under salinity stress in soybean. Bioelectromagnetics 37(7):455–470
Baki G, Siefritz F, Man HM, Weiner H, Kaldenhoff R, Kaiser W (2000) Nitrate reductase in Zea mays L. under salinity. Plant Cell Environ 23(5):515–521
Bayuelo-Jimenez JS, Craig R, Lynch JP (2002) Salinity tolerance of species during germination and early seedling growth. Crop Sci 42(5):1584–1594
Bekki A, Trinchant JC, Rigaud J (1987) Nitrogen fixation (C_2H_2 reduction) by Medicago nodules and bacteroids under sodium chloride stress. Physiol Plant 71(1):61–67
Beltagi M, Ismail MA, Mohamed FH (2006) Induced salt tolerance in common bean (*Phaseolus vulgaris* L.) by gamma irradiation. Pak J Biol Sci 9(6):1143–1148
Birben E, Sahiner UM, Sackesen C, Erzurum S, Kalayci O (2012) Oxidative stress and antioxidant defense. World Allergy Organ J 5(1):9
Bliss R, Platt-Aloia K, Thomson W (1986) Osmotic sensitivity in relation to salt sensitivity in germinating barley seeds. Plant Cell Environ 9(9):721–725
Blumwald E (2000) Sodium transport and salt tolerance in plants. Curr Opin Cell Biol 12(4):431–434
Bose J, Rodrigo-Moreno A, Shabala S (2013) ROS homeostasis in halophytes in the context of salinity stress tolerance. J Exp Bot:ert430
Bradford KJ (1995) Water relations in seed germination. Seed Dev Germination 1(13):351–396
Bray EA (1997) Plant responses to water deficit. Trends Plant Sci 2(2):48–54
Brevedan R, Egli D (2003) Short periods of water stress during seed filling, leaf senescence, and yield of soybean. Crop Sci 43(6):2083–2088
Brini F, Hanin M, Lumbreras V, Amara I, Khoudi H, Hassairi A, Pages M, Masmoudi K (2007) Overexpression of wheat dehydrin DHN-5 enhances tolerance to salt and osmotic stress in *Arabidopsis thaliana*. Plant Cell Rep 26(11):2017–2026
Bruns S, Hecht-Buchholz C (1990) Light and electron microscope studies on the leaves of several potato cultivars after application of salt at various development stages. Potato Res 33(1):33–41
Bustingorri C, Lavado RS (2011) Soybean growth under stable versus peak salinity. Sci Agric 68(1):102–108
Bustingorri C, Lavado R (2013) Soybean response and ion accumulation under sprinkler irrigation with sodium-rich saline water. J Plant Nutr 36(11):1743–1753
Bybordi A (2010) The influence of salt stress on seed germination, growth and yield of canola cultivars. Notulae Botanicae Horti Agrobotanici Cluj-Napoca 38(1):128
Byrt CS, Xu B, Krishnan M, Lightfoot DJ, Athman A, Jacobs AK, Watson-Haigh NS, Plett D, Munns R, Tester M (2014) The Na+ transporter, TaHKT1; 5-D, limits shoot Na+ accumulation in bread wheat. Plant J 80(3):516–526
Carillo P, Mastrolonardo G, Nacca F, Fuggi A (2005) Nitrate reductase in durum wheat seedlings as affected by nitrate nutrition and salinity. Funct Plant Biol 32(3):209–219
Carillo P, Mastrolonardo G, Nacca F, Parisi D, Verlotta A, Fuggi A (2008) Nitrogen metabolism in durum wheat under salinity: accumulation of proline and glycine betaine. Funct Plant Biol 35(5):412–426
Carpýcý E, Celýk N, Bayram G (2009) Effects of salt stress on germination of some maize (Zea mays L.) cultivars. Afr J Biotechnol 8(19)
Casadevall R, Rodriguez RE, Debernardi JM, Palatnik JF, Casati P (2013) Repression of growth regulating factors by the microRNA396 inhibits cell proliferation by UV-B radiation in Arabidopsis leaves. Plant Cell 25(9):3570–3583

Cassaniti C, Leonardi C, Flowers TJ (2009) The effects of sodium chloride on ornamental shrubs. Sci Hortic 122(4):586–593

Cassaniti C, Romano D, Flowers TJ (2012) The response of ornamental plants to saline irrigation water. Intech Open Access Publisher

Chang H, Siegel B, Siegel S (1984) Salinity-induced changes in isoperoxidases in taro Colocasia esculenta. Phytochemistry 23(2):233–235

Chaves M, Flexas J, Pinheiro C (2009) Photosynthesis under drought and salt stress: regulation mechanisms from whole plant to cell. Ann Bot 103(4):551–560

Cheeseman JM (1988) Mechanisms of salinity tolerance in plants. Plant Physiol 87(3):547–550

Chen G-X, Asada K (1989) Ascorbate peroxidase in tea leaves: occurrence of two isozymes and the differences in their enzymatic and molecular properties. Plant Cell Physiol 30(7):987–998

Chen D, Yu R (1995) Studies on relative salt tolerance of crops II. Salt tolerance of some main crop species. Acta Pedol Sin 33(2):121–128

Chen C, Tao C, Peng H, Ding Y (2007) Genetic analysis of salt stress responses in asparagus bean (*Vigna unguiculata* (L.) ssp. *sesquipedalis* Verdc.) J Hered 98(7):655–665

Chen L, Ren F, Zhong H, Jiang W, Li X (2009) Identification and expression analysis of genes in response to high-salinity and drought stresses in Brassica napus. Acta Biochim Biophys Sin 42(2):154–164

Chinnusamy V, Jagendorf A, Zhu J-K (2005) Understanding and improving salt tolerance in plants. Crop Sci 45(2):437–448

Cho U-H, Park J-O (2000) Mercury-induced oxidative stress in tomato seedlings. Plant Sci 156(1):1–9

Colmer T, Munns R, Flowers T (2006) Improving salt tolerance of wheat and barley: future prospects. Aust J Exp Agric 45(11):1425–1443

Comba M, Benavides M, Gallego S, Tomaro M (1997) Relationship between nitrogen fixation and oxidative stress induction in nodules of salt-treated soybean plants. Phyton

Cramer GR (2002) Sodium-calcium interactions under salinity stress. In: Salinity: environment-plants-molecules. Springer, Dordrecht, pp 205–227

Davenport R, James RA, Zakrisson-Plogander A, Tester M, Munns R (2005) Control of sodium transport in durum wheat. Plant Physiol 137(3):807–818

Deane-Drummond CE, Glass AD (1982) Studies of nitrate influx into barley roots by the use of 36ClO3− as a tracer for nitrate. 1. Interactions with chloride and other ions. Can J Bot 60(10):2147–2153

Delfine S, Alvino A, Zacchini M, Loreto F (1998) Consequences of salt stress on conductance to CO2 diffusion, Rubisco characteristics and anatomy of spinach leaves. Funct Plant Biol 25(3):395–402

Demir M, Arif I (2003) Effects of different soil salinity levels on germination and seedling growth of safflower (*Carthamus tinctorius l*). Turkish J Agric 27:221–227

Ding D, Zhang L, Wang H, Liu Z, Zhang Z, Zheng Y (2009) Differential expression of miRNAs in response to salt stress in maize roots. Ann Bot 103(1):29–38

Duarte B, Santos D, Marques J, Caçador I (2013) Ecophysiological adaptations of two halophytes to salt stress: photosynthesis, PS II photochemistry and anti-oxidant feedback–implications for resilience in climate change. Plant Physiol Biochem 67:178–188

El Naim AM, Mohammed KE, Ibrahim EA, Suleiman NN (2012) Impact of salinity on seed germination and early seedling growth of three sorghum (*Sorghum biolor* L. Moench) cultivars. Sci Technol 2(2):16–20

Elstner EF (1987) Metabolism of activated oxygen species. The biochemistry of plants: a comprehensive treatise (USA)

Farhoudi R, Sharifzadeh F, Poustini K, Makkizadeh M, Kochak Por M (2007) The effects of NaCl priming on salt tolerance in canola (*Brassica napus*) seedlings grown under saline conditions. Seed Sci Technol 35(3):754–759

Feng L, Han Y, Liu G, An B, Yang J, Yang G, Li Y, Zhu Y (2007) Overexpression of sedoheptulose-1, 7-bisphosphatase enhances photosynthesis and growth under salt stress in transgenic rice plants. Funct Plant Biol 34(9):822–834

Flagella Z, Trono D, Pompa M, Di Fonzo N, Pastore D (2006) Seawater stress applied at germination affects mitochondrial function in durum wheat (*Triticum durum*) early seedlings. Funct Plant Biol 33(4):357–366

Flores P, Botella M, Martinez V, Cerdá A (2000) Ionic and osmotic effects on nitrate reductase activity in tomato seedlings. J Plant Physiol 156(4):552–557

Flowers T (2004) Improving crop salt tolerance. J Exp Bot 55(396):307–319

Flowers T, Duque E, Hajibagheri M, McGonigle T, Yeo A (1985) The effect of salinity on leaf ultrastructure and net photosynthesis of two varieties of rice: further evidence for a cellular component of salt-resistance. New Phytol 100(1):37–43

Flowers T, Koyama M, Flowers S, Sudhakar C, Singh K, Yeo A (2000) QTL: their place in engineering tolerance of rice to salinity. J Exp Bot 51(342):99–106

Foolad M (1999) Comparison of salt tolerance during seed germination and vegetative growth in tomato by QTL mapping. Genome 42(4):727–734

Foyer CH (1993) Ascorbic acid. Antioxidants Higher Plants:31–58

Foyer CH, Lelandais M, Edwards EA, Mullineaux PM (1991) The role of ascorbate in plants, interactions with photosynthesis, and regulatory significance. Current Topics Plant Physiol (USA)

Franco JA, Fernández JA, Bañón S, González A (1997) Relationship between the effects of salinity on seedling leaf area and fruit yield of six muskmelon cultivars. Hortscience 32(4):642–644

Franco J, Bañón S, Vicente M, Miralles J, Martínez-Sánchez J (2011) Review article: root development in horticultural plants grown under abiotic stress conditions–a review. J Hortic Sci Biotechnol 86(6):543–556

Fridovich I (1986a) Biological effects of the superoxide radical. Arch Biochem Biophys 247(1):1–11

Fridovich I (1986b) Superoxide dismutases. Adv Enzymol Relat Areas Mol Biol 58(6):61–97

Fukuda A, Nakamura A, Tagiri A, Tanaka H, Miyao A, Hirochika H, Tanaka Y (2004) Function, intracellular localization and the importance in salt tolerance of a vacuolar Na+/H+ antiporter from rice. Plant Cell Physiol 45(2):146–159

Gama P, Inanaga S, Tanaka K, Nakazawa R (2007) Physiological response of common bean (*Phaseolus vulgaris* L.) seedlings to salinity stress. Afr J Biotechnol 6(2)

Ganieva RA, Allahverdiyev SR, Guseinova NB, Kavakli HI, Nafisi S (1998) Effect of salt stress and synthetic hormone polystimuline K on the photosynthetic activity of cotton (*Gossypium hirsutum*). Turk J Bot 22(4):217–222

Gao P, Bai X, Yang L, Lv D, Pan X, Li Y, Cai H, Ji W, Chen Q, Zhu Y (2011) osa-MIR393: a salinity-and alkaline stress-related microRNA gene. Mol Biol Rep 38(1):237–242

Garg B, Gupta I (1997) Saline wastelands environment and plant growth. Scientific Publishers, Jodhpur

Garratt LC, Janagoudar BS, Lowe KC, Anthony P, Power JB, Davey MR (2002) Salinity tolerance and antioxidant status in cotton cultures. Free Radic Biol Med 33(4):502–511

Genc Y, Oldach K, Verbyla AP, Lott G, Hassan M, Tester M, Wallwork H, McDonald GK (2010) Sodium exclusion QTL associated with improved seedling growth in bread wheat under salinity stress. Theor Appl Genet 121(5):877–894

Ghassemi-Golezani K, Taifeh-Noori M, Oustan S, Moghaddam M (2009) Response of soybean cultivars to salinity stress. J Food Agric Environ 7(2):401–404

Glenn EP, Brown JJ, Blumwald E (1999) Salt tolerance and crop potential of halophytes. Crit Rev Plant Sci 18(2):227–255

Gomes-Filho E, Lima CRFM, Costa JH, da Silva ACM, Lima MGS, de Lacerda CF, Prisco JT (2008) Cowpea ribonuclease: properties and effect of NaCl-salinity on its activation during seed germination and seedling establishment. Plant Cell Rep 27(1):147–157

Gorham J, Hardy C, Wyn Jones R, Joppa L, Law C (1987) Chromosomal location of a K/Na discrimination character in the D genome of wheat. TAG Theor Appl Genet 74(5):584–588

Goswami S, Kumar RR, Rai RD (2014) Heat-responsive microRNAs regulate the transcription factors and heat shock proteins in modulating thermo-stability of starch biosynthesis enzymes in wheat ('*Triticum aestivum* L.) under the heat stress. Aust J Crop Sci 8(5):697

Greenway H, Munns R (1980) Mechanisms of salt tolerance in nonhalophytes. Annu Rev Plant Physiol 31(1):149–190
Gueta-Dahan Y, Yaniv Z, Zilinskas BA, Ben-Hayyim G (1997) Salt and oxidative stress: similar and specific responses and their relation to salt tolerance in citrus. Planta 203(4):460–469
Günes A, Inal A, Alpaslan M (1996) Effect of salinity on stomatal resistance, proline, and mineral composition of pepper. J Plant Nutr 19(2):389–396
Guan R, Qu Y, Guo Y, Yu L, Liu Y, Jiang J, Chen J, Ren Y, Liu G, Tian L, Jin L, Liu Z, Hong H, Chang R, Gilliham M, Qiu L (2014) Salinity tolerance in soybean is modulated by natural variation in GmSALT3. Plant J 80:937–950
Gunes A, Inal A, Alpaslan M, Eraslan F, Bagci EG, Cicek N (2007) Salicylic acid induced changes on some physiological parameters symptomatic for oxidative stress and mineral nutrition in maize (*Zea mays* L.) grown under salinity. J Plant Physiol 164(6):728–736
Gupta O, Sharma P, Gupta R, Sharma I (2014) MicroRNA mediated regulation of metal toxicity in plants: present status and future perspectives. Plant Mol Biol 84(1–2):1–18
Ha E, Ikhajiagba B, Bamidele J, Ogic-Odia E (2008) Salinity effects on young healthy seedling of Kyllingia peruviana collected from escravos, Delta state. Glob J Environ Res 2(2):74–88
Halliwell B (1982) The toxic effects of oxygen on plant tissues. Superoxide dismutase 1:89–123
Halliwell B (1987) Oxidative damage, lipid peroxidation and antioxidant protection in chloroplasts. Chem Phys Lipids 44(2–4):327–340
Halliwell B, Gutteridge JM (1985) Free radicals in biology and medicine. Pergamon
Halliwell B, Gutteridge JM (2015) Free radicals in biology and medicine. The Clarendon Press/Oxford University Press, New York
Hauser F, Horie T (2010) A conserved primary salt tolerance mechanism mediated by HKT transporters: a mechanism for sodium exclusion and maintenance of high K+/Na+ ratio in leaves during salinity stress. Plant Cell Environ 33(4):552–565
Hernández JA, Almansa MS (2002) Short-term effects of salt stress on antioxidant systems and leaf water relations of pea leaves. Physiol Plant 115(2):251–257
Hernandez JA, Corpas FJ, Gomez M, Río LA, Sevilla F (1993) Salt-induced oxidative stress mediated by activated oxygen species in pea leaf mitochondria. Physiol Plant 89(1):103–110
Hernandez JA, Olmos E, Corpas FJ, Sevilla F, Del Rio LA (1995) Salt-induced oxidative stress in chloroplasts of pea plants. Plant Sci 105:151–167
Hernández JA, Ferrer MA, Jiménez A, Barceló AR, Sevilla F (2001) Antioxidant systems and $O_2^{·-}/H_2O_2$ production in the apoplast of pea leaves. Its relation with salt-induced necrotic lesions in minor veins. Plant Physiol 127(3):817–831
Horie T, Motoda J, Kubo M, Yang H, Yoda K, Horie R, Chan WY, Leung HY, Hattori K, Konomi M (2005) Enhanced salt tolerance mediated by AtHKT1 transporter-induced Na+ unloading from xylem vessels to xylem parenchyma cells. Plant J 44(6):928–938
Houimli SIM, Denden M, El Hadj SB (2008) Induction of salt tolerance in pepper (*Capsicum annuum*) by 24-epibrassinolide. Eur Asian J BioSciences 2:83–90
Howat D (2000) Acceptable salinity, sodicity and pH values for boreal forest reclamation. Environmental Sciences Division
Hsiao TC, Xu LK (2000) Sensitivity of growth of roots versus leaves to water stress: biophysical analysis and relation to water transport. J Exp Bot 51:1595–1616
Huang J, Redmann R (1995) Salt tolerance of Hordeum and Brassica species during germination and early seedling growth. Can J Plant Sci 75(4):815–819
Ikbal FE, Hernández JA, Barba-Espín G, Koussa T, Aziz A, Faize M, Diaz-Vivancos P (2014) Enhanced salt-induced antioxidative responses involve a contribution of polyamine biosynthesis in grapevine plants. J Plant Physiol 171(10):779–788
Imlay JA, Linn S (1988) DNA damage and oxygen radical toxicity. Science 240(4857):1302
Ingram J, Bartels D (1996) The molecular basis of dehydration tolerance in plants. Annu Rev Plant Biol 47(1):377–403
Inzé D, Van Montagu M (1995) Oxidative stress in plants. Curr Opin Biotechnol 6(2):153–158

Isla R, Aragüés R, Royo A (1998) Validity of various physiological traits as screening criteria for salt tolerance in barley. Field Crop Res 58(2):97–107

Jafarzadeh AA, Aliasgharzad N (2007) Salinity and salt composition effects on seed germination and root length of four sugar beet cultivars. Biologia 62(5):562–564

James RA, Rivelli AR, Munns R, von Caemmerer S (2002) Factors affecting CO_2 assimilation, leaf injury and growth in salt-stressed durum wheat. Funct Plant Biol 29(12):1393–1403

James RA, Davenport RJ, Munns R (2006) Physiological characterization of two genes for Na+ exclusion in durum wheat, Nax1 and Nax2. Plant Physiol 142(4):1537–1547

James RA, Blake C, Byrt CS, Munns R (2011) Major genes for Na+ exclusion, Nax1 and Nax2 (wheat HKT1; 4 and HKT1; 5), decrease Na+ accumulation in bread wheat leaves under saline and waterlogged conditions. J Exp Bot 62(8):2939–2947

Jamil M, Lee CC, Rehman SU, Lee DB, Ashraf M, Rha ES (2005) Salinity (NaCl) tolerance of Brassica species at germination and early seedling growth. Elec J Env Agricult Food Chem Title 4(4):970–976

Jamil M, Rehman S, Rha E (2007) Salinity effect on plant growth, PSII photochemistry and chlorophyll content in sugar beet (*Beta Vulgaris* L.) and cabbage (*Brassica Oleracea Capitata* L.) Pak J Bot 39(3):753–760

Kalaji M, Guo P (2008) Chlorophyll fluorescence: a useful tool in barley plant breeding programs. Photochemistry Res Prog:439–463

Kalaji H, Nalborczyk E (1991) Gas exchange of barley seedlings growing under salinity stress. Photosynthetica 25(2):197–202

Kalaji HM, Pietkiewicz S (1993) Salinity effects on plant growth and other physiological processes. Acta Physiol Plant 15(2):89–124

Kangasjärvi S, Neukermans J, Li S, Aro E-M, Noctor G (2012) Photosynthesis, photorespiration, and light signalling in defence responses. J Exp Bot:err402

Karahara I, Ikeda A, Kondo T, Uetake Y (2004) Development of the Casparian strip in primary roots of maize under salt stress. Planta 219(1):41–47

Kataria S, Baghel L, Guruprasad K (2017) Alleviation of adverse effects of ambient UV stress on growth and some potential physiological attributes in soybean (*Glycine max*) by seed pretreatment with static magnetic field. J Plant Growth Regul:1–16

Kaveh H, Nemati H, Farsi M, Jartoodeh SV (2011) How salinity affect germination and emergence of tomato lines. J Biol Environ Sci 5(15)

Kawasaki S, Borchert C, Deyholos M, Wang H, Brazille S, Kawai K, Galbraith D, Bohnert HJ (2001) Gene expression profiles during the initial phase of salt stress in rice. Plant Cell 13(4):889–905

Khajeh-Hosseini M, Powell A, Bingham I (2003) The interaction between salinity stress and seed vigour during germination of soyabean seeds. Seed Sci Technol 31(3):715–725

Khan MA, Rizvi Y (1994) Effect of salinity, temperature, and growth regulators on the germination and early seedling growth of Atriplex griffithii var. stocksii. Can J Bot 72(4):475–479

Khan MA, Weber DJ (2006) Ecophysiology of high salinity tolerant plants, vol 40. Springer Science & Business Media, New York

Khan MA, Ungar IA, Showalter AM, Dewald HD (1998) NaCl-induced accumulation of glycinebetaine in four subtropical halophytes from Pakistan. Physiol Plant 102(4):487–492

Khan MA, Ahmed MZ, Hameed A (2006) Effect of sea salt and L-ascorbic acid on the seed germination of halophytes. J Arid Environ 67(3):535–540

Khodarahmpour Z, Ifar M, Motamedi M (2012) Effects of NaCl salinity on maize (*Zea mays* L.) at germination and early seedling stage. Afr J Biotechnol 11(2):298–304

Kim JY, Kwak KJ, Jung HJ, Lee HJ, Kang H (2010) MicroRNA402 affects seed germination of *Arabidopsis thaliana* under stress conditions via targeting DEMETER-LIKE Protein3 mRNA. Plant Cell Physiol 51(6):1079–1083

Krishnamurthy A, Rathinasabapathi B (2013) Oxidative stress tolerance in plants: novel interplay between auxin and reactive oxygen species signaling. Plant Signaling & Behavior 8(10):e25761

Larcher W (2003) Physiological plant ecology: ecophysiology and stress physiology of functional groups. Springer Science & Business Media/Springer/Springer-Verlag, Berlin/Heidelberg/New York

Läuchli A, James RA, Huang CX, McCully M, Munns R (2008) Cell-specific localization of Na+ in roots of durum wheat and possible control points for salt exclusion. Plant Cell Environ 31(11):1565–1574

Lee DH, Kim YS, Lee CB (2001) The inductive responses of the antioxidant enzymes by salt stress in the rice (*Oryza sativa* L.) J Plant Physiol 158(6):737–745

Lee MH, Cho EJ, Wi SG, Bae H, Kim JE, Cho J-Y, Lee S, Kim J-H, Chung BY (2013) Divergences in morphological changes and antioxidant responses in salt-tolerant and salt-sensitive rice seedlings after salt stress. Plant Physiol Biochem 70:325–335

Liang G, Yang F, Yu D (2010) MicroRNA395 mediates regulation of sulfate accumulation and allocation in *Arabidopsis thaliana*. Plant J 62(6):1046–1057

Lin CC, Kao CH (1995) NaCl stress in rice seedlings: starch mobilization and the influence of gibberellic acid on seedling growth. Botanical Bull Academia Sinica 36

Liu H-H, Tian X, Li Y-J, Wu C-A, Zheng C-C (2008) Microarray-based analysis of stress-regulated microRNAs in *Arabidopsis thaliana*. RNA 14(5):836–843

Liu R, Sun W, Chao M, C-J JI, Wang M, YE B-P (2009) Leaf anatomical changes of Bruguiera gymnorrhiza seedlings under salt stress. J Trop Subtropical Bot 2:012

Liu Y, Yu L, Qu Y, Chen J, Liu X, Hong H, Liu Z, Chang R, Gilliham M, Qiu L, Guan R (2016) GmSALT3, which confers improved soybean salt tolerance in the field, increases leaf Cl^- exclusion prior to Na^+ exclusion but does not improve early vigor under salinity. Front Plant Sci 7:1485-1–1485-14

Long R, Li M, Zhang T, Kang J, Sun Y, Cong L, Gao Y, Liu F, Yang Q (2016) Comparative proteomic analysis reveals differential root proteins in *Medicago sativa* and Medicago truncatula in response to salt stress. Front Plant Sci 7

Longstreth DJ, Nobel PS (1979) Salinity effects on leaf anatomy consequences for photosynthesis. Plant Physiol 63(4):700–703

Maggio A, Hasegawa PM, Bressan RA, Consiglio MF, Joly RJ (2001) Review: unravelling the functional relationship between root anatomy and stress tolerance. Funct Plant Biol 28(10):999–1004

Maggio A, Raimondi G, Martino A, De Pascale S (2007) Salt stress response in tomato beyond the salinity tolerance threshold. Environ Exp Bot 59(3):276–282

Mahajan S, Tuteja N (2005) Cold, salinity and drought stresses: an overview. Arch Biochem Biophys 444(2):139–158

Mansour M, Salama K, Al-Mutawa M, Abou Hadid A (2002) Effect of NaCl and polyamines on plasma membrane lipids of wheat roots. Biol Plant 45(2):235–239

Marschner H, Rimmington G (1988) Mineral nutrition of higher plants. Plant Cell Environ 11:147–148

Martínez-Atienza J, Jiang X, Garciadeblas B, Mendoza I, Zhu J-K, Pardo JM, Quintero FJ (2007) Conservation of the salt overly sensitive pathway in rice. Plant Physiol 143(2):1001–1012

Mass EV (1990) Crop salt tolerance. Chapter 13, P. 262–304. In: Tanji KK (ed) Agricultural salinity assessment and management. ASCE Manuals and Reports on Engineering No. 71, American Society of Civil Engineers, New York

Mathur N, Singh J, Bohra S, Bohra A, Vyas A (2006) Biomass production, productivity and physiological changes in moth bean genotypes at different salinity levels. Am J Plant Physiol 1(2):210–213

Maxwell K, Johnson GN (2000) Chlorophyll fluorescence—a practical guide. J Exp Bot 51(345):659–668

Meloni D, Gulotta M, Martinez C (2008) Salinity tolerance in Schinopsis quebracho colorado: seed germination, growth, ion relations and metabolic responses. J Arid Environ 72(10):1785–1792

Memon SA, Hou X, Wang LJ (2010) Morphlogical analysis of salt stress response of pak choi. Elec J Env Agric Food Chem 9(1)

Meneguzzo S, Sgherri CL, Navari-Izzo F, Izzo R (1998) Stromal and thylakoid-bound ascorbate peroxidases in NaCl-treated wheat. Physiol Plant 104(4):735–740

Meyer G, Schmitt JM, Bohnert HJ (1990) Direct screening of a small genome: estimation of the magnitude of plant gene expression changes during adaptation to high salt. Mol Gen Genet MGG 224(3):347–356

Mian A, Oomen RJ, Isayenkov S, Sentenac H, Maathuis FJ, Véry AA (2011) Over-expression of an Na+–and K+–permeable HKT transporter in barley improves salt tolerance. Plant J 68(3):468–479

Miransari M, Smith D (2007) Overcoming the stressful effects of salinity and acidity on soybean nodulation and yields using signal molecule genistein under field conditions. J Plant Nutr 30(12):1967–1992

Miransari M, Smith D (2009) Alleviating salt stress on soybean (*Glycine max* (L.) Merr.)–Bradyrhizobium japonicum symbiosis, using signal molecule genistein. Eur J Soil Biol 45(2):146–152

Mitsuya S, Takeoka Y, Miyake H (2000) Effects of sodium chloride on foliar ultrastructure of sweet potato (*Ipomoea batatas* Lam.) plantlets grown under light and dark conditions in vitro. J Plant Physiol 157(6):661–667

Mittova V, Tal M, Volokita M, Guy M (2003) Up-regulation of the leaf mitochondrial and peroxisomal antioxidative systems in response to salt-induced oxidative stress in the wild salt-tolerant tomato species Lycopersicon pennellii. Plant Cell Environ 26(6):845–856

Moradi F, Ismail AM (2007) Responses of photosynthesis, chlorophyll fluorescence and ROS-scavenging systems to salt stress during seedling and reproductive stages in rice. Ann Bot 99(6):1161–1173

Munns R (1992) A leaf elongation assay detects an unknown growth inhibitor in xylem sap from wheat and barley. Funct Plant Biol 19(2):127–135

Munns R (1993) Physiological processes limiting plant growth in saline soils: some dogmas and hypotheses. Plant Cell Environ 16(1):15–24

Munns R (2002) Comparative physiology of salt and water stress. Plant Cell Environ 25(2):239–250

Munns R, Gilliham M (2015) Salinity tolerance of crops–what is the cost? New Phytol 208(3):668–673

Munns R, Rawson H (1999) Effect of salinity on salt accumulation and reproductive development in the apical meristem of wheat and barley. Funct Plant Biol 26(5):459–464

Munns R, Termaat A (1986) Whole-plant responses to salinity. Funct Plant Biol 13(1):143–160

Munns R, Tester M (2008) Mechanisms of salinity tolerance. Annu Rev Plant Biol 59:651–681

Munns R, Rebetzke GJ, Husain S, James RA, Hare RA (2003) Genetic control of sodium exclusion in durum wheat. Crop Pasture Sci 54(7):627–635

Munns R, James RA, Läuchli A (2006) Approaches to increasing the salt tolerance of wheat and other cereals. J Exp Bot 57(5):1025–1043

Murat C, Zampieri E, Vizzini A, Bonfante P (2008) Is the Perigord black truffle threatened by an invasive species? We dreaded it and it has happened! New Phytol 178(4):699–702

Mustard J, Renault S (2006) Response of red-osier dogwood (*Cornus sericea*) seedlings to NaCl during the onset of bud break. Botany 84(5):844–851

Netondo GW, Onyango JC, Beck E (2004) Sorghum and salinity. Crop Sci 44(3):797–805

Niazi B, Athar M, Salim M, Rozema J (2005) Growth and ionic relations of fodderbeet and seabeet under saline environments. Int J Environ Sci Technol 2(2):113–120

Niazi A, Ramezani A, Dinari A (2014) GSTF1 gene expression analysis in cultivated wheat plants under salinity and ABA treatments. Mol Biol Res Commun 3(1):9

Niu X, Zhu J-K, Narasimhan ML, Bressan RA, Hasegawa PM (1993) Plasma-membrane H+-ATPase gene expression is regulated by NaCl in cells of the halophyte Atriplex nummularia L. Planta 190(4):433–438

Ochiai K, Matoh T (2002) Characterization of the Na+ delivery from roots to shoots in rice under saline stress: excessive salt enhances apoplastic transport in rice plants. Soil Sci Plant Nutr 48(3):371–378

Oh D-H, Lee SY, Bressan RA, Yun D-J, Bohnert HJ (2010) Intracellular consequences of SOS1 deficiency during salt stress. J Exp Bot 61(4):1205–1213

Othman Y, Al-Karaki G, Al-Tawaha A, Al-Horani A (2006) Variation in germination and ion uptake in barley genotypes under salinity conditions. World J Agric Sci 2(1):11–15

Parida AK, Das AB (2004) Effects of NaCl stress on nitrogen and phosphorous metabolism in a true mangrove Bruguiera parviflora grown under hydroponic culture. J Plant Physiol 161(8):921–928

Parida AK, Das AB (2005) Salt tolerance and salinity effects on plants: a review. Ecotoxicol Environ Saf 60(3):324–349

Parida A, Das AB, Das P (2002) NaCl stress causes changes in photosynthetic pigments, proteins, and other metabolic components in the leaves of a true mangrove, *Bruguiera parviflora*, in hydroponic cultures. J Plant Biol 45(1):28–36

Parida AK, Das A, Mittra B (2004) Effects of salt on growth, ion accumulation, photosynthesis and leaf anatomy of the mangrove, *Bruguiera parviflora*. Trees 18(2):167–174

Parker MB, Gascho G, Gaines T (1983) Chloride toxicity of soybeans grown on Atlantic coast flatwoods soils. Agron J 75(3):439–443

Pasapula V, Shen G, Kuppu S, Paez-Valencia J, Mendoza M, Hou P, Chen J, Qiu X, Zhu L, Zhang X (2011) Expression of an Arabidopsis vacuolar H+−pyrophosphatase gene (AVP1) in cotton improves drought-and salt tolerance and increases fibre yield in the field conditions. Plant Biotechnol J 9(1):88–99

Passioura J (1988) Water transport in and to roots. Annu Rev Plant Physiol Plant Mol Biol 39(1):245–265

Pi E, Qu L, Hu J, Huang Y, Qiu L, Lu H, Jiang B, Liu C, Peng T, Zhao Y (2016) Mechanisms of soybean roots' tolerances to salinity revealed by proteomic and phosphoproteomic comparisons between two cultivars. Mol Cell Proteomics 15(1):266–288

Poljakoff-Mayber A, Somers G, Werker E, Gallagher J (1994) Seeds of *Kosteletzkya virginica* (Malvaceae): their structure, germination, and salt tolerance. II. Germination and salt tolerance. Am J Bot:54–59

Popova LP, Stoinova ZG, Maslenkova LT (1995) Involvement of abscisic acid in photosynthetic process in *Hordeum vulgare* L. during salinity stress. J Plant Growth Regul 14(4):211

Popova OV, Ismailov SF, Popova TN, Dietz K-J, Golldack D (2002) Salt-induced expression of NADP-dependent isocitrate dehydrogenase and ferredoxin-dependent glutamate synthase in *Mesembryanthemum crystallinum*. Planta 215(6):906–913

Prasad S, Bagali P, Hittalmani S, Shashidhar H (2000) Molecular mapping of quantitative trait loci associated with seedling tolerance to salt stress in rice (*Oryza sativa* L.) Curr Sci 78(2):162–164

Qiu Q-S, Guo Y, Dietrich MA, Schumaker KS, Zhu J-K (2002) Regulation of SOS1, a plasma membrane Na+/H+ exchanger in *Arabidopsis thaliana*, by SOS2 and SOS3. Proc Natl Acad Sci 99(12):8436–8441

Quintero FJ, Martinez-Atienza J, Villalta I, Jiang X, Kim W-Y, Ali Z, Fujii H, Mendoza I, Yun D J, Zhu J K (2011) Activation of the plasma membrane Na/H antiporter Salt-Overly-Sensitive 1 (SOS1) by phosphorylation of an auto-inhibitory C-terminal domain. Proc Natl Acad Sci 108(6):2611–2616

Rahnama A, James RA, Poustini K, Munns R (2010) Stomatal conductance as a screen for osmotic stress tolerance in durum wheat growing in saline soil. Funct Plant Biol 37(3):255–263

Rameeh V, Cherati A, Abbaszadeh F (2012) Salinity effects on yield, yield components and nutrient ions in rapeseed genotypes. J Agric Sci Belgrade 57(1):19–29

Ranathunge K, Shao S, Qutob D, Gijzen M, Peterson CA, Bernards MA (2010) Properties of the soybean seed coat cuticle change during development. Planta 231(5):1171–1188

Rao D, Giller K, Yeo A, Flowers T (2002) The effects of salinity and sodicity upon nodulation and nitrogen fixation in chickpea (*Cicer arietinum*). Ann Bot 89(5):563–570

Raul L-A, Andres O-C, Armando L-A, Bernardo M-A, Enrique T-D (2003) Response to salinity of three grain legumes for potential cultivation in arid areas. Soil Sci Plant Nutr 49(3):329–336

Redondo-Gómez S, Mateos-Naranjo E, Davy AJ, Fernández-Muñoz F, Castellanos EM, Luque T, Figueroa ME (2007) Growth and photosynthetic responses to salinity of the salt-marsh shrub *Atriplex portulacoides*. Ann Bot 100(3):555–563

Ren Z-H, Gao J-P, Li L-G, Cai X-L, Huang W, Chao D-Y, Zhu M-Z, Wang Z-Y, Luan S, Lin H-X (2005) A rice quantitative trait locus for salt tolerance encodes a sodium transporter. Nat Genet 37(10):1141–1146

Rengasamy P, Olsson K (1993) Irrigation and sodicity. Soil Res 31(6):821–837

Rodrıguez P, Torrecillas A, Morales M, Ortuno M, Sánchez-Blanco M (2005) Effects of NaCl salinity and water stress on growth and leaf water relations of *Asteriscus maritimus* plants. Environ Exp Bot 53(2):113–123

Romero-Aranda R, Soria T, Cuartero J (2001) Tomato plant-water uptake and plant-water relationships under saline growth conditions. Plant Sci 160(2):265–272
Roppolo D, De Rybel B, Tendon VD, Pfister A, Alassimone J, Vermeer JE, Yamazaki M, Stierhof Y-D, Beeckman T, Geldner N (2011) A novel protein family mediates Casparian strip formation in the endodermis. Nature 473(7347):380–383
Roxas VP, Lodhi SA, Garrett DK, Mahan JR, Allen RD (2000) Stress tolerance in transgenic tobacco seedlings that overexpress glutathione S-transferase/glutathione peroxidase. Plant Cell Physiol 41(11):1229–1234
Ruiz-Sánchez MC, Domingo R, Torrecillas A, Pérez-Pastor A (2000) Water stress preconditioning to improve drought resistance in young apricot plants. Plant Sci 156(2):245–251
Saffan E (2008) Effect of salinity and osmotic stresses on some economic plants. Res J Agric Biol Sci 4(2):159–166
Santos CV, Falcão IP, Pinto GC, Oliveira H, Loureiro J (2002) Nutrient responses and glutamate and proline metabolism in sunflower plants and calli under Na_2SO_4 stress. J Plant Nutr Soil Sci 165(3):366–372
Saqib M, Zörb C, Schubert S (2006) Salt-resistant and salt-sensitive wheat genotypes show similar biochemical reaction at protein level in the first phase of salt stress. Zeits Pflanzenernahr Bodenkunde-Journ Plant Nutrit. Soil Sci 169(4):542–548
Schuch UK, Kelly JJ (2008) Salinity tolerance of cacti and succulents. Turfgrass, Landscape and Urban IPM Research Summary
Seemann JR, Critchley C (1985) Effects of salt stress on the growth, ion content, stomatal behaviour and photosynthetic capacity of a salt-sensitive species, *Phaseolus vulgaris* L. Planta 164(2):151–162
Serraj R, Vasquez-Diaz H, Drevon J (1998) Effects of salt stress on nitrogen fixation, oxygen diffusion, and ion distribution in soybean, common bean, and alfalfa. J Plant Nutr 21(3):475–488
Serraj R, Vasquez-Diaz H, Hernandez G, Drevon J-J (2001) Genotypic difference in the short-term response of nitrogenase activity (C_2H_2 reduction) to salinity and oxygen in the common bean. Agronomie 21(6–7):645–651
Shalata A, Neumann PM (2001) Exogenous ascorbic acid (vitamin C) increases resistance to salt stress and reduces lipid peroxidation. J Exp Bot 52(364):2207–2211
Shannon MC, Grieve CM, Francois LE (1994) Whole-plant response to salinity. Plant-Environ Interact:199–244
Shao H-B, Guo Q-J, Chu L-Y, Zhao X-N, Su Z-L, Hu Y-C, Cheng J-F (2007) Understanding molecular mechanism of higher plant plasticity under abiotic stress. Colloids Surf B: Biointerfaces 54(1):37–45
Sharma T, Sen D (1989) A new report on abnormally fast germinating seeds of Haloxylon spp. An ecological adaptation to saline habitat. Curr Sci Bangalore 58(7):382–385
Sharma P, Prashat GR, Kumar A, Mann A (2016) Physiological and molecular insights into mechanisms for salt tolerance in plants. In: Innovative saline agriculture. Springer, pp 321–349
Shen C-X, Zhang Q-F, Li J, Bi F-C, Yao N (2010) Induction of programmed cell death in Arabidopsis and rice by single-wall carbon nanotubes. Am J Bot 97(10):1602–1609
Shereen A, Ansari R, Soomro A (2001) Salt tolerance in soybean (*Glycine max* L.): effect on growth and ion relations. Pak J Bot 33(4):393–402
Shi H, Ishitani M, Kim C, Zhu J-K (2000) The *Arabidopsis thaliana* salt tolerance gene SOS1 encodes a putative Na+/H+ antiporter. Proc Natl Acad Sci 97(12):6896–6901
Shi H, Quintero FJ, Pardo JM, Zhu J-K (2002) The putative plasma membrane Na+/H+ antiporter SOS1 controls long-distance Na+ transport in plants. Plant Cell 14(2):465–477
Shinwari ZK, Nakashima K, Miura S, Kasuga M, Seki M, Yamaguchi-Shinozaki K, Shinozaki K (1998) An Arabidopsis gene family encoding DRE/CRT binding proteins involved in low-temperature-responsive gene expression. Biochem Biophys Res Commun 250(1):161–170
Shriram V, Kumar V, Devarumath RM, Khare TS, Wani SH (2016) MicroRNAs as potential targets for abiotic stress tolerance in plants. Front Plant Sci 7:817
Shu S, Yuan L-Y, Guo S-R, Sun J, Yuan Y-H (2013) Effects of exogenous spermine on chlorophyll fluorescence, antioxidant system and ultrastructure of chloroplasts in *Cucumis sativus* L. under salt stress. Plant Physiol Biochem 63:209–216

Silva P, Gerós H (2009) Regulation by salt of vacuolar H+-ATPase and H+−pyrophosphatase activities and Na+/H+ exchange. Plant Signal Behav 4(8):718–726

Soussi M, Lluch C, Ocana A, Norero A (1999) Comparative study of nitrogen fixation and carbon metabolism in two chick-pea (*Cicer arietinum* L.) cultivars under salt stress. J Exp Bot 50(340):1701–1708

Srivastava A, Srivastava S, Lokhande V, D'souza S, Suprasanna P (2016) Salt stress reveals differential antioxidant and energetics responses in glycophyte (*Brassicajuncea* L.) and halophyte (*Sesuvium portulacastrum* L.). Front. Redox Homeostasis Managers in Plants under Environmental Stresses 3:20

Steiger HM, Beck E, Beck R (1977) Oxygen concentration in isolated chloroplasts during photosynthesis. Plant Physiol 60(6):903–906

Stepien P, Johnson GN (2009) Contrasting responses of photosynthesis to salt stress in the glycophyte Arabidopsis and the halophyte Thellungiella: role of the plastid terminal oxidase as an alternative electron sink. Plant Physiol 149(2):1154–1165

Steppuhn H, Falk K, Zhou R (2010) Emergence, height, grain yield and oil content of camelina and canola grown in saline media. Can J Soil Sci 90(1):151–164

Steudle E (2000) Water uptake by roots: effects of water deficit. J Exp Bot 51(350):1531–1542

Sun X, Xu L, Wang Y, Yu R, Zhu X, Luo X, Gong Y, Wang R, Limera C, Zhang K (2015) Identification of novel and salt-responsive miRNAs to explore miRNA-mediated regulatory network of salt stress response in radish (*Raphanus sativus* L.) BMC Genomics 16(1):197

Szabolcs I (1994) Soils and salinization. In: Pessarakli M (ed) Handbook of plant and crop stress, Marcel Dekker, New York, pp 3–11

Takeda S, Matsuoka M (2008) Genetic approaches to crop improvement: responding to environmental and population changes. Nat Rev Genet 9(6):444–457

Tang X, Mu X, Shao H, Wang H, Brestic M (2015) Global plant-responding mechanisms to salt stress: physiological and molecular levels and implications in biotechnology. Crit Rev Biotechnol 35(4):425–437

Tardieu F, Davies W (1993) Integration of hydraulic and chemical signalling in the control of stomatal conductance and water status of droughted plants. Plant Cell Environ 16(4):341–349

Tattini M, Gucci R, Coradeschi MA, Ponzio C, Everard JD (1995) Growth, gas exchange and ion content in *Olea europaea* plants during salinity stress and subsequent relief. Physiol Plant 95(2):203–210

Tavakkoli E, Fatehi F, Coventry S, Rengasamy P, McDonald GK (2011) Additive effects of Na+ and Cl−ions on barley growth under salinity stress. J Exp Bot 62(6):2189–2203

Tester M, Davenport R (2003) Na+ tolerance and Na+ transport in higher plants. Ann Bot 91(5):503–527

Tsugane K, Kobayashi K, Niwa Y, Ohba Y, Wada K, Kobayashi H (1999) A recessive Arabidopsis mutant that grows photoautotrophically under salt stress shows enhanced active oxygen detoxification. Plant Cell 11(7):1195–1206

Turan MA, Katkat V, Taban S (2007) Salinity-induced stomatal resistance, proline, chlorophyll and ion concentrations of bean. Int J Agric Res 2(5):483–488

Valiollah R (2013) Effect of salinity stress on yield, component characters and nutrient compositions in rapeseed (*Brassica napus* L.) genotypes. Agric Trop Subtrop 46(2):58–63

Vamerali T, Saccomani M, Bona S, Mosca G, Guarise M, Ganis A (2003) A comparison of root characteristics in relation to nutrient and water stress in two maize hybrids. In: Roots: the dynamic interface between plants and the earth. Springer, pp 157–167

Van Hoorn J, Katerji N, Hamdy A, Mastrorilli M (2001) Effect of salinity on yield and nitrogen uptake of four grain legumes and on biological nitrogen contribution from the soil. Agric Water Manag 51(2):87–98

Walker R, Sedgley M, Blesing M, Douglas T (1984) Anatomy, ultrastructure and assimilate concentrations of roots of citrus genotypes differing in ability for salt exclusion. J Exp Bot 35(10):1481–1494

Wang D, Shannon M (1999) Emergence and seedling growth of soybean cultivars and maturity groups under salinity. Plant Soil 214(1):117–124

Wang L, Zhao H, Chen D, Li L, Sun H, Lou Y, Gao Z (2016) Characterization and primary functional analysis of a bamboo NAC gene targeted by miR164b. Plant Cell Rep 35(6):1371–1383
Welbaum GE, Tissaoui T, Bradford KJ (1990) Water relations of seed development and germination in muskmelon (*Cucumis melo* L.) III. Sensitivity of germination to water potential and abscisic acid during development. Plant Physiol 92(4):1029–1037
Wise RR, Naylor AW (1987) Chilling-enhanced photooxidation evidence for the role of singlet oxygen and superoxide in the breakdown of pigments and endogenous antioxidants. Plant Physiol 83(2):278–282
Wu S-J, Ding L, Zhu J-K (1996) SOS1, a genetic locus essential for salt tolerance and potassium acquisition. Plant Cell 8(4):617–627
Wu C-A, Yang G-D, Meng Q-W, Zheng C-C (2004) The cotton GhNHX1 gene encoding a novel putative tonoplast Na+/H+ antiporter plays an important role in salt stress. Plant Cell Physiol 45(5):600–607
Xu D, Duan X, Wang B, Hong B, Ho T-HD, Wu R (1996) Expression of a late embryogenesis abundant protein gene, HVA1, from barley confers tolerance to water deficit and salt stress in transgenic rice. Plant Physiol 110(1):249–257
Xu S, Hu B, He Z, Ma F, Feng J, Shen W, Yang J (2011) Enhancement of salinity tolerance during rice seed germination by presoaking with hemoglobin. Int J Mol Sci 12(4):2488–2501
Xue D, Huang Y, Zhang X, Wei K, Westcott S, Li C, Chen M, Zhang G, Lance R (2009) Identification of QTLs associated with salinity tolerance at late growth stage in barley. Euphytica 169(2):187–196
Yadav R, Flowers T, Yeo A (1996) The involvement of the transpirational bypass flow in sodium uptake by high-and low-sodium-transporting lines of rice developed through intravarietal selection. Plant Cell Environ 19(3):329–336
Yamaguchi T, Blumwald E (2005) Developing salt-tolerant crop plants: challenges and opportunities. Trends Plant Sci 10(12):615–620
Yang Q, Chen Z-Z, Zhou X-F, Yin H-B, Li X, Xin X-F, Hong X-H, Zhu J-K, Gong Z (2009) Overexpression of SOS (Salt Overly Sensitive) genes increases salt tolerance in transgenic Arabidopsis. Mol Plant 2(1):22–31
Yasuta Y, Kokubun M (2014) Salinity tolerance of super-nodulating soybean genotype En-b0-1. Plant Production Science 17(1):32–40
Yeo A, Yeo M, Flowers T (1987) The contribution of an apoplastic pathway to sodium uptake by rice roots in saline conditions. J Exp Bot 38(7):1141–1153
Yilmaz H, Kina A (2008) The influence of NaCl salinity on some vegetative and chemical changes of strawberries (*Fragaria* x *ananssa* L.) Afr J Biotechnol 7(18)
Yousif S, Al-Saadawi I (1997) Effect of salinity and nitrogen fertilization on osmotic potential and elements accumulation in four genotypes of broad bean *Vicia faba* L. Dirasat Agric Sci 24:395–401
Yupsanis T, Moustakas M, Eleftheriou P, Damianidou K (1994) Protein phosphorylation-dephosphorylation in alfalfa seeds germinating under salt stress. J Plant Physiol 143(2):234–240
Zekri M, Parsons LR (1989) Growth and root hydraulic conductivity of several citrus rootstocks under salt and polyethylene glycol stresses. Physiol Plant 77(1):99–106
Zhang B, Wang Q (2015) MicroRNA-based biotechnology for plant improvement. J Cell Physiol 230(1):1–15
Zhang H, Han B, Wang T, Chen S, Li H, Zhang Y, Dai S (2011) Mechanisms of plant salt response: insights from proteomics. J Proteome Res 11(1):49–67
Zhao G, Ma B, Ren C (2007) Growth, gas exchange, chlorophyll fluorescence, and ion content of naked oat in response to salinity. Crop Sci 47(1):123–131
Zhao G, Yu H, Liu M, Lu Y, Ouyang B (2017) Identification of salt-stress responsive microRNAs from *Solanum lycopersicum* and *Solanum pimpinellifolium*. Plant Growth Regul:1–12
Zhu J-K (2000) Genetic analysis of plant salt tolerance using Arabidopsis. Plant Physiol 124(3):941–948

Zhu J-K (2001a) Plant salt stress. In: eLS. John Wiley & Sons, Ltd. https://doi.org/10.1002/9780470015902.a0001300.pub2

Zhu J-K (2001b) Plant salt tolerance. Trends Plant Sci 6(2):66–71

Zhu J-K, Liu J, Xiong L (1998) Genetic analysis of salt tolerance in Arabidopsis: evidence for a critical role of potassium nutrition. Plant Cell 10(7):1181–1191

Zidan M, Elewa M (1995) Effect of salinity on germination, seedling growth and some metabolic changes in four plant species (Umbelliferae). Indian J Plant Physiol 38:57–61

Chapter 2
Deploying Mechanisms Adapted by Halophytes to Improve Salinity Tolerance in Crop Plants: Focus on Anatomical Features, Stomatal Attributes, and Water Use Efficiency

Ankanagari Srinivas, Guddimalli Rajasheker, Gandra Jawahar, Punita L. Devineni, Maheshwari Parveda, Somanaboina Anil Kumar, and Polavarapu B. Kavi Kishor

Abstract Nearly 1200 million hectares of land is affected by salinity throughout the world, and it is increasing year after year. It is one of the major causes that threaten our crop productivity at a time when we need to meet our growing food demands with limited land and freshwater resources. This leaves us but to understand the complex salinity tolerance mechanisms adapted by halophytic species especially their stomatal conductance (g_s), epidermal salt bladders, and water use efficiency (WUE) and to utilize the candidate genes associated with them in crop plants for better tolerance and crop productivity.

Keywords Bulliform or motor cells · Cation channels · Epidermal bladder cells · Glycophytes · Halophytes · Lignified cells · Salt bladders · Salt hairs · Salt secretion · Secretory cells · Stomatal attributes · Stomatal patchiness · Successive cambia · Trichome patterning · Water use efficiency

Abbreviations

ABA	Abscisic acid
CAM	Crassulacean acid metabolism
CCC	Cation-chloride cotransporter
CNGC	Cyclic nucleotide-gated channels
EBC	Epidermal bladder cells
GIS	*GLABROUS INFLORESCENCE STEMS*

A. Srinivas · G. Rajasheker · G. Jawahar · P. L. Devineni
M. Parveda · S. A. Kumar · P. B. Kavi Kishor (✉)
Department of Genetics, Osmania University, Hyderabad, India

V. Kumar et al. (eds.), *Salinity Responses and Tolerance in Plants, Volume 1*,
https://doi.org/10.1007/978-3-319-75671-4_2

g_s	Stomatal conductance
HKT	High-affinity potassium transporter
KEA	Potassium-efflux antiporter
KIRC	Potassium inward-rectifying channel
KORC	Potassium outward-rectifying channel
KUP	Potassium uptake
NADPH	Nicotinamide adenine dinucleotide phosphate (reduced)
NSCC	Nonselective cation channels
PIP	Plasma membrane intrinsic protein
ROS	Reactive oxygen species
SIM	SIAMESE
SLA	Specific leaf area
SOS	Salt overly sensitive
TIP	Tonoplast intrinsic protein
WUE	Water use efficiency

2.1 Halophytes and Their Significance

Most of the water on the earth is locked up in the oceans, which is salty, and hence crop plants cannot grow under such conditions. However, some native floras grow in areas like mangrove swamps, marshes, sloughs, and seashores where soil salinity is very high (35 g of salt per kilogram of water) and complete their life cycle successfully. Such plants are known as salt plants or halophytes. Nearly, 350 halophytic species have been identified which are distributed across 256 families (Flowers et al. 2010). Understanding the molecular mechanisms underlying salt tolerance in halophytic species and identifying the candidate genes that modulate stomatal density, stomatal conducatnce (g_s), epidermal salt bladders, and water use efficiency (WUE) are vital to develop crop species with better salt stress tolerance. By the year 2050, global food grain production should match the projected population growth of nearly 9.3 billion (Millar and Roots 2012). But this target cannot be achieved by using the arable land that is currently in use. Due to scarcity of freshwater resources, in the future, farmers will have to resort to the use of brackish and saline water for irrigation purposes (Barrett-Lennard and Setter 2010). To combat this problem, we need to breed our crop plants that can grow in arable land by withstanding harsh environments. At the same time, it is hard to achieve this goal due to narrow genetic resources that are at our disposal currently (Colmer et al. 2005). Therefore, salt stress tolerance genes must be identified in extremophiles like halophytes and transferred into crop plants by utilizing genetic engineering/gene-editing technologies. A review of literature related to halophytes indicates that many differences exist between glycophytes and halophytes in traits related to anatomy of leaves, stems, stomatal density, g_s, epidermal salt bladders, WUE, and cellular processes that emphasize the ability of halophytes to tolerate high tissue concentrations of Na^+ and

Cl^-, including regulation of membrane transport, and their ability to synthesize compatible solutes to deal with reactive oxygen species (ROS) accumulation. Unfortunately, glycophytes, especially Poaceae members, do not possess high tissue tolerance since they are salt excluders. It appears that halophytes play vital roles as model plants for understanding plant salt stress tolerance and as genetic resources to improve salt tolerance in crop plants and also for revegetation of saline lands (Flowers and Colmer 2015). This review focuses mainly on the unique morpho-anatomical features like successive cambia, lignified cells, bulliform cells, salt hairs, salt bladders, and salt glands, specific stomatal attributes that majority of halophytes display, and also the key physiological mechanisms that confer salinity tolerance in halophytes and the specific genes that need to be targeted to achieve salt tolerance in crop plants.

2.2 Classification of Halophytic Species

The first classification of halophytes was carried out by Steiner in the year 1934 based on the responses of plants to the internal salt levels, i.e., salt-regulating types (excluders) and salt-accumulating types (includers). Albert and Popp (1977) classified the halophytic species especially that grow in salt marshes as physiotypes. But, they are generally classified based on the presence (black mangroves) or absence (red mangroves) of salt glands located in the leaves (Popp et al. 1993).

2.2.1 Classification Based on Plant Morphology

While Marschner (1995) noticed that some halophytes have the potential to efflux the salt from the leaf surface through salt glands (excretives), Weber (2009) observed accumulation of water within the plant leaves to counter high levels of salt toxicity (succulents).

2.2.2 Classification Based on Salt Demand

Some of the halophytes require salt for their growth, but all of them that occur in nature are not obligates. Hence, they have also been classified as obligate, facultative, and habitat-indifferent types by Sabovljevic and Sabovljevic (2007) and Cushman (2001), respectively. Obligates, mostly the Chenopodiaceae members, grow in waters that contain over 0.5% to 1% of NaCl (Ungar 1978) and display optimal growth under these conditions. On the other hand, facultative halophytes (mostly Poaceae, Cyperaceae, and Juncaceae members) grow both under salt stress and devoid of it (Sabovljevic and Sabovljevic 2007). Habitat-indifferent halophytes (*Festuca rubra*, *Agrostis stolonifera*, and *Juncus bufonius*) grow in soils without salt and also in salt-affected areas (Cushman 2001; Sabovljevic and Sabovljevic 2007).

2.2.3 Classification Based on Physiological Types

Physiological classification of halophytes is based on salt uptake and storage as described by Breckle (1995). Salt-secreting halophytes (exorecretohalophytes and endorecretohalophytes), euhalophytes (salt-diluting halophytes), and pseudohalophytes (salt-excluding) fall under this category. While the exorecretohalophytes secrete salt from salt glands located in the leaves and stems (e.g., *Limonium*, *Tamarix*, *Spartina*, and *Avicennia*), endorecretohalophytes exclude salt from epidermal bladders present on the leaves (e.g., *Atriplex*, *Chenopodium*, and *Mesembryanthemum*). On the other hand, some euhalophytes have succulent stems (e.g., *Kalidium*, *Salicornia*), while others have succulent leaves as in the case of *Suaeda* and *Salsola* (Yensen and Biel 2006; Youssef 2009; Aslam et al. 2011; Zhao et al. 2011). Pseudohalophytes include several species such as *Artemisia*, *Juncus*, and *Phragmites* (Zhao et al. 2011).

2.2.4 Classification Based on Ecological Types

Halophytes have been subdivided into different categories by van Eijk (1939) and Topa (1939). Euhalophytes are subdivided into three ecological types like mesohalophytes (grow in salt meadows, 0.5–10% NaCl, e.g., *Atriplex*, *Apocynum*, *Suaeda*), xerohalophytes (grow in salt deserts, e.g., *Tamarix*, *Kalidium*, *Suaeda*, *Zygophyllum*), and hydrohalophytes (Bucur et al. 1957; Patrut et al. 2005; Youssef 2009; Aslam et al. 2011). Hydrohalophytes are further divided into emergent halophytes (grow in salt marsh and coastal marsh habitats, e.g., *Rhizophora*, *Acanthus*, *Nypa*, *Acrostichum*, and *Salicornia*) and submerged halophytes (grow beneath the surface of seawater, e.g., *Zostera marina*, *Cymodocea*, and *Halophila*) (Hou 1982; Ramadan 1998; Zhao and Li 1999; Patrut et al. 2005).

2.3 Anatomical Changes Observed in Glycophytes and Halophytes

Two types of halophytes were described by Marius-Nicusor and Constantin (2010) based on anatomical adaptations. The first one is called as extreme halophytes and the second as mesohalophytes. Extreme halophytes have two subtypes: one is irreversible extreme halophytes (e.g., members of Chenopodiaceae such as *Petrosimonia oppositifolia*, *P. triandra*, *Salicornia europaea*, *Suaeda maritima*, and *Halimione verrucifera*), and the second is reversible halophytes. In irreversible halophytes, succulence in the form of a well-developed water and sugar storage parenchyma tissue in aerial organs has been noticed alongside the erect position of species with less developed mechanical system (Grigore 2008; Grigore et al. 2010). Succulence helps in minimizing heating of leaves and thus reduces transpiration demands (Ehleringer and Forseth 1980). Thus, succulence appears a natural strategy adapted by some of the species exposed to harsh environments.

2.3.1 Successive Cambia in Glycophytes and Halophytes

Taxa mentioned in the above paragraphs possess a rare anatomical phenomenon, i.e., secondary growth by successive cambia (a developmental oddity) in roots and stems (Marius-Nicusor and Constantin 2010). It has been pointed out that, often, these successive cambia are organized concentrically (Robert et al. 2011). Working on *Avicennia*, Robert et al. (2011) pointed out that formation of successive cambia is an ecologically important characteristic and strongly related with water-limited environments. They noticed a complex network of non-cylindrical wood patches in *Avicennia* and its increased complexity with more stressful conditions. Nearly, 85% of the woody shrub and tree species have been found with such concentric internal phloem in either dry or saline environments (Robert et al. 2011). Such successive cambia have also been noticed in several glycophytic species like *Ipomoea arborescens*, *Rivea hypocrateriformis* (Convolvulaceae), *Hebanthe eriantha* (Amaranthaceae), and several species of *Abuta* (Menispermaceae) that grow mostly in dry forests and deserts (Tamaio et al. 2009; Terrazas et al. 2011; Rajput and Marcati 2013; Rajput 2016). This infers that development of successive cambia is an adaptive tactics and it provides the necessary advantages for survival in harsh environments like salt and drought stress conditions in both glycophytic and halophytic species.

2.3.2 Other Anatomical Anomalies

In halophytic species like *Salicornia europaea*, tracheoidioblasts in its fleshy tissues have been recorded which may be involved in water storage or in giving mechanical support (Marius-Nicusor and Constantin 2010). But there are no reports on the occurrence of such tracheoidioblasts in glycophytic species. Other halophytes like *Petrosimonia oppositifolia* and *P. triandra* have foliar Kranz type of anatomy, an important feature related to C_4 photosynthesis (Marius-Nicusor and Constantin 2010). Such Kranz type of anatomy has also been noticed in several glycophytic species (e.g., *Sorghum*, maize, etc.), which exhibit better tolerance to abiotic stresses when compared to C_3 plants. This type of anatomy may help the plants to reduce water loss when exposed to harsh environments. Thus, halophytes and mangrove plants occurring in different ecological regions display plasticity in their anatomical structures unlike that of glycophytes to better adapt themselves under abiotic stress conditions.

Reversible extreme halophytes have been named so, since they can pass from much salinized areas through less salinized soils as noticed in *Atriplex tatarica*, *Atriplex littoralis*, *Atriplex prostrata*, *Bassia hirsuta*, *Camphorosma annua*, and *C. monspeliaca*. Also, these halophytes exhibit either salt hairs, or successive cambia, or Kranz type of anatomy and succulence. Mesohalophytes on the other hand are species with intermediary anatomical adaptations between extreme halophytes and glycophytes (e.g., *Aster tripolium* subsp. *pannonicus*, *Lactuca saligna*, *Scorzonera cana*, *Lepidium cartilagineum* subsp. *crassifolium*, *Lepidium latifolium*, *Lepidium perfoliatum*, *Iris halophila*, *Plantago schwarzenbergiana*, *Trifolium fragiferum*, and *Spergularia media*). Well-developed endodermis, aerenchyma, and xerophytic features such as sunken stomata and water storage tissue are the usual features in these taxa. Many of these adaptations are also related with other ecological factors like flooding, dryness, and humidity. Amphibious halophytes possess bulliform cells, which is an adaptation to temporary dry conditions of the habitats, despite the fact that these species are hygrophilous (e.g., *Scirpus maritimus*, *Carex distans*, *Juncus gerardii*, *Puccinellia distans* subsp. *limosa*, *Carex vulpina*, and *Alopecurus arundinaceus*).

2.3.3 Lignified Cells

In some of the halophytic taxa, associated with successive cambia, increased lignin content has been noticed (Marius-Nicusor and Constantin 2010) which may confer cellular resistance to a high osmotic pressure. In glycophytic species like *Morus alba*, increased epidermal thickness and changes in stomatal distribution and in xylem components of both stem and root were observed under NaCl stress conditions (Vijayan et al. 2008). In the halophytic species like *Salvadora persica*, decreased epidermal cell diameter and cortex thickness but increased thickness of hypodermal layer, pith area, and pith cell diameter were observed under salinity (Parida et al. 2016). Such morpho-anatomical modifications including stomatal density imply that these changes help the plants in osmotic adjustment and better conductivity of water under high saline conditions. In both glycophytic and halophytic species, lignin is involved in supporting and counteracting the high osmotic pressure. Also, lignified cells provide the necessary rigidity to the plant organs. On the other hand, the phloem could delay the water absorption in such plants. Thick cuticular layer covers the epidermis of stems and leaves of desert halophytic species like *Zygophyllum album* and *Nitraria retusa*. Such an anatomical feature is fully justified for better adaptation of desert halophytes exposed to adverse moisture-limited conditions (Abd Elhalim et al. 2016). In several of these plants, adaptation is achieved by succulence, development of trichomes, sunken stomata, xylem fibers, and storage materials. Further, plants with waxy cuticle in the leaves and stems minimize the loss of water under salinity and drought stress conditions. Sunken stomata with trichomes arising from the epidermis are another adapting mechanism to minimize water loss under stressful habitats. Leaves of several desert shrubs exhibit small leaf surface area, well-developed epidermal hairs covering the leaf surface, stronger cutinization, and sunken stomata. Such an array of anatomical features perhaps can reduce water loss under stressful environments (Yan et al. 2002). Identification of candidate genes associated with such phenomena would help us to isolate and validate them in glycophytic species.

2.3.4 Bulliform or Motor Cells

Members of Poaceae (Claudia and Murray 2000; Peterson 2000), Cyperaceae (Mittler 2006), and Juncaceae (Duval-Jouve 1871) have been found to have bulliform or motor cells at the foliar epidermis level. These are large, thin-walled water-containing cells seen in many halophytic and glycophytic species. These structures enable the plants to roll or curl the leaves during water stress and to reopen during favorable conditions. Under water-deficit conditions, they lose turgor and cause lamina to fold or roll inward edge to edge (Dickison 2000). Bulliform cells also occur in several glycophytic species. Grigore et al. (2010) studied ecological implications of bulliform cells and found that they are associated with salt and water stress conditions. Thus, these structural features help the plants in leaf movements during salt and drought stress (Cutler et al. 2007).

2.3.5 Salt Hairs, Salt Bladders, and Salt Glands

Plant epidermis plays a vital role in water relations, in defense, as well as in pollinator attraction. Such a wide spectrum of functions are carried out by specialized cells which differentiate from undifferentiated epidermis (Glover 2000). Glover (2000) pointed out that interaction between differentiating cells adapting different cell fates is the key to the patterning of a multifunctional tissue. Salt-secreting hairs (found in halophytes only) are vital devices in salt removal from plant shoots. In *Halimione verrucifera*, Marius-Nicusor and Constantin (2010) have observed salt-secreting hairs on the surface of leaves. Such vesicular hairs for salt excretion have also been noticed in other halophytic species such as *Atriplex tatarica* at the foliar limb. It is highly interesting to note that the microhair density increases with enhanced salinity levels in the soil (Ramadan and Flowers 2004). But Thomson et al. (1988) noticed that microhairs do not have glandular function in cereals in contrast to halophytic grasses. In cereals, microhairs are smaller in size unlike that of halophytes and hence cannot sequester Na^+ ions (Shabala 2013). Another structure is salt glands which are specialized epidermal bladder cells (EBCs) or specialized trichomes. EBCs sequester excessive Na^+ (1000-fold more when compared with epidermal cells) away from the mesophyll cells in the leaves. Halophytes that contain salt glands are generally termed as salt secretors as pointed out by Liphschitz et al. (1974) or recretohalophytes (Breckle 1990). All salt glands are epidermal in origin and in essence specialized trichomes (Esau 1965). Nearly 50% of the halophytic species contain salt bladders (Flowers and Colmer 2008) that are ten times larger than epidermal cells (Shabala and Mackay 2011) and help to store and exclude salts (Flowers and Colmer 2015; Santos et al. 2016). Secretion of salts *via* salt glands has been noticed in more than 50 species in 14 angiosperm families distributed in Caryophyllales, asterids, rosids, and grasses (Dassanayake and Larkin 2017).

2.3.6 Salt Secretion by Salt Bladders and Salt Glands

Recretohalophytes (e.g., *Chenopodium quinoa*) have developed salt glands for secreting salt out of a plant under saline environments. Knowledge of salt secretion pathways in relation to the function of salt glands and the genes associated with the phenomenon helps us in developing crop plants that can be grown in saline-affected soils. Two types of salt-secreting structures that are unique to halophytes, namely, salt bladders and salt glands, which secrete the ions out of the plants have been identified. Such structures are absent in other types of halophytes and in non-halophytic species (Shabala et al. 2014; Yuan et al. 2015). Salt bladders (modified epidermal hairs) and salt glands differ in certain aspects of their structures. Salt bladders are classified as trichomes, glandular hairs, thorns, and surface glands (Ishida et al. 2008). Single epidermal cells can function as salt bladders as in the case of *Mesembryanthemum crystallinum*, which are only modified trichomes. Salt bladders are composed of one bladder cell, with or without one or more stalk cells. Salt bladders sequester not only cations such as Na^+ but also anions like Cl^- and act as secondary epidermis to reduce the loss of water under stress (Shabala and Mackay 2011). Ramadan (1998) working on a xero-halophyte *Reaumuria hirtella* found that there is diurnal salt excretion pattern from the salt bladders, i.e., during the night and also early morning. In other words, the rejection of salts at the roots and the secretion mechanism at the shoots allow the plant to maintain its internal salt concentration at a constant level in spite of variations in soil salt content. Oh et al. (2015) and Barkla and Vera-Estrella (2015) presented a transcriptomic and metabolomic analysis of bladder cells of *M. crystallinum*, respectively. While Oh et al. (2015) demonstrated cell-type-specific responses in this plant during salt stress adaptation through transcriptomic work, Barkla and Vera-Estrella (2015) identified 352 different metabolites in bladder cells exposed to salt treatment. Pan et al. (2016) demonstrated that increasing of Na^+ ion accumulation in bladders of *Atriplex canescens* enhances salt stress tolerance.

Salt glands form stable structures and contain either bicellular or multicellular structures. Salt glands in dicotyledonous recretohalophytes are multicellular and sunken into the epidermis (Yuan et al. 2016). In *Tamarix aphylla*, 8 cells were noticed (Thomson and Platt-Aloia 1985) and, in *Limonium bicolor* (Plumbaginaceae), 16 were observed (Feng et al. 2015; Yuan et al. 2015). Interestingly, Yuan et al. (2013) noticed autofluorescence of salt glands under UV excitation, and later it has been found that this autofluorescence is localized in the cuticle of salt glands (Deng et al. 2015). It has also been considered that cuticle is an essential structure for preventing leakage of ions and for protecting the mesophyll cells from salt damage. Though salt glands exclude both cationic (Na^+, K^+, Ca^{2+}, N, Mg^{2+}, Fe^{2+}, Mn^{2+}, Si^{2+}, and Zn^{2+}) and anionic (Cl^-, Br^-, O, S, P, and C) elements (Sobrado and Greaves 2000; Ceccoli et al. 2015; Feng et al. 2015; Zouhaier et al. 2015), it has been noticed that they exclude more Na^+ and Cl^- than other types of ions (Ma et al. 2011). Into the salt glands, salt is mainly transported through plasmodesmata, vesicles, and ion transporters. However, the exact mechanism of transport of ions into salt glands and out of them is not clearly known. Transport of Na^+ in a typical leaf mesophyll cell and a secretory cell are shown in Figs. 2.1 and 2.2, respectively.

It appears that the positioning of inner most cells of the salt glands is crucial. They are usually positioned adjacent to the mesophyll cells. Such a unique structure of salt glands may help the ion transport into it. But, it is not known how salt is transported from mesophyll cells of leaves into the salt glands directionally and if the cuticle or ferulic acid present in cuticle surrounding the salt glands plays any vital role in this process (Deng et al. 2015). The secretory cells of the salt glands adjacent to the mesophyll cells possess many small vesicles as in the case of *Limonium* (Tan et al. 2010, 2013; Yuan et al. 2015; Zouhaier et al. 2015) which may play a vital role in transporting salt into the salt glands. It is also known that

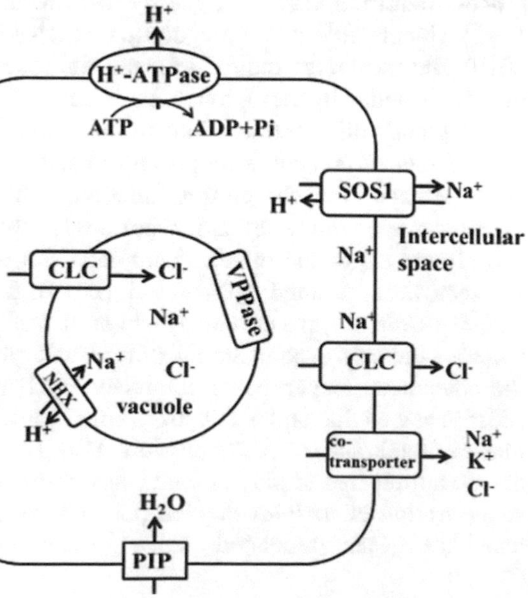

Fig. 2.1 Transport of Na^+ in leaf mesophyll cell

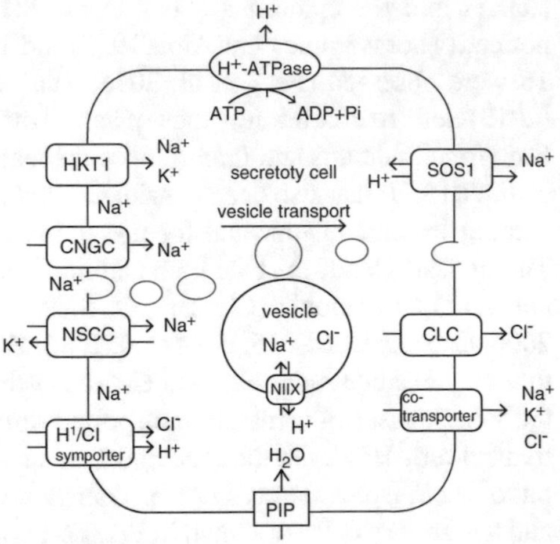

Fig. 2.2 Transport of Na^+ in a secretory cell

plasmodesmata play critical roles in transporting large molecules between cells (Overall and Blackman 1996). Many transporters like HKT1, KUP, KEA, cyclic nucleotide-gated channels (CNGC), and nonselective cation channels (NSCCs) may enhance Na^+ accumulation in cytoplasm (Flowers et al. 1977; Flowers and Colmer 2008), and a low-affinity K^+ transporter called AlHKT2;1 plays an important role in K^+ uptake during salt stress and in maintaining high K^+/Na^+ ratio (Sanadhya et al. 2015) in the cytoplasm of recretohalophyte *Aeluropus lagopoides*.

Flowers et al. (2010) pointed out that salt glands differ in structural complexity and mechanism of salt exclusion from that of salt bladders. Salt glands directly secrete salt out of the plant to the external environment by an active process requiring very high energy and are triggered by light (Dschida et al. 1992). Under salt stress, upregulation of H^+-ATPase was observed (Chen et al. 2010). But the energy required for this purpose (photoassimilates and NADPH) may be provided by mesophyll cells. Yuan et al. (2015) found that mitochondria in salt glands differentiated first in order to provide energy for salt secretion. The number of mitochondria present in salt glands is more than the mesophyll cells, perhaps to discharge this function. It has been discovered that salts are excreted along with water and, accordingly, two aquaporin genes (*PIP* and *TIP*) were found expressed in salt gland cells and water may be reabsorbed during salt removal as pointed by Tan et al. (2013). Experiments with halophytic species like *Chloris gayana* (Kobayashi et al. 2007) and *Tamarix* (Ma et al. 2011) revealed the existence of liquid flow in salt glands (possible Na^+-ATPase). On the other hand, experiments conducted by Feng et al. (2014) and Yuan et al. (2015) showed that under salt stress conditions, numerous vesicles fuse with the plasma membrane for salt exclusion. Also, H^+-ATPase and the Na^+/H^+ antiporters are anticipated to play crucial roles in the process of salt secretion and Na^+ sequestration of *Atriplex marina* (Chen et al. 2010; Tan et al. 2010). Further, candidate genes associated with salt secretion have been discovered using transcriptomics in recretohalophyte *Reaumuria trigyna*. Here genes related to ion transport have been found relevant to the salt secretion function (Dang et al. 2013, 2014). In both glycophytes and halophytes, epidermal structural adaptations include cuticle-covered pavement cells which prevent dehydration and pathogen attack. Small chloroplasts with high stroma to grana ratio in the pavement cells of *Arabidopsis thaliana* have been noticed by Barton et al. (2016). Identification and validation of genes and generation of mutants will help us in finding out their actual contribution to the upkeep and functioning of the aerial epidermis and ultimately in developing crop plants with better salt tolerance. Other structural adaptations include trichomes, nectaries, prickles, and hydathodes, and they may be made up of single or multiple cellular structures consisting several cell types (Esau 1965). Thus, these studies point out that halophytic species have complex salt tolerance mechanisms when compared to glycophytes which are being unraveled slowly.

2.3.7 Trichomes and Trichome Patterning

Trichomes can be glandular or nonglandular. Martin and Glover (2007) pointed out that the nature of the products that glandular trichomes exude affects their functions and consequently their distribution and patterning. In glycophytic species such as *Arabidopsis thaliana*, trichomes appear as large single cells (200–300 μm in length), with a cuticle and a suite of socket cells (Glover 2000). In maize, structurally three distinct types of trichomes such as macrohairs, prickle hairs, and bicellular microhairs were noticed (Martin and Glover 2007). The main function of trichomes under arid conditions is light reflectance to reduce the heat load of the leaves (Ehleringer et al. 1976; Benz and Martin 2006). While pubescent leaves of *Populus alba* reflect 50% of incident light, glabrous leaves reflect 20% of it (Johnson 1975). In the adult vegetative phase of *Arabidopsis*, the *GLABROUS INFLORESCENCE STEMS* (*GIS*) gene that affects trichome production has been identified (Gan et al. 2006). Also, biochemical and genetic changes associated with trichome initiation have been well studied (Martin and Glover 2007). But in cereals, sharp and pointed trichome structures have been observed in contrast to round, balloon-like trichome structures in halophytes (Shabala 2013). Trichomes present in halophytes (round and balloon-like) are highly efficient for sequestering Na^+ ions into them. Therefore, in the future, efforts must be made to identify target genes associated with trichome differentiation and formation patterns and for transferring them into glycophytic species through genetic manipulation besides increasing trichome density, size, and volume. It appears that trichome size (determined by the number of endoreduplications) is controlled by SIM and D-type of cyclins in *A. thaliana* (Churchman et al. 2006). But, no efforts have been made so far to transfer such genes into crop plants.

2.4 Stomatal Attributes

The evolution of stomata dates back to Silurian to Devonian, but its structure and function predominantly remained the same throughout ~400 million years of higher plant evolution (Edwards et al. 1998). Shtein et al. (2017) recently investigated stomatal cell wall composition and structural patterns associated with diverse groups of plants. They noticed kidney-shaped stomata in *Platycerium bifurcatum* (a fern with CAM photosynthesis, Polypodiaceae family), *Asplenium nidus* (a fern with C_3 photosynthesis, Aspleniaceae), *Arabidopsis thaliana* (angiosperm, dicot, belongs to Brassicaceae), and *Commelina erecta* (monocot, Commelinaceae) but dumbbell-shaped stomata in grass species like *Sorghum bicolor* (C_4, Poaceae) and *Triticum aestivum* (C_3, Poaceae). In *P. bifurcatum* and *A. nidus*, stomatal movement is passive, but in other abovementioned species, it has been found active. Shtein et al. (2017) discovered different spatial patterns of varying cellulose crystallinity, pectin, lignin, and phenolics in guard cell walls. In ferns, lignin serves wall strengthening function in guard cells; but, in angiosperms, crystalline cellulose has replaced the function of lignin. Such varying cell wall modifications clearly indicate different biochemical functions for cell walls and that they might have occurred over a period of time in response to specific environmental challenges. Unfortunately, the study group does not include any halophytic species; therefore, not much information is known about the stomatal shape, wall structure of guard cells, and stomatal movements in diverse groups of halophytes. Whether the shape of stomata or varying cellulose crystallinity has any role to play in salt stress tolerance especially in halophytes needs, however, further investigations.

2.5 Stomatal Patchiness

On a single leaf blade, different behaviors of small groups of stomata occur which is usually termed as stomatal patchiness. It varies depending on the plant species, age, leaf position, and stress conditions (Pospisilova and Santrucek 1994). Stomatal patchiness is common in plants, but not clearly understood despite its detrimental role to water use efficiency. It has been pointed out that occurrence and characteristics of patchiness are difficult to predict (Mott and Peak 2007). In xerophytes, generally large number of smaller stomata occur which may be related to increased need of g_s regulation under stress conditions. Aasamaa and Sober (2001) pointed out that shoot hydraulic conductance is correlated well with better photosynthesis, g_s, and stomatal sensitivity to enhanced leaf water potential. Further, Beyschlag and Eckstein (2001) analyzed the cause for stomatal patchiness which is a result of hydraulic interactions that produce a stomatal closure. Taking the above facts into consideration, Delgado et al. (2011) pointed out that in natural conditions, stomatal density and distribution are under selective pressure. These phenomena indicate that stomatal density may play a critical role in enhancing the adaptation of plants to salt stress/water-deficit conditions.

2.6 Stomatal Density and Stomatal Aperture

Moderate water stress had positive effects on stomatal number, but severe stress led to reduced stomatal number and size in a perennial grass, *Leymus chinensis* (Xu and Zhou 2008). Stomatal density was positively correlated with g_s, net assimilation rate of CO_2, and water use efficiency (Xu and Zhou 2008). Enhanced stomatal density was reported under stress in glycophytic species like *Triticum aestivum* (Yang and Wang 2001; Zhang et al. 2006) besides decreased stomatal size (Quarrie and Jones 1977; Spence et al. 1986). On the other hand, in Na_2SO_4-treated halophytic species *Prosopis strombulifera*, higher stomatal density, transpiration rate, and ABA levels were observed. But at high (−2.6 MPa) salinity (Na_2SO_4) levels, higher epidermal cell density with smaller stomata was noticed (Reginato et al. 2013). Under normal conditions, no change in stomatal density is usually observed, but salinity-induced reduction was recorded in salt-tolerant as well as in a C_3, halophytic species, *Chenopodium quinoa* (Shabala et al. 2013). Thus, stomatal abundance and distribution (leaf micromorphology) vary in diverse plant species. Orsini et al. (2010) noticed shortest stomata in *Thellungiella parvula* and *T. salsuginea* and the longest in *Lepidium virginicum* and *A. thaliana* compared to those in *Descurainia pinnata*. They recorded higher leaf stomatal densities in *T. salsuginea*, *L. virginicum*, and *D. pinnata* when compared with *Arabidopsis*. But, no differences in stomatal width (shorter axis) were noticed. The authors could also find a fine correlation between lower stomatal sizes and higher stomatal density. Reduced stomatal density and stomatal pore area of leaf were noticed in the halophytic species *Salvadora persica* with increasing salinity.

2.6.1 *Stomatal Aperture and Ion Accumulation in Glycophytes and Halophytes*

Unlike glycophytes, some halophytes accumulate salts in their tissues; therefore, they prevent desiccation and improve their biomass under salt stress (Greenway and Munns 1980; Flowers 1985). Plant biomass depends on stomatal movements since the stomatal apertures regulate the entry of CO_2. Halophytes should exhibit less reduction in g_s when compared to glycophytes, since they are salt stress tolerant, but the reports are conflicting. Perera et al. (1994) and Robinson (1996) reported decreased stomatal aperture in isolated epidermal strips of halophytic species like *Aster tripolium* and *Cochlearia anglica* under NaCl stress conditions. Contrarily, in glycophytes, enhanced stomatal aperture was reported by Zeiger (1983). X-ray microanalysis has revealed higher accumulation of Na^+ in epidermal and subsidiary cells (mesophyll tissue) when compared to guard cells of *A. tripolium* (Perera et al. 1997). This indicates that entry of Na^+ ions into guard cells in this halophyte is restricted by some mechanism. Perera et al. (1997) also observed higher accumulation of K^+ ions in the guard cells compared to Na^+. On the other hand, in the glycophyte *Commelina communis*, guard cells accumulate high amounts of Na^+ instead of K^+ under salt stress. Contrarily, halophytic species may substitute K^+ for Na^+ in their stomata (Shabala and Mackay 2011). Perera et al. (1997) suggested that the capacity of guard cells to restrict the intake of Na^+ ions is an important component of sodium-driven regulation of transpiration and therefore *A. tripolium* exhibits salinity tolerance. While working with *Atriplex halimus*, Martinez et al. (2005) reported that Na^+ added to the plants helps them to cope up with osmotic stress conditions as imposed by polyethylene glycol and helps in osmotic adjustment. In contrast to other halophytic species, *A. tripolium* partially closes its stomata in response to high Na^+ ion concentrations. Such a stomatal response prevents high accumulation of Na^+ within the shoot *via* control of the transpiration rate (Very et al. 1998). They demonstrated that guard cell cation channels are involved in Na^+-induced stomatal closure in halophytic species *A. tripolium.* While K^+ uptake is downregulated in this species and guard cell Na^+ is increased, it did not happen in the non-halophytic species *A. amellus*, a relative of *A. tripolium*. Tester (1988) and Thiel and Blatt (1991) observed inhibition of inward- and outward-rectifying K^+ channels by Na^+ in plant cells. Niu et al. (1993, 1996) and Tsiantis et al. (1996) observed NaCl-induced expression of plasma membrane and tonoplast H^+-ATPases. It has been documented that Na^+ helps in stomatal opening in glycophytes (Pallaghy 1970). Very et al. (1998) found out that enhanced guard cell cytosolic Na^+ in *A. tripolium* lead to delayed deactivation of KIRCs, but not in *A. amellus*. It appears that instead of K^+, Na^+ ions are used by some halophytic species to increase the turgor in guard cells under NaCl stress. Besides KIRCs, striking differences were also noticed in the depolarization-activated K^+ outward-rectifying channels (KORC) between halophytic and glycophytic species under stress. The above observations infer that stomata of glycophytic species lose their ability to close due to accumulation of Na^+ ions, but not halophytes. So, the genes associated with Na^+-induced closure of stomata need to be manipulated in glycophytes.

2.7 Water Use Efficiency (WUE) in Glycophytes and Halophytes Under Salt Stress

Higher leaf stomatal densities were noticed in *Thellungiella salsuginea*, *L. virginicum*, *D. pinnata*, and *T. parvula* compared with salt-susceptible species like *A. thaliana* (Orsini et al. 2010). Stomatal patterning was well studied in *Arabidopsis*, but not in halophytes. Orsini et al. (2010) while working with *Arabidopsis thaliana* (glycophyte) and its relatives including halophytes observed much lower whole-plant day/night transpiration rate in halophytes under both saline and nonsaline conditions in comparison with *A. thaliana*. Following salt exposure, decreased g_s was also recorded in several halophytic species by Lovelock and Ball (2002) and Boughalleb et al. (2009). Orsini et al. (2011) and Shabala et al. (2013) observed decreased stomatal density in *Chenopodium quinoa*, a halophyte. Likewise, *Distichlis spicata* (Kemp and Cunningham 1981), *Suaeda maritima* (Flowers 1985), *Kochia prostrata* (Karimi et al. 2005), *Lasiurus scindicus* (Naz et al. 2010), and all salt-tolerant halophytes display decrease in stomatal density with increasing salt stress conditions. Adolf et al. (2012) are of the opinion that reduced stomatal density or few fully opened stomata may be beneficial to plants (rather than many partially opened stomata) to optimize water productivity under salt stress. On the other hand, Omami et al. (2006) noticed more stomata per unit leaf area and larger stomatal aperture in the genotypes *Amaranthus tricolor* (C_4 species) and accession '83 when compared to *A. hypochondriacus* and *A. cruentus*. They recorded decreased plant height, leaf number, and leaf area under NaCl stress in different species of amaranth. Reduction in photosynthetic rate and g_s under salinity was also observed. Specific leaf area (SLA) was reduced under salt stress, and the extent of reduction depended on the genotype. Also, SLA and WUE were negatively correlated. Enhanced WUE was recorded at 100 mM NaCl salt stress, and it ranged from 3.9 g in *A. tricolor* to 6.7 g dry mass/kg water in *A. cruentus*. While halophytes display reduction in stomatal density, salt-sensitive lines show 20% increase in stomatal density (Shabala 2013). Thus, it appears that reduction in stomatal density plays a pivotal role in plant adaptive responses and increases WUE under saline environments (Shabala et al. 2012). *Too many mouths* (*tmm*) mutants obtained by Serna (2009) in *Arabidopsis* are the only mutants available for unraveling the molecular mechanisms better about stomatal density and aperture sizes. Unfortunately, we do not have such mutants available in halophytes or in glycophytic crop plants to understand this phenomenon well. Unless such mutants are available, crop plants cannot be targeted for reduced stomatal density and increased WUE.

2.8 Influx of Na^+ Ions into the Xylem of Glycophytes and Halophytes and Salt Stress Tolerance

Loading of Na^+ ions into xylem is highly important for salt stress tolerance. Balnokin et al. (2004) showed that *Suaeda altissima*, a succulent halophyte, accumulates more Na^+ ions in leaves than in roots. Balnokin et al. (2005) also reported sustained water potential gradient from roots to shoots that corresponds to the Na^+ ion distribution pattern among different plant parts. Halophytes need to maintain water potential gradient in the soil-root-shoot system when exposed to high levels of soil salinity. Hence, halophytes accumulate salt in their cells for maintaining appropriate levels of water potentials and to absorb water properly. Na^+ ion uptake is higher in the shoots and leaves compared to the roots (Balnokin et al. 2005). Balnokin et al.

(2005) while working with halophytic members of the family Chenopodiaceae observed that Na^+ concentration in leaves is higher compared to roots when grown under varied NaCl concentrations. In other words, this pattern of Na^+ ion accumulation did not change irrespective of the NaCl concentration in the medium. When halophytes were grown without NaCl, higher K^+ accumulation was noticed again in leaves than in roots. This distribution pattern of Na^+ and K^+ ions among organs in halophytes signifies proper absorption of water content by roots and its transport to shoots and leaves under high soil salinity (Balnokin et al. 2005).

Loading of Na^+ into xylem was compared in pea and barley and two glycophytic plants but with contrasting differences in their salinity tolerance. While salt-tolerant barley plants loaded higher Na^+ content in the xylem within 6 hours of salt stress, the entry of Na^+ ions into xylem was restricted in pea (Bose et al. 2014). Pea plants usually exclude Na^+ ions at the plasma membrane level (like all salt-susceptible lines) in contrast to that of salt-tolerant lines which use Na^+ ions for osmotic adjustment. Shabala et al. (2000) noted that higher Na^+ ion concentrations in the xylem impose high osmotic stress on the leaf mesophyll cells. Consequently, this results in the K^+ ion leakage and upregulation of caspase-like proteases and endonucleases (Hughes and Cydlowski 1999; Demidchik et al. 2010) leading to the death of the cells. Demidchik et al. (2010) reported that such root K^+ efflux conductance is activated by hydroxyl radicals in *Arabidopsis* under salt stress. Thus, the time-dependent loading of Na^+ ions into the xylem is critical for salt stress tolerance. Unfortunately, how exactly halophytes achieve such a time-dependent Na^+ loading into xylem and the signaling events leading to the loading of Na^+ ions into the xylem are not completely understood. Discovering such events is crucial for developing glycophytic crops with time-dependent loading of Na^+ ions into xylem and crop plant stress tolerance.

In glycophytic species, loading of Na^+ ions into the xylem may be mediated by a channel. Wegner and Raschke (1994) investigated voltage-dependent ion conductance in the plasmalemma of xylem parenchyma cells of barley using patch-clamp technique. They suggested that salt is released into the xylem apoplast from the xylem parenchyma cells by both anions and cations following electrochemical gradients set by the uptake of ions. In other words, Na^+ ion-permeable nonselective outward-rectifying channels exist at the xylem parenchyma interface in glycophytic species (Shabala 2013). Wegner and De Boer (1997) discovered the properties of two outward-rectifying channels in root xylem parenchyma cells. It is known that xylem parenchyma cells regulate the transpiration stream and play a role in long-distance signaling. Wegner and De Boer (1997) pointed out that KORC mediates the release of K^+ to the xylem sap and suggested a role in K^+ homeostasis and long-distance signaling. Pilot et al. (2003) discovered that ABA plays a key role in the regulation of these channels. Unfortunately, such channels in halophytic species have not yet been discovered. But, how do Na^+ ions enter the xylem cells in halophytes and glycophytes? Balnokin et al. (2005) and Lun'kov et al. (2005) observed that accumulation of Na^+ in the root xylem of halophytes is achieved through the Na^+/H^+ antiporter located at the plasma membrane of parenchymal cells adjacent to the xylem. Shi et al. (2002) found out that in glycophytes, Na^+/H^+ exchanger (SOS1) controls long-distance Na^+ transport and is highly expressed at the xylem symplast boundary of roots. Further, *SOS1* activity has been shown to be inducible under NaCl stress conditions in both glycophytic (Shi et al. 2002) and halophytic species (Oh et al. 2010). Na^+ ions may also enter into xylem through a cation-Cl^- cotransporter (CCC) as has been noticed in the case of animals (Delpire and Mount 2002). Experiments carried out by Colmenero-Flores et al. (2007) reveal high CCC expression in xylem parenchyma. It appears that not only loading of Na^+ into xylem is important but also its timing for improving salt stress tolerance.

2.9 Outlook

Halophytic species accumulate salt in their cells and hence maintain appropriate water potentials and absorb water from the soil. In other words, halophytes have well-orchestrated mechanisms in place to cope with salt stress. Unfortunately, we have been making efforts or breeding crops for salt exclusion for the past several years. But, we should try to identify the genes associated with tissue tolerance of salts and transfer such traits into crop plants for better performance like that of halophytes. Crop plants must be targeted for reduced stomatal density, increased external Na^+ sequestration using trichomes, limiting xylem Na^+ loading, and increased WUE like that of halophytes by developing mutants for such traits. Also, the distribution pattern of Na^+ and K^+ ions among organs in halophytes signifies proper absorption of water content by roots and its transport to shoots and leaves under high soil salinity which is not observed in crop plants. This character is vital, and hence, attempts must be made in the future for proper distribution pattern of these ions in the crop plant systems. However, the in-depth mechanism of halophyte salt tolerance is not known. Hence, it must be unzipped before we realize the dream of developing crop plants that can withstand soil salinities like that of halophytic species.

Acknowledgments The research activities in the laboratory of Dr. AS supported by DST-PURSE, DST-FIST, and UGC-CAS, New Delhi, are gratefully acknowledged. PBK gratefully acknowledges the CSIR, New Delhi, for providing the Emeritus Scientist Fellowship.

References

Aasamaa K, Sober A (2001) Hydraulic conductance and stomatal sensitivity to changes of leaf water status in six deciduous tree species. Biologia Plant 44:65–73. https://doi.org/10.1023/A:1017970304768

Abd Elhalim ME, Abo-Alatta OK, Habib SA, Abd Elbar Ola H (2016) The anatomical features of the desert halophytes *Zygophyllum album* L.F. and *Nitraria retusa* (Forssk.) Asch. Ann Agric Sci 61:97–104. https://doi.org/10.1016/j.aoas.2015.12.001

Adolf VI, Shabala S, Andersen MN, Razzaghi F, Jacobsen SE (2012) Varietal differences of quinoa's tolerance to saline conditions. Plant Soil 357:117–129. https://doi.org/10.1007/s11104-012-1133-7

Albert R, Popp M (1977) Chemical composition of halophytes from the Neusiedler Lake region in Austria. Ecologia (Berl) 27:157–170. https://doi.org/10.1007/bf00345820

Aslam R, Bostan N, Amen N, Maria M, Safdar W (2011) A critical review on halophytes: salt tolerant plants. J Med Plants Res 5:7108–7118. https://doi.org/10.5897/jmprx11.009

Balnokin YV, Kurkova EB, Myasoedov NA, Bukhov NG (2004) Structural and functional state of thylakoids in a halophyte *Suaeda altissima* before and after disturbance of salt-water balance by extremely high concentrations of NaCl. Russ J Plant Physiol 51:815–821. https://doi.org/10.1023/b:rupp.0000047831.85509.a6

Balnokin YV, Myasoedov NA, Shamsutdinov ZS, Shamsutdinov NZ (2005) Significance of Na^+ and K^+ for sustained hydration of organ tissues in ecologically distinct halophytes of the family Chenopodiaceae. Russ J Plant Physiol 52:779–787. https://doi.org/10.1007/s11183-005-0115-5

Barkla BJ, Vera-Estrella R (2015) Single cell-type comparative metabolomics of epidermal bladder cells from the halophyte *Mesembryanthemum crystallinum*. Front Plant Sci. https://doi.org/10.3389/fpls.2015.00435

Barrett-Lennard EG, Setter TL (2010) Developing saline agriculture: moving from traits and genes to systems. Funct Plant Biol 37(7). https://doi.org/10.1071/FPv37n7_FO

Barton KA, Schattat MH, Jakob T, Hause G, Wilhelm C, Mckenna JF, Máthé C, Runions J, Van Damme D, Mathur J (2016) Epidermal pavement cells of *Arabidopsis* have chloroplasts. Plant Physiol 171:723–726. https://doi.org/10.1104/pp.16.00608

Benz BW, Martin CE (2006) Foliar trichomes, boundary layers, and gas exchange in 12 species of epiphytic *Tillandsia* (Bromeliaceae). J Plant Physiol 163:648–656. https://doi.org/10.1016/j.jplph.2005.05.008

Beyschlag W, Eckstein J (2001) Towards a causal analysis of stomatal patchiness: the role of stomatal size variability and hydrological heterogeneity. Acta Oecologica-Int J Ecol 22:161–173. https://doi.org/10.1016/s1146-609x(01)01110-9

Bose J, Shabala L, Pottosin I, Zeng F, Velarde-Buendia A, Massart A, Poschenrieder C, Hariadi Y, Shbala S (2014) Kinetics of xylem loading, membrane potential maintenance, and sensitivity of K^+ –permeable channels to ROS: physiological traits that differentiate salinity tolerance between pea and barley. Plant Cell Environ 37:589–600. https://doi.org/10.1111/pce.12180

Boughalleb F, Denden M, Ben Tiba B (2009) Photosystem II photochemistry and physiological parameters of three fodder shrubs, *Nitraria retusa, Atriplex halimus*, and *Medicago arborea* under salt stress. Acta Physiol Plant 31:463–476. https://doi.org/10.1007/s11738-008-0254-3

Breckle SW (1990) Salinity tolerance of different halophyte types. In: Bassam NE, Dambroth M, Loughman BC (eds) Genetic aspects of plant mineral nutrition. Springer, Berlin, pp 167–175. https://doi.org/10.1007/978-94-009-2053-8_26

Breckle SW (1995) How do halophyte overcome salinity? Biology of salt tolerant plants. In: Khan MA, Ungar IA (eds) Chelsca, pp 199–213

Bucur N, Dobrescu C, Turcu GH, Lixandru GH, Tesu C, Dumbrava I, Afusoaie (1957) Contributii la studiul halofiliei plantelor din pasuni si fanete de saratura din Depresiunea Jijia-Bahlui (parten a I-a). Stud. Si Cerc. (Biol. Si St. Agric.) Acad. R.P. Romane, filial Iasi 8:277–317

Céccoli G, Ramos J, Pilatti V, Dellaferrera I, Tivano JC, Taleisnik E, Vegetti AC (2015) Salt glands in the Poaceae family and their relationship to salinity tolerance. Bot Rev 81:162–178. https://doi.org/10.1007/s12229-015-9153-7

Chen J, Xiao Q, Wu F, Dong X, He J, Pei Z, Zheng H, Nasholm T (2010) Nitric oxide enhances salt secretion and Na^+ sequestration in a mangrove plant, *Avicennia marina*, through increasing the expression of H^+-ATPase and Na^+/H^+ antiporter under high salinity. Tree Physiol 30:1570–1585. https://doi.org/10.1093/treephys/tpq086

Churchman ML, Brown ML, Kato N, Kirik V, Hulskamp M, Inze D, De Veylder L, Walker JD, Zheng Z, Oppenheimer DG, Gwin T, Churchman J, Larkin JC (2006) SIAMESE, a plant-specific cell cycle regulator, controls endoreplication onset in *Arabidopsis thaliana*. Plant Cell 18:3145–3157. https://doi.org/10.1105/tpc.106.044834

Claudia T, Murray DR (2000) Effects of elevated atmospheric [CO_2] in *Panicum* species of different photosynthetic modes (*Poaceae: Panicoideae*). In: Jacobs SWL, Everett J (eds) Grasses: systematics and evolution. CSIRO Publishing, Collingwood, pp 259–266

Colmer TD, Munns R, Flowers TJ (2005) Improving salt tolerance of wheat and barley: future prospects. Aust J Exp Agric 45(11):1425. https://doi.org/10.1071/EA04162

Colmenero-Flores JM, Martinez G, Gamba G, Vazquez N, Iglesias DJ, Brumos J, Talon M (2007) Identification and functional characterization of cation–chloride cotransporters in plants. Plant J 50:278–292. https://doi.org/10.1111/j.1365-313x.2007.03048.x

Cushman JC (2001) Osmoregulation in plants: implications for agriculture. Am Zool 414:758–769. https://doi.org/10.1093/icb/41.4.758

Cutler DF, Botha T, Stevenson DW (2007) Plant Anatomy- An applied approach. Blackwell Publishing, Australia. https://doi.org/10.1093/aob/mcn118

Dang ZH, Zheng LL, Wang J, Gao Z, Wu SB, Qi Z, Qi Z, Wang YC (2013) Transcriptomic profiling of the salt-stress response in the wild recretohalophyte *Reaumuria trigyna*. BMC Genomics 14:29. https://doi.org/10.1186/1471-2164-14-29

Dang ZH, Qi Q, Zhang HR, Yu LH, Wu SB, Wang YC (2014) Identification of salt-stress-induced genes from the RNA-Seq data of *Reaumuria trigyna* using differential-display reverse transcription PCR. Int J Genomics 381501. https://doi.org/10.1155/2014/381501

Dassanayake M, Larkin JC (2017) Making plants break a sweat: the structure, function, and evolution of plant salt glands. Front Plant Sci 8:406. https://doi.org/10.3389/fpls.2017.00406

Delgado D, Alonso-Blanco C, Fenoll C, Mena M (2011) Natural variation in stomatal abundance of *Arabidopsis thaliana* includes cryptic diversity for different developmental processes. Ann Bot 107:1247–1258. https://doi.org/10.1093/aob/mcr060

Delpire E, Mount DB (2002) Human and murine phenotypes associated with defects in cation-chloride cotransport. Annu Rev Physiol 64:803–843. https://doi.org/10.1146/annurev.physiol.64.081501.155847

Demidchik V, Cuin TA, Svistunenko D, Smith SJ, Miller AJ, Shabala S, Sokolik A, Yurin V (2010) *Arabidopsis* root K^+ –efflux conductance activated by hydroxyl radicals: single-channel properties, genetic basis and involvement in stress-induced cell death. J Cell Sci 123:1468–1479. https://doi.org/10.1242/jcs.064352

Deng Y, Feng Z, Yuan F, Guo J, Suo S, Wang B (2015) Identification and functional analysis of the autofluorescent substance in *Limonium bicolor* salt glands. Plant Physiol Biochem 97:20–27. https://doi.org/10.1016/j.plaphy.2015.09.007

Dickison WC (2000) Integrative plant anatomy. Harcourt Academic Press, New York

Dschida W, Platt-Aloia K, Thomson W (1992) Epidermal peels of *Avicennia germinans* (L.) Stearn: a useful system to study the function of salt glands. Ann Bot 70:501–509. https://doi.org/10.1093/oxfordjournals.aob.a088510

Duval-Jouve J (1871) Sur quelques tissus de *Joncées*, de *Cyperacées* et de *Graminées*. Bull Soc Bot Fr 18:231–239

Edwards D, Kerp H, Hass H (1998) Stomata in early land plants: an anatomical and ecophysiological approach. J Exp Bot 49:255–278. https://doi.org/10.1093/jexbot/49.suppl_1.255

Ehleringer JR, Forseth I (1980) Solar tracking by plants. Science 210:1094–1098. https://doi.org/10.1126/science.210.4474.1094

Ehleringer J, Björkman O, Mooney HA (1976) Leaf pubescence: effects on absorptance and photosynthesis in a desert shrub. Science 191:376–377. https://doi.org/10.1126/science.192.4237.376

Esau K (1965) Plant anatomy. Wiley, New York

Feng Z, Sun Q, Deng Y, Sun S, Zhang J, Wang B (2014) Study on pathway and characteristics of ion secretion of salt glands of *Limonium bicolor*. Acta Physiol Plant 36:2729–2741. https://doi.org/10.1007/s11738-014-1644-3

Feng Z, Deng Y, Zhang S, Liang X, Yuan F, Hao J et al (2015) K^+ accumulation in the cytoplasm and nucleus of the salt gland cells of *Limonium bicolor* accompanies increased rates of salt secretion under NaCl treatment using NanoSIMS. Plant Sci 238:286–296. https://doi.org/10.1016/j.plantsci.2015.06.021

Flowers TJ (1985) Physiology of halophytes. Plant Soil 89:41–56. https://doi.org/10.1007/978-94-009-5111-2_3

Flowers TJ, Colmer TD (2008) Flooding tolerance in halophytes. New Phytol 179:964–974. https://doi.org/10.1111/j.1469-8137.2008.02483.x

Flowers TJ, Colmer TD (2015) Plant salt tolerance: adaptations in halophytes. Ann Bot 115:327–331. https://doi.org/10.1093/aob/mcu267

Flowers TJ, Troke P, Yeo A (1977) The mechanism of salt tolerance in halophytes. Annu Rev Plant Physiol 28:89–121. https://doi.org/10.1146/annurev.pp.28.060177.000513

Flowers TJ, Galal HK, Bromham L (2010) Evolution of halophytes: multiple origins of salt tolerance in land plants. Funct Plant Biol 37:604–612. https://doi.org/10.1071/fp09269

Gan Y, Kumimoto R, Liu C, Ratcliffe O, Yu H, Broun P (2006) Glabrous inflorescence stems modulates the regulation by gibberellins of epidermal differentiation and shoot maturation in *Arabidopsis*. Plant Cell 18:1383–1395. https://doi.org/10.1105/tpc.106.041533

Glover BJ (2000) Differentiation in plant epidermal cells. J Exp Bot 51:497–505. https://doi.org/10.1093/jexbot/51.344.497

Greenway H, Munns R (1980) Mechanisms of salt tolerance in non-halophytes. Ann Rev Plant Physiol 31:149–190. https://doi.org/10.1146/annurev.pp.31.060180.001053

Grigore M (2008) Introducere în Halofitologie. Elemente de anatomie integrativ. Edit. Pim. Ia. pp 3–28

Grigore MN, Toma C, Boscaiu M (2010) Ecological implications of bulliform cells on halophytes in salt and water stress natural conditions. An. Şt. Univ., Al. I. Cuza" Iaşi, s. II, a. (Biol. Veget.) 56:5–15

Hou XY (1982) Botanical geography and chemical ingredients of dominant plant species in China. Science, Beijing. (in Chinese)

Hughes FM Jr, Cidlowski JA (1999) Potassium is a critical regulator of apoptotic enzymes *in vitro* and *in vivo*. Adv Enzyme Reg 39:157–171. https://doi.org/10.1016/s0065-2571(98)00010-7

Ishida T, Kurata T, Okada K, Wada T (2008) A genetic regulatory network in the development of trichomes and root hairs. Ann Rev Plant Biol 59:365–386. https://doi.org/10.1146/annurev.arplant.59.032607.092949

Johnson HB (1975) Plant pubescence: an ecological perspective. Bot Rev 41:233–258. https://doi.org/10.1007/bf02860838

Karimi G, Ghorbanli M, Heidari H, Nejad RAK, Assareh MH (2005) The effects of NaCl on growth, water relations, osmolytes and ion content in *Kochia prostrata*. Biol Plant 49:301–304. https://doi.org/10.1007/s10535-005-1304-y

Kemp PR, Cunningham GL (1981) Light, temperature and salinity effects on growth, leaf anatomy and photosynthesis of *Distichlis spicata* (L) green. Am J Bot 68:507–516. https://doi.org/10.2307/2443026

Kobayashi H, Masaoka Y, Takahashi Y, Ide Y, Sato S (2007) Ability of salt glands in Rhodes grass (*Chloris gayana* Kunth) to secrete Na^+ and K^+. Soil Sci Plant Nutr 53:764–771. https://doi.org/10.1111/j.1747-0765.2007.00192.x

Liphschitz N, Adiva-Shomer-Ilan, Eshel A, Waisel Y (1974) Salt glands on leaves of Rhodes grass (*Chloris gayana* Kth.) Ann Bot 38:459–462. https://doi.org/10.1093/oxfordjournals.aob.a084829

Lovelock CE, Ball MC (2002) Influence of salinity on photosynthesis of halophytes. In: Lauchli A, Luttge U (eds) Salinity: environment-plants-molecules. Springer, Dordrecht, pp 315–339. https://doi.org/10.1007/0-306-48155-3_15

Lun'kov RV, Andreev IM, Myasoedov NA, Khailova GF, Kurkova EB, Balnokin YV (2005) Functional identification of H^+-ATPase and Na^+/H^+ antiporter in the plasma membrane isolated from the root cells of salt accumulating halophyte *Suaeda altissima*. Russian J Plant Physiol 52:635–644. https://doi.org/10.1007/s11183-005-0094-6

Ma H, Tian C, Feng G, Yuan J (2011) Ability of multicellular salt glands in *Tamarix* species to secrete Na^+ and K^+ selectively. Sci China Life Sci 54:282–289. https://doi.org/10.1007/s11427-011-4145-2

Marius-Nicusor G, Constantin T (2010) A proposal for a new halophytes classification based on integrative anatomy observations. Muz. Olteniei, Craiova, Stud. şi Com., Şt. Nat 26:45–50.

Marschner H (1995) Mineral nutrition of higher plants. San Diego, Academic Press, USA. https://doi.org/10.1016/b978-0-12-473542-2.x5000-7

Martin C, Glover BJ (2007) Functional aspects of cell patterning in aerial epidermis. Curr Opin Plant Biol 10:70–82. https://doi.org/10.1016/j.pbi.2006.11.004

Martinez JP, Kinet JM, Bajji M, Lutts S (2005) NaCl alleviates polyethylene glycol-induced water stress in the halophyte species *Atriplex halimus* L. J Exp Bot 56:2421–2431. https://doi.org/10.1093/jxb/eri235

Millar J, Roots J (2012) Changes in Australian agriculture and land use: implications for future food security. Int J Agric Sustain 10:25–39

Mittler R (2006) Abiotic stress, the field environment and stress combination. Trends Plant Sci 11:15–19. https://doi.org/10.1016/j.tplants.2005.11.002

Mott KA, Peak D (2007) Stomatal patchiness and task-performing networks. Ann Bot 99:219–226. https://doi.org/10.1093/aob/mcl234

Naz N, Hameed M, Ashraf M, Al-Qurainy F, Arshad M (2010) Relationships between gas-exchange characteristics and stomatal structural modifications in some desert grasses under high salinity. Photosynthetica 48:446–456. https://doi.org/10.1007/s11099-010-0059-7

Niu X, Narasimhan ML, Salzman RA, Bressan RA, Hasegawa PM (1993) NaCl regulation of plasma membrane H^+-ATPase gene expression in a glycophyte and a halophyte. Plant Physiol 103:713–718. https://doi.org/10.1104/pp.103.3.713

Niu X, Damsz B, Kononowicz AK, Bressan RA, Hasegawa PM (1996) NaCl-induced alterations in both cell structure and tissue-specific plasma membrane H^+-ATPase gene expression. Plant Physiol 111:679–686. https://doi.org/10.1104/pp.111.3.679

Oh DH, Lee SY, Bressan RA, Yun DJ, Bohnert HJ (2010) Intracellular consequences of SOS1 deficiency during salt stress. J Exp Bot 61:1205–1213. https://doi.org/10.1093/jxb/erp391

Oh DH, Barkla BJ, Vera-Estrella R, Pantoja O, Lee SY, Bohnert HJ, Dassanayake M (2015) Cell type-specific responses to salinity-the epidermal bladder cell transcriptome of *Mesembryanthemum crystallinum*. New Phytol 207:627–644. https://doi.org/10.1111/nph.13414

Omami EN, Hammes PS, Robbertse PJ (2006) Differences in salinity tolerance for growth and water-use efficiency in some amaranth (*Amaranthus* spp.) genotypes. New Zealand J Crop Hort Sci 34:11–22. https://doi.org/10.1080/01140671.2006.9514382

Orsini F, Matilde Paino D'Urzo F, Inan G, Serra S, Oh D, Mickelbart MV, Consiglio F, Xia Li X, Jeong JC, Yun DJ, Bohnert HJ, Bressan RA, Maggio A (2010) A comparative study of salt tolerance parameters in 11 wild relatives of *Arabidopsis thaliana*. J Exp Bot 61:3787–3798. https://doi.org/10.1093/jxb/erq188

Orsini F, Accorsi M, Gianquinto G, Dinelli G, Antognoni F, Ruiz Carrasco KB, Martinez EA, Alnayef M, Marotti I, Bosi S, Biondi S (2011) Beyond the ionic and osmotic response to salinity in *Chenopodium quinoa*: functional elements of successful halophytism. Funct Plant Biol 38:818–831. https://doi.org/10.1071/fp11088

Overall RL, Blackman LM (1996) A model of the macromolecular structure of plasmodesmata. Trends Plant Sci 1:307–311. https://doi.org/10.1016/s1360-1385(96)88177-0

Pallaghy CK (1970) The effect of Ca^{2+} on the ion specificity of stomatal opening in epidermal strips of *Vicia faba*. Z Pflanzenphysiol 62:58–62

Pan Y, Guo H, Wang S, Zhao B, Zhang J, Ma Q, Yin HJ, Bao AK (2016) The photosynthesis, Na^+/K^+ homeostasis and osmotic adjustment of *Atriplex canescens* in response to salinity. Front Plant Sci 7:848. https://doi.org/10.3389/fpls.2016.00848

Parida AK, Veerabathini SK, Kumari A, Agarwal PK (2016) Physiological, anatomical and metabolic implications of salt tolerance in the halophyte *Salvadora persica* under hydroponic culture condition. Front Plant Sci 7:351. https://doi.org/10.3389/fpls.2016.00351

Patrut DI, Adelina P, Ioan C (2005) Biodiversitatea halofitelor din Campia Banatului. Edit, Eurobit, Timisoara

Perera L, Mansfield TA, Malloch AJC (1994) Stomatal responses to sodium-ions in *Aster tripolium* – a new hypothesis to explain salinity regulation in aboveground tissues. Plant Cell Environ 17:335–340. https://doi.org/10.1111/j.1365-3040.1994.tb00300.x

Perera LKRR, De Silva DLR, Mansfield TA (1997) Avoidance of sodium accumulation by the stomatal guard cells of the halophyte *Aster tripolium*. J Exp Bot 48:707–717. https://doi.org/10.1093/jxb/48.3.707

Peterson PM (2000) Systematics of the *Muhlenbergiinae* (*Poaceae: Eragrostidae*). In: Jacobs SWL, Everett J (eds) Grasses: systematics and evolution. CSIRO Publishing, Collingwood, pp 195–212

Pilot G, Gaymard F, Mouline K, Cherel I, Sentenac H (2003) Regulated expression of Arabidopsis Shaker K^+ channel genes involved in K^+ uptake and distribution in the plant. Plant Mol Biol 51:773–787. https://doi.org/10.1023/A:1022597102282

Popp M, Polania J, Weiper M (1993) Physiological adaptations to different salinity levels in mangrove. In: Lieth H, Al Masoom A (eds) Towards the rational use of high salinity tolerant plants, vol 1. Kluwer Academic Publisher, Dordrecht, pp 217–224. https://doi.org/10.1007/978-94-011-1858-3_22

Pospisilova J, Santrucek J (1994) Stomatal patchiness. Biologia Plant 36:481–510. https://doi.org/10.1007/bf02921169

Quarrie SA, Jones HG (1977) Effects of abscisic acid and water stress on development and morphology of wheat. J Exp Bot 28:192–203. https://doi.org/10.1093/jxb/28.1.192

Rajput KS (2016) Development of successive cambia and wood structure in stem of *Rivea hypocriteriformis* (Convolvulaceae). Polish Bot J 61:89–98. https://doi.org/10.1515/pbj-2016-0003

Rajput KS, Marcati CR (2013) Stem anatomy and development of successive cambia in *Hebanthe eriantha* (Poir.) Pedersen: a neotropical climbing species of the Amaranthaceae. Plant Systematic Evol 299:1449–1459. https://doi.org/10.1007/s00606-013-0807-9

Ramadan T (1998) Ecophysiology of salt excretion in the xero-halophyte *Reaumuria hirtella*. New Phytol 139:273–281. https://doi.org/10.1046/j.1469-8137.1998.00159.x

Ramadan T, Flowers TJ (2004) Effects of salinity and benzyl adenine on development and function of micro hairs of *Zea mays* L. Planta 219:639–648. https://doi.org/10.1007/s00425-004-1269-7

Reginato M, Reinoso H, Llanes AS, Luna MV (2013) Stomatal abundance and distribution in *Prosopis strombulifera* plants growing under different iso-osmotic salt treatments. American J Plant Sci 4:80–90. https://doi.org/10.4236/ajps.2013.412a3010

Robert EMR, Schmitz N, Boeren I, Driessens T, Herremans K, De Mey J, Van de Casteele E, Beeckman H, Koedam N (2011) Successive cambia: a developmental oddity or an adaptive structure? PLoS one 6(1):e16558. https://doi.org/10.1371/journal.pone.0016558

Robinson MF (1996) Sodium-induced stomatal closure in the maritime halophyte *Aster tripolium (L.)*. Lancaster University U.K, Ph.D. Thesis.

Sabovljevic M, Sabovljevic A (2007) Contribution to the coastal bryophytes of the Northern Mediterranean: are there halophytes among bryophytes? Phytol Balcanica 13:131–135

Sanadhya P, Agarwal P, Khedia J, Agarwal PK (2015) A low-affinity K^+ transporter AlHKT2; 1 from recretohalophyte *Aeluropus lagopoides* confers salt tolerance in yeast. Mol Biotechnol 57:489–498. https://doi.org/10.1007/s12033-015-9842-9

Santos J, Al-Azzawi M, Aronson J, Flowers TJ (2016) eHALOPH a database of salt-tolerant plants: helping put halophytes to work. Plant Cell Physiol 57:e10. https://doi.org/10.1093/pcp/pcv155

Serna L (2009) Cell fate transitions during stomatal development. BioEssays 31:865–873. https://doi.org/10.1002/bies.200800231

Shabala S (2013) Learning from halophytes: physiological basis and strategies to improve abiotic stress tolerance in crops. Ann Bot 112:1209–1221. https://doi.org/10.1093/aob/mct205

Shabala S, Mackay A (2011) Ion transport in halophytes. Adv Bot Res 57:151–199. https://doi.org/10.1016/b978-0-12-387692-8.00005-9

Shabala S, Babourina O, Newman I (2000) Ion-specific mechanisms of osmoregulation in bean mesophyll cells. J Exp Bot 51:1243–1253. https://doi.org/10.1093/jxb/51.348.1243

Shabala L, Mackay A, Tian Y, Jacobsen SE, Zhou DW, Shabala S (2012) Oxidative stress protection and stomatal patterning as components of salinity tolerance mechanism in quinoa (*Chenopodium quinoa*). Physiol Plant 146:26–38. https://doi.org/10.1111/j.1399-3054.2012.01599.x

Shabala L, Hariadi Y, Jacobsen SE (2013) Genotypic difference in salinity tolerance in quinoa is determined by differential control of xylem Na^+ loading and stomatal density. J Plant Physiol 170:906–914. https://doi.org/10.1016/j.jplph.2013.01.014

Shabala S, Bose J, Hedrich R (2014) Salt bladders: do they matter? Trends Plant Sci 19:687–691. https://doi.org/10.1016/j.tplants.2014.09.001

Shi HZ, Quintero FJ, Pardo JM, Zhu JK (2002) The putative plasma membrane Na^+/H^+ antiporter SOS1 controls long-distance Na^+ transport in plants. Plant Cell 14:465–477. https://doi.org/10.1105/tpc.010371

Shtein I, Shelef Y, Marom Z, Zelinger E, Schwartz A, Popper ZA, Bar-On B, Harpaz-Saad S (2017) Stomatal cell wall composition: distinctive structural patterns associated with different phylogenetic groups. Ann Bot 119:1021–1033. https://doi.org/10.1093/aob/mcw275

Sobrado MA, Greaves ED (2000) Leaf secretion composition of the mangrove species *Avicennia germinans* (L.) in relation to salinity: a case study by using total-reflection X-ray fluorescence analysis. Plant Sci 159:1–5. https://doi.org/10.1016/s0168-9452(00)00292-2

Spence RD, Wu H, Sharpe PJ, Clark K (1986) Water stress effects on guard cell anatomy and the mechanical advantage of the epidermal cells. Plant Cell Environ 9:197–202. https://doi.org/10.1111/1365-3040.ep11611639

Steiner M (1934) To the ecology of the salt march of the Nordostlichen united countries of Nordamerika. Jahrb Know Offered 81:94

Tamaio N, Cardoso-Vieira R, Angyalossy V (2009) Origin of successive cambia on stem in three species of Menispermaceae. Rev Bras Bot 32:839–848. https://doi.org/10.1590/s0100-84042009000400021

Tan WK, Lim TM, Loh CS (2010) A simple, rapid method to isolate salt glands for three-dimensional visualization, fluorescence imaging and cytological studies. Plant Methods 6:24. https://doi.org/10.1186/1746-4811-6-24

Tan WK, Lin Q, Lim TM, Kumar P, Loh CS (2013) Dynamic secretion changes in the salt glands of the mangrove tree species *Avicennia officinalis* in response to a changing saline environment. Plant Cell Environ 36:1410–1422. https://doi.org/10.1111/pce.12068

Terrazas T, Aguilar-Rodríguez S, Ojanguren CT (2011) Development of successive cambia, cambial activity, and their relationship to physiological traits in *Ipomoea arborescens* (Convolvulaceae) seedlings. Am J Bot 98:765–774. https://doi.org/10.3732/ajb.1000182

Tester M (1988) Blockade of potassium channels in the plasmalemma of *Chara corallina* by tetraethylammonium, Ba^{2+}, Na^+ and Cs^+. J Membrane Biol 105:77–85. https://doi.org/10.1007/bf01871108

Thiel G, Blatt MR (1991) The mechanism of ion permeation through K^+ channels of stomatal guard cells: voltage-dependent block by Na^+. J Plant Physiol 138:326–334. https://doi.org/10.1016/s0176-1617(11)80296-8

Thomson W, Platt-Aloia K (1985) The ultrastructure of the plasmodesmata of the salt glands of *Tamarix* as revealed by transmission and freeze-fracture electron microscopy. Protoplasma 125:13–23. https://doi.org/10.1007/BF01297346

Thomson WW, Faraday CD, Oross JW (1988) Salt glands. In: Baker DA, Hall JL (eds) Solute transport in plant cells and tissues. Longman, Harlow, pp 498–537

Topa E (1939) Vegetatia halofitelor din nordul Romaniei in legatura cu cea din restul tarii. Teza presentata la Facultatea de Stiinte din Cernauti pentru obtinerea titlului de doctor in Stiintele Naturale

Tsiantis MS, Bartholomew DM, Smith JA (1996) Salt regulation of transcript levels for the c subunit of a leaf vacuolar H^+-ATPase in the halophyte *Mesembryanthemum crystallinum*. Plant J 9:729–736. https://doi.org/10.1046/j.1365-313x.1996.9050729.x

Ungar IA (1978) Halophyte seed germination. Bot Rev 44:233–364

van Eijk M (1939) Analyse der Wirkung des NaCl auf die Entwicklung Sukkulenze und Transpiration bei *Salicornia herbacea*, sowie Untersuchungen über den Einfluss der Salzaufnahme, auf die Wurzelatmung bei *Aster tripolium*. Rec Trav Bot Neerl 36:559–657

Véry AA, Robinson MF, Mansfield TA, Sanders D (1998) Guard cell cation channels are involved in Na^+-induced stomatal closure in a halophyte. Plant J 14:509–521. https://doi.org/10.1046/j.1365-313x.1998.00147.x

Vijayan K, Chakraborti SP, Ercisli S, Ghosh PD (2008) NaCl induced morpho-biochemical and anatomical changes in mulberry (*Morus* spp.) Plant Growth Regul 56:61–69. https://doi.org/10.1007/s10725-008-9284-5

Weber DJ (2009) Adaptive mechanisms of halophytes in desert regions. In: Ashraf M, Ozturk M, Athar H (eds) Salinity and water stress. Tasks for vegetation sciences, vol 44. Springer, Dordrecht. https://doi.org/10.1007/978-1-4020-9065-3_18

Wegner LH, De Boer AH (1997) Properties of two outward-rectifying channels in root xylem parenchyma cells suggest a role in K^+ homeostasis and long distance signaling. Plant Physiol 115:1707–1719. https://doi.org/10.1104/pp.115.4.1707

Wegner LH, Raschke K (1994) Ion channels in the xylem parenchyma of barley roots – a procedure to isolate protoplasts from this tissue and a patch-clamp exploration of salt passageways into xylem vessels. Plant Physiol 105:799–813. https://doi.org/10.1104/pp.105.3.799

Xu Z, Zhou G (2008) Responses of leaf stomatal density to water status and its relationship with photosynthesis in a grass. J Exp Bot 59:3317–3325. https://doi.org/10.1093/jxb/ern185

Yan L, Li H, Liu Y (2002) The anatomical ecology studies on the leaf of 13 species in *Caragana* genus. J Arid Land Resour Environ 16:100–106

Yang HM, Wang GX (2001) Leaf stomatal densities and distribution in *Triticum aestivum* under drought and CO_2 enrichment. Acta Phytoecologica Sin 25:312–316

Yensen NP, Biel KY (2006) Soil remediation *via* salt-conduction and the hypotheses of halosynthesis and photoprotection. Tasks for Vegetation Science. Series-40. Ecophysiology of High Salinity Tolerant Plants. pp 313–344. https://doi.org/10.1007/1-4020-4018-0_21

Youssef AM (2009) Salt tolerance mechanisms in some halophytes from Saudi Arabia and Egypt. Res J Agr Biol Sci 5:623–638

Yuan F, Chen M, Leng BY, Wang B (2013) An efficient autofluorescence method for screening *Limonium bicolor* mutants for abnormal salt gland density and salt secretion. S Afr J Bot 88:110–117. https://doi.org/10.1016/j.sajb.2013.06.007

Yuan F, Lyu MA, Leng B, Zheng GY, Feng ZT, Li P, Zhu XG, Wang BS (2015) Comparative transcriptome analysis of developmental stages of the *Limonium bicolor* leaf generates insights into salt gland differentiation. Plant Cell Environ 38:1637–1657. https://doi.org/10.1111/pce.12514

Yuan F, Lyu MJA, Leng BY, Zhu XG, Wang BS (2016) The transcriptome of NaCl-treated *Limonium bicolor* leaves reveals the genes controlling salt secretion of salt gland. Plant Mol Biol 91:241–256. https://doi.org/10.1007/s11103-016-0460-0

Zeiger E (1983) The biology of stomatal guard cells. Ann Rev Plant Physiol 34:441–474. https://doi.org/10.1146/annurev.pp.34.060183.002301

Zhang YP, Wang ZM, Wu YC, Zhang X (2006) Stomatal characteristics of different green organs in wheat under different irrigation regimes. Acta Agron Sin 32:70–75

Zhao KF, Li FZ (1999) Halophytes in China. Scientific, Beijing. (in Chinese)

Zhao KF, Song J, Feng G, Zhao M, Liu JP (2011) Species, types, distribution, and economic potential of halophytes in China. Plant Soil 342:495–509. https://doi.org/10.1007/s11104-010-0470-7

Zouhaier B, Abdallah A, Najla T, Wahbi D, Wided C, Aouatef BA et al (2015) Scanning and transmission electron microscopy and X-ray analysis of leaf salt glands of *Limoniastrum guyonianum* Boiss. under NaCl salinity. Micron 78:1–9. https://doi.org/10.1016/j.micron.2015.06.001

Chapter 3
Targeting Aquaporins for Conferring Salinity Tolerance in Crops

Kundan Kumar and Ankush Ashok Saddhe

Abstract Salinity is one of the well-known abiotic stresses which affects crop productivity through imposing ion imbalance and disrupting the metabolic pathways. Soil salinity is dramatically increasing throughout the world because of climate change, rise in sea levels, excessive irrigation, and natural leaching process. To overcome this problem, many approaches were reported including selection of natural salt-tolerant variety, breeding program, and genetic-engineered plants. Membrane intrinsic proteins (MIPs; also called aquaporins) are membrane channel proteins initially discovered as water channels, but their roles in the transport of small neutral solutes, metal ions, and gasses are now well established. Based on homology and subcellular localization, plant MIPs are divided into four major subfamilies: plasma membrane intrinsic proteins (PIPs), tonoplast intrinsic proteins (TIPs), NOD26-like intrinsic proteins (NIPs), and small basic intrinsic proteins (SIPs). Besides these four subfamilies, some unique subfamilies were reported such as GlpF-like intrinsic proteins (GIPs), hybrid intrinsic proteins (HIPs), and the uncategorized X intrinsic proteins (XIPs). In plants, MIPs are involved in diverse physiological roles such as seed germination; fruit ripening; leaf, petal, and stomata movement; phloem loading and unloading; reproduction; and stress response. However, a large number of studies have suggested the involvement of MIPs in various abiotic stresses, including drought, salt, and cold stress. PIPs and TIPs have shown differential regulation pattern in roots and shoots of *Arabidopsis*, barley, and maize in salinity stress. Moreover, overexpression studies of various PIPs and TIPs in plant suggest their possible role in salt tolerance. Transcriptome analyses of citrus under salt stress showed that in addition to PIPs and TIPs, most of the NIPs, XIPs, and SIPs were differentially regulated in root tissues. In the present chapter, we discussed roles of plant aquaporins in salinity stress and exploitating the same for genetic engineering approach.

K. Kumar (✉) · A. A. Saddhe
Department of Biological Sciences, Birla Institute of Technology & Science Pilani, K K Birla Goa Campus, Goa, India
e-mail: kundan@goa.bits-pilani.ac.in

V. Kumar et al. (eds.), *Salinity Responses and Tolerance in Plants, Volume 1*,
https://doi.org/10.1007/978-3-319-75671-4_3

Keywords Abiotic stress · Salinity · Aquaporins · MIPs · PIPs · TIPs · NIPs · SIPs · XIPs · HIPs · GIPs

Abbreviations

GIPs	GlpF-like intrinsic proteins
HIPs	Hybrid intrinsic proteins
McMipA and McMipC	*Mesembryanthemum crystallinum* MIP-related genes
MIPs	Membrane intrinsic proteins
NIPs	NOD26-like intrinsic proteins
PIPs	Plasma membrane intrinsic proteins
SIPs	Small basic intrinsic proteins
TIPs	Tonoplast intrinsic proteins
XIPs	Uncategorized X intrinsic proteins

3.1 Introduction

Abiotic stresses such as drought, salinity, heat, cold, and anaerobic stress are imposing negative effect on plant growth and productivity (Cavanagh et al. 2008; Munns and Tester 2008; Chinnusamy and Zhu 2009; Mittler and Blumwald 2010; Kumar et al. 2013). Compared to other abiotic stresses, soil salinity is one of the brutal climatic factors which would impose hyperionic and osmotic stress, disrupting metabolic activities and thus limiting the productivity of crop plants (Munns and Tester 2008). Soil salinity is a global issue which affected approximately 45 million hectares of irrigated land, and about 1.5 million hectares of productive land turned into non-fertile lands (Munns and Tester 2008). Salt stress is affecting plants in several ways such as ion imbalances due to Na^+ and Cl accumulation, nutritional disorders, oxidative stress, alteration of metabolic processes, membrane disorganization, enhanced lipid peroxidation and increased production of reactive oxygen species, and reduction of cell division and expansion (Kumar et al. 2013). Recent studies have identified various adaptive responses to salinity stress at cellular, molecular, physiological, and biochemical levels. The plant molecular responses towards salt stress involve interactions and cross talks between numerous metabolic and signaling pathways. The pathways that are involved in various plants include transcription factors, photosynthesis, antioxidant mechanisms, hormone signaling, and osmolyte synthesis (Atkinson et al. 2013; Iyer et al. 2013; Prasch and Sonnewald 2013; Rasmussen et al. 2013).

Till date many approaches have been incorporated from classical to most advance techniques to develop salt-resistant plant variety. But unfortunately not a single approach fulfilled the criteria, and still researchers are hunting for the best possible approach. Here, we discussed recent advancement in the membrane intrinsic protein (aquaporins) research during salt stress. Major intrinsic proteins (MIPs) are a unique class of membrane channel proteins which are ubiquitously distributed in all kingdoms such as bacteria, archaea, protozoa, yeast, and plants (Fortin et al. 1987; Calamita et al. 1995; Mitra et al. 2000; Carbrey et al. 2001; Kozono et al. 2003; Srivastava et al. 2016; Deshmukh et al. 2016, 2017). It is mainly involved in water homeostasis and transport, in addition to a wide range of low-molecular-weight solutes across membrane such as glycerol, urea, ammonia (NH_3), methyl ammonium, hydrogen peroxide, formamide, acetamide, lactic acid, CO_2, and metalloids such as boron (B), silica (Si), arsenic (As), and antimony (Sb) (Forrest and Bhave 2007; Maurel et al. 2008; Srivastava et al. 2016; Deshmukh et al. 2016). Major intrinsic proteins (MIPs), as a broader term, are used currently to describe aquaporins because they are not only water channels but are also involved in transport of small uncharged and cation molecules (Yool et al. 1996; Forrest and Bhave 2007; Byrt et al. 2017). Plant MIP is considered as one of the largest superfamilies with more than 30 isoforms, almost three times more compared to animal MIP family members. The multiple isoforms suggest that MIPs have important roles in plant life, but the functions of some subfamilies and individuals are still unknown. For instance, *Arabidopsis thaliana*, rice (*Oryza sativa*), and maize (*Zea mays*) have more than 30 MIPs, which were further phylogenetically categorized into subfamilies. Genome-wide analysis of dicot genome predicted 35 MIPs in *Arabidopsis*, 55 in *Populus*, 55 in Chinese cabbage, 71 in upland cotton, 66 in soybean, and 41 in potato (Johanson et al. 2001; Gupta and Sankararamakrishnan 2009; Park et al. 2010; Zhang et al. 2013; Venkatesh et al. 2013; Tao et al. 2014; Deshmukh et al. 2015). Similarly monocot MIPs have 38 members in rice, 36 in maize, 41 in sorghum, 47 in banana, and 40 in barley (Chaumont et al. 2001; Forrest and Bhave 2007; Reddy et al. 2015; Hu et al. 2015; Hove et al. 2015; Deshmukh et al. 2015, 2016) (Fig. 3.1). Membrane intrinsic protein (MIP) family is categorized into seven different subfamilies based on their intracellular locations and sequence similarities such as the plasma membrane intrinsic proteins (PIPs), tonoplast intrinsic proteins (TIPs), NOD26-like intrinsic proteins (NIPs), small basic intrinsic proteins (SIPs), GlpF-like intrinsic proteins (GIPs), hybrid intrinsic proteins (HIPs), and the uncategorized X intrinsic proteins (XIPs) (Forest and Bhave 2007; Deshmukh et al. 2017; Secchi et al. 2017). The GlpF-like intrinsic protein (GIP) was reported in *Physcomitrella patens* which was homologous to bacterial glycerol channel. The HIP subfamily shared characteristic features of the PIP and the TIP subfamilies, hence called hybrid intrinsic proteins. The GIPs, XIPs, and HIPs were restricted to moss, but XIPs were reported in various dicots such tobacco, potato, tomato, soybean, and poplar (Danielson and Johanson 2008; Lopez et al. 2012) (Fig. 3.2). PIPs, NIPs, and XIPs are prevalently localized to the plasma membrane. TIP localization is confined to the vacuolar membrane also called as "tonoplast." The SIPs localized to endoplasmic reticulum (ER) were observed during the processes of posttranscriptional and translational modification (Maurel et al. 2015). However, XIPs and GIPs were localized to plasma membrane of plant cell. Three additional subfamilies have been recently reported.

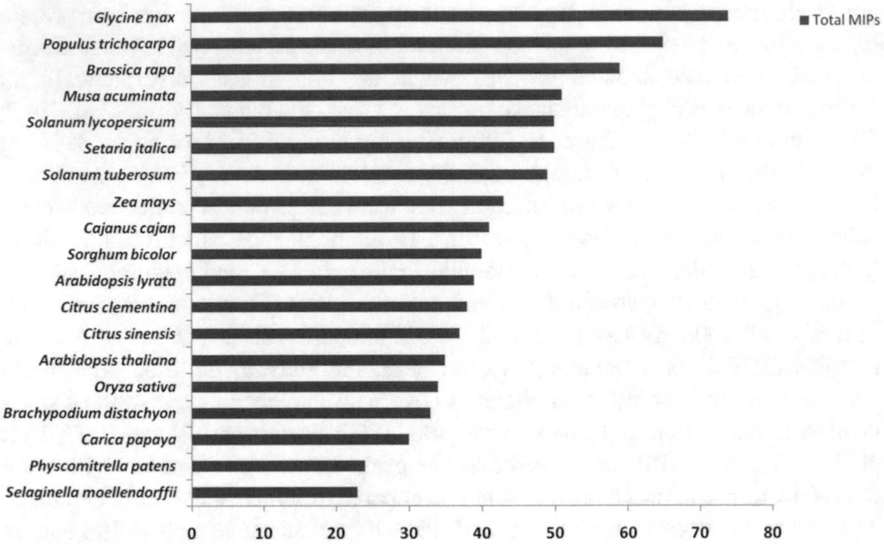

Fig. 3.1 Distribution of membrane intrinsic proteins (MIPs) into plant kingdom represented by bar diagram. X-axis values are represented as number of aquaporins in plant species and Y-axis represent plant species

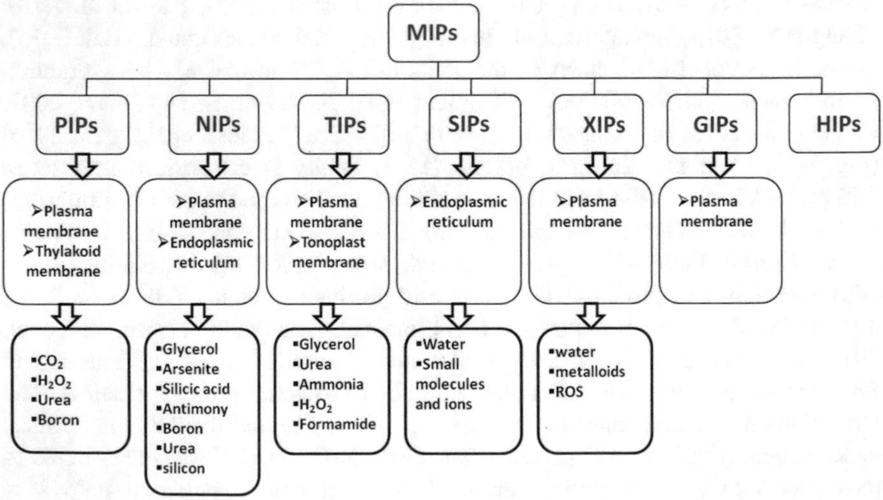

Fig. 3.2 Classification of membrane intrinsic proteins (MIPs) into seven different subfamilies based in their subcellular localization. *PIPs* plasma membrane intrinsic proteins, *TIPs* tonoplast intrinsic proteins, *NIPs* NOD26-like intrinsic proteins, *SIPs* small basic intrinsic proteins, *GIPs* GlpF-like intrinsic proteins, *HIPs* hybrid intrinsic proteins, *XIPs* uncategorized X intrinsic proteins, *ROS* reactive oxygen species

The plant AQPs are highly conserved structures, in which six membrane-spanning α-helices are linked by five short loops and N- and C-termini facing toward the cytosol. Two loops such as loops B and E consist of highly conserved Asp-Pro-Ala (NPA) motifs which play a major role in the formation of water-selective channels. Other important residues in aquaporin sequences are the ones forming the aromatic/arginine selectivity filter (ar/R). This region is formed by four residues toward the extracellular side approximately 8 Å from the NPA region. Four residues of aromatic/arginine (Ar/R) helix 2 (H2), helix 5 (H5), loop E1 (LE1), and loop E2 (LE2) regions contribute to a size-exclusion barrier. The N-terminal has AEF (Ala-Glu-Phe) or AEFXXT motif and two highly conserved NPA (asparagines-proline-alanine) motifs in the loops B and E (Forrest and Bhave 2007; Deshmukh et al. 2016). The substrate specificity of MIPs is depending on Ar/R filter and two NPA motifs which can form narrowest pore around 8 Å in diameter (Forrest and Bhave 2007; Deshmukh et al. 2015).

Plant aquaporins are considered as multifunctional proteins involved in several physiological functions such as water homeostasis, small neutral solutes, metal ions, gasses, and nutrients (boron (B), silicon (Si)) (Srivastava et al. 2016; Mosa et al. 2016; Deshmukh et al. 2017; Groszmann et al. 2017). They are regulating seed germination; growth and development; fruit ripening; leaf, petal, and stomata movement; phloem loading and unloading; reproduction; and stress response. Genetically altered plants with aquaporins (PIPs, TIPs, and NIPs) are now tested for their ability to improve plant tolerance to abiotic stresses.

3.2 Role of PIPs in Salinity Stress

The plasma membrane intrinsic proteins (PIPs) are one of the biggest subfamilies of aquaporins, and their localization is confined to plasma and thylakoid membrane. Their distribution in plants ranged from 3 in *Selaginella*, and the highest 22, was recorded in *Glycine* and *Brassica* species. They are involved in water and small molecule homeostasis and played important roles during several abiotic stresses. Multiple PIP isoforms are present in a plant species and can be further classified into two subgroups PIP1 and PIP2 (Chaumont et al. 2000; Secchi et al. 2017). The role of PIPs during salinity stress in glycophytes as well as halophytes is well studied.

The up- and downregulation pattern of PIP genes in the roots and aerial parts of the *Arabidopsis* plants subjected to abiotic stress (salt, drought, and cold stress) implied that lower or higher expression of these aquaporin genes was beneficial to keep a suitable status of water under stress conditions. Many *Arabidopsis* PIP genes including PIP1;1, PIP1;2, and PIP2;3 were upregulated, whereas PIP1;5 and PIP2;6 were downregulated during salt treatment. During salt stress, in the aerial parts, PIP2;2 and PIP2;3 were upregulated, but PIP2;6 was downregulated. Similarly, in root tissues, PIP1;1, PIP1;2, PIP1;3, and PIP2;7 showed upregulation pattern, and transcript abundance of PIP1;5 was decreased. It revealed that PIP1;2 gene was downregulated by drought in the aerial parts but was upregulated under salt stress in the roots. The lower expression of PIP1;2 gene in the aerial parts was to curb water loss and the same aquaporin gene expression increases in the roots to increase water uptake from environment and to maintain reasonable water status during salt stress (Jang et al. 2004). Further correlation between salinity stress and aquaporin response based on microarray experiment claimed that the hybridization signal of PIP1;1 and PIP1;2 was decreased with time point and subcellular relocalization was observed in both PIPs (Boursiac et al. 2005). The differential down- or upregulation of aquaporin gene expression during salt stress may play roles in limiting initial water loss during early stage of salt stress and assisting the subsequent uptake of water to maintain water homeostasis in high-cellular salt conditions. Overexpression studies of *Arabidopsis PIP1;2* in tobacco plants have shown that overexpression improves plant vigor under normal condition but has no significant effect under salt stress (Aharon et al. 2003). In rice (*Oryza sativa*), OsPIP1;3 gene was involved in water homeostasis under salt-stressed condition (Abdelkader et al. 2012). However, overexpression of either *OsPIP1;1* or *OsPIP2;2* genes in the *Arabidopsis* showed increased tolerance to salt stress (100 mM of NaCl) (Guo et al. 2006). The preliminary studies on effect of salt stress on barley seedling root tissue-based transcript level underscored the involvement of PIPs such as HvPIP1;2, HvPIP1;3, and HvPIP2;2, and their mRNA were accumulated. Accumulation of HvPIP1;2, HvPIP1;3, HvPIP1;4, HvPIP2;1, HvPIP2;2, and HvPIP2;3 was significantly reduced in response to 200 mM salt stress. These results showed that the low transcript level of six HvPIP under severe salinity stress leads the barley plants unable to manage stress and normal growth (Horie et al. 2011). Heterologous expression of banana *MaPIP1;1* in *Arabidopsis* conferred tolerance to salt and drought stress and transgenic plant showed increased primary root elongation, root hair numbers, and survival rates compared to wild type (Xu et al. 2014). Transgenic banana plants overexpressing a native PIP, MusaPIP1;2 showed high tolerance levels to various abiotic stresses such as drought, cold, and salt stresses (Sreedharan et al. 2013). Overexpression of MusaPIP2;6 enhanced the salt tolerance in transgenic banana and displayed better photosynthetic efficiency and lower membrane damage under salt-stressed conditions (Sreedharan et al. 2015). Maize PIP members ZmPIP1 and ZmPIP2 showed downregulation pattern during salt stress. However, a transient upregulation pattern was observed in ZmPIP1;1, ZmPIP1;5, and ZmPIP2;4 members, preferentially in the outer parts of the roots (Zhu et al. 2005). The effects of salinity stress on two cultivars of tomato were observed with respect to the transcript levels of the LePIP1 and LePIP2 genes which showed higher transcript accumulation in the salt-sensitive tomato cultivar than in the salt-tolerant cultivar (Zhao et al. 2015). Under salt stress, high transcript abundance of all *Brassica* BrPIP were observed except BrPIP1;1a and BrPIP1;1b. However, most of the BrPIP transcript showed initial downregulation and subsequent upregulation pattern, and highest expression was recorded at 24 h of salt stress (Kayum et al. 2017). Effect of salt stress on radish seedling PIP through an immunoblot analysis showed that the RsPIP2–1 protein level was increased by NaCl treatment (Suga et al. 2002). Osmotic and salt stress exposure to cucumber seedlings led to the decrease in hydraulic conductivity of leaves which may be attributed to downregulation of the two most highly expressed isoforms of PIPs, *CsPIP1;2* and *CsPIP2;4* (Qian et al. 2015). Constitutive overexpression of *GmPIP1;6* in soybean examined under normal and salt stress conditions resulted in enhanced leaf gas exchange, higher net assimilation, and increased growth under 100 mM salt stress (Zhou et al. 2014). The expressions of the three genes such as GhPIP1;1, GhPIP2;1, and GhPIP2;2 were significantly upregulated or downregulated under different stresses such as salt and cold stress; PEG (polyethylene glycol) treatments indicated that they were involved in modulating all these stresses (Li et al. 2009).

High-throughput analysis of the grapevine under drought and salinity stress showed that salinity significantly increased the abundance of transcripts of PIP2;1, whereas water deficit significantly reduced the transcript abundance of these genes relative to control plants (Cramer et al. 2007).

Few halophytic aquaporins have been well characterized in order to understand their roles in abiotic stresses. The ice plant (*Mesembryanthemum crystallinum*) showed that the transcript accumulation of McMipA and McMipC (MIP-related genes) correlated with turgor recovery following salt-induced water stress (Yamada et al. 1995). Overexpression of the euhalophyte *Salicornia bigelovii* SbPIP1 gene into wheat confers tolerance against salinity stress which could increase the accumulation of the osmolyte proline, decrease the MDA content, and enhance the soluble sugar biosynthesis in the early period (Yu et al. 2015). However, salt-induced genes were isolated and well characterized from true mangroves *Rhizophora apiculata* using suppression subtractive hybridization and reported highest transcript abundance of aquaporin recorded at 6 h of salt stress (Menon and Soniya 2014). A plasma membrane intrinsic protein from *Atriplex canescens* (AcPIP2) overexpressed in the *Arabidopsis* enhanced plant growth and abiotic stress (salt and drought) tolerance (Li et al. 2015). The salt stress triggered repression of PIP2;7 promoter activity which led to a significant decrease in transcript abundance within 2 h, and aquaporin internalization led to alter root cell water conductivity in the *Arabidopsis* (Pou et al. 2016). A model forage grass species such as tall fescue (*Festuca arundinacea* Schreb) and meadow fescue (*Festuca pratensis* Huds) were subjected to droughts, salt, and cold stress and performed transcript abundance of MIP family. It was revealed that high salt-tolerant genotypes showed reduction of PIP1;2 transcript level and an increase of TIP1;1 transcript abundance in both *F. arundinacea* genotypes (Pawłowicz et al. 2017). The halophyte *Sesuvium portulacastrum* SpAQP1 belongs to PIP2 subfamily and was significantly induced by NaCl treatment and inhibited by abscisic acid (ABA) treatment. Heterologous expression of SpAQP1 in yeast and tobacco enhanced the salt tolerance of yeast strains and tobacco plants under salt stress in transgenic plants (Chang et al. 2016) (Table 3.1).

Table 3.1 (A) Role of PIPs in salinity stress

Sr. No	Plant species	PIP	Role in salinity	References
1	*Arabidopsis*	AtPIP1;2 and AtPIP2;1	Role in salinity at 100 mM	Luu et al. (2012)
		PIP1;1, PIP1;2, PIP1;3, PIP1;5, PIP1;2, PIP2;2, PIP2;3, PIP2;6, PIP2;7	Role in salinity at 100 mM	Jang et al. (2004) Boursiac et al. (2005)
2	*Arabidopsis thaliana*	PIP2;7	Repress transcriptional activity under salt stress	Pou et al. (2016)
3	Rice	OsPIP1 and OsPIP2	Role in salinity at 100 mM	Guo et al. (2006)
		OsPIP1–3	Role in salinity at 150 mM	Abdelkader et al. (2012)
4	Maize	ZmPIP1;1, ZmPIP1;5, and ZmPIP2;4	Role in salinity at 100 mM	Zhu et al. (2005)
5	Barley	HvPIP1;2, HvPIP1;3, HvPIP2;1, HvPIP2;2, and HvPIP2;3	Role in salinity at 200 mM	Horie et al. (2011)
6	Banana	MaPIP1;1	Role in salinity at 350 mM	Xu et al. (2014)
		MaPIP1;2	Role in salinity at 250 mM	Sreedharan et al. (2013)
		MaPIP2;6	Role in salinity at 250 mM	Sreedharan et al. (2015)
7	*Brassica rapa*	BrPIP1;3b, 2;4b, 2;6, 2;7a, and 2;7c, BrPIP2;1, 2;2a, and 2;2b	Role in salinity at 200 mM	Kayum et al. (2017)
8	Tomato	LePIP1	Role in salinity at 100 mM	Zhao et al. (2015)
9	Cucumber	CsPIP1;2, CsPIP2;4	Role in salinity	Qian et al. (2015)
10	Soybean	GmPIP1;6	Role in salinity at 100 mM	Zhou et al. (2014)
11	Cotton	GhPIP1;1, GhPIP2;1, and GhPIP2;2	Role in salinity at 250 mM	Li et al. (2009)
12	Grape wine	VvPIP2;1, VvPIP2.2	Role in salinity at 100 mM	Cramer et al. (2006), Mohammadkhani et al. (2012)
13	Radish	RsPIP2;1	Role in salinity at 150 mM	Suga et al. (2002)
14	*Salicornia bigelovii*	SbPIP1	Role in salinity at 250 mM	Yu et al. (2015)

(continued)

Table 3.1 (continued)

Sr. No	Plant species	PIP	Role in salinity	References
15	Citrus	CsPIP	Role in salt stress	Martins et al. (2015)
16	*Festuca* species	PIP1;2	Reduced transcript level under salt stress	Pawłowicz et al. (2017)
17	*Atriplex canescens*	AcPIP2	Overexpressed in *Arabidopsis* improved tolerance against salt stress	Li et al. (2015)
18	*Sesuvium portulacastrum*	SpAQP1 (PIP2)	Heterologous expressed in yeast- and plant-conferred salinity stress tolerance	Chang et al. (2016)

(B) Role of TIPs in salinity stress

Sr. No	Plant species	TIP	Role	Reference
1	Rice	*OsTIP1;1*, *OsTIP1;2*, *OsTIP2;2*, *OsTIP4;1*, *OsTIP4;2*, and *OsTIP4;3*	Role in salinity at 150 mM NaCl	Li et al. (2008), Sakurai et al. (2005), Liu et al. (1994)
2	Maize	*ZmTIP1–1*, *ZmTIP1–2*, *ZmTIP2–1*, *ZmTIP2–2*, *ZmTIP2–4*	Role in salinity at 100 mM NaCl	Zhu et al. (2005)
3	Salicornia	*ShTIP* and *ShTIP*	Role in salinity at 300 mM NaCl	Ermawati et al. (2009)
4	*Glycine soja*	*GsTIP2;1*	Role in salinity at 100 mM NaCl	Wang et al. (2011)
5	Tomato	*TIP2;2*	Role in salinity at 200 mM NaCl	Xin et al. (2014)
6	*Panax ginseng*	*PgTIP1*	Role in salinity at 100 mM NaCl	Pang et al. (2007), Li and Cai (2015)
7	*Thellungiella salsuginea*	*TsTIP1;2*	Role in multiple stresses such as salt stress	Wang et al. (2014)
8	*Populus*	*TIP1;1*, *TIP1;2*, *TIP2;3*, and *TIP2;4*	Downregulated under salt stress	Cohen et al. (2013)
9	*M. crystallinum*	*TIP* gene *(MIP-F)*	Downregulated under salt stress	Kirch et al. (2000)
10	*Festuca* species	*TIP1;1*	Upregulate transcript abundance	Pawłowicz et al. (2017)
11	*Glycine max*	*GmTIP2;1*, *GmTIP1;7*, *GmTIP1;8*	Conferred salt stress 200 mM in transgenic *Arabidopsis*	Zhang et al. (2017)
12	Citrus	*CsTIP2;1*	Overexpression in tobacco-conferred salt stress	Martins et al. (2017)

(continued)

Table 3.1 (continued)

(C) Role of NIP, SIP, and XIP in salinity stress

Sr. No	Plant species	MIPs	Role	Reference
1	Wheat	*TaNIP*	Upregulation during salt stress	Gao et al. (2010)
2	Citrus	CsNIPs	Upregulation during salt stress	Martins et al. (2015)
3	Citrus	CsSIPs	Upregulation during salt stress	Martins et al. (2015)
4	Citrus	CsXIPs	Upregulation during salt stress	Martins et al. (2015)
5	Poplar	PtXIP2;1	Response to abiotic stresses	Lopez et al. (2012)
6	*Glycine max*	GmSIP1;3	Oxidative stress	Zhang et al. (2017)

PIP2;1 is a plasma membrane localized water channel that regulates water uptake, and it is constitutively trafficked between the plasma membrane and the trans-Golgi network (TGN) in *Arabidopsis thaliana* (Ueda et al. 2016). Under salinity stress PIP2;1 was internalized from plasma membrane into the vacuole intracellular compartments and possibly involved in decreasing the water uptake of the root (Ueda et al. 2016).

3.3 Role of TIPs in Salinity Stress

Tonoplast intrinsic proteins (TIPs) are subclass of MIPs, confined localization to the vacuolar membrane also called as "tonoplast" (Maurel et al. 2015). Identification of multiple vacuolar compartments were detected by immunofluorescence experiments in root tips and mature embryos of different plant species (Paris et al. 1996; Jauh et al. 1998, 1999; Gillespie et al. 2005; Poxleitner et al. 2006). These experiments confirmed the localization of TIP1;1 to vegetative vacuole, TIP3;1 to lytic-type vacuoles, and TIP2;1 with protein storage vacuoles. Based on *Arabidopsis* proteomic studies, TIPs have been predicted to localize to the inner envelop and thylakoids, while it has also been found that AtTIP5;1 is located to tonoplast (Maurel et al. 2015; Wudicke et al. 2014). Many reports have claimed the relocalization of aquaporins under salt stress, in *Arabidopsis* salt stress-induced relocalization of AtTIP1;1 into intravacuolar invaginations (Boursiac et al. 2005). Presence of several TIP isoforms on separate tonoplasts provided evidence for multiple, functionally different vacuolar compartments within plant cells. TIPs are ubiquitous in distribution throughout plant kingdom, and they are ranged from lowest 3 TIPs in *Selaginella moellendorffii* to highest 23 TIPs in *Glycine max* (Deshmukh et al. 2015). The *Arabidopsis* genome encodes ten TIP isoforms, further classified into five subgroups such as TIP1, TIP2, seed-specific TIP3, TIP4, and TIP5 (Gattolin et al. 2009). The main role of TIPs has been described in the permeability of water. AtTIP1;1 was identified, and its roles was proved in high water permeability. In addition to water transport, TIPs have been involved in facilitating transport of glycerol, urea, H_2O_2, $NH4^+/NH_3$, methyl ammonium, and formamide (Afzal et al. 2016). Moreover, many studies reported that TIP expression is regulated by salinity, drought, gibberellic acid (GA_3), and abscisic acid (ABA) (Afzal et al. 2016).

Specific TIP isoforms of rice, maize, and *Arabidopsis* also show differential responses to several abiotic stresses such as water stress, salt, and cold stress. Systematic analysis of TIP expression in response to abiotic stresses was conducted in *Arabidopsis* and maize (Alexandersson et al. 2005; Zhu et al. 2005). The results indicated that most TIPs were repressed by drought and salinity stress in *Arabidopsis* and maize. Rice TIP expression patterns under various abiotic stress conditions including dehydration, high salinity, and ABA during seed germination were investigated by real-time PCR.

In rice, the expression of OsTIP1;1 was increased under drought, salt stress, and exogenous ABA (Liu et al. 1994). OsTIP1;1 and OsTIP2;2 expression was repressed by chilling stress and recovered following warming. In rice, the expression of OsTIP1;1 was downregulated in response to cold stress but upregulated during response to water and salinity stress (Sakurai et al. 2005; Liu et al. 1994). Similarly, in rice, NaCl induced the expression of four TIPs (OsTIP1;1, OsTIP1;2, OsTIP2;2, and OsTIP4;3) and repressed the expression of OsTIP2;2 and OsTIP4;3 in roots (Li et al. 2009). In maize, no change in expression of ZmTIPs was observed under salt stress experiments (Zhu et al. 2005). The correlation between TIP mRNA level and the bleeding volume suggested a relationship between root water uptake and TIP expression (Sakurai et al. 2005). At the posttranscriptional level, TIP activities can be regulated by phosphorylation and glycosylation. The *Mesembryanthemum crystallinum* TIP1;2 was glycosylated and redistributed to endosomal compartments, which served to maintain osmotic balance within the cytoplasm by mediating the stress-induced uptake of specific solutes or ions into these vesicles during osmotic stresses (Vera-Estrella et al. 2004).

A tonoplast AQP gene (TsTIP1;2) from halophyte *Thellungiella salsuginea* is possibly involved in the survival mechanism under multiple stresses such as drought and salt (Wang et al. 2014). Ligaba et al. (2011) studied the expression patterns of seven MIP genes from barley under different abiotic stresses using quantitative real-time PCR (RT-PCR), indicating that abiotic stress modulates the expression of major intrinsic proteins in barley. Tomato SlTIP2;2 expressed in transgenic *Arabidopsis* could enhance the tolerance to salt stress and interact with its homologous proteins SlTIP1;1 and SlTIP2;1 (Xin et al. 2014). Wang et al. (2011) cloned the novel *Glycine soja* tonoplast intrinsic protein gene GsTIP2;1, and the overexpression of GsTIP2;1 in *Arabidopsis* repressed/reduced tolerance to salt and dehydration stress, suggesting that GsTIP2;1 might mediate stress sensitivity by enhancing water loss in plants.

TIPs are also involved in the accumulation of ions in vacuoles in response to salt stress. The overexpression of AtTIP5;1 in *Arabidopsis* resulted in the tolerance of transgenic plants to high levels of borate and influxing into vacuolar compartment (Pang et al. 2010). The overexpression of the *Panax ginseng* aquaporin, PgTIP1, in *Arabidopsis* showed significant plant growth and enhanced tolerance to salt and drought stress (Peng et al. 2007). PgTIP1 is a functional water channel protein, but the mutation of Ser^{128} abolished water channel. PgTIP1 confers salt stress tolerance to *Arabidopsis*, but mutation of Ser^{128} in PgTIP1 nullifies this phenotype and alters the expression of stress-related genes in transgenic *Arabidopsis* (Li and Cai 2015).

In *Populus* TIP1;1, TIP1;2, TIP2;3, and TIP2;4 were downregulated under salt stress in root tissues (Cohen et al. 2013). A salt-tolerant ice plant (*M. crystallinum*) showed downregulation of TIP gene (MIP-F in roots and leaves, respectively) under salt stress. In contrast, upregulation of MIP-C in the plasma membrane of roots, which might be controlled by endosome trafficking, may increase the cellular uptake of water in the plants (Kirch et al. 2000). A novel TIP homolog from *Salicornia herbacea* showed involvement in salt stress in a different way compared to that of *Arabidopsis* TIP (Ermawati et al. 2009) (Table 3.1). Heterologous expression of the citrus *CsTIP2;1* in the tobacco plant enhanced growth, antioxidant activity, and physiological adaptation under drought and salt stress (Martins et al. 2017). The bamboo aquaporin family member, *PeTIP4;1–1*, was involved in shoot growth and when overexpressed in *Arabidopsis* led to drought and salinity tolerance (Sun et al. 2017). The heterologous expression of *Glycine max* TIP2;1 in yeast and overexpression of *GmTIP2;1*, *GmTIP1;7*, and *GmTIP1;8* in *Arabidopsis* enhanced salt and drought stress tolerance. Further, it was identified that GmTIP2;1 forms homodimers as well as interacts with GmTIP1;7 and GmTIP1;8 proteins (Zhang et al. 2017a).

3.4 Role of NIPs in Salinity Stress

The NOD26-like intrinsic proteins (NIPs) are a unique subfamily of MIPs, and the archetype reported first time in soybean was nodulin 26 protein (Fortin et al. 1987). These members are expected to be involved in homeostasis of metabolites between the host and the symbiont. NIPs are not restricted to legume crops but are widely distributed in both leguminous and nonleguminous plants and showed that plant NIP functions are not limited to nodule symbiosis (Wallace et al. 2006). The lowest NIPs observed in *Physcomitrella patens* (5) and *Arabidopsis thaliana* encode 9, and *Oryza sativa* genomes encode 11, and highest 23 NIPs were recorded in *Glycine max* (Deshmukh et al. 2015). Based on the ar/R regions of aquaporins, NIPs can be divided into three distinct groups such as NIP I, NIP II, and NIP III (Rouge and Barre 2008). NIP I proteins in *Arabidopsis* have been reported to transport water, glycerol, and lactic acid; NIP II proteins are permeable to larger solutes than NIP I protein. They are involved in transport of a wide but specified range of small solutes such as glycerol, silicic acid, antimony, arsenite, boron, silicon, and urea (Wallace et al. 2006). In addition to PIPs and TIPs, being small solute and ion transporters, NIPs have also been found to be involved in salinity stress. Wheat TaNIP, an AQP gene, was identified, and it showed upregulation pattern during salt stress. Further TaNIP was cloned and overexpressed in *Arabidopsis* which could enhance tolerance of *Arabidopsis* to various abiotic stresses (Gao et al. 2010). Transcriptome analyses of citrus roots and leaves under salt stress revealed that in addition to CsPIPs and CsTIPs, most of the CsNIPs, CsXIPs, and CsSIPs are also upregulated in roots, whereas in leaves both up- and downregulation patterns were observed for some homologs of each aquaporin family except CsSIPs (Table 3.1).

3.5 Other Aquaporin Members and Their Role in Salt Stress Tolerance

Small basic intrinsic protein (SIP) is the smallest subfamily of plant MIPs; they are called small basic intrinsic proteins because of their small molecular size (compared to PIP, TIP, and NIP) and they are relatively rich in basic amino acid such as lysine. There was very scarce information available on SIPs, but, recently, some reports have revealed that SIPs were localized to the endoplasmic reticulum membrane and were related to mammalian AQP11 and AQP12 (Johanson and Gustavsson 2002; Maeshima and Ishikawa 2008). SIPs tagged with green fluorescent protein (GFP) and transiently expressed in *Arabidopsis* cells, showed subcellular localization of SIPs to ER (Ishikawa et al. 2005). SIP1;1 and SIP1;2 may function as water channels in the ER, while SIP2;1 might act as an ER channel for other small molecules or ions (Ishikawa et al. 2005). SIPs have moderate water transport activity and may also function in original pore conformation. SIPs are distributed throughout plant kingdom and their distribution ranged from three SIPs members in *Arabidopsis* and maize and two members in rice (Chaumont et al. 2001; Johanson et al. 2001; Sakurai et al. 2005). Overexpression of *Glycine max* SIP1;3 in *Nicotiana tabacum* showed a short root phenotype, growth retardation, and significant tolerance to oxidative stress both in yeast and plant systems (Zhang et al. 2017b).

However, physiological roles of SIP members except as water channel are very scarce, and no definite evidences are available on their roles in abiotic stresses. An uncategorized X intrinsic protein (XIP) is a unique class of MIPs which share less similarities with other identified aquaporin subfamilies and is recently reported in the nonvascular moss *Physcomitrella patens* (Danielson and Johanson 2008; Lopez et al. 2012). They are uncharacterized proteins, and their further molecular characterization will be required (Lopez et al. 2012). XIPs are found in protozoa, fungi, and certain land plant species, such as *Populus trichocarpa*, *Nicotiana tabacum*, *Solanum lycopersicum*, and *Solanum tuberosum* (Gupta and Sankararamakrishnan 2009; Shelden et al. 2009; Bienert et al. 2011; Lopez et al. 2012). It was initially assumed that XIPs were nonfunctional as water channels but recent studies suggested their involvement in transport of hydrophobic solutes (Danielson and Johanson, 2008; Gupta and Sankararamakrishnan, 2009). These predictions were confirmed in XIPs from three Solanales members such as *Nicotiana tabacum*, *Solanum lycopersicum*, and *Solanum tuberosum*. Poplar has largest number of XIPs (9) with significant amino acid diversity. Expression analyses using quantitative real-time PCR reported that only two PtXIP genes such as *PtXIP2;1* and *PtXIP3;2* were expressed in vegetative tissues and *PtXIP2;1* was differentially and significantly modulated in response to abiotic stresses. In situ hybridization experiments observed that the *PtXIP2;1* and *PtXIP3;2* genes showed tissue-specific high expression in poplar. Eventually, PtXIP2;1 and PtXIP3;3 were functionally characterized as being the poplar XIPs able to transport water through *Xenopus* oocyte expression assays (Lopez et al. 2012). Recently, two different splice variants (α and β) of NbXIP1;1 from *Nicotiana benthamiana* were studied and showed that NbXIP1;1α was permeable to boric acid and phosphorylated at N-terminal domain (Ampah-Korsah et al. 2016). Moreover, a mutant study of NbXIP1;1α using single amino acid substitutions in selectivity filter and deletions in loops C and D rendered water permeability (Ampah-Korsah et al. 2017). Transcriptome analyses of citrus roots and leaves under salt stress revealed that in addition to CsPIPs and CsTIPs, most of the CsNIPs, CsXIPs, and CsSIPs are also upregulated in roots, whereas in leaves both up- and downregulation patterns were observed for some homologs of aquaporin family except CsSIPs (Martins et al. 2015).

3.6 Conclusion and Future Perspective

The discovery of novel plant aquaporin family opened new research horizon and established their roles in water as well as small molecule homeostasis. The emerging research also strengthens our understanding about roles of aquaporins during plant growth, development and several biotic and abiotic stresses. In plant kingdom, MIPs are one of the biggest families divided into seven subfamilies, and multiple isoforms were recorded which help to understand their importance in diverse physiological functions as well strengthen their survival rate during adverse conditions. Moreover, most of MIP members of glycophytes as well as halophytes were well characterized and confirmed their involvement in water homeostasis along with small uncharged molecules and improve tolerance during various abiotic stresses such as drought, salt, and cold stresses. Comparatively, plant PIP and TIP members were actively involved in salt stress tolerance mechanism, but on the other hand, very few reports were available on plant NIPs, XIPs, and SIPs role in abiotic stress. Interestingly, halophytic *Salicornia* TIP homologs are involved in the salt stress tolerance and have different mechanism from glycophytic plants under salt stress. This new finding will open new research pathways and definitively help to understand salt tolerance mechanism of halophytic plants under salt stress. An aquaporin family expression analysis from tree species such as *Populus* revealed involvement of PIPs and TIPs along with few XIPs as well orchestrated under abiotic stresses. Nowadays most of aquaporin family members are cloned and well characterized from plant kingdom and established their roles under normal as well as environmental stresses, but their mechanism of providing salt tolerance is still lacking. In upcoming future, it will be very useful to explore MIP mechanism and pathways under normal developmental and stress condition. However, comparative mechanism in glycophytes and halophytes during salt stress will help us to understand different control mechanisms which will mitigate into crop plants to improve salt tolerance trait. Plant "omics" such as transcriptomics, proteomics, and metabolomics of halophyte investigation relative to commercial crops are required in order to complete understanding of mechanisms underlying salt tolerance.

Acknowledgments The work in KK lab was supported by financial assistance from Board of Research in Nuclear Sciences (37(1)/14/28/2016-BRNS/37248), India. AAS acknowledges the Senior Research Fellowship provided by University Grants Commission, India.

References

Abdelkader AF, El-khawas S, El-Din El-Sherif NA, Hassanein RA, Emam MA, Hassan RE (2012) Expression of aquaporin gene (OsPIP1-3) in salt-stressed rice (*Oryza sativa* L.) plants pre-treated with the neurotransmitter (dopamine). Plant Omics 5(6):532

Afzal Z, Howton TC, Sun Y, Mukhtar MS (2016) The roles of aquaporins in plant stress responses. J Dev Biol 4(1):9

Aharon R, Shahak Y, Wininger S, Bendov R, Kapulnik Y, Galili G (2003) Overexpression of a plasma membrane aquaporin in transgenic tobacco improves plant vigor under favorable growth conditions but not under drought or salt stress. Plant Cell 15(2):439–447

Alexandersson E, Fraysse L, Sjövall-Larsen S, Gustavsson S, Fellert M, Karlsson M et al (2005) Whole gene family expression and drought stress regulation of aquaporins. Plant Mol Biol 59(3):469–484

Ampah-Korsah H, Anderberg HI, Engfors A, Kirscht A, Norden K, Kjellstrom S, Kjellbom P, Johanson U (2016) The aquaporin splice variant NbXIP1; 1α is permeable to boric acid and is phosphorylated in the N-terminal domain. Front Plant Sci 7

Ampah-Korsah H, Sonntag Y, Engfors A, Kirscht A, Kjellbom P, Johanson U (2017) Single amino acid substitutions in the selectivity filter render NbXIP1;1α aquaporin water permeable. BMC Plant Biol 17(1):61

Atkinson NJ, Lilley CJ, Urwin PE (2013) Identification of genes involved in the response of Arabidopsis to simultaneous biotic and abiotic stresses. Plant Physiol 162(4):2028–2041

Bienert GP, Bienert MD, Jahn TP, Boutry M, Chaumont F (2011) Solanaceae XIPs are plasma membrane aquaporins that facilitate the transport of many uncharged substrates. Plant J 66(2):306–317

Boursiac Y, Chen S, Luu DT, Sorieul M, van den Dries N, Maurel C (2005) Early effects of salinity on water transport in *Arabidopsis* roots, molecular and cellular features of aquaporin expression. Plant Physiol 139(2):790–805

Byrt CS, Zhao M, Kourghi M, Bose J, Henderson SW, Qiu J, Gilliham M, Schultz C, Schwarz M, Ramesh SA, Yool A (2017) Non-selective cation channel activity of aquaporin AtPIP2;1 regulated by Ca^{2+} and pH. Plant Cell Environ 40:802–815

Calamita G, Bishai WR, Preston GM, Guggino WB, Agre P (1995) Molecular cloning and characterization of AqpZ, a Water Channel from Escherichia coli. J Biol Chem 270(49):29063–29066

Carbrey JM, Bonhivers M, Boeke JD, Agre P (2001) Aquaporins in Saccharomyces: characterization of a second functional water channel protein. PNAS 98(3):1000–1005

Cavanagh C, Morell M, Mackay I, Powell W (2008) From mutations to MAGIC: resources for gene discovery, validation and delivery in crop plants. Curr Opin Plant Biol 11(2):215–221

Chang W, Liu X, Zhu J, Fan W, Zhang Z (2016) An aquaporin gene from halophyte *Sesuvium portulacastrum*, SpAQP1, increases salt tolerance in transgenic tobacco. Plant Cell Rep 35(2):385–395

Chaumont F, Barrieu F, Jung R, Chrispeels MJ (2000) Plasma membrane intrinsic proteins from maize cluster in two sequence subgroups with differential aquaporin activity. Plant Physiol 122(4):1025–1034

Chaumont F, Barrieu F, Wojcik E, Chrispeels MJ, Jung R (2001) Aquaporins constitute a large and highly divergent protein family in maize. Plant Physiol 125(3):1206–1215

Chinnusamy V, Zhu JK (2009) Epigenetic regulation of stress responses in plants. Curr Opin Plant Biol 12(2):133–139

Cohen D, Bogeat-Triboulot MB, Vialet-Chabrand S, Merret R, Courty PE, Moretti S, Bizet F, Guilliot A, Hummel I (2013) Developmental and environmental regulation of Aquaporin gene expression across Populus species: divergence or redundancy? PLoS One 8(2):e55506

Cramer GR, Ergül A, Grimplet J, Tillett RL, Tattersall EA, Bohlman MC, Vincent D, Sonderegger J, Evans J, Osborne C, Quilici D (2007) Water and salinity stress in grapevines: early and late changes in transcript and metabolite profiles. Funct integr Genomic 7(2):111–134

Danielson JÅ, Johanson U (2008) Unexpected complexity of the aquaporin gene family in the moss *Physcomitrella patens*. BMC Plant Biol 8(1):45

Deshmukh RK, Nguyen HT, Belanger RR (2017) Aquaporins: dynamic role and regulation. Frontiers in Plant Sci 8:1420

Deshmukh RK, Sonah H, Bélanger RR (2016) Plant Aquaporins: genome-wide identification, transcriptomics, proteomics, and advanced analytical tools. Front Plant Sci 7

Deshmukh RK, Vivancos J, Ramakrishnan G, Guérin V, Carpentier G, Sonah H, Labbé C, Isenring P, Belzile FJ, Bélanger RR (2015) A precise spacing between the NPA domains of aquaporins is essential for silicon permeability in plants. Plant J 83(3):489–500

Ermawati N, Liang YS, Cha JY, Shin D, Jung MH, Lee JJ, Lee BH, Han CD, Lee KH, Son D (2009) A new TIP homolog, ShTIP, from *Salicornia* shows a different involvement in salt stress compared to that of TIP from *Arabidopsis*. Biol plantarum 53(2):271–277

Forrest KL, Bhave M (2007) Major intrinsic proteins (MIPs) in plants: a complex gene family with major impacts on plant phenotype. Funct integr genomic 7(4):263

Fortin MG, Morrison NA, Verma DP (1987) Nodulin-26, a peribacteroid membrane nodulin is expressed independently of the development of the peribacteroid compartment. Nucleic Acids Res 15(2):813–824

Gao Z, He X, Zhao B, Zhou C, Liang Y, Ge R, Shen Y, Huang Z (2010) Overexpressing a putative aquaporin gene from wheat, TaNIP, enhances salt tolerance in transgenic *Arabidopsis*. Plant Cell Physiol 51(5):767–775

Gattolin S, Sorieul M, Hunter PR, Khonsari RH, Frigerio L (2009) In vivo imaging of the tonoplast intrinsic protein family in *Arabidopsis* roots. BMC Plant Biol 9(1):133

Gillespie J, Rogers SW, Deery M, Dupree P, Rogers JC (2005) A unique family of proteins associated with internalized membranes in protein storage vacuoles of the Brassicaceae. Plant J 41(3):429–441

Groszmann M, Osborn HL, Evans JR (2017) Carbon dioxide and water transport through plant aquaporins. Plant Cell Environ 40(6):938–961

Gupta AB, Sankararamakrishnan R (2009) Genome-wide analysis of major intrinsic proteins in the tree plant *Populus trichocarpa*: characterization of XIP subfamily of aquaporins from evolutionary perspective. BMC Plant Biol 9(1):134

Guo L, Wang ZY, Lin H, Cui WE, Chen J, Liu M, Chen ZL, Qu LJ, Gu H (2006) Expression and functional analysis of the rice plasma-membrane intrinsic protein gene family. Cell Res 16(3):277–286

Horie T, Kaneko T, Sugimoto G, Sasano S, Panda SK, Shibasaka M, Katsuhara M (2011) Mechanisms of water transport mediated by PIP aquaporins and their regulation via phosphorylation events under salinity stress in barley roots. Plant Cell Physiol 52(4):663–675

Hove RM, Ziemann M, Bhave M (2015) Identification and expression analysis of the barley (*Hordeum vulgare L.*) aquaporin gene family. PLoS One 10(6):e0128025

Hu W, Hou X, Huang C, Yan Y, Tie W, Ding Z, Wei Y, Liu J, Miao H, Lu Z, Li M (2015) Genome-wide identification and expression analyses of aquaporin gene family during development and abiotic stress in banana. Int J Mol Sci 16(8):19728–19751

Ishikawa F, Suga S, Uemura T, Sato MH, Maeshima M (2005) Novel type aquaporin SIPs are mainly localized to the ER membrane and show cell-specific expression in Arabidopsis thaliana. FEBS Lett 579(25):5814–5820

Iyer NJ, Tang Y, Mahalingam R (2013) Physiological., biochemical and molecular responses to a combination of drought and ozone in *Medicago truncatula*. Plant Cell Environ 36(3):706–720

Jang JY, Kim DG, Kim YO, Kim JS, Kang H (2004) An expression analysis of a gene family encoding plasma membrane aquaporins in response to abiotic stresses in *Arabidopsis thaliana*. Plant Mol Biol 54(5):713–725

Jauh GY, Fischer AM, Grimes HD, Ryan CA, Rogers JC (1998) δ-Tonoplast intrinsic protein defines unique plant vacuole functions. PNAS 95(22):12995–12999

Jauh GY, Phillips TE, Rogers JC (1999) Tonoplast intrinsic protein isoforms as markers for vacuolar functions. Plant Cell 11(10):1867–1882

Johanson U, Gustavsson S (2002) A new subfamily of major intrinsic proteins in plants. Mol Biol Evol 19(4):456–461

Johanson U, Karlsson M, Johansson I, Gustavsson S, Sjövall S, Fraysse L, Weig AR, Kjellbom P (2001) The complete set of genes encoding major intrinsic proteins in *Arabidopsis* provides a framework for a new nomenclature for major intrinsic proteins in plants. Plant Physiol 126(4):1358–1369

Kayum MA, Park JI, Nath UK, Biswas MK, Kim HT, Nou IS (2017) Genome-wide expression profiling of aquaporin genes confer responses to abiotic and biotic stresses in Brassica Rapa. BMC Plant Biol 17(1):23

Kirch HH, Vera-Estrella R, Golldack D, Quigley F, Michalowski CB, Barkla BJ, Bohnert HJ (2000) Expression of water channel proteins in *Mesembryanthemum crystallinum*. Plant Physiol 123(1):111–124

Kozono D, Ding X, Iwasaki I, Meng X, Kamagata Y, Agre P, Kitagawa Y (2003) Functional expression and characterization of an archaeal aquaporin AqpM from *Methanothermobacter marburgensis*. J Biol Chem 278(12):10649–10656

Kumar K, Kumar M, Kim SR, Ryu H, Cho YG (2013) Insights into genomics of salt stress response in rice. Rice 6(1):27

Li DD, Wu YJ, Ruan XM, Li B, Zhu L, Wang H, Li XB (2009) Expressions of three cotton genes encoding the PIP proteins are regulated in root development and in response to stresses. Plant Cell Rep 28(2):291–300

Li GW, Peng YH, Yu X, Zhang MH, Cai WM, Sun WN, Su WA (2008) Transport functions and expression analysis of vacuolar membrane aquaporins in response to various stresses in rice. J Plant Physiol 165(18):1879–1888

Li J, Cai W (2015) A ginseng PgTIP1 gene whose protein biological activity related to ser 128 residue confers faster growth and enhanced salt stress tolerance in *Arabidopsis*. Plant Sci 234:74–85

Li J, Yu G, Sun X, Liu Y, Liu J, Zhang X, Jia C, Pan H (2015) AcPIP2, a plasma membrane intrinsic protein from halophyte Atriplex Canescens, enhances plant growth rate and abiotic stress tolerance when overexpressed in Arabidopsis Thaliana. Plant Cell Rep 34(8):1401–1415

Ligaba A, Katsuhara M, Shibasaka M, Djira G (2011) Abiotic stresses modulate expression of major intrinsic proteins in barley (Hordeum vulgare). C R Biol 334(2):127–139

Liu Q, Umeda M, Uchimiya H (1994) Isolation and expression analysis of two rice genes encoding the major intrinsic protein. Plant Mol Biol 26(6):2003–2007

Lopez D, Bronner G, Brunel N, Auguin D, Bourgerie S, Brignolas F, Carpin S, Tournaire-Roux C, Maurel C, Fumanal B, Martin F, Sakr S, Label P, Julien JL, Gousset-Dupont A, Venisse JS (2012) Insights into Populus XIP aquaporins: evolutionary expansion, protein functionality, and environmental regulation. J Exp Bot 63(5):2217–2230

Luu D-T, Martinière A, Sorieul M, Runions J, Maurel C (2012) Fluorescence recovery after photobleaching reveals high cycling dynamics of plasma membrane aquaporins in Arabidopsis roots under salt stress. Plant J 69(5):894–905

Maeshima M, Ishikawa F (2008) ER membrane aquaporins in plants. Pflügers Arch-EJP 456(4):709–716

Martins CD, Pedrosa AM, Du D, Gonçalves LP, Yu Q, GmitterJr FG, Costa MG (2015) Genome-wide characterization and expression analysis of major intrinsic proteins during abiotic and biotic stresses in sweet orange (*Citrus sinensis L. Osb.*) PLoS One 10(9):e0138786

Martins CP, Neves DM, Cidade LC, Mendes AF, Silva DC, Almeida AA, Coelho-Filho MA, Gesteira AS, Soares-Filho WS, Costa MG (2017) Expression of the citrus CsTIP2;1 gene improves tobacco plant growth, antioxidant capacity and physiological adaptation under stress conditions. Planta 245(5):951–963

Maurel C, Boursiac Y, Luu DT, Santoni V, Shahzad Z, Verdoucq L (2015) Aquaporins in plants. Physiol Rev 95(4):1321–1358

Maurel C, Verdoucq L, Luu DT, Santoni V (2008) Plant aquaporins: membrane channels with multiple integrated functions. Annu Rev Plant Biol 59:595–624

Menon TG, Soniya EV (2014) Isolation and characterization of salt-induced genes from Rhizophora apiculata Blume, a true mangrove by suppression subtractive hybridization. Curr Sci 107(4):650–655

Mitra BN, Yoshino R, Morio T, Yokoyama M, Maeda M, Urushihara H, Tanaka Y (2000) Loss of a member of the aquaporin gene family, aqpA affects spore dormancy in Dictyostelium. Gene 251(2):131–139

Mittler R, Blumwald E (2010) Genetic engineering for modern agriculture: challenges and perspectives. Annu Rev Plant Biol 61:443–462

Mohammadkhani N, Heidari R, Abbaspour N, Rahmani F (2012) Growth responses and aquaporin expression in grape genotypes under salinity. Iran J Plant Physiol 2:497–507

Mosa KA, Kumar K, Chhikara S, Musante C, White JC, Dhankher OP (2016) Enhanced boron tolerance in plants mediated by bidirectional transport through plasma membrane intrinsic proteins. Sci Rep 6(1)

Munns R, Tester M (2008) Mechanisms of salinity tolerance. Annu Rev Plant Biol 59:651–681

Pang Y, Li L, Ren F, Lu P, Wei P, Cai J, Xin L, Zhang J, Chen J, Wang X (2010) Overexpression of the tonoplast aquaporin AtTIP5;1 conferred tolerance to boron toxicity in *Arabidopsis*. J Genet Genomics 37(6):389–397

Paris N, Stanley CM, Jones RL, Rogers JC (1996) Plant cells contain two functionally distinct vacuolar compartments. Cell 85(4):563–572

Park W, Scheffler BE, Bauer PJ, Campbell BT (2010) Identification of the family of aquaporin genes and their expression in upland cotton (*Gossypium hirsutum L.*) BMC Plant Biol 10(1):142

Pawłowicz I, Rapacz M, Perlikowski D, Gondek K, Kosmala A (2017) Abiotic stresses influence the transcript abundance of PIP and TIP aquaporins in *Festuca* species. J Appl Genet 4:1–5

Peng Y, Lin W, Cai W, Arora R (2007) Overexpression of a Panax ginseng tonoplast aquaporin alters salt tolerance, drought tolerance and cold acclimation ability in transgenic *Arabidopsis* plants. Planta 226(3):729–740

Pou A, Jeanguenin L, Milhiet T, Batoko H, Chaumont F, Hachez C (2016) Salinity-mediated transcriptional and post-translational regulation of the *Arabidopsis* aquaporin PIP2;7. Plant Mol Biol 92(6):731–744

Poxleitner M, Rogers SW, Lacey Samuels A, Browse J, Rogers JC (2006) A role for caleosin in degradation of oil-body storage lipid during seed germination. Plant J 47(6):917–933

Prasch CM, Sonnewald U (2013) Simultaneous application of heat, drought, and virus to *Arabidopsis* plants reveals significant shifts in signaling networks. Plant Physiol 162(4):1849–1866

Qian Z-J, Song J-J, Chaumont F, Ye Q (2015) Differential responses of plasma membrane aquaporins in mediating water transport of cucumber seedlings under osmotic and salt stresses. Plant Cell Environ 38(3):461–473

Rasmussen S, Barah P, Suarez-Rodriguez MC, Bressendorff S, Friis P, Costantino P, Bones AM, Nielsen HB, Mundy J (2013) Transcriptome responses to combinations of stresses in *Arabidopsis*. Plant Physiol 161(4):1783–1794

Reddy PS, Rao TS, Sharma KK, Vadez V (2015) Genome-wide identification and characterization of the aquaporin gene family in *Sorghum bicolor* (L.) Plant Gene 1:18–28

Rougé P, Barre A (2008) A molecular modeling approach defines a new group of Nodulin 26-like aquaporins in plants. Biochem Biophys Res Commun 367(1):60–66

Sakurai J, Ishikawa F, Yamaguchi T, Uemura M, Maeshima M (2005) Identification of 33 rice aquaporin genes and analysis of their expression and function. Plant Cell Physiol 46(9):1568–1577

Secchi F, Pagliarani C, Zwieniecki MA (2017) The functional role of xylem parenchyma cells and aquaporins during recovery from severe water stress. Plant Cell Environ 40(6):858–871

Shelden MC, Howitt SM, Kaiser BN, Tyerman SD (2009) Identification and functional characterisation of aquaporins in the grapevine, *Vitis vinifera*. Funct Plant Biol 36(12):1065–1078

Sreedharan S, Shekhawat UK, Ganapathi TR (2015) Constitutive and stress-inducible overexpression of a native aquaporin gene (MusaPIP2;6) in transgenic banana plants signals its pivotal role in salt tolerance. Plant Mol Biol 88(1–2):41–52

Sreedharan S, Shekhawat UK, Ganapathi TR (2013) Transgenic banana plants overexpressing a native plasma membrane aquaporin MusaPIP1;2 display high tolerance levels to different abiotic stresses. Plant Biotechnol J 11(8):942–952

Srivastava AK, Penna S, Nguyen DV, Tran LS (2016) Multifaceted roles of aquaporins as molecular conduits in plant responses to abiotic stresses. Crit Rev Biotechnol 36(3):389–398

Suga S, Komatsu S, Maeshima M (2002) Aquaporin isoforms responsive to salt and water stresses and phytohormones in radish seedlings. Plant Cell Physiol 43(10):1229–1237

Sun H, Li L, Lou Y, Zhao H, Yang Y, Wang S, Gao Z (2017) The bamboo aquaporin gene PeTIP4;1–1 confers drought and salinity tolerance in transgenic *Arabidopsis*. Plant Cell Rep 36(4):597–609

Tao P, Zhong X, Li B, Wang W, Yue Z, Lei J, Guo W, Huang X (2014) Genome-wide identification and characterization of aquaporin genes (AQPs) in Chinese cabbage (*Brassica rapa ssp. pekinensis*). Mol Gen Genomics 289(6):1131–1145

Ueda M, Tsutsumi N, Fujimoto M (2016) Salt stress induces internalization of plasma membrane aquaporin into the vacuole in Arabidopsis thaliana. Biochem Biophys Res Commun 474(4):742–746

Venkatesh J, Yu JW, Park SW (2013) Genome-wide analysis and expression profiling of the *Solanum tuberosum* aquaporins. Plant Physiol Biochem 73:392–404

Vera-Estrella R (2004) Novel regulation of Aquaporins during osmotic stress. Plant Physiol 135(4):2318–2329

Wallace IS, Choi WG, Roberts DM (2006) The structure, function and regulation of the nodulin 26-like intrinsic protein family of plant aquaglyceroporins. Biochimica et Biophysica Acta (BBA)-Biomembranes 1758(8):1165–1175

Wang X, Li Y, Ji W, Bai X, Cai H, Zhu D et al (2011) A novel Glycine soja tonoplast intrinsic protein gene responds to abiotic stress and depresses salt and dehydration tolerance in transgenic Arabidopsis thaliana. J Plant Physiol 168(11):1241–1248

Wang L-L, Chen A-P, Zhong N-Q, Liu N, Wu X-M, Wang F, Yang C-L, Romero MF, Xia G-X (2014) The Thellungiella salsuginea tonoplast aquaporin TsTIP1;2 functions in protection against multiple abiotic stresses. Plant Cell Physiol 55(1):148–161

Wudick MM, Luu DT, Tournaire-Roux C, Sakamoto W, Maurel C (2014) Vegetative and sperm cell-specific aquaporins of *Arabidopsis* highlight the vacuolar equipment of pollen and contribute to plant reproduction. Plant Physiol 164(4):1697–1706

Xin S, Yu G, Sun L, Qiang X, Xu N, Cheng X (2014) Expression of tomato SlTIP2;2 enhances the tolerance to salt stress in the transgenic Arabidopsis and interacts with target proteins. J Plant Res 127(6):695–708

Xu Y, Hu W, Liu J, Zhang J, Jia C, Miao H, Xu B, Jin Z (2014) A banana aquaporin gene, MaPIP1;1, is involved in tolerance to drought and salt stresses. BMC Plant Biol 14(1):59

Yamada S, Katsuhara M, Kelly WB, Michalowski CB, Bohnert HJ (1995) A family of transcripts encoding water channel proteins: tissue-specific expression in the common ice plant. Plant Cell 7(8):1129–1142

Yool AJ, Stamer WD, Regan JW (1996) Forskolin stimulation of water and cation permeability in aquaporin1 water channels. Science 30:1216–1218

Yu GH, Zhang X, Ma HX (2015) Changes in the physiological parameters of -transformed wheat plants under salt stress. Int J Genomics 2015:1–6

Zhang DY, Ali Z, Wang CB, Xu L, Yi JX, Xu ZL, Liu XQ, He XL, Huang YH, Khan IA, Trethowan RM (2013) Genome-wide sequence characterization and expression analysis of major intrinsic proteins in soybean (*Glycine max L.*) PLoS One 8(2):e56312

Zhang DY, Kumar M, Xu L, Wan Q, Huang YH, Xu ZL, He XL, Ma JB, Pandey GK, Shao HB (2017a) Genome-wide identification of major intrinsic proteins in Glycine soja and characterization of GmTIP2;1 function under salt and water stress. Sci Rep 7

Zhang D, Huang Y, Kumar M, Wan Q, Xu Z, Shao H, Pandey GK (2017b) Heterologous expression of GmSIP1;3 from soybean in tobacco showed and growth retardation and tolerance to hydrogen peroxide. Plant Sci 263:210–218

Zhao YY, Yan F, Hu LP, Zhou XT, Zou ZR, Cui LR (2015) Effects of exogenous 5-aminolevulinic acid on photosynthesis, stomatal conductance, transpiration rate, and PIP gene expression of tomato seedlings subject to salinity stress. Genet Mol Res 14(2):6401–6412

Zhou L, Wang C, Liu R, Han Q, Vandeleur RK, Du J, Tyerman S, Shou H (2014) Constitutive overexpression of soybean plasma membrane intrinsic protein GmPIP1;6 confers salt tolerance. BMC Plant Biol 14(1):181

Zhu C, Schraut D, Hartung W, Schäffner AR (2005) Differential responses of maize MIP genes to salt stress and ABA. J Exp Bot 56(421):2971–2981

Chapter 4
Strategies to Mitigate the Salt Stress Effects on Photosynthetic Apparatus and Productivity of Crop Plants

Sonia Mbarki, Oksana Sytar, Artemio Cerda, Marek Zivcak, Anshu Rastogi, Xiaolan He, Aziza Zoghlami, Chedly Abdelly, and Marian Brestic

Abstract Soil salinization represents one of the major limiting factors of future increase in crop production through the expansion or maintaining of cultivation area in the future. High salt levels in soils or irrigation water represent major environmental concerns for agriculture in semiarid and arid zones. Recent advances in research provide great opportunities to develop effective strategies to improve crop salt tolerance and yield in different environments affected by the soil salinity. It was clearly demonstrated that plants employ both the common adaptative responses and the specific reactions to salt stress. The review of research results presented here may be helpful to understand the physiological, metabolic, developmental, and other reactions of crop plants to salinity, resulting in the decrease of biomass

S. Mbarki
National Research Institute of Rural Engineering, Water and Forests (INRGREF), B.P N°10 Rue Hedi Karray 2080 Ariana, Tunisia

Laboratory of Plant Extremophile Plants, Center of Biotechnology of Borj Cedria BP 901, Hammam-Lif 2050, Tunisia

O. Sytar (✉) · M. Zivcak · M. Brestic (✉)
Department of Plant Physiology, Slovak University of Agriculture, Nitra, Slovakia
e-mail: oksana.sytar@gmail.com; Marian.brestic@uniag.sk

A. Cerda
Department of Geography, Universitat de Valencia, Valencia, Spain

A. Rastogi
Department of Meteorology, Poznan University of Life Sciences, Poznan, Poland

X. He
Jiangsu Academy of Agricultural Sciences (JAAS), Nanjing, China

A. Zoghlami
Institut National de la Recherche Agronomique de Tunisie(INRAT), Rue Hedi Karray Ariana 2049, Tunisia

C. Abdelly
Laboratory of Extremophile Plants, Centre of Biotechnology of Borj-Cedria, Hammam-Lif, Tunisia

V. Kumar et al. (eds.), *Salinity Responses and Tolerance in Plants, Volume 1*,
https://doi.org/10.1007/978-3-319-75671-4_4

production and yield. In addition, the chapter provides an overview of modern studies on how to mitigate salt stress effects on photosynthetic apparatus and productivity of crop plants with the help of phytohormones, glycine betaine, proline, polyamines, paclobutrazol, trace elements, and nanoparticles. To understand well these effects and to discover new ways to improve productivity in salinity stress conditions, it is necessary to utilize efficiently possibilities of promising techniques and approaches focused on improvement of photosynthetic traits and photosynthetic capacity, which determines yield under salt stress conditions.

Keywords Salinity · Photosynthesis apparatus · Yield · Adaptative response

Abbreviations

ABA	Abscisic acid
APX	Ascorbate peroxidase
BRs	Brassinosteroids
CAT	Catalase
DW	Dry weight
EBL	24-Epibrassinolide
FW	Fresh weight
GPX	Guaiacol peroxidase
JA	Jasmonic acid
MeJA	Jasmonate
MDHAR	Monodehydroascorbate reductase
MDA	Malonic dialdehyde
NPs	Nanoparticles
Pn	Photosynthetic rate
PAs	Polyamines
RWC	Relative water content
SA	Salicylic acid
SOD	Superoxide dismutase
WUE	Water use efficiency

4.1 Introduction

The progressive soil degradation, especially soil salinization, will represent one of the major obstacles to increase the crop production through the expansion of cultivation area in the future (Munns and Gilliham 2015). In many regions of the world, where precipitation is insufficient to leach soluble salts from the root zone, high salt levels in soils or irrigation water represent major environmental concerns for agriculture, the severity of which can increase in conditions of climate change (Lachhab et al. 2013).

Salinity of soil is one of the main abiotic stresses limiting the growth of crops (Munns and Tester 2008). The high salt concentration in the root zone can be natural or induced by agricultural activities such as irrigation with low-quality water or the use of certain fertilizers (Bartels and Nelson 1994). Nearly 400 million hectares of land is affected by salinization, 80% of which are of natural origin and 20% of anthropogenic origin (FAO 2015).

Globally, no less than 10 million hectares of agricultural land is abandoned annually (Berthomieu et al. 1988) due to the acclimatization over time of small quantities of salts contained in the irrigation water. About 15% of the cultivated land has an excess of salt (Berthomieu et al. 1988), and large quantities of water are of very poor quality (Aissaoui and Reffas 2007). Globally, salinity affects more than 6% of the land; in case of irrigated lands, it is over 40% (Chaves et al. 2009).

In salt zones covering 16 million hectares (Hamdy 1999), plants are often subjected to strong sunlight and low rainfall. In these areas, salinity is not only related to climatic conditions but also to the often poorly controlled use of irrigation. Therefore, the high evaporation demand and low infiltration due to precipitation lead to the accumulation of salt on the soil surface (Gucci et al. 1997). Soils in affected areas contain high concentrations of soluble salts, mainly NaCl, but also Na_2SO_4, $CaSO_4$, and KCl (Hachicha 2007).

The presence of salts in the soil causes a limitation and decline in yield in many regions, in which the salt concentration of the soil solution exceeds 100 mM, inhibiting the growth of plants (Shahbaz and Ashref 2013). Salt accumulation in soils induces changes in plant physiology and metabolism. It affects germination, seedling growth, vegetative phase, flowering, and fruiting leading to decreased yields and quality of production (Zid and Grignon 1991; Vicente et al. 2004; Parida and Das 2005).

Salt tolerance has been broadly studied in numerous plant species and varieties and in halophytes to understand the mechanisms developed for their adaptation (Abdelly 2006; Messedi et al. 2004; Slama et al. 2017; Ben Hamed et al. 2013; Flowers and Colmer 2015). Salt tolerance is a complex trait that involves a set of mechanisms in plants (Lachhab et al. 2013). Different studies have shown that cultivation of specific plant species or varieties may improve productivity of marginal areas affected by salinity. This can be important, especially in conditions where complementary irrigations are often carried out with water containing high concentrations of soluble ions. In many cases, this has been done without taking into account the tolerance of the different varieties, resulting in poor crop production. Therefore, the breeding for better salt tolerance in crops has become a critical requirement for the future of agriculture in arid and semiarid regions (Owens 2001).

The identification of salt-tolerant genetic resources would certainly contribute to crop improvements, subsequently supporting the agricultural production of salinized areas or areas irrigated with brackish water. In this respect, the knowledge on different adaptative mechanisms, morphological, physiological, biochemical, and other strategies, with which different plants cope with the challenges of the salt stress is critical for success in crop improvement. Although the topic of salt tolerance is very broad and complex, this chapter will be specifically focused on the mechanisms contributing to the protection of the photosynthetic processes against detrimental effects of the salt stress.

4.2 Origin of Soil Salinization

Salts are composed of the different mineral elements of the soil, and some of them represent the essential nutrients needed for plant life. Their concentrations can become very high due to natural processes and/or poor management. In any part of the world, when evaporation exceeds precipitation, salts tend to be accumulated in soils, leading to increased concentrations (Hachicha 2007). Also, salinity increases due to poor drainage and poor water use in irrigation.

Salty soils are rich in soluble salts that affect plant growth and productivity (Mermoud 2006; Hachicha et al. 2017). Generally, in soils where drainage is weak or absent due to their impermeability, salts accumulate in the superficial horizons.

Salinity is a common phenomenon in arid and semiarid regions. Although identification of the natural processes that lead to salinization is essential for understanding the status of salt in any habitat, identification of the origin of salinity and its progression in the wild is important to cope with this constraint. Soluble salts can have three major sources.

4.2.1 Marine Origin

The salts of marine origin are transported and deposited in the continents in three ways. Cyclic salts, brought to the continents in the form of salty spray, are solubilized by rainwater, then redistributed, and transported to the oceans by drainage. Infiltration salts, brought to coastal habitats by seawater infiltration, are the main local source. Fossil salts, precipitated for a long time in certain localities, are led to the rhizosphere or to the surface of the earth by capillarity.

4.2.2 Lithogenic Sources

Some sedimentary rocks contain high levels of chloride and sulfate. The extent of salinization of groundwater and soils depends on the rate of alteration of these sedimentary rocks, which, in turn, varies considerably by region and climate zone. Generally, these rocks release more sulfate than chloride. The weathering of these rocks also releases significant amounts of carbonates, calcium, and magnesium. It

Table 4.1 The areas of salt-affected soils in different parts of the world (FAO 2008)

		Saline soils		Sodic soils	
Region	Total area (M ha)	M ha	%	M ha	%
Africa	1.899	39	2.0	34	1.8
Asia, the Pacific, and Australia	3.107	195	6.3	249	8.0
Europe	2.011	7	0.3	73	3.6
Latin America	2.039	61	3.0	51	2.5
Near East	1.802	92	5.1	14	0.8
North America	1.924	5	0.2	15	0.8
Total	12.781	397	3.1	434	3.4

was documented that these rocks may represent important sources for the local formation of saline lands (Siadat et al. 1997; FAO 2000).

4.2.3 *Anthropogenic Sources*

During the last centuries, large quantities of soluble salts accumulated in soils due to human activity. In some areas, poor-quality irrigation with intense evaporation in arid and semiarid regions and the use of inappropriate cultivation practices have led to the accumulation of large quantities of salt in the upper layers of the soil ground (Qadir et al. 2008). Following the development of the industry and the extensive use of fuels, another source of soluble salts has been added. The use of urban water leads to an increase in salinity. Similarly, the reuse of wastewater causes considerable salinization of groundwater (Shakir et al. 2017).

Saline soils are characterized by the predominance of Ca and Mg ions, as well as Na. The presence of salts leads to an increase in the osmotic pressure of the circulating solution, making it difficult for the plant roots to absorb water and nutritive elements and may cause the plasmolysis of the root cells. The excessive presence of certain anions or cations may affect the absorption and development of some plant species, while other ions (Cl, B, and Na) may also be toxic.

In Table 4.1 is the situation presented by FAO about the total area of saline soil in the world (FAO 2008).

4.3 Effects of High Salinity on Plants

Accumulation of electrolytic solutes in water and soil in higher concentrations is stressful to many plants. As stated before, saline soils occur in different regions of the world, although the majority of their occurrence is in semiarid and arid regions. There, as a result of insufficient downward percolation and lack of salt leaching, the soluble minerals can accumulate in the root zone. In hot and dry areas, the meager

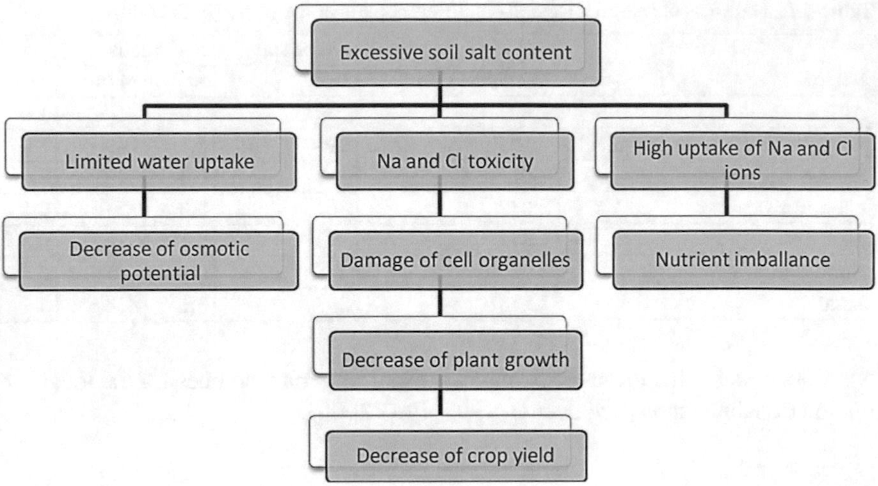

Fig. 4.1 Overview of salt stress effects on plants (Adapted from Evelin et al. 2009)

rainfall is accompanied by high temperature and low air humidity, leading to excessive evaporation. Moreover, in many places, such as river valleys, these conditions are associated with the presence of salt-bearing sediments, and shallow, brackish groundwater, resulting in gradual conversion of rich and fertile soils into barren saline soils (Hillel and Vlek 2005).

Presence of high salt concentrations in soil or water leads to osmotic stress, which is one of the most severe stress factors worldwide, with particular importance for numerous regions located in semiarid and arid climate zones. Figure 4.1 briefly illustrates the different ways by which the salinity affects plant functions. In addition to physical (osmotic) effects of soluble particles limiting water uptake by roots, the salinity effects are associated also with toxicity of individual ions and nutritional imbalance or interactions of these factors (Ashraf and Harris 2004).

4.3.1 Effect of Salinity on Germination and Emergence

Salt stress can limit plant growth, by changing the balance between availability and needs. The scarcity of rains in many regions of the world accentuates salinization of irrigated land and makes them unsuitable for cultivation and abandoned (Yan et al. 2015). Salinity is a limiting factor in agriculture (Qadir et al. 2014). Generally salinity causes a decrease in soil hydraulic conductivity and root aeration and an increase in resistance to root penetration. Moreover, roots meet greater difficulty in suction of water and absorption of nutritional elements.

The salinity results in most of plants in reduction of growth and development (Munns and Tester 2008). A large number of alterations occur in plants (e.g., osmotic adjustments and various reactions) which affect the various organs from the roots to the stems and leaves and which may actually prevent plant growth. This adverse effect is translated by physiological, morphological, molecular, and biochemical changes which negatively affect plant growth and production (Qadir et al. 2014).

Dynamics of germination depends on genetic predispositions and health status, but it is strongly influenced by environmental conditions, including soil water availability (Gutterman 1993). Abdelly (2006) showed that most plants are more sensitive to salinity during their germination and emergence, when the harmful osmotic or toxic effects of salts are very direct and strong.

Germination of seeds represents often the critical step in the establishment of crop canopy, and thus, it can determinate successful agricultural production. Indeed, under salt stress, a late development favors the accumulation of toxic ions that can kill plants before the end of their development cycle (Munns 2002). Salt tolerance can therefore be evaluated by the precocity of germination.

The response to salt of plant species depends on several variables, starting with the species itself and specific genotypes, but also on salt concentration, growing conditions, and developmental stage of the plant. The decrease of the soil osmotic potential prevents the imbibition of the seed following a decrease in enzymatic activities and a strong absorption of Na^+ compared to K^+, leading to embryonic toxicity and delay in metabolic processes (Pires et al. 2017). Percentage of germination under salt stress depends on the level and duration of salinity applied (Fig. 4.2). In addition to final percentage, the salinity influences also the germination dynamics and vitality of the seedlings (Wang et al. 2011).

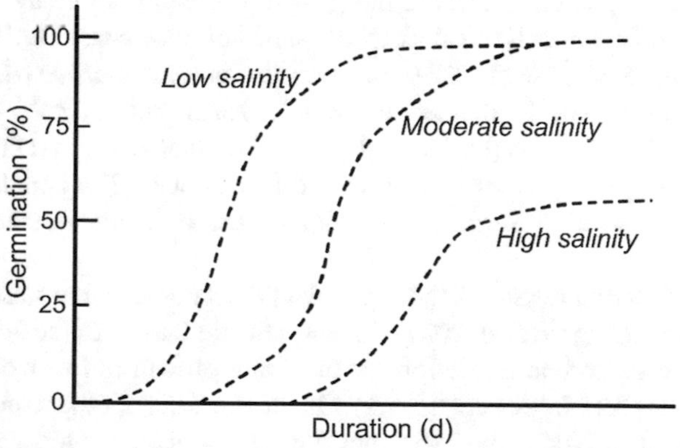

Fig. 4.2 Relationship between rate of germination and time after sowing at different salinity levels (Modified from Ibrahim 2016)

4.3.1.1 Plant Responses on the Different Levels of the Salt Stress

The salinity of soils and waters is induced by the presence of too high concentrations of ions, especially Na^+ and Cl^-. Salinity has three effects, reduces water potential, affects ionic homeostasis, and induces toxicity. At the same time, salt stress affects by different ways different levels of plant organization (Fig. 4.3) and various plant organs (Fig. 4.4).

Excess salt induces osmotic and ionic stress (Yan et al. 2015). The fast response to salt stress is the slowdown of leaf expansion that ceases at high concentrations (Wang and Nil 2000) and affects negatively plant growth (Hernández et al. 1995). The reduction in growth has been shown to be correlated with the salt concentration and the osmotic potential of soils (Flowers and Colmer 2015). The harmful effect of salt is remarkable and causes the death of plants or reduces their productivity. However, growth reduction occurs in most of plant species, but a level of tolerance or sensitivity varies widely among species. For example, for *Raphanus sativus* (radish) plants, 80% of the growth reduction under salt stress is attributed to loss of leaf area (Chartzoulakis and Klapaki 2000).

The action of salinity on the growth leading to both an imbalanced nutritional value of essential ions and a high uptake of toxic ions by the plant (Munns 2002) is connected with stress inducing low osmotic potential of soil solution (Munns and Tester 2008; Yan et al. 2013). Growth reduction in Poaceae can be attributed to an excessive uptake of Na^+ ions (Tester and Davenport 2003, Gu et al. 2016). In tomato, salinity significantly reduces the mass of aerial parts, the number of leaves, the height of plants, and the plant root area and length (Mohammad et al. 1998).

However, in other obligatory halophytes, growth is increased with the salinity of the medium. These plants require a certain concentration of salt to express their maximum growth potential. In fact, the growth optimum is obtained at 200 mM NaCl in *Salicornia rubra* (Khan et al. 2010) and at 50 mM NaCl in *Alhagi pseudalhagi* (Fabaceae) (Kurban et al. 1999). In a non-excretory mangrove (*Bruguiera parviflora*), optimal growth is obtained at 100 mM NaCl, but 500 mM NaCl may be often lethal for these species (Parida et al. 2004). Optimal growth is achieved at 50% seawater in *Rhizophora mucronata* (Aziz and Khan 2001). The beneficial effect of salt was also observed in *Sesuvium portulacastrum* and *Batis maritima* (Messedi et al. 2004).

Growth of different organs of the same plant does not have the same degree sensitivity to salinity (Negrão et al. 2017). In fact, salt stress reduces the growth of aerial parts by decreased carbon allocation for the foliar growth in favor of root growth (Yeo et al. 1991; Huez-López et al. 2011). On the other hand, other studies (Hajlaoui et al. 2010, Lv et al. 2015) report the opposite; the roots are the most affected.

According to Munns and Rawson (1999), the effect of salinity usually results in a reduction in vegetative growth (reduction in height, number of tillers and leaves) which is a function of division and cell elongation. The growth of shoots is more sensitive to salts than the roots (Fig. 4.4).

Salinity affects the growth of plant roots (Kamiab et al. 2012; Rewald et al. 2012). It have shown that salt stress increases the PR/PA ratio. Indeed, plants main-

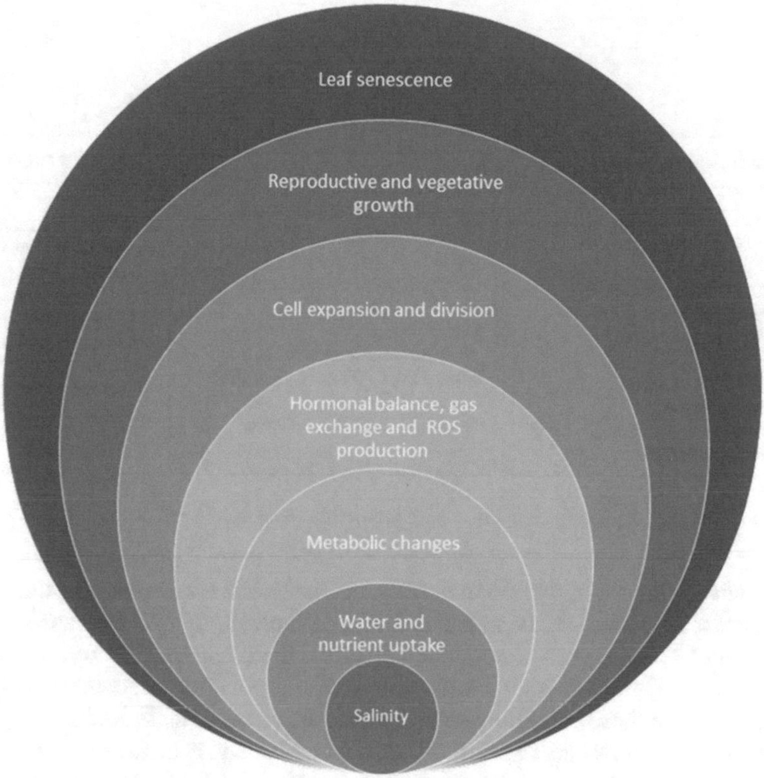

Fig. 4.3 Changes in plant growth under salinity effects (Modified from de Oliveira et al. 2013)

Fig. 4.4 Effect of salt stress on the different organs of plants

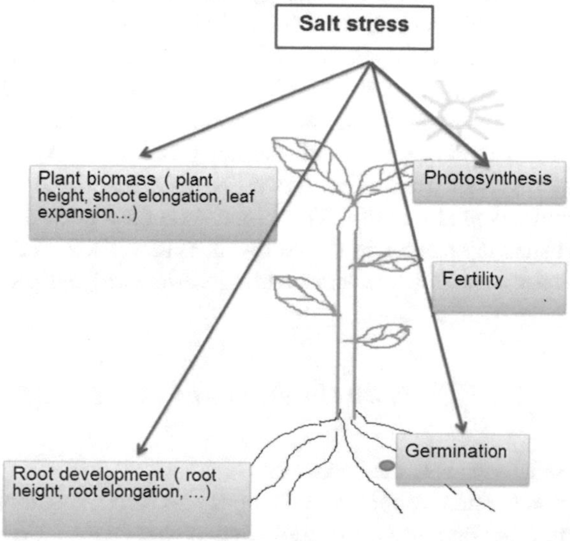

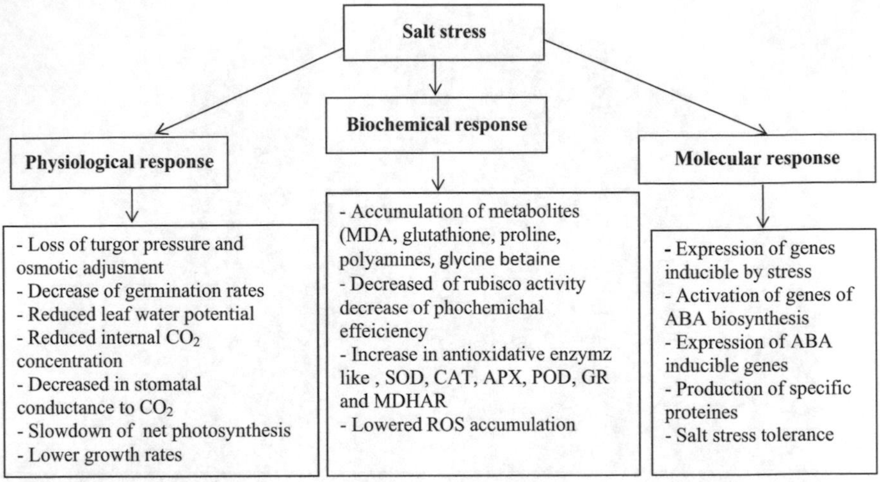

Fig. 4.5 Plant physiological, biochemical, and molecular response to salt stress

tain relatively high root growth under high saline stress; the increase in the PR/PA ratio that follows seems to be associated with an increase of their tolerance to salt.

Kafkai (1991) suggests that under salt stress, the plant spends more photosynthetic energy to maintain high water status and for production of roots for reduction of water loss. In these conditions, it seems that the arrest of leaf growth is triggered by hormones (Munns and Tester 2008) and a significant proportion of assimilates is then relocated to root growth (Fig. 4.5). This is one of the key anatomical responses to osmotic stress in many species, whose adaptative nature is evident since an increase conditions, it seems that the arrest of leaf growth is triggered by of the root mass/mass ratio of the canopy maximizes the water absorption area in decreasing the evaporation surface (Munns 2002).

The mechanisms of salt tolerance include the changes of low as well as of high complexity. The low complex mechanisms are associated with the expression of particular biochemical pathways. On the other hand, the mechanisms of high complexity modify the activities of key processes of energy metabolism, such as photosynthesis and respiration, as well as processes related to water uptake and transpiration and stomatal activity. The highly complex responses involve also the processes related to changes in the cell wall, cytoskeleton, and plasma membrane, related to water use efficiency, and securing the key cell elements (Botella et al. 1994).

4.3.2 *Effects on the Anatomy of the Leaf*

Salinity causes the increased thickness of epidermis, palisade and spongy parenchyma, and mesophyll, cell length, and cell diameter of the leaves of *Atriplex*, *Faba*, and *Gossypium* (Longstreth and Nobel 1979). However, it was shown that in mangroves (*Bruguiera parviflora*), the thickness of the epidermis and mesophyll, and the intercellular space in the leaves, is reduced (Parida et al. 2004). In wheat, area of flag leaf was significantly lowered in conditions of salt stress, significantly limiting yield (Ahmad et al. 2005). Salt causes increased vacuolation, development of the endoplasmic reticulum and mitochondria, vesicle formation and tonoplast fragmentation, and cytoplasmic degradation by mixing the matrices of cytoplasm and vacuole in the potato leaves (Mitsuya et al. 2000). It induces a reduction in intercellular spaces and number of chloroplasts in potato (Bruns and Hecht-Buchholz 1990) and stoma density in tomato (Romero-Aranda et al. 2001).

4.3.3 Plant Mineral Nutrition

In salt stress conditions, the mineral nutrition of the plant is disturbed. Indeed, the Na^+ ions disrupt cation (K^+, Ca^{2+}) absorption (Haouala et al. 2007). At the root level, Na^+ moves Ca^{2+} from cell walls (Zhu 2002). For example, in the isolated cell walls of barley roots, Na^+ and Ca^{2+} are competing for the same site absorption, while K^+ is fixed on other sites. As a result, the K^+/Na^+ ratio at the surface of cells depends on competition with Na^+/Ca^{2+} (Stassart et al. 1981).

4.3.4 Effect of Salinity on Agronomic Yield

Components of crop performance, such as the number of tillers per plant, the number of ears, the number of spikelets per ear, and the grain weight, are usually influenced by the presence of salinity stress. Ahmad et al. (2005) have shown that all performance parameters are reduced under the salinity and that the higher the salinity, the higher yield decrease is observed.

The barley plants under salt stress during heading or differentiation of the ear have shown the reduced plant height and leaf area (Singh et al. 1994), as well as decrease in stem weight, stem length, and shoot dry matter. The number of spikelets per spike is decreased, as well as the number of grains. Salinity has a detrimental effect on the remobilization of reserves during the filling of the grains. Salinity decreases yield more often by reducing the number of spikes bearing the spikelets, ear weight, and 1000-seed weight (Munns and Rawson 1999). Mass and Grieve (1990) observed the changes in the final capacity of spike, associated with lower number of spikelets and grains per spike, as well as a decrease in spike length. Salt stress led to an increase of the phyllochron. In addition, the number of leaves on main stem was lower, and a vegetative period of stem meristem was shorter (Mass and Nieman 1978).

In case of rice, screening of cultivars for salt tolerance has been based mostly on growth rate and grain yield (Kafi et al. 2013). In this respect, it was shown that yield under salt stress was well correlated with panicle weight, number, height, and tiller number. The significant variability in salt resistance and yields in salt stress conditions was observed among families of rice, providing promising background for salt stress studies and breeding in this crop (Souleymane et al. 2017).

The effect of irrigation with saline water was tested by Lamsal et al. (1999) in wheat, observing decreases of grain yield and of all yield components and growth parameters, including flag leaf area, plant height, grain per spike, number of spikes per plant, grain weight per spike, 1000 grain weight, as well as total dry matter accumulation.

4.3.4.1 Photosynthetic Responses and Acclimations on Salt Stress

Photosynthesis is strongly involved in plant productivity (reduction of production of biomass, leaves) and nutrient flows in plants. Salinity affects the physiological activity of the leaf, particularly photosynthesis, which is the main cause of reduced plant productivity (Alem et al. 2002). The adaptation of photosynthesis under salt stress is presented in Fig. 4.6.

In specific environments, the resistance to drought and salinity are important traits determining the yield of main crops (Munns and Tester 2008; Munns, et al. 2010). The phenotypic effects of these two stresses are often very similar, and, therefore, the similar screening methods and tools can be used for both stresses. Important and relatively fast response to salinity is stomatal closure, associated partly with the osmotic effect of salts on the capacity of roots to absorb the soil water (Munns and Tester 2008). Insufficient uptake of the water due to salinity is denoted as "chemical drought" (Munns et al. 2010). In the early stages of the stress, the decrease of photosynthetic rate is caused by stomatal closure. The standard measurements of stomatal conductance or screening based on photosynthetic parame-

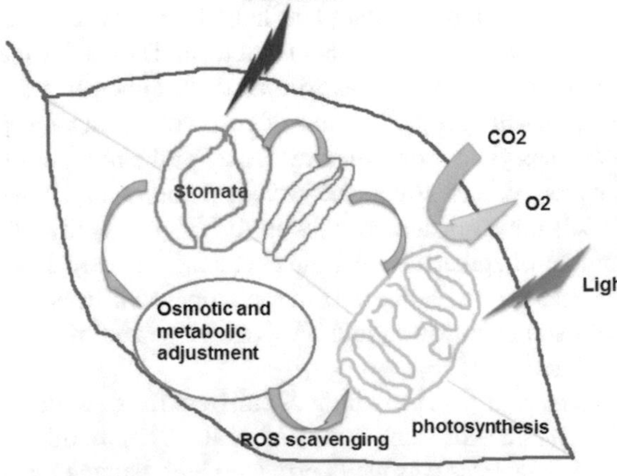

Fig. 4.6 Photosynthesis adaptation to salt stress

ters are usually quite slow, and the reproducibility is often low (Munns et al. (2010). Therefore, there is still a need for fast and reliable methods of monitoring photosynthesis in salt stress conditions. Chlorophyll fluorescence measurements can be used in plant phenotyping and breeding programs to monitor different biotic and abiotic stresses including mineral deficiencies, soil salinity, and pathogenic diseases (Brestic and Zivcak 2013; Kalaji et al. 2017).

Under moderate salinity, the photosynthetic efficiency can reach the values similar to the control, but under high salinity, photosynthesis is significantly inhibited, as it was found in *Desmostachya bipinnata* (L.) Staph. (Asrar et al. 2017). At high concentrations of salt, the photosynthetic pigments, photochemical quenching, and electron transport rate were significantly decreased, whereas at moderate salt stress, the decrease was not observed. The content of MDA increased at high salinity, documenting excessive accumulation of ROS. Under increasing salinity treatments, a subsequent decrease of Rubisco content was also observed. The proteins of photosynthetic complexes were overexpressed (D1, AtpA, PetD) or remained unaffected (PsbO) under moderate salinity but decreased at higher concentrations, except of AtpA. Severe salt stress caused damages to PSII photochemistry and downregulation of chloroplast proteins connected with biochemical limitations in *D. bipinnata* (Asrar et al. 2017). The rate of photosynthesis increases for low levels of salinity and decreases in high level, without modification on stomatal conductance *Bruguiera parviflora* (Parida et al. 2004).

Salinity stress induces for the majority of plants a reduction in production of biomass essentially due to a decrease in photosynthesis (Kalaji et al. 2018). These changes are mostly connected with changes of carboxylation processes, but not photophosphorylation, which are most affected by salt stress. The response of plants to salt stress is heavily dependent on genotype. The salt stress has both short- and long-term effects on photosynthetic processes. The early effects can be observed after a few hours or days of exposure of the plants to salinity, and there can be a full cessation of carbon uptake for a few hours. The long-term effect appears after several days of salt treatment. The reduction of photosynthesis is caused by the accumulation of salt in the leaves (Munns and Termaat 1986). Whereas some studies indicate a reduction in photosynthesis by salinity (Chaudhuri and Choudhuri 1997; Soussi et al. 1998; Romero-Aranda et al. 2001; Kao et al. 2001), others show that carbon assimilation is not affected by salt stress or, in some cases, also enhanced by moderate salt concentrations (Rajesh et al. 1998; Kurban et al. 1999).

In the mulberry tree, the net assimilation of CO_2 (PN), the stomatal conductance (gs), and the rate of transpiration (E) are reduced under salt stress, whereas the intercellular concentration of CO_2 (Ci) is increased (Agastian et al. 2000). In *Bruguiera parviflora*, PN increases at low salinity but decreases at high salinity, whereas at low salinity, it remains comparable to that of control plants and falls at high concentrations (Parida et al. 2004).

Many studies showed, however, a major role of salt stress in limiting osmotic conductance and reducing reactive oxygen species (ROS) and main enzymes of detoxification of ROS. According to Munns and Tester (2008), the reduction in photosynthesis is related to the decrease in leaf water potential, which is responsible for the closure of stomata (Allen et al. 1985), which causes the reduction of photosynthesis and stomatal conductance (Orcutt and Nilsene 2000).

Salt stress reduces chlorophyll content and increases respiration, but has no significant effect on carotenoid levels in alfalfa (Khavari-Nejad and Chaparzadeh 1998). In *Atriplex lentiformis*, uptake of CO_2 and the ratio of Rubisco to phosphoenolpyruvate carboxylase (PEPC) activity decreased under salt stress. Phosphoenolpyruvate carboxylase (PEPC) activity increases linearly with salinity, whereas that of ribulose bisphosphate carboxylase/oxygenase (Rubisco) is not changing significantly (Zhu and Meinzer 1999). The reduction in photosynthesis is caused also by a decrease in stomatal conductance which restricts the access of CO_2 for Calvin-Benson cycle (Brugnoli and Bjorkman 1992). Stomata closure minimizes transpiration and affects light capture by chloroplasts and photosystems leading to impaired activity of these organelles (Iyengar and Reddy 1996).

In the process of photosynthesis, two key complex events occur: light reactions, in which light energy is converted into ATP and NADPH and oxygen is evolved, and light-independent, dark reactions, in which CO_2 is fixed into carbohydrates by utilizing both of the products of light reactions (Allakhverdiev et al. 2002).

Pigment analysis showed that salt stress resulted in a significant decrease in chlorophyll a content, whereas the content of chlorophyll *b* remained unaffected. It appeared that chlorophyll b tolerated more the salinity than chlorophyll a. However, total carotenoid contents were increased. For some crops carotenoid concentration can be increased in conditions with low salinity and on the other hand under high salinity can be significantly decreased.

CO_2 assimilation rate (Chaves et al. 2011; Kalaji et al. 2011; Kanwal et al. 2011; Chen et al. 2015; Hniličková 2017; Dąbrowski et al. 2017) and photosynthetic oxygen evolution can be also inhibited (Dąbrowski et al. 2016). Increases in Fo have been attributed to physical separation of the PSII from associated pigment antennae and decreases in the number of active RCs (Strasser et al. 2010). In the present study, the salt-induced increase in Fo indicates that some active RCs were inactivated by salt stress. This inactivation of RCs by salt stress was further evidenced by the increase in the ABS/RC value and the decrease in the values of RC/CSO and TRO/ABS after salinity treatment. Specific chlorophyll a fluorescence transient parameters derived from OJIP test (φEo, φPo, ψO, RC/CS, RC/ABS, PIABS, and PICS) were decreased under salt stress, while dVo/dto(Mo), Vj, and φDo were increased. The decrease of ETRmax and yield and the change of chlorophyll a fluorescence transients showed that salt stress had an important influence on photosynthesis. These results indicated that the effects of salinity stress on photosynthesis may depend on the inhibition of electron transport and the inactivation of the reaction centers, but this inhibition may occur in the electron transport pathway at the photosystem II (PSII) donor and acceptor sites (Mehta et al. 2010). Salt stress can damage active reactive centers of PSII and destroys the oxygen-evolving complex (OEC) and impairs the electron transfer capacity on the donor side of PSII (Misra et al. 2001; Mehta et al. 2010; Sun et al. 2016; Kan et al. 2017; Kalaji et al. 2018).

The major consequences of salt stress on the main photosynthetic parameters in the various crop plants are presented in Table 4.2.

Table 4.2 Salt stress photosynthetic adaptative responses of different plant species

Plant species	Salt stress	Photosynthetic parameters measured	Adaptative response to salt stress	References
Cakile maritima	0, 100, 200, 300, and 500 m*M* NaCl	RGR, ROS, Φ_{PSII}, DW, F_m, F_o, F_s, F_v/F_m, NPQ, qP	Adaptative response to salt stress Net photosynthetic rate, stomatal conductance, maximum quantum efficiency of PSII, and quantum yield were stimulated in the 100–200 mM NaCl range Higher salinity adversely affected gas exchange and changed PSII functional characteristics, resulting in a reduction of net photosynthetic rate per leaf area unit	Debez et al. (2008)
Atriplex portulacoides	0, 200, 400, 800, and 1000 mM NaCl	RGR, ROS, SOD, APX, ΦPSII, DW, Fm, Fo, Fs, Fv/Fm, NPQ, qP	The maintaining photosynthetic activity, photosynthetic pigment contents, and preserving PSII functional integrity, in conjunction with the accumulation of natural bioactive compounds (e.g., polyphenols, anthocyanins, and proline) and the active contribution of enzymatic antioxidant defenses	Benzarti et al. (2012)
Atriplex hortensis	0, 5,15 g/l NaCl	The plant height, root system length, leaf area, Chl content	Improved tolerance to salt stress may be accomplished by decline in growth and photosynthetic activity	Sai Kachout et al. (2009)
Suaeda salsa	2, 4, 6, and 8 g/kg NaCl	FW, DW, height, total nitrogen (TN), and total carbon (TC)	FW, DW, and height were promoted at lower salinity treatments but reduced at higher salinity treatments, while TN and TC contents were kept stable with increasing salinity level Nitrogen addition could significantly mitigate the deleterious effects of salt stress	Jia et al. (2017)
Aeluropus littoralis	0, 200, and 400 mM NaCl	Proteomic analysis	The reduction of proteins related to photosynthesis and induction of proteins involved in glycolysis, tricarboxylic acid (TCA) cycle, and energy metabolism	Azri et al. (2009)
Mediterranean *Limonium* sp. *L. santapolense*, *L. girardianum*, *L. narbonense*, *L. virgatum*	0, 200, 400, or 800 mM NaCl	DW, ion content, pigment content-free proline, glycine betaine, total soluble sugars, analysis of soluble carbohydrates	The efficient transport of Na^+ and Cl^- to the leaves and their compartmentalization in vacuoles The accumulation of fructose and proline, as the main physiological osmolytes responsible for cellular osmotic adjustment The activation of K^+ transport from the roots to the leaves in the presence of high salt concentrations	Al Hassan et al. (2017)

(continued)

Table 4.2 (continued)

Plant species	Salt stress	Photosynthetic parameters measured	Adaptative response to salt stress	References
Limonium bicolor (bag.) Kuntze	0, 100, 200, and 300 mM NaCl	DW, water content, net CO_2 assimilation rate (PN), stomatal conductance (gs), transpiration rate of MDA, content of total phenolic and flavonoids	Adaptation resistance mechanisms involving an increased number of salt glands, enhanced activities of antioxidant enzymes, and an accelerated accumulation of secondary metabolites	Wang et al. (2016)
Zea mays L. Maize	0,100, 200, 300 mM NaCl	Prompt Chl fluorescence (PF), delayed Chl fluorescence (DF), modulated 820 nm reflection	Decreasing the number of active PSII reaction centers Impaired the connectivity between independent PSII units, destroyed the oxygen-evolving complex, and limited electron transport beyond the primary quinone acceptor Q_A^-. The photochemical activity of PSII is higher in salt stress than that of PSI	Kan et al. (2017a, b)
Paulownia sp.	Saline soils (EC = 1.6; 6.3;14 $mS.m^{-2}$)	Chlorophyll pigments Chlorophyll *a* fluorescence (PAM fluorimeter)	The Q_A^-reoxidation and photochemical quenching increase The stimulation of the linear electron transport Increased ETR and qP in presence of salt enhanced the ability of the photosynthetic apparatus to maintain QA in the oxidized state and an increase in the proportion of the "open" PSII reaction centers	Stefanov et al. (2016)
Paspalum vaginatum Swartz cv.	0, 75, 150, 300, and 600 mM NaCl	FW, DW, Chl fluorescence (mini-PAM)	Maximum efficiency of PSII photochemistry declined only in the highest salinity. A full recovery of the fluorescence parameters upon rewatering. The rapid osmotic adjustment capacity was associated with higher sugar levels	Pompeiano et al. (2016)

According to Munns and Tester (2008), the reduction of photosynthesis is related to decreased leaf water potential, which is responsible for stomatal closure (Price and Hendry 1991; Allen et al. 1985), which causes the reduction of stomatal conductance (Orcutt and Nilsen 2000). The diffusion of CO inside the stomata then becomes limited, and its binding at the chloroplast level consequently decreases; the regeneration of RuBP (ribulose bisphosphate) becomes limited.

4.3.5 *Effects of Salt Stress on Ultrastructure of Chloroplasts*

The diffusion of CO_2 inside the stomata then becomes limited, and its binding at the level of chloroplasts decreases (Graan and Boyer 1990); consequently the regeneration of RuBP (ribulose bisphosphate) becomes limited (Gimenez et al. 1992). Stomatic control and regulation involve cell turgor but also root signals, such as abscisic acid (ABA) (Zhang and Davies 1989). Cellular turgor occurs more or less directly in the chloroplast: directly by maintaining the volume of the chloroplast and indirectly by its effect on the stomatal opening, which controls the conductance and conditions the use of energy photochemical (ATP, NADPH) in chloroplasts (Gupta and Berkowitz 1987).

Salt causes disorganization of the thylakoid structure, increases the number and size of plastoglobules, and reduces starch ranges (Hernández et al. 1995, 1999). In mesophyll potato cells, thylakoid membranes are distended, and most are altered under severe salt stress (Mitsuya et al. 2000). Salinity reduces the number and thickness of the stack of thylakoids by grana (Bruns and Hecht-Buchholz 1990a, b). In NaCl-treated tomato plants, chloroplasts are assembled, cell membranes are deformed and wavy, and the structure of grana and thylakoids is affected (Khavari-Nejad and Mostofi 1998). Ultrastructural changes of chloroplasts under salt stress are well studied in *Eucalyptus microcorys*. These changes include the appearance of large starch grains, dilation of thylakoid membranes, almost complete absence of grana, and development of mesophyll cells (Keiper et al. 1998).

4.3.5.1 Effect of Salinity on Reproduction Process

Salinity reduces the growth rate of the plant and its reproductive organs (Hu et al. 2017). They studied the effect of salinity on the physiology of reproduction; they found that the number of pollen in two different types of barley cultivars was reduced from 24 to 37%. Studies by Munns and Rawson (1999) on the effect of salt accumulation in the meristem of barley on reproduction and development show that short periods of salt stress during organogenesis may have irreversible consequences on the fertility of the ear; it causes the abortion of the ovaries.

It was observed higher floret fertility contributed to higher seed set and grain yields in tolerant genotypes, whereas higher spikelet sterility led to poor seed set and lower grain yields in sensitive to salt stress genotypes. It is important to do screening at reproductive stage for morphological traits like floret fertility are thus more useful to identify plant genotypes tolerant to salinity stress (Rao et al. 2008).

Floral phenology, pollen quality, and seed set of *Plantago crassifolia* plants were optimal in plants grown in non-saline conditions. Same positive tendency was observed in the presence of 100 mM NaCl. But progressive reduction of pollen fertility, seed set, and seed viability has been observed by higher salt concentrations (Boscaiu et al. 2005).

4.3.5.2 Symptoms of Toxicity Connected with Ionic and Nutritional Balance in Plants

Saline solutions impose ionic and osmotic stress in plants. The effects of this stress can be observed at different levels. In sensitive plants, the growth of aerial parts and to a lesser extent that of roots is rapidly reduced. This reduction phenomenon appears to be independent of the tissue Na^+ concentration but would rather be a response to the osmolarity of the culture medium (Munns 2002). The specific toxicity of Na^+ ions is related to the accumulation of these ions in the leaf tissues and leads to necrosis of the aged leaves. Generally, this necrosis begins with the tip and the edges to finally invade the entire leaf. The reduction in growth is due to a reduction in leaf life, and thus there will be a reduction in growth and productivity (Munns 1993 and 2002). In saline soils, Na^+ ions induce deficiency in other elements (Silberbush et al. 2005). The effects of Na^+ are also the result of deficiency in other nutrients and interactions with other environmental factors, such as drought, which increase the problems of Na^+ toxicity. The excess of Na^+ ions inhibits the uptake of other nutrients either by competition at the sites of the root cell plasma membrane transporters or by inhibition of root growth by an osmotic effect. Thus, the absorption of water and the limitation of the nutrients essential for the growth and the development of the microorganisms of the soil can be inhibited.

Leaves are more sensitive to Na^+ ions than roots because these ions accumulate more in the aerial parts than in the roots. These can regulate the concentration of Na^+ ions by their export either to the aerial parts or to the ground. The metabolic toxicity of Na^+ is mainly related to its competition with K^+ at sites essential for cell function. Thus, more than 50 enzymes are activated by K^+ ions; Na^+ ions cannot replace K^+ in these functions (Bhandal and Malik 1988; Tang et al. 2015; Gu et al. 2016). For that, a high concentration of Na^+ can affect the functioning or the synthesis of several enzymes. In addition, protein synthesis requires high K^+ concentrations for tRNA binding on ribosomes (Blaha et al. 2000) and probably for other ribosome functions (Wyn Jones et al. 1979). The disruption of protein synthesis by the high concentration of Na^+ represents the major toxic effect of Na^+ ions. Osmotic stress could occur following an increase in Na^+ concentration at leaf apoplasm (Oertli 1968). This result was verified by microanalyses (R-X) of Na^+ concentration in apoplasm of rice leaves (Flowers et al. 1991). The presence of high concentrations of Na^+ in the cells allows the plant to maintain its water potential lower than that of the soil to maintain its turgor and water absorption capacity. This leads to an increase in osmotic concentration either by absorption of solutes from the soil or by synthesis of compatible solutes. The former, usually Na^+ and Cl^-, are toxic, while the latter are compatible but energetically expensive for the plant.

4.4 Specific Adaptative Strategies of Plants Under Salt Stress

Several studies have shown that plants adapted to salt stress use one or more mechanisms to mitigate the effect of NaCl such as (i) the Na^+ reabsorption by transfer cells or vascular parenchyma (Karray-Bouraoui 1995); (ii) the compartmentalization of ions between the organs (roots/aerial parts), tissues (epidermis/mesophyll), or cellular compartments (vacuole/cytoplasm) (Cheeseman 1988); and (iii) the dilution of Na^+ by the material produced growing leaves (Tester and Davenport 2003).

The mechanism by which a plant tolerates salt is complex, and it differs from species to species (Munns and Tester 2008; Bueno et al. 2017). The main effects of salt stress and development of plant adaptation of various crop plants are presented in Table 4.3.

Salt affects seed germination through osmotic effects (Khan et al. 2000), ion toxicity, or a combination of the two (Khan and Ungar 1998; Munns and Tester 2008). Osmotic stress can result in inhibition of water uptake that is essential for enzyme activation, breakdown, and translocation of seed reserves (Ashraf and Foolad 2007; Munns and Tester 2008). Furthermore, ionic stress can inhibit critical metabolic steps in dividing and expanding cells and may be toxic at high concentrations (Munns and Tester 2008). The excess Na^+ and Cl^- have the potential to affect plant cell enzymes, resulting in reduced energy production and some physiological processes (Munns and Tester 2008; Morais et al. 2012). The degree of salt tolerance varies among plant species and, for a given species, also at different developmental stages (Ahmad et al. 2013; Bueno et al. 2017). Some plants have adapted to grow in high-salinity environments due to the presence of different mechanisms in them for salt tolerance, such plants are known as salt-tolerant plants or halophytes (Ahmad et al. 2013), but a large majority of plant species grown in non-saline areas are salt-sensitive (glycophytes). Glycophytic and halophytic species differ greatly in their tolerance to salt stress (Munns and Tester 2008).

With the degree of salinity of soil solution, glycophytes in general are exposed to changes in their morphophysiological (Bennaceur et al. 2001) and biochemical (Grennan 2006) behavior. So the plants react to these variations in salinity in the biotope to set off resistance mechanisms. Among these mechanisms, the osmotic adjustment plays a vital role in the resistance or the tolerance of the plant to stress (Munns 2002). The plant will have to synthesize organic solutes to adjust its water potential. A salinity adaptation strategy consists of synthesizing osmoprotective agents, mainly amino compounds and sugars, and to the accumulated in the cytoplasm and organelles (Ashraf and Foolad 2007; Chen et al. 2010; Sengupta and Majumder 2010; Yan et al. 2013). Identification and understanding of plant tolerance mechanisms to salinity therefore have a clear interest in varietal improvement. The objective of this study is to determine certain morphological and physiological criteria allowing early identification of saline-tolerant plants.

Table 4.3 Salt stress adaptative responses of different plant species

Plant species	Conditions of salt stress	Parameters measured	Adaptative response to salt stress (mechanism implicated in salt tolerance)	References
Cakile maritima	50, 100, 200, 300, 400, or 500 mM NaCl which progressively adjusted with increasing NaCl concentrations 50 mM step per day	Germination rate, RGR, fresh weight (FW), dry weight (DW), ionic status (K+, Na+, Cl-, K=/Na+, Mg+, Ca+)	Growth activity maintained up to 500 mM NaCl Preservation of the biomass production Ability to maintain the tissue water status The efficiency of selective K+ uptake Na+ utilization by the plant for osmotic adjustment	Debez et al. (2004) Megdiche et al. (2007)
Crithmum maritimum	0, 50, and 200 mM which progressively adjusted with increasing NaCl concentrations 50 mM step per day	Growth parameters, MDA, POD, SOD, catalase, peroxidase	Keeping of convenient tissue water supply Selective accumulation of K^+ versus Na^+ High antioxidant enzyme activities (SOD-CAT-POD), preventing toxic accumulation of AOS	Ben Amor et al. (2005) Ben Hamed et al. (2007)
Gypsophila oblanceolata Bark.	0, 50, 100, 150, 300 mM NaCl	Germination, antioxidant activities of enzymes/isoenzymes (SOD, CAT and POX, MDA)	Different antioxidant metabolism to the different salt concentration	Sekmen et al. (2012)
Cakile maritima (halophyte) *Arabidopsis thaliana* (glycophyte)	100 mM and 400 mM NaCl (daily increase by 50 mM NaCl or 10 mM)	DW, H_2O_2, antioxidant enzyme activities, MDA	Increase in H2O2, antioxidant activities and MDA in halophyte compared to the glycophyte plant	Ellouzi et al. (2011) Ellouzi et al. (2014)
Limonium latifolium, *Matricaria maritima*, *Crambe maritima*	0, 100, 200, or 400 mM NaCl	H_2O_2 assay, MDA, Na accumulation in plant tissues, CAT activity, APX activity, GR activity	Results showed that *L. latifolium*, *M. maritima*, and *C. maritima* used different antioxidant enzymes to cope with salinity The most tolerant species (*L. latifolium*) uses GR to protect from stress-induced ROS	Ben Hamed et al. (2014)
Triticum aestivum L. (wheat) *Hordeum vulgare* L. (barley)	100 mM NaCl	K, Ca, Na, DW	Na^+ uptake and transport increased under NaCl presence Specific osmotic adjustment of barley is distinguished by a good ability to sequester Na^+ inside the vacuole	Ben Ahmed et al. (2010)

Sesuvium portulacastrum	0, 100, 200, 300, 400, and 500 mM NaCl incorporated with Hoagland's nutrient solution	*PN*, transpiration rate (*E*), water use efficiency (WUE), antioxidant enzyme activity (CAT, GR, SOD, APX, and GPX)	Na^+ uptake and transport increased under NaCl presence	Muchate et al. (2016)
Hordeum vulgare L. (barley)	0, 50, 100, 150, and 200 mM NaCl	Germination, the radicle breaching, and the coleoptile's emergence	NaCl, at doses up to 200 mM, slows the rate of barley germination without affecting their ability to germinate During the phase of coleoptile's emergence, barley cultivars show greater sensitivity to NaCl than during radicle breaching	Abdi et al. (2016)
Sarcocornia fruticosa (L.)	10 (control), 60, 100, 200, and 300 mM NaCl	DW, biomass allocation, tissue water content	Increase Cl– and Na+ uptake used instead of organic osmolytes Increase of succulence in shoots at higher salt concentrations and the ability to maintain a lower K^+/Na^+ ratio and higher K-Na selectivity in all organs	García-Caparrós et al. (2017)
Barbarea verna, *Capsella bursa-pastoris*, *Hirschfeldia incana*, *Lepidium densiflorum*, *Malcolmia triloba*, etc.	Starting 25 days after sowing (DAS), plants watered with 150 mM NaCl for 30 days	Germination, root elongation, leaf water relation, Na^+, K^+, stomatal size and density	The control of transpirational water flux (i.e., via stomatal and/or aquaporin regulation) that is associated with ion loading and accumulation	Orsini et al. (2010)
Plantago crassifolia L.	100, 200, 300, 400, or 500 mm NaCl	Germination, seedling recovery, plant growth, proline content, soluble sugar fraction	Proline may still play an important role in the protection against high salinity "Osmolytes" work primarily through oxidative detoxification	Vicente et al. (2004)
Atriplex halimus L., *Atriplex nummularia* L.	0, 300, 600, 800, and 1000 mM NaCl; 0, 300, 600, 800, and 1000 mM KCl	Leaf tissue water, leaf area, (Na^+, Ca_2^+, K^+, Mg_2+), proline concentration	High absorption and transport of Na+ to shoots and its use for osmotic adjustment The efficiency of the vacuolar compartmentalization of Na^+ ions The salt tolerance depends upon the chemical composition of the salts in the soil solution	Belkheiri and Mulas (2013)
Matricaria recutita L.	0, 6, 9, and 12 dS m^{-1}NaCl	POX activity, oil and chamazulene content, flavonoid compounds	Increasing of POX activity, flavonoid content, and chamazulene content Root growth stimulation at 9 ds m^{-1} NaCl Na compartmentation in root at 12 dS m^{-1} NaCl	Askari-Khorasgani et al. (2017)

4.4.1 Morphological and Anatomical Adaptations of Plants

Native plants in saline and desert environments have developed, over time, certain traits that give them the ability to develop in these stressful conditions. These traits are often of morphological and anatomical types. However, scientific studies have been done on the development and role of these morpho-anatomical features and also studies with other adaptative reactions of plants to the salt stress.

The most obvious morphological adaptations of plants in saline desert habitats are the reduction in leaf size and the number of stomata per unit leaf area, the increase in succulence and cuticle thickness of the leaf, and the formation of a layer of wax (Gale 1975; Mass and Nieman 1978). These adaptations play a crucial role in the conservation of water for the development of the plant in saline conditions. More recently, it has been shown that there is variability among the *Cenchrus ciliaris* ecotypes that allows it to withstand severe salinity conditions especially during drought periods (Hameed et al. 2015).

These are the most visible adaptations found throughout the plant (Dickison 2000). At the leaf level, there are certain structures that allow the plant to secrete excess salt. The most important are the secretory trichomes (*Atriplex* spp.) and the salt glands found in many plants of desert flora and coastal habitats. The latter are characteristic of a few families, including Poaceae, Avicenniaceae, Acanthaceae, Frankeniaceae, Plumbaginaceae, and Tamaricaceae (Mauseth 1988; Thomson et al. 1988, Marcum and Murdoch 1994).

Desert plants have usually succulent stems characterized by a well-developed water storage tissue in the cortex and marrow (Lyshede 1917; Dickison 2000). A multilayered epidermis may have thick walls, covered with thick cuticle surmounted by wax. For example, at *Anabasis* sp. the epidermis is formed of 8 to 11 layers. The stem of *Salicornia fruticosa* consists of a single cortex and an epidermis with a single thin-walled cell base. The palisadic and parenchymal tissue of photosynthetic organs is used for water storage (Fahn 1990).

The roots of xerohalophytes reduced the cortex to shorten the distance between the epidermis and the stele (Wahid 2003). The Caspari band is larger in dry-habitat plants than in mesophytes (Wahid 2003). In saline environment plants, the endoderm and exoderm represent barriers of variable resistance to the radial flow of water and ions, from the cortex to the stele (Hose et al. 2001; Taiz and Zeiger 2002).

4.4.2 Salt Tolerance Mechanisms

The physiological characterization of plant tolerance to salinity results from processes that allow the plant to absorb water and mineral salts from substrates with low hydric potential but also to live by accepting the important presence of the sodium in its tissues; halophytes, which accumulate the most sodium (Elzam and Epstein 1969, Guerrier 1984a), are distinguished by a strong capacity development of organic compounds (Briens and Larhe 1982), these two factors allowing the maintenance of a high internal osmotic pressure which favors the water exchange between the external and cellular compartments (Guerrier 1984b).

All plants do not react in the same way to salt stress; according to their production of biomass in the presence of salt, four main trends have been discerned: The first is true halophyte, whose production of biomass is stimulated by the presence of salt. These plants (*Atriplex* sp., *Salicornia* sp., *Sueda* sp.) present extensive adaptations and are naturally favored by soil salinity. Optional halophytes can show a slight increase in biomass at levels low in salts: *Plantago maritima* and *Aster tripolium*. Nonresistant halophytes can support low concentrations of salts: *Hordeum* sp. Glycophytes or halophytes can be sensitive to the presence of salts: *Phaseolus vulgaris* (Cheeseman 2015).

The adaptation reaction of various glycophytes to different salt stress treatments is presented in Table 4.4.

According to Munns et al. (2006), tolerance of cereals to salinity depends on variability genetics such as some species that resist this type of abiotic stress than others. In particular, the toxic effect of salts is less pronounced in common wheat than in durum wheat. This character is conferred by the presence of Kna1, a gene responsible for the exclusion of sodium. In addition barley can grow normally under conditions considered as limiting. Indeed, in addition to the exclusion of sodium, the barley plant uses another salinity tolerance mechanism that manifests itself by the imprisonment of salts in a very specific compartment in the leaf. This not only spares their effects toxic but also counteracts the osmotic pressure of the soil (Munns and Tester 2008).

4.4.3 Ion Exclusion and Compartmentalization Mechanisms of Plant Tolerance to Salinity

According to Berthomieu (1988), the plant prevents salt from rising up to leaves. A first barrier exists at the level of the endoderm, an inner layer of the root. However, this barrier can be interrupted, especially during emergence ramifications of the root. Other mechanisms limit the passage of salt from the roots to leaves, but the genes that govern them are still largely unknown.

It is also indicated that the exclusion capacity of Na^+ and/or Cl^- stems is good correlated with the degree of salt tolerance. Maintaining a low concentration of Na in the leaves may be due to an exclusion mechanism that causes an accumulation of Na in the roots, avoiding excessive translocation to the stems; but it can also be linked to a high mobility of this element in the phloem. However, some physiological measures concord to suggest the existence of an active expulsion of cytoplasmic sodium apoplasm or to the vacuole, thus protecting enzymatic equipment from the cytoplasm in aerial organs (Greenway and Munns 1980).

Table 4.4 Adaptative responses to salt stress in different glycophytes

Plant species	Salt stress	Photosynthetic parameters measured	Adaptative response to salt stress	References
Matricaria chamomilla L. (German chamomile)	0, 50, 100, 150 mM NaCl	Na, K, Fe, soluble carbohydrates, proline	The difference in growth at the salinity treatments can be attributed to differences in ion transfer rates to the leaves and proline and carbohydrate accumulation	Heidari and Sarani (2012)
Amaranthus cruentus L.	5 g L^{-1} NaCl	DW, pigment content photosynthesis rate, transpiration, antioxidant enzyme activities	5 g L^{-1} of NaCl does not affect the yield as well as its pigment content, photosynthesis, transpiration rate, stomatal conductance, or antioxidant enzyme activities	Quin et al. (2013)
Brassica rapa L.	0, 50, 100, and 150 mmol NaCl	DW; FW; shoot length; root length; chlorophyll *a*, *b*, *a* + *b*; proline contents; relative water content (RWC)	Increasing proline content in leaves Low plant dry mass at high salt concentrations High salinity levels decrease normal photosynthetic and other biological processes	Jan et al. (2017)
Tomato landrace	0, 300, 450, and 600 mM NaCl	FW, DW, leaf gas exchange, CO_2 assimilation rate, transpiration rate, stomatal conductance (g_s), intercellular CO_2 concentration (C_i), total soluble sugars, total antioxidant capacity	Changes in total antioxidant capacity and leaf pigment content that emphasized the occurrence of modifications in the photosynthetic apparatus according to salt gradient The more efficient assimilate supply and an integrated root protection system provided by sugars and antioxidants can explain the significantly higher root/shoot ratio	Moles et al. (2016)
Triticum aestivum L. (wheat)	200 mmol NaCl	RWC photosynthetic rate (Pn) and stomatal conductance (gs)	Under salt stress, the salt-resistant cv. YN19 had higher efficiency in photosynthetic electron transport, hence maintaining higher photosynthetic rate under salt stress, compared with the salt-sensitive cv. JM22	Sun et al. (2016)
Lolium perenne L. (ryegrass)	0, 0.15, and 0.30 M NaCl	JIP test parameters	JIP test estimated higher potential photosynthetic efficiency (vitality) during salt stress conditions in the most tolerant variety	Dąbrowski et al. (2016)

Glycine max L. (Merr.) (soybean)	0.59% NaCl kg^{-1} 1.29% NaCl kg^{-1} soil	Gas exchange parameters (PN, gs, Ci, E, Fv/Fm, and ETR)	Regression analysis indicated that there has been found extremely significantly positive correlation between GY and P_N in field condition P_N parameter can be used as physiological index for field resistance of soybean to salt stress	He et al. (2016)
Beta vulgaris sp. *L.* (sugar beet)	50 mM NaCl applied in 50 mM increments each day over 6 days until a final level of 300 mM	ΦPSII, Rubisco activity, chlorophyll and protein contents, enzyme activities, phosphoenolpyruvate carboxylase (PEPC) activity, chloroplast metabolite level	The contribution of chloroplasts and the extra-chloroplast space to salinity tolerance via metabolic adjustment	Hossain et al. (2017)
Amaranthus cruentus L.	0, 25, 50, 75, and 100 mM of NaCl. NaCl was added gradually (25 mM of NaCl day^{-1}) to avoid osmotic shock	Plant height, DW, proline, sugar, amino acid, carbohydrates	Reduction in the relative water content and membrane integrity suggest a low ability of experimental cultivar to adjust osmotically under salt stress The K^+/Na^+ ratio abruptly decreased in 25 mM of NaCl, suggesting an ionic imbalance, which may partially explain the salt-induced growth reduction	Menezes et al. (2017)
Chrysanthemum paludosum	0, 50, 100, 150, and 200 mM NaCl	Plant growth parameters, ion leakage, leaf relative water content (RWC), leaf chlorophyll concentration (SPAD-502)	The water content and cell turgor may relatively be preserved under saline conditions An increase in ion concentrations (Na^+ and Cl^-) induced from salinity largely contributed to a decrease in leaf osmotic potential, and this adaptation allows plants to keep cell turgor by continuing water uptake	Yasemin et al. (2017)
Oryza sativa subsp. indica cv	200 mM NaCl	Cytokinin oxidase activity	Previous studies have shown that an enhanced CK content improves stress tolerance ability. Previous studies have shown that an enhanced CK content improves stress tolerance ability. Enhanced cytokinin activities improves stress tolerance ability of rice	Joshi et al. (2017)

(continued)

Table 4.4 (continued)

Plant species	Salt stress	Photosynthetic parameters measured	Adaptative response to salt stress	References
Matricaria Recutita L. (chamomile)	2, 6, 9, and 12 dS m^{-1} NaCl	Nutrient uptake, plant productivity, essential oil and chamazulene percentage	The differential responses to productivity and salt resistance were attributed to the genetic variation, higher root-to-shoot ratios, and compartmentalization of sodium in roots of the Shiraz and Ahvaz genotypes, leading to better nutrient uptake and balance, while the nutrient composition was relatively in the same range for all genotypes. The differential responses to productivity and salt resistance were attributed to the genetic variation, higher root-to-shoot ratios, and compartmentalization of sodium in roots, leading to better nutrient uptake and balance	Askari-Khorasgani et al. (2017)

An organism can hardly exclude the Na^+ completely of its tissues. At the plants, one of the best-known salinity tolerance strategies is compartmentalization ions (Na^+, Cl^-) in excess in the tissues. This controlled redistribution is mainly in vacuoles (Niu et al. 1995) and possibly at the whole-plant scale, in the oldest or least sensitive organs (Cheeseman 1988a, b; Munns 1993).

To be controlled, the movement of ions through the membranes involves an active transport, energy consumer, who uses different carriers (in variable density) to the surface of cell membranes (Orcutt and Nelen 2000; Tyerman and Skerret 1999; Al-Khateeb 2006). Once vacuolated, the excess Na^+ contributes to the osmotic adjustment without altering the process metabolic rate (Levitt 1980; Yeo 1983, 1998).

The best way to maintain a low cytoplasmic concentration in Na+ is to compartmentalize this ion in the vacuole. This intracellular compartmentalization can be associated with succulence, which increases the volume of vacuoles in which Na+ ions accumulate. These ions are pumped into the vacuole before being concentrated in the cytoplasm. Ion pumping is provided by Na^+/H^+ antiports. The difference in pH is restored by H+-ATPase and pyrophosphatases (Blumwald 2000). Anti-Na^+/H^+ activity was increased following the addition of Na^+ in wheat roots (Garbarino and DuPont 1989), tomato (Wilson and Shannon 1995), and sunflower (Ballesteros et al. 1997). Stimulation of activity is greater in the tolerant species, *Plantago maritima*, than in susceptible species, *Plantago* media (Staal et al. 1991). In sensitive rice, however, salinity does not induce the activity of anti-Na^+ /H^+ antibodies in tonoplasts (Fukuda et al. 1998). Salinity also induces the activity of H^+ vacuolar pumps in tolerant and sensitive plants (Hasegawa et al. 2000). To maintain the osmotic balance between the vacuole and the cytoplasm, there will be synthesis of compatible organic compounds. Generally these compatible compounds protect the biochemical reactions against high concentrations of inorganic compounds (Shomer et al. 1991). These compatible compounds are neutral and highly soluble and contain secondary metabolites such as glycine betaine, proline, and sucrose (Hu and Schmidhalter 2000). There is a close correlation between the synthesis of these organic compounds and tolerance to salinity or drought. This correlation has been noted in maize (Saneoka et al. 1995). These compounds appear to be very effective in maintaining a negative osmotic potential in the cytoplasm and in protecting proteins and ribosomes against the deleterious effects of Na^+ ions. Although organic compounds give the plant some tolerance to salinity, they need to be accompanied by strong regulation of Na^+ pumps.

4.4.4 Tolerance of Halophyte Plants to Salinity

Plants in salt-exposed (e.g., costal) environments have certainly acquired characteristics to adapt to soils whose chemical composition varies in time and space, depending on salinity and associated stress (Ben Hamed et al. 2013). Therefore, the adaptations of these coastal plants are complex and different. Nowadays, many authors study halophytes as plant object with developed mechanisms of salt resistance. Halophytes are plants which can be productive under stress (Ben Hamed et al. 2013; Llanes et al. 2013; Slama et al. 2015).

Salinity tolerance reflects the ability of plant halophytes to grow and complete their life cycle in environments containing soluble salts at high concentrations. Halophytes are characterized by low morphological and taxonomic diversity. At the same time, halophytes require salt for optimal growth. The high concentrations of the ions in the tissues of halophytes suggest that their metabolic process may be more tolerant to salt stress compared to the glycophytic metabolism.

4.4.5 Regulation of Salt Loads in Aerial Parts

4.4.5.1 Salt Release in Aerial Parts

Soluble substances pass from the inside of the leaves to the outside where they are accumulated. These substances are released from the leaves through the epidermis and stored in the cuticle. It is called pseudo-secretion (Klepper and Barrs 1968). The sodium is the most flexible element among osmotic cations. However, chlorine is the most flexible element among anions (Tukey et al. 1958; Tukey and Morgan 1962). The release of electrolytes is highly dependent on the plant water status. The salts released from the leaves of *Atriplex* sp. normally present half of the leaf contents. The phenomenon of salt release in coastal areas where mangroves and other native halophytes grow is based on the desalination process with halophyte participation.

4.4.5.2 Guttation

Salt glands are not the only structures through which salt is removed from the plant. Hydathodes are structures which can also eliminate water. Guttation is a common phenomenon in young leaves. The liquid obtained is not a direct secretion of the xylem sap. In fact, the content of nutrient ions in the xylem sap liquid is much lower compared to the sap liquid from the other plant tissues.

Guttation liquid contains mainly calcium, carbonate, sodium, and silicate. In some plant species, hydathodes can function as salt glands and participate in the selective removal of ions. The secretion of salt in this way may be important in young halophytes that develop under humid conditions.

4.4.5.3 Elimination of Organs Saturated with Salt

It is a phenomenon that can eliminate a large amount of salt in halophytes. In some species such as *Juncus maritimus* or *Juncus gerardii*, leaves fall after being loaded with unwanted ions. Some succulent plants such as *Halocnemum* or *Salicornia* get rid of part of the cortex. This part of the cortex released large amount of salt, which allows the plant to survive (Chapman 1960).

4.4.5.4 Remobilization of Salt

The substances accumulated in the aerial parts can be transported by the phloem vessels to the roots and after into the rhizosphere. This has been verified for sodium (Cooil et al. 1965). A similar process has been observed for *Suaeda monoica* and *Salicornia europaea* (Von Willert 1968). Salt recirculation also exists in non-excretory mangroves (Scholander et al. 1962; Atkinson et al. 1967).

4.4.5.5 Accumulation of Salt in Secretory Epidermic Hairs

In some species, epidermic hairs remove salt from sensitive sites of leaf mesophyll. The epidermic hairs work for a short time, but they are very effective. Salt accumulation by vesicle traffic under salt stress is widely known in semi-halophytic and halophytic species of the family Chenopodiaceae. For example, in *Atriplex*, salt hairs or trichomes are formed by two cells: a small basal cell and a large vesicle (Osmond et al. 1969, Mozafar and Goodin 1970). The basal cell has a high structural similarity with the cells of the salt glands. It is characterized by a dense cytoplasm rich in mitochondria, endoplasmic reticulum, and numerous small vesicles. It differs from glandular cells by the presence of chloroplasts.

4.4.5.6 Excretion of Salt

Salt excretion is the most important mechanism which supports halophyte resistance to salinity (Waisel 1972). It allows plants to eliminate excess salt and prevents excessive buildup without reaching toxic levels inside the tissues. It is typical for several halophyte species: Convolvulaceae, Frankeniaceae, Poaceae, Primulaceae, Tamaricaceae, Avicenniaceae, and Plumbaginaceae.

4.4.6 *Salt-Secreting Structures*

4.4.6.1 Trichomes

The secretory trichomes of salt are typical for *Atriplex* spp. These are vesicles that emerge at the leaf surface. They consist of a large secretory cell or vesicle at the apex or vesicle from one or a few cells on a pedicel (Smaoui 1971; Dickison 2000). These cells have mitochondria, dictyosomes, ribosomes, endoplasmic reticulum, and a large flattened nucleus. Chloroplasts are found also but rudimentary or partially developed. In the secretory cell or vesicle is present a large vacuole, while the cell on the pedicle contains the several small vacuoles (Osmond et al. 1969). A symplasmic continuity exists between the mesophyll cells and the secretory cell, for the movements of the ions. The outer walls of vesicular and pedicellar cells are cutinized, while the inner walls are not (Thomson and Platt-Aloia 1979). The salts are externally released by the removal of the leaf.

4.4.6.2 Salt Glands

The glands of herbaceous plants are usually bicellular and formed of an apical and basal cell. They can be submerged, semi-sunken, extended outside the epidermis (Liphschitz and Waisel 1974; Marcum and Murdoch 1994), or lying on the leaf surface in parallel lines on the ridges (Marcum et al. 1998). In dicotyledons (Fig. 4.7), salt glands are multicellular, consisting of basal cells and secretory cells. The number of cells can vary from 6 to 40 depending on the plant genus (Fahn 1990). For example, in *Tamarix* spp., salt glands are composed of two basal and internal collecting cells and six external secretory cells (Mauseth 1988). The glands of *Avicennia* and *Glaux* contain several secretory cells surmounting a discoidal basal cell (Rozema and Riphagen 1977).

At the ultrastructural level, glandular cells in dicotyledonous herbaceous plants contain some lipid bodies, a large flattened nucleus, an endoplasmic reticulum, ribosomes, several mitochondria, rudimentary plastids, and small vacuoles (Thomson 1975). The glands are covered with an elongated cuticle at the level of the excretory cell to form a collecting chamber, in which the salt accumulates before being excreted outside. There is continuity between the basal cell and the apical cell and between the basal cell and the mesophyll cells, thanks to plasmodesmata (Zeigler and Lüttge 1967).

Fig. 4.7 Salt excretion on adaxial leaf side of *Aeluropus littoralis* plants treated with 0, 200, 400, 600, and 800 mM NaCl observed with a magnifying glass (×4) (**a**). Note that leaves were more or less rolled. SEM micrographs of salt crystals observed on adaxial leaf surface from 400 mM NaCl-treated plants (**b**). Adaxial leaf surface of 400 mM NaCl-treated plant which was washed and observed 2 h later. The appearance of salts indicates location of salt glands (**c**)

4.4.7 Osmotic Adjustment

The one of the main physiological traits of tolerance to environmental stress is the osmotic adjustment. This one is realized, thanks to an accumulation of osmoregulatory compounds that may be ions such as K^+ and Cl^- or organic compounds such as soluble sugars (fructose, glucose, trehalose, raffinose, fructans) and certain amino acids (proline, glycine betaine, β-alaninebetaine, proline betaine) leading to a reduction of the osmotic potential, thus allowing the maintenance of the turgor potential (Zivcak et al. 2016). The accumulation of these compounds has been evident in several plant species subjected to salt stress. It varies in large proportions depending on the species, the stage of development, and the degree of salinity. The differences in solute accumulation (free amino acids, proline, and total soluble sugars) between control plants and plants subjected to salt stress are very important. This phenomenon allows the maintenance of many physiological functions (photosynthesis, transpiration, growth, etc.) and can intervene at all stages of plant development. It allows protection of membranes and enzymatic systems especially in organs where proline appears to play a role in maintaining cytosol-vacuole pressures and pH regulation (El Hassani et al. 2008).

4.4.8 Regulation of Growth by Phytohormones Under Salt Stress

They have been shown to have physiological responses to various stresses such as drought or salinity and have similar characteristics. They cause a whole increase in the ABA concentration in the aerial part or a reduction in concentrations in cytokinin (Itai 1999). According to Zhu (2001), the reduction of growth is an adaptative capacity necessary for survival of a plant exposed to abiotic stress. Indeed, this delay in plant development can support energy accumulation and resources to limit the effects of stress before that the imbalance between the inside and outside of the plant body does not increase until a threshold where the damage is irreversible. To illustrate this trend, in nature, growth is inversely correlated with salt stress resistance of a species or variety (Zhu 2001). In more control of growth by hormonal signals, the reduction of growth results from the expenditure of resources in adaptation strategies and cytosol-vacuole and pH regulation (Hassani et al. 2008).

4.4.9 Mechanism of Membrane Control Under Salt Stress

Adaptation to salt stress is also taking place at the level of membrane cell (plasma membrane, tonoplast). The qualitative and quantitative modification of aquaporins (transmembrane proteins) is, for example, a process capable of modifying the water conductivity of the plant and promoting restricting water movements (Yeo 1998). In terms of ion transport, the salinity resistance strategy is qualitative and quantitative. The selectivity of the ions at the input constitutes the component which is defined from the different recent membrane transporters (Na^+/H^+). In the diffusion facilitated as in the active transport, the membrane proteins can be very specific to certain solutes. Nevertheless, several solutes can compete for the same transport protein (Na^+ and K). From a point of quantitative view, Na+ membrane permeability and activity, quantity, and sensitivity of the membrane Na^+/H^+ antiports evolve to adapt to sodium stress at long term (Niu et al. 1995; Tyerman and Skerrett 1999).

4.4.10 The Biological Compounds, Trace Elements Useful for Mitigation of Salt Stress Effects

4.4.10.1 Jasmonic Acid and Salicylic Acid

Generally, under stressful conditions such as salinity stress, plants employ multiple mechanisms to increase their tolerance (Borsani et al. 2003). One of the adaptative plant responses to salt stress is the production of phytohormones such as abscisic acid, salicylic acid (SA), and jasmonates that might be involved in the alleviation of salinity stress (Wang et al. 2001; Yoon et al. 2009). It was observed that SA reduced salt stress injuries via enhancing antioxidant enzyme activities. Antagonistic effects of abscisic acid and jasmonates on salt stress-inducible transcripts have been found (Moons et al. 1997).

Plant hormones such as methyl jasmonate (MeJA) and jasmonic acid (JA) have an ameliorating effect on different plant species under salt stress (Yoon et al. 2009; Manan et al. 2016). Jasmonates play a role of cellular regulators in the response to stress factors, such as salt, drought, and heavy metal (Anjum et al. 2011; Poonam et al. 2013; Qiu et al. 2014).

Jasmonic acid (JA) and methyl jasmonate (MeJA), which is the methyl ester of JA, are natural plant growth regulators, involved in regulation of the morphological, biochemical, and physiological processes in plants. Their exogenous plant treatment under conditions of high salinity can support the development of biomass yield (Sheteawi 2007).

The several studies have shown that methyl jasmonate can diminish the inhibitory effect of NaCl on photosynthesis rate and can enhance the growth and development of plants (Hristova and Popova 2002; Javid et al. 2011). Application of JA after the stress decreased the adverse effect of high salinity on photosynthesis and growth of barley (Tsonev et al. 1998). In addition, exogenous pretreatment of JA could ameliorate salt-stressed rice seedlings, particularly in salt-sensitive cultivars, and could decrease sodium concentration remarkably (Kang et al. 2005).

4.4.10.2 Brassinosteroids

One of the promising options to mitigate the detrimental effects of salt stress is the exogenous application of plant protectants such as brassinosteroids (BRs) (Vardhini and Anjum 2015). BRs belong to plant polyhydroxysteroids, which are important for the regulation of plant growth and development.

The ability to protect the cellular structures was documented for 24-epibrassinolide (EBL), which reduced damage to membrane lipids, and hence a low MDA concentration. MDA, a product of polyunsaturated fatty acid decomposition, is used as a marker to assess the lipid peroxidation in plasmalemma or organelle membranes, which typically occurs in stress conditions (Sharma et al. 2012). The peroxidation of lipids disturbs the bilayer structure, affecting the membrane fluidity, permeability, bilayer thickness, and other membrane properties due to oxidative damage of lipids and membrane proteins. This may alter ion gradients, strongly influencing the metabolic processes. The lipid peroxidation caused by salt stress is probably one of the major reasons inhibiting plant growth.

In perennial ryegrass exposed to salt stress, it was observed that the treatment by exogenous brassinosteroids led to higher K^+, Ca^{2+}, and Mg^{2+} content and lower Na^+/K^+ ratio (Sun et al. 2015). The exogenous brassinosteroid application led also to upregulation of antioxidant enzyme (SOD, CAT, and APX) activity, keeping the level of plant hormones at a physiologically favorable level and an increase of proline and ion content (K^+, Ca^{2+}, and Mg^{2+}). Exogenous brassinosteroids could prevent the nutritional imbalance and ion toxicity under salt stress (Wu et al. 2017). The main effects of brassinosteroid use have been presented in Table 4.5.

4.4.10.3 Amino Acids: Glycine Betaine and Proline

Amino acids glycine betaine and proline are effectively used for exogenous treatment to mitigate salt stress effects on plants (Sobahan et al. 2012; Li et al. 2014). Improvements of salt tolerance can be partly attributed to more favorable water status, as well as to activity of antioxidative enzymes in leaves, especially of peroxidase. It was found that mitigative effects of exogenous proline in salt stress conditions are more efficient than exogenous application of betaines (Hoque et al. 2007). The proline compared to the betaine can directly scavenge superoxide ($O_2^{\bullet}$) or hydrogen peroxide (H_2O_2) and induce an increase of antioxidant enzyme activities (Demiral 2004; Ashraf and Foolad 2007; Nawaz and Ashraf 2010). At the same time, the unequal reaction of antioxidative enzymes has been observed under different level of salt concentrations. Treatment of 5 mM proline significantly reduced POX3 activity, which resulted in modulating salinity stress compared to 200 mM concentration (Varjovi et al. 2016). Therefore, it is obvious that the effect of proline in plants exposed to salinity is specific, at least to some extent.

The role of proline and also betaine in maintaining the plant water status under salt stress is important, as the initial slowdown of plant growth after salt imposition is a result of salt osmotic effects (Munns and Tester 2008; Yang and Lu 2006). The

Table 4.5 Use of exogenous brassinosteroids to mitigate the salt stress effects on plants

Plant species	Mitigation effects on development of salt stress	References
Cucumis sativus L.	Increasing the contents of free proline and soluble sugar Increasing the activity of antioxidative enzymes (SOD, POX, CAT) Effective protection of membrane from salt stress	Shang et al. (2006)
Pisum sativum L.	Increasing FW and DW, seedling height, enhanced photosynthesis rate (Pn), stomatal conductance (gs), total chlorophyll contents (Chl), proline contents Increasing the activity of antioxidative enzymes (SOD, POX, CAT) Increasing the activity of nitrate reductase activity (NRA) and nitrite reductase activity (NiRA)	Shahid et al. (2011)
Oryza sativa L.	Improvement in the growth, level of protein, and proline content Increasing the activity of antioxidative enzymes (SOD, APX, CAT, GPX, and MDHAR)	Sharma et al. (2012) Özdemir et al. (2004)
Brassica juncea L.	Enhancing level of pigments and photosynthetic parameters Improving membrane stability index and relative water content Increasing the activity of antioxidative enzymes (SOD, POX, CAT)	Ali et al. (2008)
Capsicum annuum L.	Increasing relative growth rate and water use efficiency Improving photosynthesis by increasing stomatal conductance	Samira et al. (2012)
Lolium perenne L.	Decrease of electrolyte leakage (EL), malondialdehyde (MDA), and H_2O_2 contents Enhancing the leaf relative water content (RWC), proline, soluble sugar, and soluble protein content Reducing the accumulation of Na^+ and increased K^+, Ca^{2+}, and Mg^{2+} contents	Sun et al. (2015)

water-retaining ability can enhance salt tolerance by preventing too high concentration of ions (Romero-Aranda et al. 2006).

In rice, it was found that exogenous application of glycine betaine and proline may suppress the Na^+ uptake from the apoplast, preventing the detrimental effects of salts (Yang and Lu 2005; Sobahan et al. 2012). Proline inhibits opening of stomata, which keeps the transpiration and Na^+ uptake low. Moreover, in rice treated with betaine, the cells of the root tip and root cap produced numerous vacuoles, playing a role of storage vessels for Na^+ (Rahman et al. 2002). Thus, the specific functions of proline and betaine can contribute to improvements of salinity tolerance.

The use of exogenous proline can balance grain and straw yields under increased salinity levels. The production of grain and straw yields under salt stress conditions after the use of exogenous proline was kept on the significant level. Foliar application of proline and betaine decreased the sodium content and uptake by plants. Thus, it can be concluded that the exogenous application of proline and betaine may mitigate significantly the salt stress effects in crop plants (Siddique et al. 2015; Athar et al. 2015).

4.4.10.4 Polyamines

Polyamines (PAs) are abundant compounds, present in plant cells in concentrations from 10 μM to 10 mM (Roychoudhury et al. 2011). They represent low-molecular-weight, straight-chain, aliphatic amines, including the diamine putrescine (Put^{2+}), triamine spermidine (Spd^{3+}), and tetramine spermine (Spm^{4+}), and are involved in various biochemical and physiological processes related to the regulation of plant growth and development (Puyang et al. 2015). Thanks to their polycationic nature at physiological pH, these compounds can interact with membrane phospholipids, proteins, nucleic acids, and constituents of cell walls, which stabilize these molecules (Roychoudhury et al. 2011).

In the last period, the role of polyamines as second messengers was investigated, especially in response to environmental stresses like osmotic stress, salinity, drought, heat, mineral nutrient deficiency, heavy metals, pH variation, UV irradiation, etc. (Liu et al. 2015). It was documented that exogenous spermidine application diminished the oxidative stress induced by salinity, leading to lower MDA, H_2O_2, and $O_2^{\cdot-}$ concentrations in cultivars of bluegrass. Results indicated that exogenous spermidine treatment is able to improve quality of turfgrass, thanks to promoting the tolerance to salinity by eliminating the oxidative damages and upregulating activities of antioxidative enzymes directly or through gene expression (Puyang et al. 2015).

Spermidine efficiently alleviated the inhibitory role of alkaline ions on plant growth and inhibited related oxidative stress (Zhang et al. 2015). At the same time, the exogenous spermidine treatment had positive effects on nitrogen metabolism and activity of its enzymes in tomato seedlings under salt stress (Zhang et al. 2013). Exogenous spermidine application helps tomato seedlings to overcome salinity stress by regulation of protective mechanism of plant cells, including activating of detoxification, which may protect the cellular structures from oxidative damage under salinity stress. Exogenous spermidine is also able to increase salt tolerance of *Panax ginseng* by upregulation of scavenging enzyme activities, which eliminates the oxidative impairment (Parvin et al. 2014).

4.4.10.5 Paclobutrazol

The enhancement of salt tolerance in plants can be achieved through exogenous application of plant growth regulators with specific effects on the content of key plant phytohormones and signal molecules (Kishor et al. 2009; Hu et al. 2017). Paclobutrazol [(2RS,3RS)-1-(4-chlorophenyl)-4,4-dimethyl-2-(1,2,4-triazol-1-yl) pentan-3-ol] is a triazole fungicide which regulates the plant growth mostly by antagonizing the hormone gibberellin (Hajihashemi et al. 2006; Kishor et al. 2009). Paclobutrazol and other triazole compounds are synthetic plant growth regulators, which cause different physiological responses in plants, such as increasing content of chlorophylls, promoting net photosynthesis, regulating cytokinin biosynthesis, inhibiting ABA biosynthesis, reducing free-radical damage, and enhancing the peroxidase and SOD activities and proline content (Kishor and Sreenivasulu 2014; Khunpon et al. 2017).

Paclobutrazol has also morphological effects on the leaf thickness, cuticle and epidermis, palisade layer, and spongy layer. It can reduce the diameter of xylem vessels; however the phloem elements had shown an increased diameter (Tehranifar et al. 2009). Exogenous treatment has also stimulated effect on root growth and decreasing shoot growth (Banon et al. 2003; Nivedithadevi et al. 2012). Some authors suggest that the paclobutrazol minimizes the absorption of toxic ions such as Na^+ and Cl^-, which eliminates the negative effects of NaCl. Moreover, the experimental results supported the role of paclobutrazol in upregulating the K^+ uptake (Kishor and Sreenivasulu 2014; Hu et al. 2017). We can conclude that paclobutrazol can improve important plant responses and increase crop production with a consequent benefit to saline agriculture.

4.4.10.6 Trace Elements and Nanoparticles

During the last few years, rapid advances of nanotechnology are associated with release of different types of nanoparticles. Some of them may be accumulated in soil or natural environment with negative effects on biota (Alharby et al. 2016; Yassen et al. 2017). The authors have reported the detrimental effects of nanoparticles (usually at relatively high concentration) on plant health. However, there is an evidence about the positive effects of nanoparticles, which is achieved at relatively low concentrations. This provides the scope for possible agricultural applications of nanoparticles (Siddiqui et al. 2014; Siddiqui and Al-Whaibi 2014; Askary et al. 2016).

One of the promising ways is the use of nanofertilizers. Applying the nutrients in the form of nanoparticles improves the nutrient use efficiency, with low risk of toxicity for soil microbiota and roots. Moreover, such a way of application reduces the frequency of the application and prevents the risk of overdosage. Hence, the potential of nanotechnology to support the sustainable farming is high, including developing countries (Naderi and Danesh-Shahraki 2013; Yassen et al. 2017).

The second way of application relates to exogenous use of trace elements and nanoparticles to mitigate stress effects by influencing some specific plant processes (Zhao et al. 2012; Rico et al. 2013; Rossi et al. 2016). For example, the zinc treatment led to lower MDA and H_2O_2 concentration in tissues in the experimental plants under salt stress, which was associated with upregulation of total APX, CAT, POD, and PPO activities under salt stress (Weisany et al. 2012). Decreasing of lipid peroxidation and proline contents under salinity by applying Fe2O3NPs has been found in the peppermint plants. The appropriate concentration of iron nanoparticles can be used for stress resistance of the peppermint (Askary et al. 2017). Fathi et al. (2017) and Soliman et al. (2015) have demonstrated the positive influence of Zn and Fe and their NPs in stress conditions. Nanoparticles were more efficient than other tested forms of these micronutrients. It can be caused by their size, shape, distribution, and other physical characteristics. Latef et al. (2017) reported that priming of seeds with ZNPs is a useful strategy to increase the salt tolerance of lupine plants. The most efficient was concentration of ZnO NPs 60 mg L^{-1}.

It has been also shown that exogenous nanoparticles such as cerium oxide nanoparticles (CeO_2-NPs) positively influence plant growth and production under normal growth conditions. Depending on soil moisture content, CeO_2-NPs supported photosynthesis, which led to increase of water use efficiency (WUE), especially in water-restricted conditions (Cao et al. 2017). Under salinity, it was found that CeO_2-NPs application led to improved plant growth and physiological responses of canola, improving the salt stress responses. However, the stress effects were not fully alleviated by CeO_2-NPs (Rossi et al. 2016).

Adding SiO_2 nanoparticles was found to be able to improve germination and seedling early growth under salinity stress (Sabaghnia and Janmohammadi 2014; Siddiqui and Al-Whaibi 2014). In similar, nano-silicon (N-Si) was shown to improve seed germination, plant growth, and photosynthesis under environmental stresses in tomato (Almutairi 2016a, b).

Also in the case of application of AgNPs, the alleviative effects in conditions of salt stress were found, including positive influence on seed germination, growth of roots, and thus the overall growth and dry mass increase in tomato seedlings under NaCl stress (Almutairi 2016a). The combined application of AgNPs and salinity increased the soluble sugars and proline contents. On the other hand, it decreased catalase activity and increased peroxidase activity compared to the respective AgNP treatments alone. AgNPs enhanced the salt tolerance in wheat, but the long-term response of AgNPs under salt stress needs further investigation.

El-Sharkawy et al. (2017) have demonstrated that application of K nanoparticles in alfalfa may be more efficient than the use of conventional fertilizers, as the nutrition can be more adequate and this way of application may prevent the negative effects of salt stress in some specific conditions.

The abovementioned results suggest that the application of different nanoparticles is a promising strategy to stimulate the plant tolerance to salt stress. According to the many researchers, engineered nanoparticles have a great chance of getting into agricultural lands (Delfani et al. 2014; Benzone et al. 2015; Liu et al. 2015; Liu and Lai 2015; Mastronardi et al. 2015; Rastogi et al. 2017). We report that a common industrial nanoparticle could in fact have a positive impact on crops. Modern nanofertilizers are expected to contribute to the improvement of crop growth, photosynthesis, and tolerance to environmental stress, which will result to better nutrient and water use efficiency and yield increase.

4.5 Conclusion

Nowadays, the advances in research aimed at salt stress effects on plants at different levels, described broadly in this chapter, provide great opportunities to develop effective strategies to improve crop tolerance and yield in different environments affected by the soil salinity. It was clearly demonstrated that plants employ both the common adaptative responses and the specific reactions to salt stress. Presented data may be helpful to understand the physiological, metabolic, developmental, and other reactions of crop plants to salinity, resulting in the decrease of biomass production and yield. In addition, the chapter provides an overview of modern studies on how to mitigate salt stress effects on photosynthetic apparatus and productivity of crop plants with the help of phytohormones, glycine betaine, proline, polyamines, paclobutrazol, trace elements, and nanoparticles. Plant production in saline agriculture can avoid or, at least, diminish the negative salt effects with use of different approaches and tools, which can have an economic impact worldwide but, especially, in most endangered developing countries.

References

Abdelly C (2006) Caractérisation des halophytes pour le dessalement des sols salins et letraitement des eaux salines. Rapport d'activités 2007. Centre de biotechnologie à la technopoledeBorj-Cedria, Tunisie, pp 28–31

Abdi N, Wasti S, Slama A, Ben Salem M, El Faleh M, Mallek-Maalej E (2016) Comparative study of salinity effect on some tunisian barley cultivars at germination and early seedling growth stages. J Plant Physiol Pathol 4(3):1–9. https://doi.org/10.4172/2329-955X.1000151

Agastian P, Kingsley SJ, Vivekanandan M (2000) Effect of salinity on photosynthesis and biochemical characteristics in mulberry genotypes. Photosynthetica 38:287–290

Ahmad M, Niazi BH, Zaman B, Athar M (2005) Varietals differences in agronomic performance of six wheat varieties grown under saline field environment. Int J Environ Sci Technol 2(1):49–57

Ahmad M, Zahir Zahir A, Nazli F, Akram F, Arshad MKM (2013) Effectiveness of halo-tolerant, auxin producing pseudomonas and rhizobium strains to improve osmotic stress tolerance in mung bean (*Vigna radiata* L.) Braz. J Microbiol 44(4):1341–1348

Aissaoui HS, Reffas S (2007) Effet de stress salin sur la productivité de populations sahariennes locales de la luzerne (*Medicago sativa* L.), Université Kasdi Merbah Ouargla.

Al Hassan M, Chaura J, Donat-Torres MP, Boscaiu M, Vicente O (2017) Antioxidant responses under salinity and drought in three closely related wild monocots with different ecological optima. AoB Plants 9(2):1–20. https://doi.org/10.1093/aobpla/plx009

Alem C, Labhilili M, Brahmi K, Jlibene M, Nasrallah N, Filali-Maltouf A (2002) Adaptations hydrique et photosynthétique du blé dur et du blé tendre au stress salin. C R Biologies 325(11):1097–1109

Alharby HF, Metwali EMR, Fuller MP, Aldhebiani AY (2016) Impact of application of zinc oxide nanoparticles on callus induction, plant regeneration, element content and antioxidant enzyme activity in tomato (*Solanum lycopersicum* Mill.) under salt stress. Arch Biol Sci 68(4):723–735

Ali B, Hayat S, Fariduddin Q, Ahmad A (2008) 24-Epibrassinolide protects against the stress generated by salinity and nickel in *Brassica juncea*. Chemosphere 72:1387–1392

Al-Khateeb SA (2006) Effect of calcium/sodium ratio on growth and ion relations of alfalfa (*Medicago sativa* L.) seedling grown under saline condition. J Agron 5(2):175–181

Allakhverdiev SI, Nishiyama Y, Miyairi S, Yamamoto H, Inagaki N, Kanesaki Y, Murata N (2002) Salt stress inhibits the repair of photodamaged photosystem II by suppressing the transcription and translation of psbA genes in Synechocystis. Plant Physiol 130:1443–1453

Allen SG, Dobrenz AK, Scharnhorst M, Stoner JEA (1985) Heritability of NaCl tolerance in germinating alfalfa seeds. Agron J 77:99–105

Almutairi ZM (2016a) Influence of silver nano-particles on the salt resistance of tomato (*Solanum lycopersicum* L.) during germination. Int J Agri Biol 18(2):449–457. https://doi.org/10.17957/IJAB/15.0114

Almutairi ZM (2016b) Effect of nano-silicon application on the expression of salt tolerance genes in germinating tomato (*Solanum lycopersicum* L.) seedlings under salt stress. Plant Omics Journal 9(1):106–114

Anjum SA, Xie X, Wang L et al (2011) Morphological, physiological and biochemical responses of plants to drought stress. Afr J Agr Res 6:2026–2032

Ashraf M, Foolad MR (2007) Roles of glycine betaine and proline in improving plant abiotic stress resistance. Environ Exp Bot 59:206–216

Ashraf M, Harris PJC (2004) Potential biochemical indicators of salinity tolerance in plants. Plant Sci 166:3–6

Askari-Khorasgani O, Emadi S, Mortazaienezhad F, Pessarakli M (2017) Differential responses of three chamomile genotypes to salinity stress with respect to physiological, morphological, and phytochemical characteristics. J Plant Nutr 40(18):2619–2630

Askary M, Talebi SM, Amini F, Ali D, Bangan B (2017) Effects of iron nanoparticles on *Mentha piperita* L. under salinity stress. Biologija 63(1):65–75

Askary M, Talebi SM, Amini F, Bangan ADB (2016) Effect of NaCl and iron oxide nanoparticles on *Mentha piperita* essential oil composition. Environ Exp Biol 14:27–32. https://doi.org/10.22364/eeb.14.05

Asrar H, Hussain T, Midhat S, Hadi S, Gul B, Nielsen BL, Khan MA (2017) Salinity induced changes in light harvesting and carbon assimilating complexes of *Desmostachya bipinnata* (L.) staph. Environ Exp Bot 135:86–95

Athar H-U-R, Zafar ZU, Ashraf M (2015) Glycinebetaine improved photosynthesis in canola under salt stress: evaluation of chlorophyll fluorescence parameters as potential indicators. J Agro Crop Sci 201:428–442. https://doi.org/10.1111/jac.12120

Atkinson MR, Findlay CP, Hope AB, Pitman MG, Saddler HDW, West KR (1967) Salt regulation in the mangroves *Rhizophora mucronata* Lam. and *Aegialitis annulata* R. Br Aust J Biol Sci 20:589–599

Aziz I, Khan MA (2001) Experimental assessment of salinity tolerance of Ceriops tagal seedlings and saplings from the Indus delta, Pakistan. Aqua Bot 70(3):259–268

Azri W, Chambon C, Herbette S, Brunel N, Coutand C, Leplé JC, Ben Rejeb I, Ammar S, Julien JL, Roeckel-Drevet P (2009) Proteome analysis of apical and basal regions of poplar stems under gravitropic stimulation. Physiol Plant 136:193–208. https://doi.org/10.1111/j.1399-3054.2009.01230.x

Ballesteros E, Blumwald E, Donaire JP, Belver A (1997) Na^+/H^+ antiport activity in tonoplast vesicles isolated from sunflower roots induced by NaCl stress. Physiol Plant 99:328–334. https://doi.org/10.1111/j.1399-3054.1997.tb05420.x

Banon S, Ochoa J, Martinez JA, Fernandez JA, Franco JA, Sanchez-Blanco MJ, Alarcon JJ, Morales MA (2003) Paclobutrazol as an aid to reducing the effects of salt stress in *Rhamnus alaternus* plants. Acta Hortic 609:263–268

Barhoumi Z, Djebali W, Smaoui A, Chaïbi W, Abdelly C (2007) Contribution of NaCl excretion to salt resistance of Aeluropus littoralis (Willd) Parl. J Plant Physiol 164(7):842–850

Bartels D, Nelson D (1994) Approaches to improve stress tolerance using molecular genetics. Plant Cell Environ 17:659–667. https://doi.org/10.1111/j.1365

Belkheiri O, Mulas M (2013) The effects of salt stress on growth, water relations and ion accumulation in two halophyte *Atriplex* species. Environ Exp Bot 86:17–28

Ben Ahmed C, Ben Rouina B, Sensoy S, Boukhriss S, Ben Abdullah F (2010) Exogenous proline effects on photosynthetic performance and antioxidant defense system of young olive tree. J Agricult Food Chem 58:416–422

Ben Amor N, Ben Hamed K, Debez A, Grignon C, Abdelly C (2005) Physiological and antioxidant responses of the perennial halophyte Crithmum maritimum to salinity. Plant Sci 168:889–899

Ben Hamed K, Ellouzi H, Talbi-Zribi O, Hessini K, Slama I, Ghnaya T, Munné Bosch S, Savouré A, Abdelly C (2013) Physiological response of halophytes to multiple stresses. Funct Plant Biol 40(9):883–896. https://doi.org/10.1071/FP13074

Ben Hamed KB, Castagna A, Salem E, Ranieri A, Abdelly C (2007) Sea fennel (*Crithmum maritimum* L.) under salinity conditions: a comparison of leaf and root antioxidant responses. Plant Growth Regul 53:185–194. https://doi.org/10.1007/s10725-007-9217-8

Ben Hamed KB, Chibani F, Abdelly C, Magne C (2014) Growth, sodium uptake and antioxidant responses of coastal plants differing in their ecological status under increasing salinity. Biologia 69(2):193–201. https://doi.org/10.2478/s11756-013-0304-1

Bennaceur M, Rahmoune C, Sdiri H, Meddhi ML, Selmi M (2001) Effet du stress salin sur la germination, la croissance et la production en grains de quelques variétés maghrébines de blé. Sècheresse 12(3):167–174

Benzarti M, Ben Rejeb K, Debez A, Messedi D, Abdelly C (2012) Photosynthetic activity and leaf antioxidative responses of *Atriplex portulacoides* subjected to extreme salinity. Acta Physiol Plant 34:1679–1688

Benzon HRL, Rubenecia MRU, Ultra VU, Lee SC (2015) Nano-fertilizer affects the growth, development, and chemical properties of rice. IJAAR 7(1):105–117

Berthomieu P, Conejero G, Nublat A, Brackenbury WJ, Lambert C, Savio C, Uozumi N, Oik S, Yamada K, Cellier F, Gosti F, Simonneau T, Essah PA, Tester M, Véry AA, Sentenac H, Bhandal IS, Malik CP (1988) Potassium estimation, uptake, and its role in the physiology and metabolism of flowering plants. Int Rev Cytol 110:205–254

Bhandal IS, Malik CP (1988) Potassium estimation, uptake, and its role in the physiology and metabolism of flowering plants. Int Rev Cytol 110:205–254

Blaha G, Stelzl U, Spahn CM, Agrawal RK, Frank J, Nierhaus KH (2000) Preparation of functional ribosomal complexes and effect of buffer conditions on tRNA positions observed by cryoelectron microscopy. Methods Enzymol 317:292–309

Blumwald E (2000) Sodium transport and salt tolerance in plants. Curr Opin Cell Biol 12(4):431–434

Borsani O, Valpuesta V, Botella MA (2003) Developing salt tolerant plants in a new century: a molecular biology approach. Plant Cell Tissue Organ 73:101–115

Boscaiu M, Estrelles E, Soriano P, Vicente O (2005) Effects of salt stress on the reproductive biology of the halophyte *Plantago crassifolia*. Biol Plant 49:141–143

Botella MA, Quesada MA, Kononowicz A, Bressan RA, Hasegawa PM, Valpuesta V (1994) Characterization and *in situ* localization of a salt induced tomato peroxidase gene. Plant Mol Biol 25:105–114

Brestic M, Zivcak M (2013) PSII fluorescence techniques in drought and high temperature stress signal measurement of crop plants: protocols and applications. In: Rout GR, Das AB (eds) Molecular stress physiology of plants. Springer, India, pp 87–113. https://doi.org/10.1007/978-81-322-0807-5_4

Briens M, Larher F (1982) Osmoregulation in halophytic higher plants: a comparative study of soluble carbohydrates, polyols, betaines and free proline. Plant Cell Environ 5:287–292

Brugnoli E, Björkman O (1992) Growth of cotton under continuous salinity stress: influence on allocation pattern, stomatal and non-stomatal components of photosynthesis and dissipation of excess light energy. Planta 187(3):335–347

Bruns S, Hecht-Buchholz C (1990) Light and electron microscope studies on the leaves of several potato cultivars after application of salt at various development stages. Potato Res 33:33–41. https://doi.org/10.1007/BF02358128

Bueno M, Lendínez ML, Aparicio C, Cordovilla MP (2017) Germination and growth of *Atriplex prostrata* and *Plantago coronopus*: two strategies to survive in saline habitats. Flora 227:56–63

Cao Z, Rossi L, Stowers C, Zhang W, Lombardini L, Ma X (2017) The impact of cerium oxide nanoparticles on the physiology of soybean (*Glycine max* (L.) Merr.) under different soil moisture conditions. Environ Sci Pollut Res Int 25(1):930–939. https://doi.org/10.1007/s11356-017-0501-5

Chapman VJ (1960) Salt marshes and salt deserts of the world. Leonard Hill, London

Chartzoulakis K, Klapaki G (2000) Response of two greenhouse pepper hybrids to NaCl salinity during different growth stages. Sci Hortic 86:247–260

Chaudhuri K, Choudhuri M (1997) Effects of short-term NaCl stress on water relations and gas exchange of two jute species. Biol Plant 40:373. https://doi.org/10.1023/A:1001013913773
Chaves M, Flexas J, Pinheiro C (2009) Photosynthesis under drought and salt stress: regulation mechanisms from whole plant to cell. Ann Bot 103:551–560
Chaves MM, Earl HJ, Flexas J, Loreto F, Medrano H (2011) Photosyn- thesis under water stress, flooding and salinity. In: Terrestrial photosynthesis in a changing environment. Cambridge University Press pp 49–104
Cheeseman JM (1988) Mechanisms of salinity tolerance in plants. Plant Physiol 87(3):547–550
Cheeseman JM (2015) The evolution of halophytes, glycophytes and crops, and its implications for food security under saline conditions. New Phytol 206(2):557–570. https://doi.org/10.1111/nph.13217
Chen Q, Goldstein I, Jiang W (2010) Payoff complementarities and financial fragility: evidence from mutual fund outflows. J Financ Econ 97(2):239–262
Chen TW, Kahlen K, Stützel H (2015) Disentangling the contributions of osmotic and ionic effects of salinity on stomatal, mesophyll, biochemical and light limitation to photosynthesis. Plant Cell Environ 38:1528–1542. https://doi.org/10.1111/pce.12504
Cooil BJ, de la Fluente RK, de la Pena RS (1965) Absorption and transport of sodium and potassium in squash. Plant Physiol 40:625–633
Dąbrowski P, Baczewska AH, Pawluśkiewicz B, Paunov M, Alexantrov V, Goltsev V, Kalaji MH (2016) Prompt chlorophyll a fluorescence as a rapid tool for diagnostic changes in PSII structure inhibited by salt stress in *Perennial ryegrass*. J Photochem Photobiol B 157:22–31. https://doi.org/10.1016/j.jphotobiol.2016.02.001
Dąbrowski P, Kalaji MH, Baczewska AH, Pawluśkiewicz B, Mastalerczuk G, Borawska-Jarmułowicz B, Paunov M, Goltsev V (2017) Delayed chlorophyll a fluorescence, MR 820, and gas exchange changes in perennial ryegrass under salt stress. J Lumin 183:322–333. https://doi.org/10.1016/j.jlumin.2016.11.031
De Oliveira VP, Marques EC, de Lacerda CF, Prisco JT, Gomes-Filho E (2013) Physiological and biochemical characteristics of Sorghum bicolor and Sorghum sudanense subjected to salt stress in two stages of development. Afr J Agric Res 8:660–670
Debez A, Hamed KB, Grignon C, Abdelly C (2004) Salinity effects on germination, growth, and seed production of the halophyte *Cakile maritima*. Plant Soil 262:179–189
Debez A, Koyro HW, Grignon C, Abdelly C, Huchzermeyer B (2008) Relationship be-tween the photosynthetic activity and the performance of *Cakile maritima* after long-term salt treatment. Physiol Plant 133:373–385
Delfani M, Firouzabadi MB, Farrokhi N, Makarian H (2014) Some physiological responses of black-eyed pea to iron and magnesium nanofertilizers. Commun Soil Sci Plant Anal 45:530–540
Demiral IT (2004) Does exogenous glycinebetaine affect antioxidative system of rice seedlings under NaCl treatment? J Plant Physiol 161:1089–1100
Dickison WC (2000) Integrative plant anatomy, 1st edn. Harcount Academic, San Diego
Ellouzi H, Hamed KB, Cela J, Munné-Bosch S, Abdelly C (2011) Early effects of salt stress on the physiological and oxidative status of *Cakile maritima* (halophyte) and *Arabidopsis thaliana* (glycophyte). Physiol Plant 142:128–143. https://doi.org/10.1111/j.1399-3054.2011.01450.x
Ellouzi H, Hamed KB, Hernández I, Cela J, Müller M, Magné C, Abdelly C, Munné-Bosch S (2014) A comparative study of the early osmotic, ionic, redox and hormonal signaling response in leaves and roots of two halophytes and a glycophyte to salinity. Planta 240(6):1299–1317. https://doi.org/10.1007/s00425-014-2154-7
El-Sharkawy MS, El-Beshsbeshy TR, Mahmoud EK, Abdelkader NI, Al-Shal RM, Missaoui AM (2017) Response of alfalfa under salt stress to the application of potassium sulfate nanoparticles. Am J Plant Sci 8:1751–1773. https://doi.org/10.4236/ajps.2017.88120
Elzam OE, Epstein E (1969) Salt relations of two grass species differing in salt tolerance. II kinetics of the absorption of K, Na and Cl by their excised roots. Agrochimica 13:196–206
Evelin H, Kapoor R, Giri B (2009) Arbuscular mycorrhizal fungi in alleviation of salt stress: a review. Ann Bot 104:1263–1280

Fahn A (1990) Plant anatomy, 4th edn. Pergamon Press, New York

FAO (2000) Global network on integrated soil management for sustainable use of salt-affected soils. Country Specific Salinity Issues— Iran.FAO, Rome. Available at http://www.fao.org/ag/agl/agll/spush/degrad.asp?country¼iran

FAO (2015) FAO cereal supply and demand brief. http://www.fao.org/worldfoodsituation/csdb/en/

FAO/IIASA/ISRIC/ISS-CAS/JRC (2008) Harmonized world soil database (version 1.0). FAO, Rome

Fathi A, Zahedi M, Torabian S (2017) Effect of interaction between salinity and nanoparticles (Fe_2O_3 and ZnO) on physiological parameters of *Zea mays* L. J Plant Nutr 40(19):2745–2755. https://doi.org/10.1080/00103624.2013.863911

Flowers TJ, Colmer TD (2015) Plant salt tolerance: adaptations in halophytes. Ann Bot 115(3):327–331

Flowers TJ, Hajibagheri MA, Yeo AR (1991) Ion accumulation in the cell walls of rice plants growing under saline conditions: evidence for the Oertli hypothesis. Plant, Cell and Environ 14:319–325

Fukuda A, Yazaki Y, Ishikawa T, Koike S, Tanaka Y (1998) Na^+ /H^+ antiporter in tonoplast vesicles from rice roots. Plant Cell Physiol 39:196–201

Gale J (1975) The combined effect of environmental factors and salinity on plant growth. In: Book: plants in saline environ, pp 186–192. https://doi.org/10.1007/978-3-642-80929-3_12

Garbarino J, Dupont FM (1989) Rapid induction of na/h exchange activity in barley root tonoplast. Plant Physiol 89(1):1–4

García-Caparrós P, Llanderal A, Pestana M, Correia PJ, Lao MT (2017) *Lavandula multifida* response to salinity: growth, nutrient uptake, and physiological changes. J Plant Nutr Soil Sci 180:96–104. https://doi.org/10.1002/jpln.201600062

Gimenez C, Mitchell VJ, Lawlor DW (1992) Regulation of photosynthetic rate of two sunflower hybrids under water stress. Plant Physiol 98(2):516–524

Graan T, Boyer JS (1990) Very high CO2 partially restores photosynthesis in sunflower at low leaf water potentials. Planta 181:378–384

Greenway H, Munns R (1980) Mechanisms of salt tolerance in nonhalophytes. Annu Rev Plant Physiol 31:149–190

Grennan AK (2006) Abiotic stress in rice. An "omic" approach. Plant Physiol 140(4):1139–1141

Gu MF, Li N, Long XH, Brestic M, Shao HB, Li J, Mbarki S (2016) Accumulation capacity of ions in cabbage (*Brassica oleracea* L.) supplied with sea water. Plant Soil Environment 62(7):314–320. https://doi.org/10.17221/771/2015-PSE

Gucci R, Lombardini L, Tattini M (1997) Analysis of water relations in leaves of two olive (Olea Europaea) cultivars differing in tolerance to salinity. Tree Physiol 17:13–21

Guerrier G (1984a) Relations entre la tolérance ou la sensibilité à la salinité lors de la germination des semences et les composantes de la nutrition en sodium. Biol Plant 26:22–28. https://doi.org/10.1007/BF02880421

Guerrier G (1984b) Selectivité de fixation du sodium au niveau des embryons et des jeunes plantes sensible or tolerant au NaCl. Can J Bot 62:1791–1792

Gupta A, Berkowitz GA (1987) Osmotic adjustment, symplast volume, and nonstomatally mediated water stress inhibition of photosynthesis in wheat. Plant Physiol 85:1040–1047

Gutterman Y (1993) Seed germination in desert plants. Adaptations of desert organisms. Springer-Verlag, Berlin

Hachicha M (2007) Les sols salés et leur mise en valeur en Tunisie. Science et changements planétaires/Sécheresse 18(1):45–50

Hachicha M, Kahlaoui B, Khamassi N, Misle E, Jouzdan O (2017) Effect of electromagnetic treatment of saline water on soil and crops Journal of the Saudi Society of Agricultural Sciences (2017) In Press, Corrected Proof, Available on line 25 March 2016

Hajihashemi S, Kiarostami K, Enteshari S, Saboora A (2006) The effects of salt stress and paclobutrazol on some physiological parameters of two salt-tolerant and salt-sensitive cultivars of wheat. Pakistan J Biol Sci 9(7):1370–1374

Hajlaoui H, El Ayeb N, Garrec JP, Denden M (2010) Differential effects of salt stress on osmotic adjustment and solutes allocation on the basis of root and leaf tissue senescence of two silage maize (*Zea mays* L.) varieties. Ind Crop Prod 31(1):122–130

Hamdy A (1999) Active damping of vibrations in elevator cars. J Struct Control 6:53–100. https://doi.org/10.1002/stc.4300060105

Hameed A, Gulzar S, Aziz I, Hussain T, Gul B, Khan MA (2015) Effects of salinity and ascorbic acid on growth, water status and antioxidant system in a perennial halophyte. AoB Plants 7:plv004. https://doi.org/10.1093/aobpla/plv004

Haouala F, Ferjani H, Ben El Hadj S (2007) Effet de la salinité sur la répartition des cations (Na^+, K^+ et $Ca2^+$) et du chlore (Cl^-) dans les parties aériennes et les racines du ray-grass anglais et du chiendent. Biotechnol Agron Soc Environ 11(3):235–244

Hasegawa PM, Bressan RA, Zhu JK, Bohnert HJ (2000) Plant cellular and molecular responses to high salinity. Annu Rev Plant Physiol Plant Mol Biol 51:463–499

Hassani A, Dellal A, Belkhodja M, Kaid-Harche M (2008) Effect of salinity on water and some osmolytes in barley (*Hordeum vulgare*). Eur J Sci Res 23:61–69

He S, Schulthess AW, Mirdita V, Zhao Y, Korzun V, Bothe R, Ebmeyer E, Reif JC, Jiang Y (2016) Genomic selection in a commercial winter wheat population. Theor Appl Genet 129(3):641–651. https://doi.org/10.1007/s00122-015-2655-1

Heidari M, Sarani S (2012) Growth, biochemical components and ion content of chamomile (*Matricaria chamomilla* L.) under salinity stress and iron deficiency. J Saudi Soc Agric Sci 11(1):37–42

Hernández JA, Olmos E, Corpas FJ, Sevilla F, del Río LA (1995) Salt-induced oxidative stress in chloroplasts of pea plants. Plant Sci 105(2):151–167

Hernández JA, Campillo A, Jiménez A, Alarcón JJ, Sevilla E (1999) Response of antioxidant systems and leaf water relations to NaCl stress in pea plants. New Phytol 141:241–251

Hillel D, Vlek PLG (2005) The sustainability of irrigation. Adv Agron 87:55–84. https://doi.org/10.1016/S0065-2113(05)87002-6

Hniličková H, Hnilička F, Martinková J, Kraus K (2017) Effects of salt stress on water status, photosynthesis and chlorophyll fluorescence of rocket. Plant Soil Environ 63:362–367

Hoque MA, Okuma E, Banu MNA, Nakamura Y, Shimoishi Y, Murata Y (2007) Exogenous proline mitigates the detrimental effects of salt stress more than exogenous betaine by increasing antioxidant enzyme activities. J Plant Physiol 164:553–561

Hose E, Clarkson DT, Steudle E, Schreiber L, Hartung W (2001) The exodermis: a variable apoplastic barrier. J Exp Bot 52(365):2245–2264

Hossain MS, Persicke M, AI ES, Kalinowski J, Dietz KJ (2017) Metabolite profiling at the cellular and subcellular level reveals metabolites associated with salinity tolerance in sugar beet. J Exp Bot 68(21–22):5961–5976

Hristova V, Popova L (2002) Treatment with methyl jasmonate alleviates the effects of paraquat on photosynthesis in barley plants. Photosynthetica 40:567. https://doi.org/10.1023/A:1024356120016

Hu Y, Schmidhalter U (2000) A two-pinhole technique to determine distribution profiles of relative elemental growth rates in the growth zone of grass leaves. Aust J Plant Physiol 27:1187–1190

Hu Y, Yu W, Liu T, Shafi M, Song L, Du X, Huang X, Yue Y, Wu J (2017) Effects of paclobutrazol on cultivars of Chinese bayberry (*Myrica rubra*) under salinity stress. Photosynthetica 55(3):443–453. https://doi.org/10.1007/s11099-016-0658-z

Huez-López MA, Ulery April L, Samani Z, Picchioni G, Flynn RP (2011) Response of chile pepper (*Capsicum annuum* L.) to salt stress and organic and inorganic nitrogen sources: i. Growth and yield. Trop Subtrop Agroecosyst 14:137–147

Ibrahim EA (2016) Seed priming to alleviate salinity stress in germinating seeds. J Plant Physiol 192:38–46. https://doi.org/10.1016/j.jplph.2015.12.011

Itai C (1999) Role of phytohormones in plant responses to stresses. In: Lerner HR (ed) Plant responses to environmental stress. From phytohormones to genome reorganization. Marcel Dekker, New York, pp 287–301

Iyengar ERR, Reddy MP (1996) Photosynthesis in highly salt-tolerant plants. In: Pessaraki M (ed) Handbook of photosynthesis. Marcel Dekker, New York, pp 897–909

Jan SA, Bibi N, Shinwari KS, Rabbani MA, Ullah S, Qadir A, Khan N (2017) Impact of salt, drought, heat and frost stresses on morpho-biochemical and physiological properties of Brassica species: An updated review. J Rural Dev Agric 2(1):1–10

Javid MG, Sorooshzadeh A, Moradi F, Sanavy SAMM, Allahdadi I (2011) The role of phytohormones in alleviating salt stress in crop plants. Aust J Crop Sci 5:726–734

Jia J, Bai J, Gao H, Wen X, Zhang G, Cui B, Liu X (2017) In situ soil net nitrogen mineralization in coastal salt marshes (*Suaeda salsa*) with different flooding periods in a Chinese estuary. Ecol Indic 73:559–565

Joshi R, Sahoo KK, Tripathi AK, Kumar R, Gupta BK, Pareek A, Singla-Pareek SL (2017) Knockdown of an inflorescence meristem-specific cytokinin oxidase–OsCKX2 in rice reduces yield penalty under salinity stress condition. Plant Cell Environ https://doi.org/10.1111/pce.12947

Kachout SS, Mansoura AB, Leclerc JC, Jaffel K, Rejeb MN, Ouerghi Z (2009) Effects of heavy metals on antioxidant activities of *Atriplex hortensis* and *Atriplex rosea*. J Appl Bot Food Qual 83(1):37–43

Kafi M, Shariat JM, Moayedi A (2013) The sensitivity of grain sorghum (*Sorghum bicolor* L.) developmental stages to salinity stress: an integrated approach. J Agric Sci Tech 15(4):723–736

Kafkai U (1991) Root growth under stress. Plant roots: the hidden half. Marcel Dekker, New York, USA, pp 375–391

Kalaji H, Rastogi A, Živčák M, Brestic M et al (2018) Prompt chlorophyll fluorescence as a tool for crop phenotyping: an example of barley landraces exposed to various abiotic stress factors. Photosynthetica. https://doi.org/10.1007/s11099-018-0766-z

Kalaji HM, Govindjee BK et al (2011) Effects of salt stress on photosystem II efficiency and CO2 assimilation of two Syrian barley landraces. Environ Exp Bot 73:64–72

Kalaji MH, Goltsev V, Zuk-Golaszewska B, Zivcak M, Brestic M (2017) Chlorophyll fluorescence: understanding crop performance—basics and applications. Taylor and Francis, p 222. ISBN 9781498764490

Kamiab F, Talaie A, Javanshah A, Khezri M, Khalighi A (2012) Effect of long-term salinity on growth, chemical composition and mineral elements of pistachio (*Pistacia vera* cv. Badami-Zarand) rootstock seedlings. Annals Biol Res 3(12):5545–5551

Kan X, Ren J, Chen T, Cui M, Li C, Zhou R, Zhang Y, Liu H, Deng D, Yin Z (2017) Effects of salinity on photosynthesis in maize probed by prompt fluorescence, delayed fluorescence and P700 signals. Environ Exp Bot 140:56–64

Kang DJ, Seo YJ, Lee JD, Ishii R, Kim KU, Shin DH, Park SK, Jang SW, Lee IJ (2005) Jasmonic acid differentially affects growth, ion uptake and abscisic acid concentration in salt-tolerant and salt-sensitive rice cultivars. J Agron Crop Sci 191:273–282

Kanwal H, Ashraf M, Shahbaz M (2011) Assessment of salt tolerance of some newly developed and candidate wheat (Triticum Aestivum L.) cultivars using gas exchange and chlorophyll fluorescence attributes. Pak J Bot 43:2693–2699

Kao RR, Gravenor MB, McLean AR (2001) Modelling the national scrapie eradication programme in the UK. Math Biosci 174:61–76

Karray-Bouraoui N (1995) Analyse des facteurs responsables de la tolérance au stress salin chez une céréale hybride le triticale: croissance, nutrition et métabolisme respiratoire. Thèse Doc. Univ, Tunis

Kavi Kishor PB, Sreenivasulu N (2014) Is proline accumulation per se correlated with stress tolerance or is proline homeostasis a more critical issue? Plant Cell Environ 37:300–311. https://doi.org/10.1111/pce.12157

Keiper FJ, Chen DM, De Filippis LF (1998) Respiratory, photosynthetic and ultrastructural changes accompanying salt adaptation in culture of *Eucalyptus microcorys*. J Plant Physiol 152(4-5):564–573

Khan AM, Ungar IA (1998) Germination of the salt tolerant shrub Suaeda fruticosa from. Pakistan: salinity and temperature responses. Seed Sci Technol 26:657–667

Khan MA, Ungar IA, Showalter AM (2000) The effect of salinity on the growth, water status, and ion content of a leaf succulent perennial halophyte *Suadea fruticosa* (L.) Forssk. J Arid Environ 45:73–84

Khan N, Syeed S, Masood A, Nazar R, Iqbal N (2010) Application of salicylic acid increases contents of nutrients and antioxidative metabolism in mungbean and alleviates adverse effects of salinity stress. Int J Plant Biol 1(1):1–8

Khavari-Nejad RA, Chaparzadeh N (1998) The effects of NaCl and $CaCl_2$ on photosynthesis and growth of alfalfa plants. Photosynthetica 35(3):461–466

Khavari-Nejad R, Mostofi Y (1998) Effects of NaCl on photosynthetic pigments, saccharides, and chloroplast ultrastructure in leaves of tomato cultivars. Photosynthetica 35:151. https://doi.org/10.1023/A:1006846504261

Khunpon B, Chaum S, Faiyue B, Uthaibutra J, Saengnil K (2017) Influence of paclobutrazol on growth performance, photosynthetic pigments, and antioxidant efficiency of Pathumthani 1 rice seedlings grown under salt stress. ScienceAsia 43:70–81. https://doi.org/10.2306/scienceasia1513-1874.2017.43.070

Kishor A, Srivastav M, Dubey AK, Singh AK, Sairam RK, Pandey RN, Dahuja A, Sharma RR (2009) Paclobutrazol minimises the effects of salt stress in mango (*Mangifera indica* L.) J Hortic Sci Biotechnol 84(4):459–465

Klepper B, Barrs HD (1968) Effects of salt secretion on psychrometric determinations of water potential of cotton leaves. Plant Physiol 43(7):1138–1140

Kurban H, Saneoka H, Nehira K, Adilla R, Premachandra GS, Fujita K (1999) Effect of salinity on growth, photosynthesis and mineral composition in leguminous plant *Alhagi pseudoalhagi* (Bieb.) Soil Sci Plant Nutr 45(4):851–862

Lachhab I, Louahlia S, Laamarti M, Hammani K (2013) Effet d'un stress salin sur la germination et l'activité enzymatique chez deux génotypes de *Medicago sativa*. IJIAS 3(2):511–516

Lamsal K, Paudyal GN, Saeed M (1999) Model for assessing impact of salinity on soil water availability and crop yield. Agr Water Manage 41:57–70

Latef AAHA, Alhmad MFA, Abdelfattah KE (2017) The possible roles of priming with zno nanoparticles in mitigation of salinity stress in lupine (*Lupinus termis*) plants. J Plant Growth Regul 36(1):60–70. https://doi.org/10.1007/s00344-016-9618-x

Levitt J (1980) Responses of plant to environmental stress water, radiation, salt and other stresses. Academic Press, New York

Li M, Guo S, Xu Y, Meng Q, Li G, Yang X (2014) Glycine betaine mediated potentiation of HSP gene expression involves calcium signaling pathways in tobacco exposed to NaCl stress. Physiol Plant 150(1):63–75

Liphschitz N, Waisel Y (1974) Existence of salt glands in various genera of Gramineae. New Phytol 73(3):507–513

Liu JH, Wang W, Wu H, Gong X, Moriguchi T (2015) Polyamines function in stress tolerance: from synthesis to regulation. Front Plant Sci 6:827. https://doi.org/10.3389/fpls.2015.00827

Liu R, Lai R (2015) Potentials of engineered nanoparticles as fertilizers for increasing agronomic productions. Sci Total Environ 51:131–139

Llanes A, Bertazza G, Palacio G, Luna V (2013) Different sodium salts cause different solute accumulation in the halophyte *Prosopis strombulifera*. Plant Biol 15:118–125

Longstreth DJ, Nobel PS (1979) Salinity effects on leaf anatomy. Plant Physiol 63(4):700–703

Lv YC, Xu G, Sun JN, Brestič M, Živčák M, Shao HB (2015) Phosphorus release from the soils in the Yellow River Delta: dynamic factors and implications for eco-restoration. Plant Soil Environ 61(8):339–343. https://doi.org/10.17221/666/2014-PSE

Lyshede OB (1917) Studies on the mucilaginolls cells in lhl.': k ar ol' .YparlO()'Slislis jilipes. PIOllfa 133:255–260

Manan MM, Ibrahim NA, Aziz NA, Zulkifly HH, Al-Worafi YM, Long CM (2016) Empirical use of antibiotic therapy in the prevention of early onset sepsis in neonates: a pilot study. Arch Med Sci 12:603–613. https://doi.org/10.5114/aoms.2015.51208

Marcum KB, Anderson SJ, Engelk MC (1998) Salt gland ion secretion/A salinity tolerance mechanism among five Zoysiagrass species. Crop Sci 38:806–810

Marcum KB, Murdoch CL (1994) Salinity tolerance mechanisms of six C_4 turfgrasses. J Amer Soc Hort Sci 119(4):779–784

Mass EV, Grieve CM (1990) Spike and leaf development in salt stressed wheat. Crop Sci 30:1309–1313

Mass EV, Nieman RH (1978) Physiology of plant tolerance to salinity. In: Jung GA (ed) Crop tolerance to suboptimal land conditions. Amer. Soc. Agron. Spec. Publ, USA, pp 277–299

Mastronardi E, Tsae P, Zhang X, Monreal CM, DeRosa MC (2015) Strategic role of nanotechnology in fertilizers: potential and limitations. In: Rai M, Ribeiro C, Mattoso L, Duran N (eds) Nanotechnologies in food and agriculture. Springer, Berlin

Mauseth JD (1988) Plant Anatomy. The Benjamin/Cummings Publishing Co, Inc, California

Megdiche W, Amor BN, Debez A, Hessini K, Ksouri R, Zuily-Fodil Y, Abdelly C (2007) Salt tolerance of the annual halophyte *Cakile maritima* as affected by the provenance and the developmental stage. Acta Physiol Plant 29:375–384

Mehta P, Jajoo A, Mathur M, Bharti S (2010) Chlorophyll a fluorescence study revealing effects of high salt stress on photosystem II in wheat leaves. Plant Physiol Biochem 48:16–20

Menezes RV, Azevedo Neto AD, Oliveira Ribeiro M, Cova AMW (2017) Growth and contents of organic and inorganic solutes in amaranth under salt stress. Pesq Agropec Trop 47(1):22–30

Mermoud A (2006) Cours de physique du sol : Maîtrise de la salinité des sols. Ecole polytechnique fédérale de Lausanne, p 23

Messedi D, Labidi N, Grignon C, Abdelly C (2004) Limits imposed by salt to the growth of the halophyte *Sesuvium portulacastrum*. JPNSS 167(6):720–725

Misra AN, Srivastava A, Strasser RJ (2001) Utilization of fast chlorophyll a fluorescence technique in assessing the salt/ion sensitivity of mung bean and brassica seedlings. J Plant Physiol 158:1173–1181

Mitsuya S, Takeoka Y, Miyake H (2000) Effects of sodium chloride on foliar ultrastructure of sweet potato (*Ipomoea batatas* Lam.) plantlets grown under light and dark conditions in vitro. J Plant Physiol 157(6):661–667

Mohammad M, Shibli R, Ajlouni M, Nimri L (1998) Tomato root and shoot responses to salt stress under different levels of phosphorus nutrition. J Plant Nutr 21(8):1667–1680

Moles TM, Pompeiano A, Huarancca Reyes T, Scartazza A, Guglielminetti L (2016) The efficient physiological strategy of a tomato landrace in response to short-term salinity stress. Plant Physiol Biochem 109:262–272. https://doi.org/10.1016/j.plaphy.2016.10.008

Moons A, Prinsen E, Bauw G, Montagu MV (1997) Antagonistic effects of abscisic acid and jasmonates on salt stress-inducible transcripts in rice roots. Plant Cell 9:2243–2259

Morais MC, Panuccio MR, Muscolo A, Freitas H (2012) Salt tolerance traits increase the invasive success of *Acacia longifolia* in Portuguese coastal dunes. Plant Physiol Biochem 55:60–65

Mozafar A, Goodin JR (1970) Vesiculated hairs: a mechanism for salt tolerance in *Atriplex halimus* L. Plant Physiol 45:62–65

Muchate NS, Nikalje GC, Rajurkar NS, Suprasanna P, Nikam TD (2016) Physiological responses of the halophyte *Sesuvium portulacastrum* to salt stress and their relevance for saline soil bioreclamation. Flora 224:96–105

Munns R (1993) Physiological processes limiting plant growth in saline soils: some dogmas and hypotheses. Plant Cell Environ 16:15–24

Munns R (2002) Comparative physiology of salt and water stress. Plant Cell Environ 25:239–250

Munns R, Gilliham M (2015) Salinity tolerance of crops what is the cost? New Phytol 208(3):668–673

Munns R, James RA, Sirault XR, Furbank RT, Jones HG (2010) New phenotyping methods for screening wheat and barley for beneficial responses to water deficit. J Exp Bot 61(13):3499–3507. https://doi.org/10.1093/jxb/erq199

Munns R, James RA, Läuchli A (2006) Approaches to increasing the salt tolerance of wheat and other cereals. J Exp Bot 57(5):1025–1043

Munns R, Rawson HM (1999) Effect of salinity on salt accumulation and reproductive development in the apical meristem of wheat and barley. Funct Plant Biol 26(5):459–464

Munns R, Termaat A (1986) Whole plant responses to salinity. Aust J Plant Physiol 13:143–160

Munns R, Tester M (2008) Mechanisms of salinity tolerance. Annu Rev Plant Biol 59:651–681
Naderi MR, Danesh-Shahraki A (2013) Nanofertilizers and their roles in sustainable agriculture. Int J Agric Crop Sci 5:2229–2232
Nawaz K, Ashraf M (2010) Exogenous application of glycinebetaine modulates activities of antioxidants in maize plants subjected to salt stress. J Agri Crop Sci 196:28–37. https://doi.org/10.1111/j.1439-037X.2009.00385.x
Negrão S, Schmöckel SM, Tester M (2017) Evaluating physiological responses of plants to salinity stress. Ann Bot 119(1):1–11
Niu X, Bressan RA, Hasegawa PM, Pardo JM (1995) Ion homeostasis in NaCl stress environments. Plant Physiol 109:735–742
Nivedithadevi D, Somasundaram R, Pannerselvam R (2012) Effect of abscisic acid, paclobutrazol and salicylic acid on the growth and pigment variation in *Solanum trilobatum* (I). Int J Drug Dev Res 4(3):236–246
Oertli JJ (1968) Extracellular salt accumulation a possible mechanism of salt injury in plants. Agrochimica 12:461–469
Orcutt DM, Nilsen ET (2000) The physiology of plants under stress: soil and biotic factors. JohnWiley and Sons, New York
Orcutt DM, Nilsene T (2000) Physiology of plants under stress. John Wiley & Sons Inc., New York, NY, USA
Orsini F, Cascone P, De Pascale S, Barbieri G, Corrado G, Rao R, Maggio A (2010) Systemin-dependent salinity tolerance in tomato: evidence of specific convergence of abiotic and biotic stress responses. Physiolo Plant 138:10–21. https://doi.org/10.1111/j.1399-3054.2009.01292.x
Osmond CB, Lüttge U, West KR, Pallaghy CK, Shacher-Hill B (1969) Ion absorption in *Atriplex* leaf tissue. II. Secretion of ions to epidermal bladders. Aust J Biol Sci 22:797–814
Owens S (2001) Salt of the earth. Genetic engineering may help to reclaim agricultural land lost due to salinisation. EMBO Rep 2:877–879
Özdemir F, Bor M, Demiral T, Türkan I (2004) Effects of 24-epibrassinolide on seed germination, seedling growth, lipid peroxidation, proline content and antioxidative system of rice (*Oryza sativa* L.) under salinity stress. Plant Growth Regul 42:203–211. https://doi.org/10.1023/B:GROW.0000026509.25995.13
Parida AK, Das AB (2005) Salt tolerance and salinity effect on plants: a review. Ecotoxicol Environ Saf 60:324–349
Parida AK, Das AB, Mittra B, Mohanty P (2004) Salt-stress induced alterations in protein profile and protease activity in the mangrove *Bruguiera parviflora*. Zeitschrift für Naturforschung C 59(5-6):408–414
Parvin S, Lee OR, Sathiyaraj G, Khorolragchaa A, Kim YJ, Yang DC (2014) Spermidine alleviates the growth of saline-stressed ginseng seedlings through antioxidative defense system. Gene 537(1):70–78. https://doi.org/10.1016/j.gene.2013.12.021
Pires RMO, Leite DG, Santos HO, Souza GA, Von Pinho EVR (2017) Physiological and enzymatic alterations in sesame seeds submitted to different osmotic potentials. Genet Mol Res 16(3). https://doi.org/10.4238/gmr16039425
Pompeiano A, Di Patrizio E, Volterrani M, Scartazza A, Guglielminetti L (2016) Growth responses and physiological traits of seashore paspalum subjected to short-term salinity stress and recovery. Agric Water Manag 163:57–65
Poonam T, Tanushree B, Sukalyan C (2013) Water quality indicesimportant tools for water quality assessment: a review. Int J Adv Chem (IJAC) 1(1):15–28
Price AH, Hendry GAF (1991) Iron-catalysed oxygen radical formation and its possible contribution to drought damage in nine native grasses and three cereals. Plant Cell Environ 14:477–484. https://doi.org/10.1111/j.1365-3040.1991.tb01517.x
Puyang X, An M, Han L, Zhang X (2015) Protective effect of spermidine on salt stress induced oxidative damage in two Kentucky bluegrass (*Poa pratensis* L.) cultivars. Ecotoxicol Environ Saf 117:96–106. https://doi.org/10.1016/j.ecoenv.2015.03.023
Qadir M, Quillérou E, Nangia V, Murtaza G, Singh M, Thomas RJ, Noble AD (2014) Economics of salt-induced land degradation and restoration. Nat Res Forum 38(4):282–295

Qadir M, Qureshi AS, Cheraghi SAM (2008) Extent and characterisation of salt-affected soils in Iran and strategies for their amelioration and management. Land Degrad Dev 19(2):214–227

Qiu Z, Guo J, Zhu A, Zhang L, Zhang M (2014) Exogenous jasmonic acid can enhance tolerance of wheat seedlings to salt stress. Ecotoxicol Environ Saf 104:202–208

Quin L, Guo S, Ai W, Tang Y, Cheng Q, Chen G (2013) Effect of salt stress on growth and physiology in amaranth and lettuce: implications for bioregenerative life support system. Adv Space Res 51(3):476–482

Rahman MS, Miyake H, Takeoka Y (2002) Effects of exogenous glycinebetaine on growth and ultrastructure of salt-stressed rice seedlings (*Oryza sativa* L.) Plant Prod Sci 5:33–44

Rajesh A, Arumugam R, Venkatesalu V (1998) Growth and photosynthetic characteristics of *Ceriops roxburghiana* under NaCl stress. Photosynthetica 35:285. https://doi.org/10.1023/A:1006983411991

Rao PS, Mishra B, Gupta SR, Rathore A (2008) Reproductive stage tolerance to salinity and alkalinity stresses in rice genotypes. Plant Breed 127:256–261. https://doi.org/10.1111/j.1439-0523.2007.01455.x

Rastogi A, Zivcak M, Sytar O, Kalaji HM, He X, Mbarki S, Brestic M (2017) Impact of metal and metal oxide nanoparticles on plant: a critical review. Front Chem 5:78. https://doi.org/10.3389/fchem.2017.00078

Rewald B, Raveh E, Gendler T, Ephrath JE, Rachmilevitch S (2012) Phenotypic plasticity and water flux rates of *Citrus* root orders under salinity. J Exp Bot 63(7):2717–2727. https://doi.org/10.1093/jxb/err457

Rico CM, Morales MI, McCreary R, Castillo-Michel H, Barrios AC, Hong J, Tafoy A, Lee WY, Varela-Ramirez A, Peralta-Videa JR, Gardea-Torresdey JL (2013) Cerium oxide nanoparticles modify the antioxidative stress enzyme activities and macromolecule composition in rice seedlings. Environ Sci Technol 47(24):14110–14118. https://doi.org/10.1021/es4033887

Romero-Aranda MR, Jurado O, Cuartero J (2006) Silicon alleviates the deleterious salt effect on tomato plant growth by improving plant water status. J Plant Physiol 163(8):847–855

Romero-Aranda R, Soria T, Cuartero J (2001) Tomato plant-water uptake and plant-water relationships under saline growth conditions. Plant Sci 160(2):265–272

Rossi L, Zhang W, Lombardini L, Ma X (2016) The impact of cerium oxide nanoparticles on the salt stress responses of *Brassica napus* L. Environ Pollut 219:28–36

Roychoudhury A, Basu S, Sengupta DN (2011) Amelioration of salinity stress by exogenously applied spermidine or spermine in three varieties of indica rice differing in their level of salt tolerance. J Plant Physiol 168:317–328

Rozema J, Riphagen J (1977) Physiology and ecologic relevance of salt secretion by the salt gland of *Glaux maritima* L. Oecologia 29:349–357

Sabaghnia N, Janmohammadi M (2014) Effect of nano-silicon particles application on salinity tolerance in early growth of some lentil genotypes. Ann UMCS Biol 69:39–55

Saneoka H, Nagasaka C, Hahn DT, Yang WJ, Premachandra GS, Joly RJ, Rhodes D (1995) Salt tolerance of glycinebetaine-deficient and -containing maize lines. Plant Physiol 107:631–638

Scholander PF, Hammel HT, Hemmingson ED, Garey W (1962) Salt balance in mangroves. Plant Physiol 37:722–729

Sekmen AH, Turkan I, Tanyolac ZO, Ozfidan C, Dinc A (2012) Different antioxidant defense responses to salt stress during germination and vegetative stages of endemic halophyte *Gypsophila oblanceolata* bark. Environ Exp Bot 77:63–76

Sengupta S, Majumder AL (2010) *Porteresia. coarctata* (Roxb.) Tateoka, a wild rice: a potential model for studying salt-stress biology in rice. Plant Cell Environ 33:526–542

Shahbaz M, Ashraf M (2013) Improving salinity tolerance in cereals. Crit Rev Plant Sci 32(4):237–249

Shahid MA, Pervez MA, Balal RM, Mattson NS, Rashid A, Ahmad R, Ayyub CM, Abbas T (2011) Brassinosteroid (24-epibrassinolide) enhances growth and alleviates the deleterious effects induced by salt stress in pea (*Pisum sativum* L.) Aust J Crop Sci 5:500–510

Shakir E, Zahraw Z, Al-Obaidy AHM (2017) Environmental and health risks associated with reuse of wastewater for irrigation. Egypt J Pet 26(1):95–102

Shang Q, Song S, Zhang Z, Guo S (2006) Exogenous brassinosteroid induced salt resistance of cucumber (*Cucumis sativus* L.) seedlings. Sci Agric Sinica 39:1872–1877

Sharma I, Bhardwaj R, Pati PK (2012) Mitigation of adverse effects of chlorpyrifos by 24-epibrassinolide and analysis of stress markers in a rice variety Pusa Basmati-1. Ecotoxicol Environ Saf 85:72–81

Sheteawi AS (2007) Improving growth and yield of salt stressed soybean by exogenous application of jasmonic acid and ascobin. Int J Agric Biol 3:473–478

Shomer I, Frenkel H, Polinger C (1991) The existence of a diffuse electric layer at cellulose fibril surfaces and its role in the swelling mechanism of parenchyma plant cell walls. Carbohydr Polym 16:199–210

Siadat H, Bybordi M, Malakouti MJ (1997) Salt-affected soils of Iran: a country report. international symposium on sustainable management of salt-affected soils in the arid ecosystem, Cairo

Siddique AB, Islam R, Hoque A, Hasan M, Tanvir RM, Mahir UM (2015) Mitigation of salt stress by foliar application of proline in rice. Univers J Agric Res 3:81–88. https://doi.org/10.13189/ujar.2015.030303

Siddiqui MH, Al-Whaibi MH (2014) Role of nano-SiO_2 in germination of tomato (*Lycopersicum esculentum* seeds mill.) Saudi J Biol Sci 21:13–17

Siddiqui MH, Al-Whaibi MH, Faisal M, Al Sahli AA (2014) Nano-silicon dioxide mitigates the adverse effects of salt stress on *Cucurbita pepo* L. Environ Toxicol Chem 33:2429–2437

Silberbush M, Ben-Asher J, Ephrath JE (2005) A model for nutrient and water flow and their uptake by plants grown in a soilless culture. Plant Soil 271(1-2):309–319

Singh SB, Singh BB, Singh M (1994) Effect of kinetin on chlorophyll, nitrogen and proline in mung bean under saline conditions. Indian J Plant Physiol 37:37–39

Slama I, Abdelly C, Bouchereau A, Flowers T, Savoure A (2015) Diversity, distribution and roles of osmoprotective compounds accumulated in halophytes under abiotic stress. Ann Bot 115:433–447. https://doi.org/10.1093/aob/mcu239

Slama I, M'Rabet R, Ksouri R, Talbi O, Debez A, Abdelly C (2017) Effects of salt treatment on growth, lipid membrane peroxidation, polyphenol content, and antioxidant activities in leaves of *Sesuvium portulacastrum* L. Arid Land Res Manag 31(4):404–417

Smaoui MA (1971) Differentiation des trichomes chez *Atriplex halimus* L. CR Acad Sci Paris Ser D 273:1268–1271

Sobahan MA, Akter N, Ohno M, Okuma E, Hirai Y, Mori IC, Nakamura Y, Murata Y (2012) Effects of exogenous proline and glycinebetaine on the salt tolerance of rice cultivars. Biosci Biotechno Biochem 76(8):1568–1570. https://doi.org/10.1271/bbb.120233

Soliman AS, El-feky SA, Darwish E (2015) Alleviation of salt stress on *Moringa peregrina* using foliar application of nanofertilizers. J Hortic For 7(2):36–47

Souleymane O, Hamidou M, Salifou M, Manneh B, Danquah E, Ofori K (2017) Genetic improvement of rice (*Oryza sativa*) for salt tolerance: a review. Inter J Advanc Res Botany 3:22–33. https://doi.org/10.20431/2455-4316.0303004

Soussi M, Ocana A, Lluch C (1998) Effects of salt stress on growth, photosynthesis and nitrogen fixation in chickpea (*Cicer arietinum* L.) J Expt Bot 49:1329–1337

Staal M, Maathuis FJM, Elzenga JTM, Overbeek JHM, Prins HBA (1991) Na^+/H^+ antiport activity in tonoplast vesicles from roots of the salt-tolerant *Plantago maritima* and the salt-sensitive *Plantago media*. Physiol Plant 82:179–184. https://doi.org/10.1111/j.1399-3054.1991.tb00078.x

Stassart JM, Neirinckx L, De Jaegere R (1981) The interactions between monovalent cations and calcium during their adsorption on isolated cell walls and absorption by intact barley roots. Ann Bot 47(5):647–652

Stefanov M, Yotsova E, Rashkov G, Ivanova K, Markovska Y, Apostolova EL (2016) Effects of salinity on the photosynthetic apparatus of two *Paulownia* lines. Plant Physiol Biochem 101:54–59. https://doi.org/10.1016/j.plaphy.2016.01.017

Strasser RJ, Tsimilli-Michael M, Qiang S, Goltsev V (2010) Simultaneous in vivo recording of prompt and delayed fluorescence and 820-nm reflection changes during drying and after rehydration of the resurrection plant Haberlea rhodopensis. Biochimica et Biophysica Acta (BBA) - Bioenergetics 1797:1313–1326

Sun S, An M, Han L, Yin S (2015) Foliar application of 24-epibrassinolide improved salt stress tolerance of perennial ryegrass. Hortscience 50:1518–1523

Sun ZW, Ren LK, Fan JW, Li Q, Wang KJ, Guo MM, Wang L, Li J, Zhang GX, Yang ZY, Chen F, Li XN (2016) Salt response of photosynthetic electron transport system in wheat cultivars with contrasting tolerance. Plant Soil Environ 62:515–521

Taiz L, Zeiger E (2002) Plant physiology, 3rd edn. Sinauer Associates, Sunderland, MA, 690 pp

Tang X, Mu X, Shao H, Wang H, Brestic M (2015) Global plant-responding mechanisms to salt stress: physiological and molecular levels and implications in biotechnology. Crit Rev Biotechnol 35(4):425–437. https://doi.org/10.3109/07388551.2014.889080

Tehranifar A, Jamalian S, Tafazoli E, Davarynejad GH, Eshghi S (2009) Interaction effects of paclobutrazol and salinity on photosynthesis and vegetative growth of strawberry plants. Acta Hortic 842:821–824. https://doi.org/10.17660/ActaHortic.2009.842.181

Tester M, Davenport R (2003) Na+ tolerance and Na+ transport in higher plants. Ann Bot 91(5):503–527

Thomson WW (1975) The structure and function of salt glands. In: Poljakoff-Mayber A, Gale J (eds) Plants in saline environments. Springer, Heidelberg, pp 118–146

Thomson WW, Faraday CD, Oross JW (1988) Salt glands. In: Baker DA, Hall JL (eds) Solute transport in plant cells and tissues. Longman Scientific & Technical, Essex, UK, pp 498–537

Thomson WW, Platt-Aloia K (1979) Ultrastructural transitions associated with the development of the bladder cells of the trichomes of *Atriplex*. Cytobios 25:105–114

Tsonev TD, Lazova GN, Stoinova ZG, Popova LP (1998) A possible role for jasmonic acid in adaptation of barley seedling to salinity stress. J Plant Growth Regul 17:153–159

Tukey HB Jr, Tukey HB, Wittwer SH (1958) Loss of nutrients by foliar leaching as determined by radioisotopes. P Am Soc Hortic Sci 71:496–506

Tukey HB, Morgan JV (1962) The occurrence of leaching from above ground plant parts and the nature of the material leached. 16th Intern Hort Congr Brussels:153–160

Tyerman SD, Skerrett IM (1999) Root ion channels and salinity. Sci Hort 78:175–235. https://doi.org/10.1016/s0304-4238(98)00194-0

Vardhini BV, Anjum NA (2015) Brassinosteroids make plant life easier under abiotic stresses mainly by modulating major components of antioxidant defense system. Front Environ Sci 2:67

Varjovi MB, Valizadeh M, Vahed MM (2016) Effect of salt stress and exogenous application of proline on some antioxidant enzymes activity in barley cultivars seedling. Biological forum. An International Journal 8(2):34–41

Vicente O, Boscaiu M, Naranjo MÁ, Estrelles E, Bellés JM, Soriano P (2004) Responses to salt stress in the halophyte *Plantago crassifolia* (*Plantaginaceae*). J Arid Environ 58(4):463–481

Von Willert DJ (1968) Tagesschwankungen des Ionengehalts in *Salicornia europaea* in Abhängigkeit vom Standort und von der Überflutung. Ber Dtsch Bot Ges Bd 81(10):442–449

Wahid A (2003) Physiological significance of morpho-anatomical features of halophytes with particular reference to Cholistan flora. Int J Agric Biol 5:207–212

Waisel Y (1972) Biology of halophytes. Academic Press, New York

Wang L, Li W, Yang H, Wu W, Ma LI, Huang T, Wang X (2016) Physiological and biochemical responses of a medicinal halophyte *Limonium bicolor* (Bag.) kuntze to salt-stress. Pak J Bot 48(4):1371–1377

Wang W, Wang R, Yuan Y, Du N, Guo W (2011) Effects of salt and water stress on plant biomass and photosynthetic characteristics of tamarisk (*Tamarix chinensis* Lour.) seedlings. Afr J Biotechnol 10:17981–11789

Wang Y, Nil N (2000) Changes in chlorophyll, ribulose biphosphate carboxylase-oxygenase, glycine betaine content, photosynthesis and transpiration in *Amaranthus* Tricolor leaves during salt stress. J Hortic Sci Biotechnol 75:623–627

Wang YY, Mopper S, Hasenstein KH (2001) Effects of salinity on endogenous ABA, IAA, JA, and SA in *Iris hexagona*. J Chem Ecol 27:327–342

Weisany W, Sohrabi Y, Heidari G, Siosemardeh A, Golezani KG (2012) Changes in antioxidant enzymes activity and plant performance by salinity stress and zinc application in soybean (*Glycine max* L.) Plant OMICS 5(2):60–67

Wilson C, Shannon MC (1995) Salt-induced Na^+/H^+ antiport in root plasma membrane of a glycophytic and halophytic species of tomato. Plant Sci 107:147–157

Wu W, Zhang Q, Ervin EH, Yang Z, Zhang X (2017) Physiological mechanism of enhancing salt stress tolerance of perennial ryegrass by 24-epibrassinolide. Front Plant Sci 8:1017. https://doi.org/10.3389/fpls.2017.01017

Wyn Jones RG, Brady CJ, Speirs J (1979) Ionic and osmotic relations in plant cells. In: Laidman DL, Wyn Jones RG (eds) Recent advances in the biochemistry of cereals. Academic Press, London, pp 63–103

Yan K, Wu C, Zhang L, Chen X (2015) Contrasting photosynthesis and photoinhibition in tetraploid and its autodiploid honeysuckle (*Lonicera japonica* Thunb.) under salt stress. Front Plant Sci 6:227. https://doi.org/10.3389/fpls.2015.00227

Yan P, Shao HB, Shao C, Chen P, Zhao S, Brestic M, Chen X (2013) Physiological adaptive mechanisms of plant grown in saline soil and implications for sustainable saline agriculture in coastal zone. Acta Physiologia Plantarum:2867–2878. https://doi.org/10.1007/s11738-013-1325-7

Yang X, Lu C (2006) Effects of exogenous glycinebetaine on growth, CO2 assimilation, and photochemistry of maize plants. Photosystem II. Physiol Plant 127(4):593–602

Yang X, Lu C (2005) Photosynthesis is improved by exogenous glycinebetaine in salt-stressed maize plants. Physiol Plant 124:343–352. https://doi.org/10.1111/j.1399-3054.2005.00518.x

Yasemin S, Koksal N, Özkaya A, Yener M (2017) Growth and physiological responses of '*Chrysanthemum paludosum*' under salinity stress. J Biol Environ Sci 11(32):59–66

Yassen A, Abdallah E, Gaballah M, Zaghloul S (2017) Role of silicon dioxide nano fertilizer in mitigating salt stress on growth, yield and chemical composition of cucumber (*Cucumis sativus* L.) Int J Agric Res 12:130–135. https://doi.org/10.3923/ijar.2017.130.135

Yeo A (1998) Molecular biology of salt tolerance in the context of whole-plant physiology. J Exp Bot 49(323):915–929

Yeo AR (1983) Salinity resistance: physiologies and prices. Physiol Plant 58:214–222. https://doi.org/10.1111/j.1399-3054.1983.tb04172.x

Yeo AR, Lee AS, Izard P, Boursier PJ, Flowers TJ (1991) Short-and long-term effects of salinity on leaf growth in rice (*Oryza sativa* L.) J Exp Bot 42(7):881–889

Yoon JY, Hamayun M, Lee SK, Lee IJ (2009) Methyl jasmonate alleviated salinity stress in soybean. J Crop Sci Biotechnol 12:63–68

Zhang J, Davies WJ (1989) Abscisic acid produced in dehydrating roots may enable the plant to measure the water status of the soil. Plant Cell Environ 12:73–81

Zhang Y, Hu XH, Shi Y, Zou ZR, Yan F, Zhao YY, Zhang H, Zhao JZ (2013) Beneficial role of exogenous spermidine on nitrogen metabolism in tomato seedlings exposed to saline–alkaline stress. J Am Soc Hortic Sci 138(1):38–49

Zhang Y, Zhang H, Zou ZR, Liu Y, Hu XH (2015) Deciphering the protective role of spermidine against saline–alkaline stress at physiological and proteomic levels in tomato. Phytochemistry 110:13–21. https://doi.org/10.1016/j.phytochem.2014.12.021

Zhao L, Peng B, Hernandez-Viezcas JA, Rico C, Sun Y, Peralta-Videa JR, Tang X, Niu G, Jin L, Varela-Ramirez A, Zhang JY, Gardea-Torresdey JL (2012) Stress response and tolerance of *Zea mays* to CeO_2 nanoparticles: cross talk among H_2O_2, heat shock protein and lipid peroxidation. ACS Nano 6:9615–9622. https://doi.org/10.1021/nn302975u

Zhu JK (2001) Plant salt tolerance. Trends Plant Sci 6(2):66–71

Zhu JK (2002) Salt and drought stress signal transduction in plants. Annu Rev Plant Biol 53: 247–273

Zhu J, Meinzer CF (1999) Efficiency of C4 photosynthesis in *Atriplex lentiformis* under salinity stress. Aust J Plant Physiol 26:79–86

Zid E, Grignon C (1991) Les tests de sélection précoce pour la résistance des plantes aux stress. Cas des stress salin et hydrique. L'amélioration des plantes pour l'adaptation aux milieux arides. Ed. John Libbey. Eurotext, Paris, pp 91–108

Ziegler H, Lüttge U (1967) Die Salzdrüsen von *Limonium vulgare*. Planta 74:1–17. https://doi.org/10.1007/BF0038516

Zivcak M, Brestic M, Sytar O (2016) Osmotic adjustment and plant adaptation to drought stress. In: Hossain MA (ed) Drought stress tolerance in plants, vol 1. Springer International Publishing, Switzerland, pp 105–143

Chapter 5
Potassium Uptake and Homeostasis in Plants Grown Under Hostile Environmental Conditions, and Its Regulation by CBL-Interacting Protein Kinases

Mohammad Alnayef, Jayakumar Bose, and Sergey Shabala

Abstract Abiotic stresses impose major penalties on plant growth and agricultural crop production. Understanding the mechanisms by which plants perceive these abiotic stresses, and the subsequent signal transduction that activates their adaptive responses, is therefore of vital importance for improving plant stress tolerance in breeding programs. Among the plethora of second messengers employed by plant cells, calcineurin B–like proteins (CBLs) and CBL-interacting protein kinases (CIPKs) have emerged as critical components of the signal transduction pathways and regulators of plant ionic homeostasis under stress conditions. This chapter summarizes the current knowledge on interaction between CIPKs and K^+ transport systems, and the role of the former in regulating cell ionic relations and K^+ homeostasis in plants grown under adverse environmental conditions.

Keywords Potassium channels · Transporters · Homeostasis · Stress tolerance · Signal transduction · Calcineurin B–like proteins (CBL) · CBL-interacting protein kinases (CIPKs) · Calcium signature · Membrane depolarization · Gene expression · Abiotic stress

M. Alnayef · S. Shabala (✉)
School of Land and Food, University of Tasmania, Hobart, TAS, Australia
e-mail: Sergey.Shabala@utas.edu.au

J. Bose
ARC Centre of Excellence in Plant Energy Biology, School of Agriculture, Food and Wine, Waite Research Institute, The University of Adelaide, Glen Osmond, SA, Australia

V. Kumar et al. (eds.), *Salinity Responses and Tolerance in Plants, Volume 1*,
https://doi.org/10.1007/978-3-319-75671-4_5

Abbreviations

ABA	Abscisic acid
AKT	*Arabidopsis* potassium transporter
$[Ca^{2+}]_{cyt}$	Cytosolic concentration of calcium
CaM	Calmodulin
CBL	Calcineurin B–like protein
CIPK	CBL-interacting protein kinase
CDPK	Ca^{2+}-dependent protein kinase
CML	CaM-like protein
CPDK	Calcium-dependent protein kinase
E_k	Equilibrium potential
GORK	Guard cells outward-rectifying potassium channel
HAK	High-affinity potassium transporter
KAT	Inward-rectifying Shaker-like potassium channel
KC1	Silent Shaker-like potassium channel
KUP	K^+ uptake permease
mRNA	Messenger RNA
NADPH	Reduced nicotinamide adenine dinucleotide phosphate
NO	Nitric oxide
PCD	Programmed cell death
PEG	Polyethylene glycol
PLP	Pyridoxal-5′-phosphate
PP2C	2C-Type protein phosphatase
RBOH	Respiratory burst oxidase homologue
ROS	Reactive oxygen species
SKOR	Stelar outward-rectifying potassium channel
SNO1	Sensitive to nitric oxide 1
SPIK	Shaker pollen inward K^+ channel
TPK	Tandem-pore potassium channel
V-ATPase	Vacuolar adenosine triphosphate

5.1 Introduction

Abiotic stresses such as salt, drought, heat, cold, and flooding impose major penalties on plant growth, resulting in estimated annual losses in crop and fiber production in excess of US$120 billion (http://www.fao.org/docrep/008/y5800e/Y5800E06.htm). Understanding the physiological and molecular mechanisms by which plants perceive these abiotic stresses, and the subsequent signal transduction that activates their adaptive responses, is therefore of vital importance for improving plant stress tolerance in breeding programs.

Among the plethora of known signals, calcium and reactive oxygen species (ROS) play vital roles as second messengers in plant responses to biotic and abiotic stress. Under normal conditions, the cytosolic concentration of calcium ($[Ca^{2+}]_{cyt}$) is maintained at a nanomolar level, generally in the range of 100–200 nM (Bush 1995), whereas it is in the millimolar range (1–10 mM) in extracellular and intracellular Ca^{2+} stores. Under adverse environmental conditions, this cytosolic concentration can be higher—up to micromolar level. Plants not only store the excess Ca^{2+} (either in the apoplast or in the lumen intracellular organelles) in order to avoid Ca^{2+} toxicity but also use it as a signal transductor in response to adverse environmental conditions (Kader and Lindberg 2010; Perochon et al. 2011; Reddy and Reddy 2004). In response to an environmental stress, the increase in the cytosolic Ca^{2+} concentration (the so-called Ca^{2+} signature) is a result of the activation of Ca^{2+} channels, pumps, and transporters located in the plasma membrane or in other organelle membranes, resulting in the downstream response of targeted genes (Tuteja and Mahajan 2007). The peak, magnitude, frequency, and duration of the Ca^{2+} signature play a vital role in encoding specific information for plants under various conditions (Whalley and Knight 2013). The Ca^{2+} signature must be either of low amplitude or transient. The latter can be single (a spike), double (biphasic), or multiple (oscillations) (Tuteja and Mahajan 2007). Stress-induced calcium signatures are highly specific and may differ even between ecotypes within the same species (Schmöckel et al. 2015). These authors showed that, 200 mM NaCl resulted in a peak $[Ca^{2+}]_{cyt}$ of 800–855 nM, while under isotonic osmotic stress generated by sorbitol, the peak magnitude ranged between 73 and 750 nM, depending on the particular ecotype of *Arabidopsis* (Schmöckel et al. 2015). Interestingly, these results showed that in the case of the Col-0 ecotype, the double (biphasic) Ca^{2+} signatures occurred between 30–120 s after salt treatment and lasted approximately 30 s, while the C24 had only one peak (a spike). This suggested that the salt signaling pathway is triggered by NaCl and initiates a signal cascade causing Ca^{2+} influx into the cytosol to create a Ca^{2+} signature (Schmöckel et al. 2015).

Another well-known signal involved in plant sensing of hostile environment are reactive oxygen signals. ROS are not always toxic molecules that need to be detoxified, and ROS waves are required by plants for growth, development, programmed cell death (PCD), and biotic and abiotic responses (Gechev et al. 2006). Importantly, ROS and Ca^{2+} signaling networks are strongly overlapped and interact, providing another layer of signal specificity encoding. For example, both cytosolic Ca^{2+} and ROS signals were essential for maintenance of the optimal cytosolic K^+/Na^+ ratio in salt-grown plants (Ma et al. 2012a). Among other reasons, the reported cross talk between Ca^{2+} and ROS originates from the fact that many Ca^{2+}-permeable channels are ROS activated (Demidchik et al. 2003). *Arabidopsis AtrbohD* and *AtrbohF* mutants lacking functional NADPH [reduced nicotinamide adenine dinucleotide phosphate] oxidases—a major source of the apoplastic ROS generation—lacked the capacity for H_2O_2 production and showed a salt-sensitive phenotype resulting from increased Na^+ accumulation and lower K^+ content (Ma et al. 2012a). The salt tolerance of these plants was partially rescued by exogenous H_2O_2 application. Respiratory burst oxidase homologues (RBOHs) are synergistically activated by binding of Ca^{2+} to EF-hand motifs and also by phosphorylation (Ogasawara et al. 2008).

Plants have evolved a specific set of proteins that bind Ca^{2+}, with or without helix–loop–helix EF-hand motifs. Three major categories of EF-hand proteins in plants are the calmodulins (CaMs) and CaM-like proteins (CMLs), Ca^{2+}-dependent protein kinases (CDPKs), and calcineurin B–like proteins (CBLs) (DeFalco et al. 2010). Some of these Ca^{2+} binding proteins such as calcium-dependent protein kinases (CPDKs), when bound to Ca^{2+}, can immediately regulate downstream proteins (Schulz et al. 2013), while others—the CMLs and CBLs—bind to other proteins to affect their downstream target.

Over the last decade, cytosolic K^+ has emerged as a key determinant of plants' ability to adapt to a broad range of biotic and abiotic factors (Anschutz et al. 2014; Shabala and Pottosin 2014; Shabala et al. 2016). While the early focus of the research in this field was on structure–function studies and understanding of the molecular mechanisms underlying K^+ transport, more recent studies of post-translational regulation of potassium transport systems have moved to the center stage of K^+ transport research (Anschutz et al. 2014). This has highlighted the role of cytosolic potassium as a signaling molecule and as one of the "master switches" enabling plant transition from normal metabolism to a "hibernated state" during the first hours after the stress exposure and then to a recovery phase (Shabala and Pottosin 2014; Demidchik 2014; Shabala 2017). The current paradigm implies that all known abiotic stresses trigger a substantial disturbance in K^+ homeostasis and provoke a feedback control on K^+ channels and transporter expression and post-translational regulation of their activity, optimizing K^+ absorption and usage and, at the extreme end, assisting the programmed cell death. The specific details of this control remain elusive but are known to include both Ca^{2+} and ROS signaling (Shabala et al. 2016).

Over the last decade, calcineurin B–like proteins have attracted a lot of attention as important regulators of plant ionic homeostasis (Li et al. 2009)—the role fulfilled by CBLs interacting with a certain class of protein kinases. CBL-interacting protein kinases (CIPKs) are a unique family of serine/threonine protein kinases, present in plants. There is evidence from various studies that biotic and abiotic stresses such as salinity, drought, cold, and high pH induce different expression patterns and subcellular localization of CBL sensors and CIPK kinases (Kolukisaoglu et al. 2004; Li et al. 2009; Manik et al. 2015). The focus of this chapter is on the interaction between CIPKs and K^+ transport systems, and the role of the former in regulating cell ionic relations and K^+ homeostasis in plants grown under adverse environmental conditions. First, major K^+ transport systems present in plant roots are described, and their roles in plant adaptation to hostile conditions are revealed.

5.2 Plant K^+ Channels

Different types of K^+ channel have been identified as being involved in K^+ transport in plants. These are Shaker-type K+ channels and tandem-pore K^+ channels (TPKs), and all types have orthologues in animals (Véry and Sentenac 2003). The first plant K^+ channels identified through functional complementation of yeast mutant strains defective for K^+ uptake, were members of the Shaker family (Kv-like) (Schachtman et al. 1992; Sentenac et al. 1992). In the *Arabidopsis* genome, nine Kv-like Shaker proteins have been identified, seven of which are highly selective K^+ channels (Very and Sentenac 2002). On the basis of their rectification properties, these have been grouped into three known subfamilies: four are inward rectifying (KAT1 [inward-rectifying Shaker-like potassium channel 1], KAT2, AKT1 [*Arabidopsis* potassium transporter 1], and SPIK [Shaker pollen inward K^+ channel]), two are outward rectifying (GORK [guard cells outward-rectifying potassium channel] and SKOR [Stelar outward-rectifying potassium channel]), one is a silent K^+ channel (KC1 [silent Shaker-like potassium channel]), and two are weak or unknown rectifying channels (AKT2/3 and AKT5) (Pilot et al. 2003). The Shaker channels are voltage gated; thus, changes in membrane polarization, in response to the presence of K^+ in millimolar concentrations, regulate the functions of these channels (Lebaudy et al. 2007). The inward-rectifying channels are activated by membrane hyperpolarization (with negative E_k [equilibrium potential] values of about −80 and −100 mV), so their gating is almost independent of the predominant potassium concentration on both sides of the plasma membrane (Hedrich et al. 2011). However, since the outward-rectifying channels are activated by membrane depolarization, their gating is strongly dependent on the extracellular K^+ concentration (Hedrich et al. 2011; Lebaudy et al. 2007).

Among all of these channels, the AKT1 channel arguably plays the greatest role in root K^+ acquisition and thus mediates K^+ homeostasis under stress conditions. Accordingly, the following sections are focused on expression patterns and regulation modes of AKT1.

5.2.1 *Expression Pattern of AKT1*

In *Arabidopsis*, AKT1 (locus ID At2g26650) is localized in the plasma membrane of the root, and through the use of gene promoter analysis it was found that *AKT1* gene promoter activity was strong along the mature root, relatively weak at the root tip, and restricted in both the cap and the epidermis (Lagarde et al. 1996). In rice plants, transcriptional analysis showed the presence of *OsAKT1* transcripts in the root for K^+ uptake (Golldack et al. 2003). The highest levels of such transcripts were found in the epidermis, followed by the exodermis, endodermis, and pericycle (Golldack et al. 2003). In leaves, the *OsAKT1* transcript was found in the mesophyll cells and cells neighboring the metaxylem vessels, suggesting that OsAKT1 plays a role in K^+ transport and distribution. Moreover, *OsAKT1* was also expressed in phloem cells, suggesting its possible role in phloem loading in rice plants (Golldack et al. 2003).

5.2.2 Regulation of AKT1 Channels

Responses to environmental stimuli such as light, abscisic acid (ABA), and salt stress regulate the Shaker K^+ channels' activity both at transcriptional and post-translational levels (Chérel 2004). For example, a study of the signaling pathway of nitric oxide (NO) in the *Xenopus* oocyte system and in root protoplasts of *Arabidopsis* showed that NO negatively regulated the activity of AKT1 via upregulation of the activity of the vitamin B_6 salvage pathway gene, SNO1 [Sensitive to nitric oxide 1]. This caused an accumulation of B6 and pyridoxal-5'-phosphate (PLP), which in turn suppressed the activity of AKT1. Such suppression reduced K^+ uptake by the root (Xia et al. 2014). Several studies have shown that the activity of these channels is regulated by protein–protein interactions (Chérel 2004), which can be grouped into two types.

The first type are interactions that form heteromeric channels. For example, a heteromerization study showed that *AtKC1* modulates the activity of the root K^+ inward rectifier, AKT1, under a K^+-deficient condition, by a combination of two effects. The first one is the integration of *AtKC1* into the root inward K^+ rectifier, which reduces K^+ efflux and results in membrane hyperpolarization by shifting the voltage-dependent gating of AKT1 to a more negative membrane potential. The second one involves reduction of the relative cord conductance by *AtKC1*. These two factors inhibit K^+ efflux while allowing K^+ uptake at negative membrane potential from the reversal potential for K^+ (Geiger et al. 2009).

The second type are interactions involving genuine regulatory proteins (Chérel 2004). Several studies have shown that *AKT1* is upregulated by CBL–CIPK protein complexes in *Arabidopsis* (Xu et al. 2006) and in rice plants, where the calcineurin B–like protein OsCBL1 interacts with the protein kinase OsCIPK23 in the plasma membrane, phosphorylating AKT1-mediated K^+ uptake by the roots of the plants (Li et al. 2014). A recent study showed that both AtKC1 and CIPK23 synergistically modulate AKT1 activity under a low-K^+ condition (Wang et al. 2016).

5.2.3 Role of AKT1 in Plant Growth Under Various K^+ and NH_4^+ Conditions

NH_4^+ and K^+ are both monovalent cations, which influence each other both positively and negatively within the low- to high-affinity K^+ transport range (Szczerba et al. 2008). In rice, the growth of rice seedlings and the accumulation of K^+ in plant tissue under the high-affinity K^+ transport condition (0.02 mM K^+) were both significantly reduced in the presence of NH_4^+. However, when the rice plants were grown in NH_4^+, an increase in the external K^+ concentration caused them to show greater influx, translocation, and tissue accumulation of K^+, accompanied by a reduction in NH_4^+ influx, resulting in a significant improvement in their growth (Szczerba et al. 2008). One of the possible explanations for this is that under a low-affinity K^+ condition, NH_4^+ competes with K^+ for transport via the non-AKT1 component of the K^+ transport system (Spalding et al. 1999). It was demonstrated that the AKT1 component of K^+ permeability was between 55% and 63% in the wild type under a low K^+ concentration of between 10 and 100 μM, and in the absence of NH_4^+. However, in the presence of NH_4^+, seed germination, seedling growth, and K^+ uptake were all inhibited in the *akt1* mutant plants (Spalding et al. 1999). Moreover, membrane depolarization was also observed when plants were grown in the presence of NH_4^+ (Spalding et al. 1999). Such depolarization is known to activate outward-rectifying channels such as GORK, resulting in more K^+ efflux. Under a 1.5 mM external K^+ condition, K^+ influx was not affected by the presence of NH_4^+, while K^+ influx was stimulated by the addition of NH_4^+ in the low-affinity K^+ range (40 mM K^+). However, under high nutrient supply, plant growth declined because of the energy consumption required to remove NH_4^+ and K^+ (Britto and Kronzucker 2002; Szczerba et al. 2008). An electrophysiological study concluded that AKT1 encodes the inward-rectifying channel responsible for K^+ uptake from root cells (Hirsch et al. 1998). The role of AKT1 in plant nutrition, studied by using reverse genetic screening, showed that in a medium containing ≤100 μM K^+ concentration and in the presence of NH_4^+, *akt1* function in *Arabidopsis* plants was disrupted, causing significant growth inhibition in comparison with the wild type. This suggested that NH_4^+ inhibits non-AKT1 K^+ uptake pathways, thereby making growth dependent on AKT1 (Hirsch et al. 1998). In contrast, the growth of the *AKT1* mutant plants was slightly reduced at 1 mM K^+ in comparison with the wild type. Moreover, $^{86}Rb^+$ tracer flux analysis revealed that loss of function of *AKT1* reduced the plants' ability to take up K^+ from the medium (Hirsch et al. 1998). These results indicate that the AKT1 channel mediates K^+ uptake in a growth medium containing a concentration of 10 μM K^+ (Hirsch et al. 1998).

Other studies showed that when the plants were grown in 0.5 mM and 5 μM of K^+, the K^+ content of the shoots and roots was significantly reduced but without a significant change in the plant biomass. However, this K^+ deficiency did not change the expression of the *AKT1* gene. This is consistent with the theory that low-affinity transporters mediate K^+ uptake in the millimolar range (Lagarde et al. 1996). A combination study of the function of AKT1 and AKT2 in roots of *Arabidopsis* plants, using knockout mutation and electrophysiological assays, showed that a reduction in the K^+ permeability of the plasma membrane of root cells resulted in growth impairment in *akt1* seedlings. Moreover, in cotyledons AKT1 was a more major contributor to K^+ uptake than AKT2 (Dennison et al. 2001). Additionally, under a low-K^+ condition (10–100 μM range) and in the absence of NH_4^+, the growth of wild-type plants increased as the external K^+ concentration increased, indicating that in the absence of NH_4^+, their growth rate was restricted by K^+ deficiency. Mutations in *akt1* and *akt2* nonsignificantly affected the growth rate of the plants under both of the above conditions. However, when NH_4^+ was added to the growth medium, the growth rate of the *akt1* mutant plants was inhibited, but that of the *akt2* mutant plants was not. This negative effect of NH_4^+ on the plant growth rate was mitigated by the addition of more K^+ to the growth medium (Dennison et al. 2001). The need for plasma membrane hyperpolarization between −180 and −240 mV increased under a low-K^+ condition (10–100 μM) to allow the inward-rectifying channels to mediate K^+ uptake (Geiger et al. 2009). Under a low-K^+ condition, the loss of function of one of the essential root K^+ channels such as AKT1 resulted in growth impairment (Geiger et al. 2009). A study involving a single *athak5* mutation and an *athak5akt1* double mutation in *Arabidopsis* plants showed that the growth of the *athak5* mutant plants was inhibited at 10 μM but not at 20 μM K^+, while both the *athak5* and *akt1* double-mutant plants failed to grow at 100 μM K^+ and experienced growth inhibition at 450 μM K^+. These results suggested that AtHAK5 and AKT1 are the major transporters in the high-affinity K^+ transport system, and that the *AtHAK5* gene is the most important component of the non-AKT1 pathway (Pyo et al. 2010).

Desbrosses et al. (2003) found that the root hairs of the *akt1-1* mutant did not elongate when the plants were grown at ≥10 mM. It was concluded that AKT1 is required for root hair elongation in the 50–100 mM range of K^+ concentration. However, under a zero K^+ concentration, the *akt1* mutant plants developed longer root hairs. Both of these results led to the conclusion that AKT1 is a component of the low-affinity K^+ uptake system in plants.

5.2.4 Role of AKT1 in Na^+ Uptake Under a Low-K^+ Condition

It has been suggested that the AKT1 channel is involved in Na^+ uptake under a low-K^+ condition (Golldack et al. 2003; Spalding et al. 1999). This suggestion arose from the finding that under a low-K^+ (10 μM K^+) condition, the growth rate of *akt1* mutant plants increased by 119% when the external Na^+ concentration was increased to 1000 μM Na^+, while this same growth rate was not stimulated by Na^+ at 100 μM K^+ (Spalding et al. 1999). A transcriptional study of the *AKT1* gene in salt-sensitive rice (IR29) and in salt-tolerant lines (BK and Pokkali) under a 150 mM NaCl condition showed that while *OsAKT1* transcripts disappeared from the exodermis of the salt-tolerant lines, their levels remained unchanged in the endodermis and increased in the exodermis in the salt-sensitive genotype. These findings point to *OsAKT1* being regulated differently in the various salt-dependent rice genotypes (Golldack et al. 2003). Moreover, this regulation of *AKT1* does not significantly affect potassium homeostasis under salt stress. Under 100 μM K^+ to low-millimole K^+, the salt-tolerant lines selectively excluded Na^+ while maintaining cytosolic K^+. However, at low-micromole K^+, these salt-tolerant and salt-sensitive lines accumulated similar amounts of Na^+, suggesting a clear correlation between *OsAKT* expression and whole-plant Na^+ selectivity (Golldack et al. 2003).

5.2.5 *Role of AKT1 in Adaptation to Osmotic Stress and Stomatal Movement*

Potassium is considered to be a major inorganic osmoticum in plants, playing an important role in osmotic adjustment under drought conditions (Wang et al. 2013). K^+ uptake is partially mediated by voltage-gated K^+ channels located in the cellular plasma membrane (Maathuis et al. 1997). Under water stress conditions, and with a 1.4 mM K^+ concentration in a hydroponic system, shoot and root K^+ content was lower in *akt1* mutant plants than in the wild type, while at 10 mM K^+, the *akt1* plants did not show any K^+-dependent phenotypes. This result suggested that under 1.4 mM external K^+, the non-AKT1 components were not able to compensate for the loss of function of *akt1,* resulting in lower K^+ content (Nieves-Cordones et al. 2012). One of the essential adaptations that plants need to develop in order to survive drought stress is rapid stomatal closure to prevent dehydration (Wang et al. 2013). K^+ inward/outward flux from guard cells via voltage-gated channels plays a crucial role in stomatal movement. Four inward rectifiers (KAT1, AKT1, AtKC1, and AKT2/3) and one outward rectifier (GORK) have been identified in the protoplasts of guard cells in wild-type *Arabidopsis* (Szyroki et al. 2001). Moreover, patch clamp analysis on enzymatically isolated guard cell protoplasts showed that KAT1 is the dominant inward K^+ channel, being responsible for 79% of K^+ conductance, while the other inward rectifiers collectively deal with the remaining 21%. This led to the conclusion that inward-rectifying K^+ currents in guard cells are mediated not only by KAT1 homomers but also by the AKT1, AKT2, and AKT2/3 K^+ channels (Szyroki et al. 2001).

Nieves-Cordones et al. (2012) reported that in both hydroponic and soil systems, *akt1* mutant plants sustained a significantly lower percentage of water loss and experienced a lower transpiration rate, at 10 mM K^+ and under water stress conditions. Additionally, the efficiency of stomatal closure in the *akt1* mutant plants was higher in response to ABA, suggesting that disruption of *akt1* enhanced the plants' response to water stress.

5.3 The KT/HAK/KUP K^+ Transporter Family

The members of the *KT/HAK/KUP* family have been identified firstly as K^+ uptake permeases (KUPs) in bacteria K^+ (Schleyer and Bakker 1993) and secondly as high-affinity potassium transporters (HAKs) in fungi (Bañuelos et al. 1995). Several homologues have also been identified in plants such as *Arabidopsis* (Maser et al. 2001), rice (Bañuelos et al. 2002; Gupta et al. 2008; Yang et al. 2009), maize (Zhang et al. 2012), and poplar (He et al. 2012). The ubiquitous presence of KT/HAK/KUP transporter genes in ancestral plant genomes implies that they play important roles in K^+ acquisition during plant growth, development, and adaptation to stress (Grabov 2007). A considerable number of studies have shown that KT/HAK/KUP transporter genes are involved in both low- and high-affinity K^+ transport, which overlap with the activity of K^+ channels (Bañuelos et al. 2002). The existence of different genes with similar K^+ affinity can be explained by the fact that KT/HAK/KUP transporter genes mediate active K^+–H^+ transport (Rodriguez-Navarro 2000), while the other transporters mediate uniport transport to maintain electrochemical equilibrium (Bañuelos et al. 2002). A study of the phylogenetic tree of *Arabidopsis* identified 13 genes in the *KT/HAK/KUP* family that exhibit strong similarities (Maser et al. 2001). One of these genes, *AtHAK5,* has been identified as a high-affinity K^+ transporter in *Arabidopsis* (Rodríguez-Navarro and Rubio 2006). In the rice genome, 27 genes have been identified, distributed among eight chromosomes. Twenty of these—including *OsHAK2*, *OsHAK3*, *OsHAK7*, and *OsHAK9*—have been shown to be localized in the plasma membrane, while the rest are located in four different organelles: *OsHAK10* in the tonoplant, *OsHAK14* in the mitochondrial inner membrane, *OsHAK15* in the chloroplast thylakoid membrane, and AtKUP/HAK/K12 in the chloroplast membrane (Gupta et al. 2008). However, on the basis of a phylogenetic analysis of 13 *Arabidopsis* and 27 rice genes, the *KT/HAK/KUP* family can be divided into four clusters, each of which includes genes that are localized in different subcellular membranes (Gupta et al. 2008). A study on *cis*-elements, which play an important role in gene regulation, showed that the upstream region of most *OsHAK* genes contains Ca^{2+}-responsive *cis*-elements. As these may be associated with the key role of Ca^{2+}, as a second messenger in response to environmental stimuli, Ca^{2+} is able to both activate and deactivate the *OsHAK* genes by regulating its *cis*-elements (Gupta et al. 2008).

5.3.1 Expression Pattern of the KT/HAK/KUP Family in Plants

A study on the expression of *OsHAK1* and *OsHAK7* in the shoots and roots of rice plants under high- and low-K^+ conditions showed that although *OsHAK1* expression was found in the whole plant, with the highest expression identified in the root, *OsHAK7* was expressed only in the shoots and roots. This suggested that OsHAK1 belongs to the high-affinity transporters in rice roots (Bañuelos et al. 2002). An in situ hybridization analysis showed that *OsHAK1* was mainly expressed in the root tips—specifically, in their meristematic regions. The OsHAK1 messenger RNA (mRNA) signal indicated that *OsHAK1* was expressed in the epidermal cells and vascular cells, while in the shoots it was detected in the apical meristem and in the cells on the inner side of the vascular bundle of leaf sheaths, as well as at the conjunction of the roots and shoots, and in the stem (Chen et al. 2015). Another study, which investigated the expression of 26 *OsHAK* genes in the roots of rice seedlings, detected 14 genes that were involved in either K^+ uptake or some other related function (Gupta et al. 2008). In *Arabidopsis*, evidence of *OsHAK5* expression was found in the epidermis of the main and lateral roots and, to a lesser extent, in the vasculature of the main roots (Gierth et al. 2005). In contrast, overexpression of *OsHAK5* in rice plants increased the net K^+ influx rate by 2.6-fold under a 0.1 mM K^+ condition, suggesting that OsHAK5 plays a role in root K^+ acquisition under a K^+-deficient condition. Moreover, the fact that strong expression of *OsHAK5* was identified in the xylem parenchyma and phloem of root vascular tissue, particularly under a K^+-deficient condition, suggested that OsHAK5 may be involved in the distribution of K^+ transport between the roots and shoots (Yang et al. 2014).

5.3.2 HAK Genes Mediate High-Affinity K^+ Uptake in Plant Roots

Several studies have indicated that HvHAK1 functions as a high-affinity K^+ transporter in *Arabidopsis* plants (Mangano et al. 2008; Santa-María et al. 1997). When the plants were grown in media containing 1 mM KCl, overexpression of *HvHAK1* resulted in slightly higher Rb^+ uptake, which was more significant in plants under the K^+-limiting condition than in the control plants (Fulgenzi et al. 2008). When the effect that knocking out *OsHAK1* had on rice plants grown under two K^+ levels—0.1 and 1 mM K^+—was studied by Chen et al. (2015), the results showed that the K^+ uptake rates in the roots of *Oshak1* mutants were about 50% and 70%, respectively, with the total K^+ uptake per plant being 15–20% and 30–35%, respectively, and the K^+ acquisition being 15% and 35%, respectively, of the values observed in wild-type plants. Thus, while the *Oshak1* mutant plants showed significant growth impairment under both conditions, this was more pronounced under the low-K^+ condition (Chen et al. 2015). Moreover, knocking out *Oshak1* limited cell expansion, resulting in smaller mutant plants (Chen et al. 2015). Another study highlighted the key role played by *AtHAK5* in both seedling establishment and plant growth under a low-K^+ condition (Pyo et al. 2010). *Athak5* mutant plants showed lower root and shoot K^+ content than the wild type when *Arabidopsis* plants were grown under K^+-deficient conditions (Gierth et al. 2005; Nieves-Cordones et al. 2010; Pyo et al. 2010). It was consistently found that under a K^+-deficient condition, overexpression of *AtHAK5* enhanced the growth of *Arabidopsis* plants by causing them to produce a greater number and greater density of lateral roots, and higher dry weights of shoots and roots, than the wild type (Adams et al. 2014). A comparison study of the growth of overexpressed *OsHAK5* and *Oshak5* mutant plants and their wild type showed that when the plants were grown in a low-K^+ condition, expression of *OsHAK5* was rapidly upregulated in the plasma membrane of the epidermal cells of the rice root, resulting in increased acquisition. However, *Oshak5* mutant plants accumulated only 60% of the K^+ accumulated by their wild type (Yang et al. 2014). In contrast, when grown in a medium containing 1 mM K^+, overexpressed *OsHAK5* plants did not display a significant difference in their growth, in comparison with the wild type. The suggestion from this is that OsHAK5 was involved in K^+ acquisition in the rice plants under the low-K^+ condition (Yang et al. 2014).

5.3.3 Function of HAK Genes in K^+ Uptake Under Salt Stress Conditions

According to Chen et al. (2015), *OsHAK1* expression is regulated differently by salt stress, depending on the external K^+ concentration. For example, under a normal-K^+ condition, the addition of Na^+ upregulated *OsHAK1* expression, while under a low-K^+ condition, *OsHAK1* expression was reduced. This resulted in a lower K^+ uptake rate in the wild type, while *Oshak1* mutant plant uptake was completely blocked. This study concluded that all high-affinity K^+ transport systems in rice plants—including OsHAK1, OsHAK5, and OsAKT1—are salt sensitive (Chen et al. 2015). The results of a study by Fulgenzi et al. (2008) showed that the *HvHAK1* transcript was significantly upregulated by 100 mM NaCl under a 1 mM K^+ condition, resulting in significant K^+ uptake after 6 h of salt induction. In *Arabidopsis* plants grown under a low-K^+ condition, while the expression of *AtHAK5* was upregulated, accompanied by a high rate of K^+ uptake by the root, *AtHAK5* expression was significantly reduced by the addition of Na^+ to the medium (Nieves-Cordones et al. 2010). One possible explanation for this is that Na^+ induced depolarization of the membrane potential, which in turn regulated the activity of the *AtHAK5* gene (Nieves-Cordones et al. 2008). However, the differences in the Na^+ concentration between the overexpressed *HAK5* lines and the wild type was insignificant, regardless of whether they were grown under control, K^+-deficiency, or salt stress conditions. Furthermore, since no morphological phenotype was found in the overexpressed lines under salt stress, this suggested that *AtHAK5* is not involved in Na^+ uptake in *Arabidopsis* plants (Adams et al. 2014; Nieves-Cordones et al. 2010). In contrast, Yang et al. (2014) reported that even under normal-K^+ conditions, *OsHAK5* expression was vastly upregulated by salinity. Thus, overexpressed *OsHAK5* mutant plants were salt tolerant and *Oshak5* mutant plants were salt sensitive, in comparison with the wild type, under a 100 mM NaCl condition. This suggests that OsHAK5 improves rice salt tolerance by enhancing the shoot K^+/Na^+ ratio. Moreover, the presence of OsHAK5 was detected in mesophyll cells and phloem tissue (Yang et al. 2014). In general, improvement of plant K^+ acquisition and growth under salt conditions, through enhancement of the transcript levels of genes that encode high-affinity potassium transporters, is an important target for plant breeders (Nieves-Cordones et al. 2010).

5.3.4 Role of HAK Genes in K^+ Transport

Significant expression levels of *OsHAK1* and *OsHAK5* were found in the xylem parenchyma and phloem of root vascular tissue, suggesting a role for OsHAK1 and OsHAK5 in K^+ transport from the root to the aerial parts (Chen et al. 2015; Yang et al. 2014). The root K^+ content of overexpressed *OsHAK5* plants, in comparison with the wild type, was 30% and 60% lower under 0.3 mM K^+ and K^+-free conditions, respectively. While under a normal-K^+ condition (1 mM), overexpression of *OsHAK5* did not contribute to the transport of K^+ from the roots to the shoots (Yang et al. 2014), under the low-K^+ condition, OsHAK5 and OsHAK1 both played a significant role in the xylem loading (Chen et al. 2015; Yang et al. 2014), with the K^+ content of the xylem sap being 20–25% higher in the overexpressed line and lower in the *oshak5* mutant line than in the wild type. This leads to the conclusion that under a normal condition, other K^+ channels and K^+ diffusion via the apoplast may play an essential role in K^+ transport from the roots to the shoots, while under a low-K^+ condition, OsHAK5 mediates K^+ transport (Yang et al. 2014).

5.4 Role of CIPKs in Regulation of K^+ Homeostasis

5.4.1 Role of CIPKs in Osmotic Stress Signaling

The initial responses to water stress include a rapid reduction in the leaf growth rate, followed by partial or complete stomatal closure. These result in reductions in the rates of transpiration and photosynthesis. Moreover, water stress inhibits root growth, and if the water stress is long term, ultimately plant growth and yield are reduced (Neumann 2011). In a bid to survive under such adverse environmental conditions, plants have evolved numerous responsive genes to restore cellular homeostasis. Various signaling proteins such as transcription factors, protein kinases and phosphatases play important signal transduction roles under abiotic stress conditions by activating adaptive downstream responses that promote plant growth (Golldack et al. 2014).

In the rice genome, 15 genes in the CIPK family were induced by drought stress in a study by Xiang et al. (2007). Of these, the *OsCIPK12* gene was the one shown to play a positive role in rice drought tolerance. This was because *OsCIPK12*-overexpressing mutant plants had the ability to accumulate higher proline and soluble sugar content, thus exhibiting better survival rates under drought stress than the wild type (Xiang et al. 2007). Another gene, *OsCIPK23,* was also shown to mediate rice drought tolerance (Yang et al. 2008). In this study, suppression of *OsCIPK23* expression significantly reduced seed set and increased rice plants' hypersensitivity to drought stress. It was also noted in the same study that expression of several drought stress–related genes was consistently induced when the *OsCIPK23* gene was overexpressed in rice plants.

In *Arabidopsis*, *CIPK9* expression was upregulated rapidly after 1 h, reaching a maximum at both 6 and 12 h of exposure to osmotic stress (mannitol), and then falling after 24 h. However, phenotypic analysis of both seedlings and adult plants under drought stress did not reveal any significant differences between *CIPK9* mutant plants and wild-type plants (Pandey et al. 2007). CBL1 and CBL9 interact with CIPK23 in plasma membranes in vivo, and their function was detected in various tissues, including guard cells and root hairs (Cheong et al. 2007). In a study using reverse genetic screening, loss of function of *cipk23* was found to reduce transpiration by regulating stomatal movement in an ABA-dependent manner. Moreover, *cipk23* mutant plants suffered a significant degree of growth impairment, accompanied by a reduction in the efficiency of K^+ uptake by the roots, when they were grown in K^+-deficient conditions (Cheong et al. 2007). Similar morphological and physiological responses have been observed in *akt1* mutant plants subjected to long-term water stress (Nieves-Cordones et al. 2012). Both of these studies suggested that the CBL1 and 9-CIPK23 complex negatively regulated AKT1—the K^+ transporter in root and stomatal guard cells—resulting in increased plant sensitivity to water stress (Cheong et al. 2007; Nieves-Cordones et al. 2012). Another study characterized the expression of *CIPK* genes in maize under water stress, showing that five *ZmCIPK* genes (*ZmCIPK1*, *-3*, *-8*, *-17*, and *-18*) were regulated by polyethylene glycol (PEG), $CaCl_2$, ABA, and H_2O_2, and that their expression was affected by ABA and H_2O_2, in an organ-dependent manner (Tai et al. 2013).

5.4.2 Role of CIPKs in Plant Adaptation to K^+ Deficiency

As an essential macronutrient for plant growth and development, potassium has two main functions in plant cells: a biophysical one such as in osmoregulation, and a biochemical one such as in protein synthesis and enzyme activation (Hakerlerler et al. 1997; Leigh and Wyn Jones 1984). The concentration of K^+ in the soil solution varies widely within a range of 0.1–6 mM. Larger quantities of K^+, however, can be accumulated by the plant and may constitute between 2% and 10% of plant dry weight (Leigh and Wyn Jones 1984). Potassium deficiency symptoms appear when potassium constitutes less than 10% of plant dry weight. The cytoplasmic K^+ concentration is, however, maintained at approximately 100 mM (Leigh and Wyn Jones 1984), while the vacuole retains a high K^+ concentration range of 20–500 mM for use under low-K^+ conditions (Walker et al. 1996). This variability in the vacuolar K^+ concentration reflects the potassium status of the plant. Under sufficient supply, the vacuole retains up to 200 mM for maintenance of the cytoplasmic K^+ concentration. Under K^+ deficiency conditions, however, it has been reported that vacuolar K^+ drops to 10–20 mM, a concentration at which it is no longer possible to maintain cytosolic K^+ (Leigh and Wyn Jones 1984; Walker et al. 1998). A significant number of selective and nonselective channels and transporters, localized in the plasma membrane and tonoplast, confer K^+ uptake efflux and long-distance transport (Shabala and Pottosin 2014). K^+ uptake functions can be performed either by a high-affinity mechanism, which is conferred by their transport function at a low external K^+ concentration of less than 0.2 mM (Epstein et al. 1963; Maathuis and Sanders 1994), or by a low-affinity mechanism, which is conferred by K^+ channels at a high external K^+ concentration of above 0.2 mM (Epstein et al. 1963).

It has been reported that the activities of ion channels and transporters are regulated to a major extent by phosphorylation and dephosphorylation (Lee et al. 2007; Ragel et al. 2015). Low-affinity, inward-rectifying K^+ channels such as the

Arabidopsis K^+ transporter AKT1 have been shown to be regulated by CBL–CIPK complexes (Lee et al. 2007; Li et al. 2006). Of these, *CIPK23*, *-6*, and *-16* phosphorylated and activated AKT1 in a BCL1-dependent manner, with *CIPK23* being the more effective gene. In contrast, the 2C-type protein phosphatase (PP2C), interacting with CIPK23, dephosphorylated and inactivated the AKT1 channel in the plant root (Lan et al. 2011; Lee et al. 2007). A model of the signaling pathway by which plants respond to low-K^+ conditions has been proposed. According to this model, under low-K^+ conditions, the calcium signature activates CBL and CBL2 calcium sensors, which interact and activate the protein kinase *CIPK23*. The CBL–CIPK23 complex phosphorylates and activates AKT1, thereby enhancing K^+ uptake in plants (Li et al. 2006). It has been consistently shown that loss of function of *cipk23*, *cbl1*, or *cbl9* leads to increased plant hypersensitivity to low-K^+ conditions (Xu et al. 2006). Since *OsAKT1* is modulated by the OsCBL1-OsCIPK23 complex, loss of function of either *Oscipk23* or *Osakt1* under a low-K^+ condition causes the appearance of similar K^+ deficiency symptoms (Li et al. 2014). Under a low-K^+ condition, CIPK23 is purported to activate the high-affinity K^+ transporter HAK5. Under this condition (10 μM), CBL1, -8, -9, and -10 Ca^{2+} sensors activate CIPK23, which phosphorylates the N terminus of *HAK5,* upregulating its activity; this results in an increase in K^+ influx (Ragel et al. 2015). Another study demonstrated that the CBL4–CIPK6 interaction modulates the activity of the AKT2 K^+ channel in plant cells by translocating AKT2 from the endoplasmic reticulum membrane to the plasma membrane and enhancing AKT2 activity in oocytes (Held et al. 2011). As this translocation is dependent upon dual lipid modifications of CBL4 (by myristoylation and palmitoylation), the Ca^{2+} sensor modulates K^+ channel activity in a kinase interaction–dependent and/or phosphorylation-independent manner (Held et al. 2011). The results of transcriptome analysis of the rice root response to potassium deficiency showed that eight *OsCIPK* genes (*OsCIPK2*, *-6*, *-9*, *-10*, *-14*, *-15*, *-23*, and *-26*) were upregulated, while two others (*OsCIPK29* and *-31*) were downregulated (Ma et al. 2012b). The transcript of *CIPK9*, a gene that is responsive to abiotic stress conditions, was induced in both the root and shoot of *Arabidopsis* under a low-K^+ condition (Pandey et al. 2007). This was also detected under a normal-K^+ condition (MS medium) (Liu et al. 2013). Although both K^+ uptake and K^+ content in the mutant plants remained unaffected by a low-K^+ condition (≤20 μM), the disruption in the function of *Oscipk9* increased K^+ plant hypersensitivity to the K^+-deficient condition (Pandey et al. 2007). In addition, *cipk9* mutant plants did not display any phenotypic change, in comparison with the wild type, under both salt and osmotic stress, suggesting that CIPK9 has a function related specifically to K^+. The double mutation in the tonoplast-localized CBL proteins *cbl2* and *cbl3* resulted in stunted plant growth with leaf tip necrosis and reduced root development, seed set, seed weight, and fatty acid content (Eckert et al. 2014; Tang et al. 2012); these are typical symptoms of nutrient deficiencies under normal growth conditions. Moreover, the double mutant *cbl2cbl3* showed hypersensitivity to an excessive level of external ions and reacted by altering its ionic profile to one that was more specific to K^+ accumulation. At the same time, this caused a defect to occur in the activity of vacuolar adenosine triphosphate (V-ATPase), which has been shown to play a role in Na^+ sequestration in the vacuoles of plant cells (Tang et al. 2012). In contrast, another study showed that overexpression of *CIPK9*, *CBL2*, and *CBL3* in *Arabidopsis* plants under a low-K^+ condition decreased their chlorophyll and K^+ content in comparison with the wild type, resulting in a low-K^+-sensitive phenotype (Liu et al. 2013). When grown under a low-K^+ condition, both *cbl3* and *cipk9* mutants consistently showed a similar low-K^+-tolerant phenotype, which still had green shoots and higher chlorophyll and K^+ content than the wild type (Liu et al. 2013). Moreover, this study established that both *CBL3* and *CBL2* interact with *CIPK9* to upregulate its expression in the tonoplast. Further analysis revealed strong expressions of *CBL2*, *CBL3*, and *CIPK9* in root vascular bundles. The root/shoot K^+ ratio was also much lower in the *cbl3* and *cipk9* mutants than in wild-type plants, under a low-K^+ condition, suggesting that these genes have a role in K^+ translocation (Liu et al. 2013). However, the regulatory mechanism of the CBL3–CIPK9 complex in

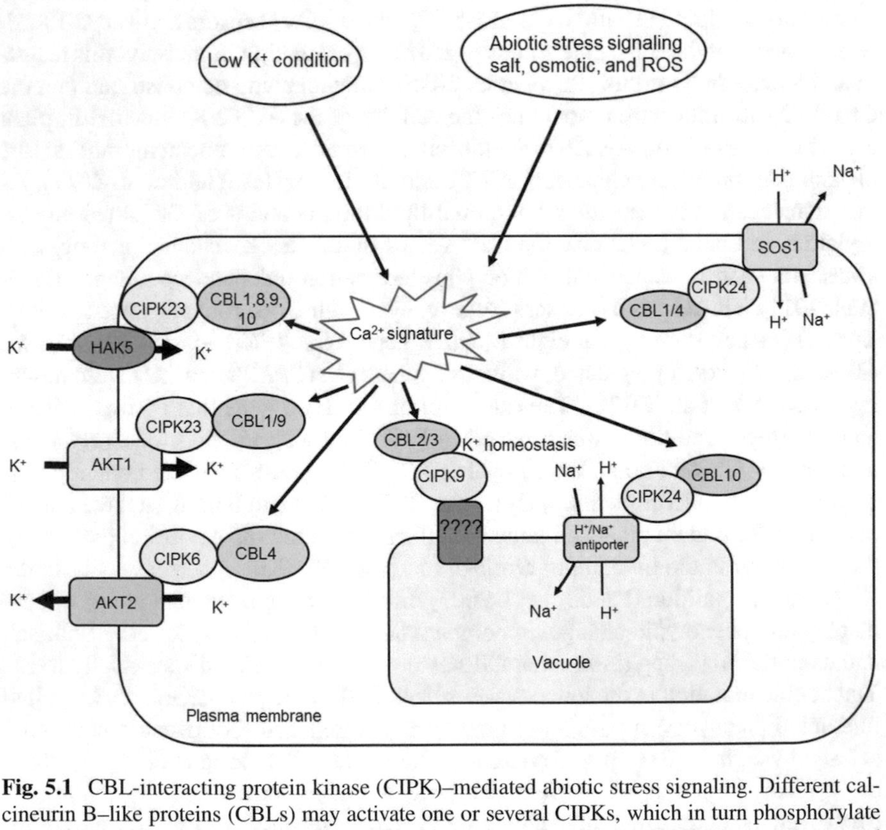

Fig. 5.1 CBL-interacting protein kinase (CIPK)–mediated abiotic stress signaling. Different calcineurin B–like proteins (CBLs) may activate one or several CIPKs, which in turn phosphorylate and activate different types of ion channels and transporters localized at the plasma membrane and the tonoplast of plant cells, to maintain ion homeostasis. *AKT Arabidopsis* K^+ transporter 1, *AKT2 Arabidopsis* K^+ transporter 2, *HAK5* high-affinity potassium transporter 5, *SOS1* salt overly sensitive 1 (an Na^+/H^+ antiporter). (Adapted from Zhu et al. (2013))

response to abiotic stress is not well understood, because neither the targeted proteins of the K^+ channel nor the K^+ transporters have been identified. In addition, when the interactions between CIPK9, K^+ transporters, and channels such as AKT1, AKT2, HAK5, SKOR, and TPK1 were tested, the results were negative (Liu et al. 2013; Pandey et al. 2007). Therefore, further studies are clearly needed to identify the role of CIPK9 in K^+ homeostasis.

5.5 Prospects and Conclusions

The potential roles of CIPKs in ion homeostasis in response to abiotic stresses such salinity, drought, and K^+ deficiency are summarized in Fig. 5.1. However, there is very little doubt that this model is incomplete and missing many other key players. In addition to CBL and CIPK, many other factors are known to control the activities of these plasma membrane–based and organelle-based ion transporters. These interactions should certainly be incorporated into the model. Interactions between CIPK and transport for other ions (specifically, Na and Ca) also need to be understood. Last, but not least, stress sensing and adaptation are known to be highly cell and tissue specific (Dinneny 2010; Shabala et al. 2016). Thus, CIPK signaling and control of ion transporters should always be considered in this context.

Acknowledgements This work was supported by the Australian Research Council and Australia–India Strategic Research Funding grants to Sergey Shabala.

References

Adams E, Diaz C, Matsui M, Shin R (2014) Overexpression of a novel component induces HAK5 and enhances growth in *Arabidopsis*. ISRN Bot 2014:9. https://doi.org/10.1155/2014/490252

Anschütz U, Becker D, Shabala S (2014) Going beyond nutrition: regulation of potassium homoeostasis as a common denominator of plant adaptive responses to environment. J Plant Physiol 171(9):670–687

Bañuelos MA, Klein RD, Alexander-Bowman SJ, Rodriguez-Navarro A (1995) A potassium transporter of the yeast *Schwanniomyces occidentalis* homologous to the Kup system of *Escherichia coli* has a high concentrative capacity. EMBO J 14:3021–3027

Bañuelos MA, Garciadeblas B, Cubero B, Rodríguez-Navarro A (2002) Inventory and functional characterization of the HAK potassium transporters of rice. Plant Physiol Biochem 130:784–795. https://doi.org/10.1104/pp.007781

Britto DT, Kronzucker HJ (2002) NH_4^+ toxicity in higher plants: a critical review. J Plant Physiol 159:567–584. https://doi.org/10.1078/0176-1617-0774

Bush DS (1995) Calcium regulation in plant cells and its role in signaling. Annu Rev Plant Biol 46:95–122

Chen G et al (2015) Rice potassium transporter OsHAK1 is essential for maintaining potassium-mediated growth and functions in salt tolerance over low and high potassium concentration ranges. Plant Cell Environ 38:2747–2765. https://doi.org/10.1111/pce.12585

Cheong YH et al (2007) Two calcineurin B–like calcium sensors, interacting with protein kinase CIPK23, regulate leaf transpiration and root potassium uptake in *Arabidopsis*. Plant J 52:223–239. https://doi.org/10.1111/j.1365-313X.2007.03236.x

Chérel I (2004) Regulation of K^+ channel activities in plants: from physiological to molecular aspects. J Exp Bot 55:337–351. https://doi.org/10.1093/jxb/erh028

DeFalco TA, Bender KW, Snedden WA (2010) Breaking the code: Ca^{2+} sensors in plant signalling. Biochem J 425:27–40

Demidchik V, Shabala SN, Coutts KB, Tester MA, Davies JM (2003) Free oxygen radicals regulate plasma membrane Ca^{2+}- and K^+-permeable channels in plant root cells. J Cell Sci 116:81–88

Demidchik V (2014) Mechanisms and physiological roles of K^+ efflux from root cells. J Plant Physiol 171(9):696–707. https://doi.org/10.1016/j.jplph.2014.01.015

Dennison KL, Robertson WR, Lewis BD, Hirsch RE, Sussman MR, Spalding EP (2001) Functions of AKT1 and AKT2 potassium channels determined by studies of single and double mutants of *Arabidopsis*. Plant Physiol 127:1012–1019

Desbrosses G, Josefsson C, Rigas S, Hatzopoulos P, Dolan L (2003) AKT1 and TRH1 are required during root hair elongation in *Arabidopsis*. J Exp Bot 54:781–788

Dinneny JR (2010) Analysis of the salt-stress response at cell-type resolution. Plant Cell Environ 33(4):543–551. https://doi.org/10.1111/j.1365-3040.2009.02055.x

Eckert C, Offenborn JN, Heinz T, Armarego-Marriott T, Schultke S, Zhang C (2014) The vacuolar calcium sensors CBL2 and CBL3 affect seed size and embryonic development in *Arabidopsis thaliana*. Plant J 78:146–156. https://doi.org/10.1111/tpj.12456

Epstein E, Rains DW, Elzam OE (1963) Resolution of dual mechanisms of potassium absorption by barley roots. Proc Natl Acad Sci 49:684–692. https://doi.org/10.1073/pnas.49.5.684

Fulgenzi FR, Peralta ML, Mangano S, Danna CH, Vallejo AJ, Puigdomenech P, Santa-María GE (2008) The ionic environment controls the contribution of the barley HvHAK1 transporter to potassium acquisition. Plant Physiol 147:252–262. https://doi.org/10.1104/pp.107.114546

Gechev TS, Van Breusegem F, Stone JM, Denev I, Laloi C (2006) Reactive oxygen species as signals that modulate plant stress responses and programmed cell death. Bioessays 28:1091–1101. https://doi.org/10.1002/bies.20493

Geiger D et al (2009) Heteromeric AtKC1. AKT1 channels in *Arabidopsis* roots facilitate growth under K^+-limiting conditions. J Biol Chem 284:21288–21295. https://doi.org/10.1074/jbc.M109.017574

Gierth M, Maser P, Schroeder JI (2005) The potassium transporter AtHAK5 functions in K^+ deprivation–induced high-affinity K^+ uptake and AKT1 K^+ channel contribution to K^+ uptake kinetics in *Arabidopsis* roots. Plant Physiol 137:1105–1114. https://doi.org/10.1104/pp.104.057216

Golldack D, Quigley F, Michalowski CB, Kamasani UR, Bohnert HJ (2003) Salinity stress–tolerant and –sensitive rice (*Oryza sativa L.*) regulate AKT1-type potassium channel transcripts differently. Plant Mol Biol 51:71–81

Golldack D, Li C, Mohan H, Probst N (2014) Tolerance to drought and salt stress in plants: unraveling the signaling networks. Front Plant Sci 5:151. https://doi.org/10.3389/fpls.2014.00151

Grabov A (2007) Plant KT/KUP/HAK potassium transporters: single family—multiple functions. Ann Bot 99:1035–1041. https://doi.org/10.1093/aob/mcm066

Gupta M et al (2008) KT/HAK/KUP potassium transporters gene family and their whole-life cycle expression profile in rice (*Oryza sativa*). Mol Genet Genomics 280:437–452. https://doi.org/10.1007/s00438-008-0377-7

Hakerlerler H, Oktay M, Eryüce N, Yagmur B (1997) Effect of potassium sources on the chilling tolerance of some vegetable seedlings grown in hotbeds. In: Johnston AE (ed) Food security in the WANA region, the essential need for balanced fertilization. International Potash Institute, Horgen, pp 317–327

He C, Cui K, Duan A, Zeng Y, Zhang J (2012) Genome-wide and molecular evolution analysis of the poplar KT/HAK/KUP potassium transporter gene family. Ecol Evol 2:1996–2004

Hedrich R, Anschütz U, Becker D (2011) Biology of plant potassium channels. In: Murphy SA, Schulz B, Peer W (eds) The plant plasma membrane. Springer Berlin Heidelberg, Berlin, pp 253–274. https://doi.org/10.1007/978-3-642-13431-9_11

Held K et al (2011) Calcium-dependent modulation and plasma membrane targeting of the AKT2 potassium channel by the CBL4/CIPK6 calcium sensor/protein kinase complex. Cell Res 21:1116–1130. http://www.nature.com/cr/journal/v21/n7/suppinfo/cr201150s1.html

Hirsch RE, Lewis BD, Spalding EP, Sussman MR (1998) A role for the AKT1 potassium channel in plant nutrition. Science 280:918–921

Kader MA, Lindberg S (2010) Cytosolic calcium and pH signaling in plants under salinity stress. Plant Signal Behav 5:233–238

Kolukisaoglu Ü, Weinl S, Blazevic D, Batistic O, Kudla J (2004) Calcium sensors and their interacting protein kinases: genomics of the *Arabidopsis* and rice CBL–CIPK signaling networks. Plant Physiol 134:43–58. https://doi.org/10.1104/pp.103.033068

Lagarde D, Basset M, Lepetit M, Conejero G, Gaymard F, Astruc S, Grignon C (1996) Tissue-specific expression of *Arabidopsis* AKT1 gene is consistent with a role in K^+ nutrition. Plant J 9:195–203. https://doi.org/10.1046/j.1365-313X.1996.09020195.x

Lan W-Z, Lee S-C, Che Y-F, Jiang Y-Q, Luan S (2011) Mechanistic analysis of AKT1 regulation by the CBL–CIPK–PP2CA interactions. Mol Plant 4:527–536. https://doi.org/10.1093/mp/ssr031

Lebaudy A, Very AA, Sentenac H (2007) K^+ channel activity in plants: genes, regulations and functions. FEBS Lett 581:2357–2366. https://doi.org/10.1016/j.febslet.2007.03.058

Lee SC et al (2007) A protein phosphorylation/dephosphorylation network regulates a plant potassium channel. Proc Natl Acad Sci U S A 104:15959–15964. https://doi.org/10.1073/pnas.0707912104

Leigh RA, Wyn Jones RG (1984) A hypothesis relating critical potassium concentrations for growth to the distribution and function of this ion in the plant cell. New Phytol 97:1–13. https://doi.org/10.1111/j.1469-8137.1984.tb04103.x

Li L, Kim B-G, Cheong YH, Pandey GK, Luan S (2006) A Ca^{2+} signaling pathway regulates a K^+ channel for low-K response in *Arabidopsis*. Proc Natl Acad Sci 103:12625–12630. https://doi.org/10.1073/pnas.0605129103

Li R, Zhang J, Wei J, Wang H, Wang Y, Ma R (2009) Functions and mechanisms of the CBL–CIPK signaling system in plant response to abiotic stress. Prog Nat Sci 19:667–676. https://doi.org/10.1016/j.pnsc.2008.06.030

Li J, Long Y, Qi GN, Xu ZJ, Wu WH (2014) The Os-AKT1 channel is critical for K^+ uptake in rice roots and is modulated by the rice CBL1–CIPK23 complex. Plant Cell 26:3387–3402. https://doi.org/10.1105/tpc.114.123455

Liu L-L, Ren H-M, Chen L-Q, Wang Y, Wu W-H (2013) A protein kinase, calcineurin B–like protein-interacting protein kinase9, interacts with calcium sensor calcineurin B–like protein3 and regulates potassium homeostasis under low-potassium stress in *Arabidopsis*. Plant Physiol 161:266–277. https://doi.org/10.1104/pp.112.206896

Ma L, Zhang H, Sun L, Jiao Y, Zhang G, Miao C, Hao F (2012a) NADPH oxidase AtrbohD and AtrbohF function in ROS-dependent regulation of Na^+/K^+ homeostasis in *Arabidopsis* under salt stress. J Exp Bot 63:305–317. https://doi.org/10.1093/jxb/err280

Ma T-L, Wu W-H, Wang Y (2012b) Transcriptome analysis of rice root responses to potassium deficiency BMC. Plant Biol 12:1–13. https://doi.org/10.1186/1471-2229-12-161

Maathuis FJ, Sanders D (1994) Mechanism of high-affinity potassium uptake in roots of *Arabidopsis thaliana*. Proc Natl Acad Sci U S A 91:9272–9276

Maathuis FJ, Ichida AM, Sanders D, Schroeder JI (1997) Roles of higher plant K^+ channels. Plant Physiol 114:1141–1149

Mangano S, Silberstein S, Santa-María GE (2008) Point mutations in the barley HvHAK1 potassium transporter lead to improved K^+-nutrition and enhanced resistance to salt stress. FEBS Lett 582:3922–3928. https://doi.org/10.1016/j.febslet.2008.10.036

Manik SMN, Shi S, Mao J, Dong L, Su Y, Wang Q, Liu H (2015) The calcium sensor CBL–CIPK is involved in plant's response to abiotic stresses. Int J Genomics 2015:10. https://doi.org/10.1155/2015/493191

Maser P et al (2001) Phylogenetic relationships within cation transporter families of *Arabidopsis*. Plant Physiol 126:1646–1667

Neumann PM (2011) Recent advances in understanding the regulation of whole-plant growth inhibition by salinity, drought and colloid stress. Adv Bot Res 57:33–48

Nieves-Cordones M, Miller AJ, Aleman F, Martinez V, Rubio F (2008) A putative role for the plasma membrane potential in the control of the expression of the gene encoding the tomato high-affinity potassium transporter HAK5. Plant Mol Biol 68:521–532. https://doi.org/10.1007/s11103-008-9388-3

Nieves-Cordones M, Aleman F, Martinez V, Rubio F (2010) The *Arabidopsis thaliana* HAK5 K^+ transporter is required for plant growth and K^+ acquisition from low K^+ solutions under saline conditions. Mol Plant 3:326–333. https://doi.org/10.1093/mp/ssp102

Nieves-Cordones M, Caballero F, Martínez V, Rubio F (2012) Disruption of the *Arabidopsis thaliana* inward-rectifier K^+ channel AKT1 improves plant responses to water stress. Plant Cell Physiol 53:423–432. https://doi.org/10.1093/pcp/pcr194

Ogasawara Y et al (2008) Synergistic activation of the *Arabidopsis* NADPH oxidase AtrbohD by Ca^{2+} and phosphorylation. J Biol Chem 283:8885–8892. https://doi.org/10.1074/jbc.M708106200

Pandey GK, Cheong YH, Kim B-G, Grant JJ, Li L, Luan S (2007) CIPK9: a calcium sensor-interacting protein kinase required for low-potassium tolerance in *Arabidopsis*. Cell Res 17:411–421

Perochon A, Aldon D, Galaud J-P, Ranty B (2011) Calmodulin and calmodulin-like proteins in plant calcium signaling. Biochimie 93:2048–2053. https://doi.org/10.1016/j.biochi.2011.07.012

Pilot G, Pratelli R, Gaymard F, Meyer Y, Sentenac H (2003) Five-group distribution of the shaker-like K^+ channel family in higher plants. J Mol Evol 56:418–434. https://doi.org/10.1007/s00239-002-2413-2

Pyo YJ, Gierth M, Schroeder JI, Cho MH (2010) High-affinity K^+ transport in *Arabidopsis*: AtHAK5 and AKT1 are vital for seedling establishment and postgermination growth under low-potassium conditions. Plant Physiol 153:863–875. https://doi.org/10.1104/pp.110.154369

Ragel P et al (2015) The CBL-interacting protein kinase CIPK23 regulates HAK5-mediated high-affinity K^+ uptake in *Arabidopsis* roots. Plant Physiol 169:2863–2873. https://doi.org/10.1104/pp.15.01401

Reddy VS, Reddy ASN (2004) Proteomics of calcium-signaling components in plants. Phytochemistry 65:1745–1776. https://doi.org/10.1016/j.phytochem.2004.04.033

Rodriguez-Navarro A (2000) Potassium transport in fungi and plants. Biochim Biophys Acta (BBA) Rev Biomembr 1469:1–30

Rodríguez-Navarro A, Rubio F (2006) High-affinity potassium and sodium transport systems in plants. J Exp Bot 57:1149–1160. https://doi.org/10.1093/jxb/erj068

Santa-María GE, Rubio F, Dubcovsky J, Rodriguez-Navarro A (1997) The HAK1 gene of barley is a member of a large gene family and encodes a high-affinity potassium transporter. Plant Cell 9:2281–2289

Schachtman DP, Schroeder JI, Lucas WJ, Anderson JA, Gaber RF (1992) Expression of an inward-rectifying potassium channel by the *Arabidopsis* KAT1 cDNA. Science 258:1654–1658

Schleyer M, Bakker EP (1993) Nucleotide sequence and 3′-end deletion studies indicate that the K^+-uptake protein kup from *Escherichia coli* is composed of a hydrophobic core linked to a large and partially essential hydrophilic C terminus. J Bacteriol 175:6925–6931

Schmöckel SM, Garcia AF, Berger B, Tester M, Webb AA, Roy SJ (2015) Different NaCl-induced calcium signatures in the *Arabidopsis thaliana* ecotypes Col-0 and C24. PloS One 10:e0117564

Schulz P, Herde M, Romeis T (2013) Calcium-dependent protein kinases: hubs in plant stress signaling and development. Plant Physiol 163:523–530. https://doi.org/10.1104/pp.113.222539

Sentenac H, Bonneaud N, Minet M, Lacroute F, Salmon JM, Gaymard F, Grignon C (1992) Cloning and expression in yeast of a plant potassium ion transport system. Science 256:663–665
Shabala S, Pottosin I (2014) Regulation of potassium transport in plants under hostile conditions: implications for abiotic and biotic stress tolerance. Physiol Plant 151:257–279. https://doi.org/10.1111/ppl.12165
Shabala L, Zhang JY, Pottosin I, Bose J, Zhu M, Fuglsang AT, Velarde-Buendia A, Massart A, Hill CB, Roessner U, Bacic A, Wu HH, Azzarello E, Pandolfi C, Zhou MX, Poschenrieder C, Mancuso S, Shabala S (2016) Cell-Type-Specific H+-ATPase Activity in Root Tissues Enables K^+ Retention and Mediates Acclimation of Barley (Hordeum vulgare) to Salinity Stress. Plant Physiol 172(4):2445–2458. https://doi.org/10.1104/pp.16.01347
Shabala S (2017) Signalling by potassium: another second messenger to add to the list? J Exp Bot 68(15):4003–4007. https://doi.org/10.1093/jxb/erx238
Spalding EP, Hirsch RE, Lewis DR, Qi Z, Sussman MR, Lewis BD (1999) Potassium uptake supporting plant growth in the absence of AKT1 channel activity: inhibition by ammonium and stimulation by sodium. J Gen Physiol 113:909–918. https://doi.org/10.1085/jgp.113.6.909
Szczerba MW, Britto DT, Ali SA, Balkos KD, Kronzucker HJ (2008) NH_4^+-stimulated and -inhibited components of K^+ transport in rice (*Oryza sativa* L.) J Exp Bot 59:3415–3423. https://doi.org/10.1093/jxb/ern190
Szyroki A et al (2001) KAT1 is not essential for stomatal opening. Proc Natl Acad Sci U S A 98:2917–2921. https://doi.org/10.1073/pnas.051616698
Tai F, Wang Q, Yuan Z, Yuan Z, Li H, Wang W (2013) Characterization of five CIPK genes expressions in maize under water stress. Acta Physiol Plant 35:1555–1564. https://doi.org/10.1007/s11738-012-1197-2
Tang RJ, Liu H, Yang Y, Yang L, Gao XS, Garcia VJ (2012) Tonoplast calcium sensors CBL2 and CBL3 control plant growth and ion homeostasis through regulating V-ATPase activity in *Arabidopsis*. Cell Res 22:1650–1665. https://doi.org/10.1038/cr.2012.161
Tuteja N, Mahajan S (2007) Calcium signaling network in plants: an overview. Plant Signal Behav 2:79–85
Very AA, Sentenac H (2002) Cation channels in the *Arabidopsis* plasma membrane. Trends Plant Sci 7:168–175
Véry AA, Sentenac H (2003) Molecular mechanisms and regulation of K^+ transport in higher plants. Annu Rev Plant Biol 54:575–603. https://doi.org/10.1146/annurev.arplant.54.031902.134831
Walker DJ, Leigh RA, Miller AJ (1996) Potassium homeostasis in vacuolate plant cells. Proc Natl Acad Sci 93:10510–10514
Walker DJ, Black CR, Miller AJ (1998) The role of cytosolic potassium and pH in the growth of barley roots. Plant Physiol 118:957–964
Wang M, Zheng Q, Shen Q, Guo S (2013) The critical role of potassium in plant stress response. Int J Mol Sci 14:7370–7390
Wang X-P et al (2016) AtKC1 and CIPK23 synergistically modulate AKT1-mediated low potassium stress responses in *Arabidopsis*. Plant Physiol 170:2264–2277. https://doi.org/10.1104/pp.15.01493
Whalley HJ, Knight MR (2013) Calcium signatures are decoded by plants to give specific gene responses. New Phytologist 197:690–693. https://doi.org/10.1111/nph.12087
Xia J et al (2014) Nitric oxide negatively regulates AKT1-mediated potassium uptake through modulating vitamin B_6 homeostasis in *Arabidopsis*. Proc Natl Acad Sci 111:16196–16201. https://doi.org/10.1073/pnas.1417473111
Xiang Y, Huang Y, Xiong L (2007) Characterization of stress-responsive CIPK genes in rice for stress tolerance improvement. Plant Physiol 144:1416–1428. https://doi.org/10.1104/pp.107.101295
Xu J, Li HD, Chen LQ, Wang Y, Liu LL, He L (2006) A protein kinase, interacting with two calcineurin B–like proteins, regulates K^+ transporter AKT1 in *Arabidopsis*. Cell 125:1347–1360. https://doi.org/10.1016/j.cell.2006.06.011

Yang W, Kong Z, Omo-Ikerodah E, Xu W, Li Q, Xue Y (2008) Calcineurin B–like interacting protein kinase OsCIPK23 functions in pollination and drought stress responses in rice (*Oryza sativa* L.). J Genet Genomics 35:531–543, s531–532. https://doi.org/10.1016/s1673-8527(08)60073-9

Yang Z, Gao Q, Sun C, Li W, Gu S, Xu C (2009) Molecular evolution and functional divergence of HAK potassium transporter gene family in rice (*Oryza sativa* L.) J Genet Genomics 36:161–172. https://doi.org/10.1016/s1673-8527(08)60103-4

Yang T et al (2014) The role of a potassium transporter OsHAK5 in potassium acquisition and transport from roots to shoots in rice at low potassium supply levels. Plant Physiol 166:945–959. https://doi.org/10.1104/pp.114.246520

Zhang Z, Zhang J, Chen Y, Li R, Wang H, Wei J (2012) Genome-wide analysis and identification of HAK potassium transporter gene family in maize (*Zea mays* L.) Plant Mol Biol Report 39:8465–8473. https://doi.org/10.1007/s11033-012-1700-2

Zhu S, Zhou X, Wu X, Jiang Z (2013) Structure and function of the CBL–CIPK Ca^{2+}-decoding system in plant calcium signaling. Plant Mol Biol Report 31:1193–1202. https://doi.org/10.1007/s11105-013-0631-y

Chapter 6
Plant Hormones: Potent Targets for Engineering Salinity Tolerance in Plants

Abdallah Atia, Zouhaier Barhoumi, Ahmed Debez, Safa Hkiri, Chedly Abdelly, Abderrazak Smaoui, Chiraz Chaffei Haouari, and Houda Gouia

Abstract Climate change has intensified the frequency and severity of many abiotic stresses. Soil salinity is a major abiotic stress affecting crop productivity worldwide. This leads to significant yield reductions, which have been reported in major cereal species such as wheat, maize, and barley. Meanwhile, the global human population is expected to rise above 9 billion by 2050. Average living standards are also increasing, with impacts on food consumption. Thus, there is an urgent need to further increase crop productivity. To meet these goals, it is imperative to develop new crops that have improved resistance to salt stress. Many plant scientists now believe that modern biotechnological approaches such as molecular breeding and genetic engineering offer the possibility to achieve these goals. Plant hormones play an important physiological role in regulating plant growth and development, and in coordinating plant responses to environmental conditions. These hormones—

A. Atia (✉)
Department of biology, College of science, King Khalid University, Abha, Saudi Arabia

Research Unit, Nutrition and Nitrogen Metabolism and Stress Protein, Department of Biology, Faculty of Sciences of Tunis, Campus Universitaire El Manar I, Tunis, Tunisia

Laboratory of Extremophile Plants, Centre of Biotechnology of Borj-Cedria, Hammam-Lif, Tunisia

Z. Barhoumi
Department of biology, College of science, King Khalid University, Abha, Saudi Arabia

Laboratory of Extremophile Plants, Centre of Biotechnology of Borj-Cedria, Hammam-Lif, Tunisia

S. Hkiri · C. C. Haouari · H. Gouia
Research Unit, Nutrition and Nitrogen Metabolism and Stress Protein, Department of Biology, Faculty of Sciences of Tunis, Campus Universitaire El Manar I, Tunis, Tunisia

A. Debez · C. Abdelly · A. Smaoui
Laboratory of Extremophile Plants, Centre of Biotechnology of Borj-Cedria, Hammam-Lif, Tunisia

V. Kumar et al. (eds.), *Salinity Responses and Tolerance in Plants, Volume 1*,
https://doi.org/10.1007/978-3-319-75671-4_6

including abscisic acid, gibberellins, auxins, cytokinins, ethylene, brassinosteroids, and jasmonates—are very important for providing adaptive responses under salt stress. Plant hormones may prove to be important metabolic engineering targets for producing abiotic stress–tolerant crop plants. This chapter discusses the physiological roles of plant hormones in salinity responses and recent success in engineering plant hormones for salinity tolerance in plants.

Keywords Abscisic acid · Auxins · Adaptive responses to salt stress · Brassinosteroids · Cytokinins · Crop productivity · Ethylene · Gibberellins · Genetic engineering · Jasmonates · Plant growth and development · Plant hormones · Soil salinity · Salinity tolerance

Abbreviations

12-OPDA	12-Oxo phytodienoic acid
13-HPOT	13-Hydroperoxy-9,11,15-octadecatrienoic acid
α-LeA	α-Linolenic acid
AAO	Abscisic acid–aldehyde oxidase
ABA	Abscisic acid
ACC	Aminocyclopropane-1-carboxylic acid
ACO	Aminocyclopropane-1-carboxylic acid oxidase
ACS	Aminocyclopropane-1-carboxylic acid synthase
ADP	Adenosine diphosphate
AHK	*Arabidopsis* histidine kinases receptor
AHP	Histidine-containing phospho-transfer protein
AO	Allene oxide
AOC	Allene oxide cyclase
AOS	Allene oxide synthase
APX	Ascorbate peroxidase
ARR	*Arabidopsis* response regulator
ATP	Adenosine triphosphate
BIN2	Brassinosteroid-insensitive 2
BL	Brassinolide
BR	Brassinosteroid
BZR1	Brassinazole-resistant 1
CAT	Catalase
CK	Cytokinin
CKX	Cytokinin oxidase/dehydrogenase
CRF	Cytokinin response factor
cZ	*Cis*-zeatin
EBR	24-Epibrassinolide (EBL)

ECe	Electrical conductivity of a saturated soil extract
ERF	Ethylene response factor
ET	ethylene
FMO	Flavin monooxygenase
GA	Gibberellin
HSD	11-b-Hydroxysteroid dehydrogenase
IAA	Indole-3-acetic acid.
IAM	Indole-3-acetamide
IAOx	Indole-3-acetaldoxime
iPA	Isopentenyl adenine
IPA	Indole-3-pyruvic acid
IPP	Isopentyl diphosphate
IPT	Adenosine phosphate isopentenyl transferase
JA	Jasmonate/jasmonic acid
JA-ACC	Jasmonoyl aminocyclopropane-1-carboxylic acid
JA-Ile	Jasmonoyl isoleucine
LOX	13-Lipoxygenase
MAPK	Mitogen-activated protein kinase
MCSU	Molybdenum cofactor sulfurase
MDA	Malondialdehyde
MeJA	Methyl jasmonate
NCED	9-*Cis*-epoxycarotenoid dioxygenase
NR	Nitrate reductase
RES	Reactive electrophile species
ROS	Reactive oxygen species
SA	Salicylic acid
S-AdoMet	*S*-adenosyl-methionine
SnRK2.4	Serine/threonine protein kinase gene
SOD	Superoxide dismutase
TAM	Tryptamine
tZ	*Trans*-zeatin
Z	Zeatin
ZEO	Zeaxanthin oxidase
ZEP	Zeaxanthin epoxidase
ZR	Zeatin riboside

6.1 Introduction

Environmental stresses—including cold, drought, and salinity—are the most severe agricultural problems affecting plant growth and crop yield (Mao et al. 2010). Climate change has intensified the frequency and severity of many abiotic stresses such as salinity, with significant yield reductions reported in major cereal species such as wheat, maize, and barley (Jewell et al. 2010). Salt-affected soils impact nearly 10% of the earth's land surface (950 million hectares) and 50% of all irrigated land (230 million hectares) in the world (Ruan et al. 2010). Salinity is distributed largely throughout coastal salt marshes and inland desert sands. These have arisen naturally through mineral weathering, which leads to the release of soluble salts such as sodium salts, chlorides, magnesium salts, sulfates, and carbonates (Munns and Tester 2008). On irrigated land, secondary salinization occurs when irrigation practices cause salt accumulation and tree clearing on agricultural land causes water tables to rise and concentrates salts in the root zone (Rengasamy 2006). Soils are classified as saline when the electrical conductivity of a saturated soil extract (ECe) is 4 dS/m or more, which is equivalent to approximately 40 mM NaCl and generates an osmotic pressure of approximately 0.2 MPa. This salt concentration significantly reduces the yield of most crops (Munns and Tester 2008). The global annual losses in agricultural production from salt-affected land are in excess of US$120 billion (Flowers et al. 2010). Meanwhile, the global human population is expected to rise above 9 billion by 2050. Average living standards are also increasing, with impacts on food consumption, the demand for grain for livestock sustenance and, ultimately, agricultural land use. Consequently, an average annual increase in cereal production of 44 million metric tons is required (Cominelli et al. 2012). Therefore, the needs for prevention of crop losses and production of more food and feed to meet the demands of ever-increasing human populations have gained more and more attention (Wani et al. 2017). Investigating how salinity affects plant growth and development at the physiological and molecular levels is an important approach to increase the productivity of crops, because salinity causes widespread crop losses throughout the world (Flowers et al. 2010; Cominelli et al. 2012; Wani et al. 2016).

Plant hormones are a group of small molecules that play important physiological roles in regulating plant growth and development and coordinating plant responses to environmental conditions. They act either close to or remote from their site of synthesis to perform their function (Colebrook et al. 2014; Kumar et al. 2016; Raja et al. 2017). Recently, plant hormones have received great attention. In fact, they are of great significance for plants' stress responses and adaptation. Usually, salicylic acid (SA), jasmonates (JAs), and ethylene are associated with plant defense, whereas gibberellins (GAs), auxins, brassinosteroids (BRs), and cytokinins (CKs) are associated with plant development. For many years, abscisic acid (ABA) was considered the key hormone that regulates plant responses to abiotic stresses (Kazan 2015). However, it is becoming increasingly evident that all plant hormones can play important roles under abiotic stress conditions (Raja et al. 2017; Wani et al. 2017). Plants are able to regulate and coordinate both growth and/or stress tolerance by variations in hormone production, distribution, or signal transduction, which consequently promote survival or escape from environmental stress (Colebrook et al. 2014).

Currently, under the increasingly environmentally challenging conditions, there is an urgent need to further increase crop productivity. Thus, to meet this goal, it is imperative to develop new crops that have improved water use efficiency and improved resistance against water scarcity and salinity (Cominelli et al. 2012). This goal might be attained by conventional plant breeding approaches. In fact, using the traditional methods of crossing and selecting progeny, breeders have produced new varieties with improved ability to resist stresses (Cominelli and Tonelli 2010). However, in recent decades, many plant scientists have concluded that modern biotechnological approaches such as molecular breeding and genetic engineering offer the possibility to achieve these goals in a shorter time (Cominelli et al. 2012).

This chapter focus on the physiological roles of plant hormones in salinity response and discusses recent success in engineering of plant hormones for salinity tolerance in plants.

6.2 Abscisic Acid

Abscisic acid (ABA) is a type of metabolite known as an isoprenoid or terpenoid (Vishwakarma et al. 2017). Various enzymes are involved in utilizing β-carotene to synthesize ABA. ABA was first isolated and identified from cotton balls in 1960. Conversion of β-carotene to ABA is mediated via a number of steps (Vishwakarma et al. 2017) catalyzed by the enzymes zeaxanthin oxidase (ZEO), 9-*cis*-epoxycarotenoid dioxygenase (NCED), ABA-aldehyde oxidase (AAO), and molybdenum cofactor sulfurase (MCSU) (Tuteja 2007). Under normal conditions, ABA plays important roles in plant growth and development (Basu and Rabara, 2017). ABA regulates seed maturation, seed germination, seedling growth, flowering, and stomatal movements (Finkelstein et al. 2002; Wani et al. 2017). Seeds imbibition leads to rapidly decreased ABA content in the seeds, which facilitates germination (Ali-Rachedi et al. 2004). Exposure of seeds to ABA during germination arrests radicle elongation and growth, but reversibly (Finkelstein et al. 2002; Atia et al. 2009). ABA is well known as a hormone that enables plants to survive severe environmental conditions such as salt and drought stresses (Raghavendra et al. 2010; Basu and Rabara, 2017). Salinity causes osmotic stress and water deficit, increasing the endogenous ABA level (Gupta and Huang 2014). The elevated ABA hormone level helps plants to close their stomata and accumulate numerous proteins and osmoprotectants for osmotic adjustment (Ruy and Chao 2015). Accumulation of ABA can mitigate the inhibitory effect of salinity on photosynthetic activity, growth (Gupta and Huang 2014), and many other physiological functions, including mineral nutrition and translocation of assimilates. Following salt stress, ABA accumulation lead to accumulation of K^+, Ca^{2+}, and compatible solutes such as proline and sugars in root cells, which may counteract the uptake of Na^+ and Cl^- (Gurmani et al. 2011).

ABA is a vital cellular signal that modulates the expression of a number of salt responsive genes (Fig. 6.1). In fact, ABA induces the expression of genes encoding ion transporters and hemostasis, thus enhancing ion-selective absorption and contributing to the transfer of Na^+ from the cytoplasm to the vacuole or its discharge from the plant. For instance, in *Hordeum vulgare*, ABA controls the allocation of toxic Na^+ ions in the vacuole by controlling the expression of two genes—*HVP1*

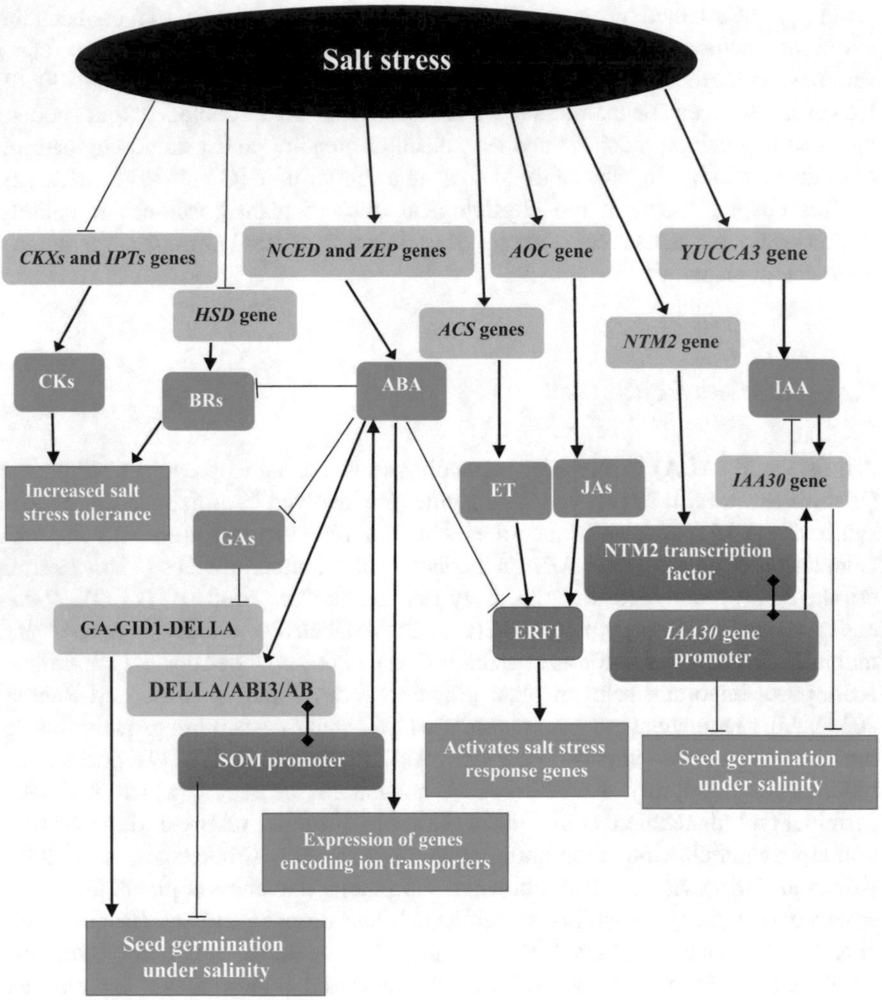

Fig. 6.1 Schematic model summarizing the roles of plant hormones in regulation of salt stress tolerance in plants. See text for additional details

and *HVP10*—for vacuolar H^+-inorganic pyrophosphatase, and expression of *HvVHA-A* for the catalytic subunit (subunit A) of vacuolar H^+-ATPase (Fukuda and Tanaka 2006). Yang et al. (2014) reported that ABA controls ATPase activities across the plasma membrane and tonoplast, and improves proton pump activity, which provides more power for Na^+/H^+ antiport, thus simultaneously enhancing selective absorption of K^+. Recently, Pons et al. (2013) showed that in rice, ABA enhances H^+ pumping under salt conditions. This positive effect occurs across the tonoplast and is more pronounced in the more tolerant rice lines. The authors showed that strong activation of Na^+/H^+ antiport activity occurs under salinity across the tonoplast and the plasma membrane. The authors concluded that under saline conditions, ABA seems to synergistically affect H^+ pumping and antagonistically affect Na^+ extrusion. Furthermore, ABA activates the expression of genes encoding ROS-scavenging enzymes—including superoxide dismutase (SOD), catalase (CAT), and ascorbate peroxidase (APX)—and increases the activity of these enzymes (Jiang and Zhang 2002; Raja et al. 2017). This alleviates the effect of salt-induced ROS accumulation and prevents cell damage. Since plant hormones such as ABA are key regulators of responses to salinity, hormone metabolism and signaling processes are potent targets for engineering salinity tolerance in plants (Kumar et al. 2016). ABA is a vital cellular signal that modulates the expression of a number of salt-responsive and water deficit–responsive genes (Basu and Rabara 2017). Salinity stress rapidly increases ABA accumulation following the activation of biosynthesis-related genes such as ZEO, NCED, and AAO. Currently, genes for the key enzymes involved in ABA synthesis are used in plant transformation to improve plant stress tolerance. For instance, NCED is the key enzyme in the ABA biosynthesis pathway. The transformation of *Nicotiana tabacum* by SgNCED from *Stylosanthes guianensis* has been shown to improve salinity tolerance in transgenic plants. This is accompanied by an increase in ABA biosynthesis and decreases in stomatal conductance, transpiration, and photosynthesis. Transgenic plants have been reported to exhibit increased activities of SOD, CAT, and APX, and high production of H_2O_2 and NO (Zhang et al. 2008) (Table 6.1). The transformation of *Arabidopsis thaliana* by MhNCED3 from *Malus hupehensis* Rehd has been shown to improve salinity tolerance in transformed plants. They displayed greater endogenous ABA content, nitric oxide generation rates, AtNIA1 transcript levels, and nitrate reductase (NR) activity than wild-type plants (Zhang et al. 2015) (Table 6.1). Zeaxanthin epoxidase (ZEP) is an enzyme in ABA biosynthesis and in the xanthophyll cycle. This enzyme is required for an initial step in ABA synthesis from isopentyl diphosphate (IPP) and β-carotene. Transformation of *A. thaliana* by the AtZEP gene improved the response to salinity in transgenic plants, which exhibited more vigorous growth than wild-type plants under high-salt and drought treatments (Park et al. 2008) (Table 6.1). *N. tabacum* plants overexpressing the MsZEP gene from *Medicago sativa* showed high water content, lower malondialdehyde (MDA) content, and induction of two salt tolerance genes, DREB and P5CS (Zhang et al. 2015) (Table 6.1). *AtLOS5* encoding MCSU is a key regulator of ABA biosynthesis. The transformation of *Zea mays* by this gene from *A. thaliana* improved K^+ retention in host plants and maintained a high cytosolic K^+/Na^+ ratio under salt stress (Zhang et al. 2016) (Table 6.1).

The ABA hormone controls the induction of ROS-scavenging activity through signaling processes. Signal molecules such as H_2O_2 and NO are involved in ABA-induced stomatal closure and gene expression, and the activities of antioxidant enzymes. ABA-induced H_2O_2 production mediates NO generation, which in turn activates mitogen-activated protein kinase (MAPK) and results in upregulation of the expression and activities of antioxidant enzymes (Zhang et al. 2008). In *Oryza sativa*, expression of the OsC3H47 gene, which encodes a signal protein belonging to the CCCH zinc finger family, is induced by drought, NaCl, and ABA. Overexpression of OsC3H47 significantly enhanced tolerance of salt stresses

Table 6.1 Examples of salt resistance genes implicated in plant hormone metabolism or plant hormone signaling, identified by functional analysis

Salt resistance gene	Function	Source species	Transformed species	Physiological responses of transformed plants under salinity	Reference
SgNCED1	NCED is the key enzyme in the ABA biosynthesis pathway	*Stylosanthes guianensis*	*Nicotiana tabacum*	Plant transformation resulted in improved ABA biosynthesis, which led to lower levels of stomatal conductance, transpiration, and photosynthesis. Transgenic plants also exhibited increased activities of SOD, CAT, and APX and production of H_2O_2 and NO, as well as tolerance of drought and salinity stresses	Zhang et al. (2008)
MhNCED3	NCED is the key enzyme in the ABA biosynthesis pathway	*Malus hupehensis* Rehd	*Arabidopsis thaliana*	Transgenic plants were more tolerant of salinity than wild-type plants. They displayed greater endogenous ABA content, nitric oxide generation, AtNIA1 transcript levels, and NR activity than wild-type plants	Zhang et al. (2015)
AtZEP	ZEP is an enzyme involved in ABA biosynthesis and in the xanthophyll cycle	*Arabidopsis thaliana*	*Arabidopsis thaliana*	Transgenic lines exhibited more vigorous growth under high-salt and drought treatments than wild-type plants and had much higher expression of the endogenous stress-responsive genes RD29A and Rab18 than wild-type plants under salt stress	Park et al. (2008)
Arabidopsis CBF1	cDNA encoding *CBF1*	*Arabidopsis thaliana*	*Lycopersicon esculentum*	Transformation enhanced tolerance of chilling, water deficit, and salt stress	Lee et al. (2003)
At*YUCCA6*	YUCCA, a Flavin monooxygenase–like protein, is involved in the IAA biosynthesis pathway	*Arabidopsis thaliana*	Poplar (*Populus alba* × *Populus glandulosa*)	Transgenic plants exhibited tolerance of osmotic stress, which was associated with reduced ROS levels	Ke et al. (2015)
AtHSD	HSD1 gene encoding a protein with homology to animal HSD	*Arabidopsis thaliana*	*Brassica napus*	Transgenic plants showed increases in growth and seed yield, as well as increased tolerance of saline stress and reduced seed dormancy	Li et al. (2007)

AtLOS5	*AtLOS5* encoding MCSU is a key regulator of ABA biosynthesis	*Arabidopsis thaliana*	*Zea mays*	Transgenic plants maintained a high cytosolic K^+/Na^+ ratio under salt stress and had higher leaf water potential and turgor, correlating with greater biomass accumulation under salt stress	Zhang et al. (2016)
MsZEP	Zeaxanthin epoxidase gene	*Medicago sativa*	*Nicotiana tabacum*	Transgenic plants had higher leaf water content and lower MDA content. They had higher expression levels of two salt tolerance genes, DREB and P5CS	Zhang et al. (2015)
TaSnRK2.4	SNF1-type serine/threonine protein kinase	*Triticum aestivum*	*Arabidopsis thaliana*	Transgenic *Arabidopsis* showed more tolerance of drought, salt, and freezing stresses than wild-type plants. They also showed decreased water loss, higher relative water content, strengthened cell membrane stability, improved photosynthesis potential, and a significant increase in osmotic potential	Mao et al. (2010)
OsC3H47	Signal protein belonging to the CCCH zinc finger family	*Oryza sativa*	*Oryza sativa*	Overexpression of OsC3H47 significantly enhanced tolerance of drought and salt stresses in rice seedlings, indicating that OsC3H47 plays important roles in post-stress recovery	Wang et al. (2015)
EIN3	Ethylene-Insensitive 3	*Arabidopsis thaliana*	*Arabidopsis thaliana*	Transgenic plants overexpressing EIN3 (EIN3ox) showed high salt tolerance during germination when compared with the Columbia ecotype (Col-0), a single ein3 mutant allele (ein3-1), and an ein3 and eil1 double mutant (ein3-1eil1-1)	Lin et al. (2013)

(continued)

Table 6.1 (continued)

Salt resistance gene	Function	Source species	Transformed species	Physiological responses of transformed plants under salinity	Reference
ERF1	Transcription factors regulating expression of ethylene-dependent genes	*Arabidopsis thaliana*	*Arabidopsis thaliana*	Overexpression of ERF1 increased drought and salt stress tolerance in transformed plants compared with wild-type plants	Cheng et al. (2013)
SlIPT3	Isopentenyltransferase enzyme involved in the first step in CK biosynthesis	*Solanum lycopersicum* L.	*Solanum lycopersicum* L.	35S::SlIPT3 tomato plants showed high salt tolerance, high accumulation of different CK metabolites, high photosynthetic pigment levels, and a higher K^+/Na^+ ratio than wild-type tomato	Žižková et al. (2015)
TαAOC1	AOC enzyme	*Triticum aestivum*	*Arabidopsis thaliana*	Transgenic plants showed high JA levels and increased salt sensitivity in comparison with wild-type plants. This transformation also suppressed expression of the stress-responsive genes AtRAB18, AtCBF1, and AtCBF3	Chen et al. (2014)
TaAOC1	AOC enzyme	*Triticum aestivum*	*Arabidopsis thaliana*	Transformation increased JA content and improved salt tolerance in *Arabidopsis*	Zhao et al. (2014)
TaOPR1	12-Oxo-phytodienoic acid reductase	*Triticum aestivum*	*Arabidopsis thaliana*	Transformation enhanced root growth in the presence of salinity, promoted ABA synthesis, and reduced MDA and ROS levels in *Arabidopsis*	Dong et al. (2013)
OsCYP94C2b	Gene encoding an OsCYP94C1-related enzyme that catalyzes conversion of JA-Ile to an inactive form	*Oryza sativa*	*Oryza sativa*	Transgenic plants showed high salt tolerance and a high survival rate	Kurotani et al. (2015)

ABA abscisic acid, *AOC* allene oxide cyclase, *APX* ascorbate peroxidase, *CAT* catalase, *CBF1* CRT/DRE binding factor 1, *cDNA* complementary DNA, *CK* cytokinin, *HSD* 11-b-hydroxysteroid dehydrogenase, *IAA* indole-3-acetic acid, *JA* jasmonate/jasmonic acid, *JA-Ile* jasmonoyl isoleucine, *MCSU* molybdenum cofactor sulfurase, *MDA* malondialdehyde, *NCED* 9-*cis*-epoxycarotenoid dioxygenase, *NR* nitrate reductase, *ROS* reactive oxygen species, *SOD* superoxide dismutase, *ZEP* zeaxanthin epoxidase in rice seedlings by decreasing ABA sensitivity (Wang et al. 2015) (Table 6.1). This gene may play an important role in ABA feedback and post-transcription processes leading to salinity tolerance (Wang et al. 2015). The serine/threonine protein kinase gene *SnRK2.4* is implicated in ABA signaling. Mao et al. (2010) showed that transgenic *Arabidopsis* overexpressing this gene exhibited enhanced tolerance of abiotic stress, including salinity. This was accompanied by a decrease in water loss, strengthened cell membrane stability, improvement of photosynthetic potential, and a significant increase in osmotic potential (Table 6.1).

6.3 Gibberellins

Gibberellins (GAs) are a large group of tetracyclic diterpenoid carboxylic acids. GAs play a major role in developmental processes in plants, including seed germination, leaf expansion, stem elongation, flower and trichome initiation, pollen maturation, and fruit development (Wani et al. 2016). GAs were first identified in the pathogenic fungus *Gibberella fujikuroi*—the causal agent of the 'foolish-seedling' disease of rice, causing excessive elongation of infected plants (Yabuta and Sumiki 1938). Since that discovery, 136 GAs have been identified in plants, fungi, and bacteria. Most GAs are precursors of bioactive forms or deactivated metabolites, and only a few (GA1, GA3, GA4, and GA7) have biological activity (Yamaguchi 2008). Interestingly, there is increasing evidence that they play vital roles in abiotic stress response and adaptation. GAs act as antagonists to ABA. GAs enhance seed germination under salt stress through different mechanisms such as by inducing synthesis of some enzymes and stimulating H^+-ATPase activity across the tonoplast (Yang et al. 2014). Under low salinity, GA_3 reduces the stomatal resistance of leaves, accelerates transpiration, and increases water use efficiency, thus improving the salt tolerance of plants (Yang et al. 2014). In tomato plants, application of GA decreased stomatal resistance and improved water use efficiency under low salinity (Maggio et al. 2010). Treatment with GAs may increase crop growth and yield under saline conditions. The positive effect of GA_3 is related to multiple effects such as decreased activity of ribonuclease and increases in reducing sugars, activity of enzymatic antioxidants, and protein synthesis (Fahad et al. 2014b).

During the past decade, most of the components of the GA signaling pathway have been identified. They include the GA receptor GIBBERELLIN INSENSITIVE DWARF1 (GID1), the DELLA growth inhibitors (DELLAs), the F-box proteins SLEEPY1 (SLY1) and SNEEZY (SNZ) in *A. thaliana*, and GIBBERELLIN INSENSITIVE DWARF2 (GID2) in *O. sativa* (Achard and Genschik 2009). The current model of GA action proposes that DELLA proteins induce plant growth inhibition whereas GAs stimulate growth by overcoming DELLA-mediated growth inhibition (Achard and Genschik 2009) (Fig. 6.1). In fact, the interaction of DELLA proteins with ABI3 and ABI5 led to the discovery of the regulatory mechanism governed by the balance between ABA and GA hormones. DELLA, ABI3, and ABI5 form a protein complex that binds the promoter and activates the transcription of target genes such as SOMNUS (SOM)—a C3H-type zinc finger—which negatively regulates seed germination (Fig. 6.1). According to the current model, unfavorable conditions such as drought and salinity increase ABA accumulation and decrease GA content. This allow formation of DELLA/ABI3/ABI5 complexes on the SOM promoter and activates its transcription (reviewed by Davière and Achard 2016). Interestingly, because SOM also activates ABA biosynthesis and represses GA biosynthesis (Kim et al. 2008), SOM might form a positive feedback that inhibits seed germination under unfavorable conditions such as high salinity (Lim et al. 2013). However, how do GAs suppress the negative effect of ABA on germination? Recently, the characterization of the GA-insensitive dwarfism gid1-1 mutant allele in rice led to the discovery of the GA receptor GID1 (Ueguchi-Tanaka et al. 2005). In brief, the perception of GA binding to GID1 allows the formation of the GA-GID1-DELLA complex (Fig. 6.1). This leads to degradation of the DELLAs by 26S proteasome, overcoming their growth-restraining effects (Davière and Achard 2016). Hence, the genes controlling the synthesis of these proteins may be potent targets for utilization in transgenic engineering to improve salt stress responses in cultivated plants.

6.4 Ethylene

Ethylene is a simple gaseous hormone, which plays multiple roles in regulation of plant growth and development, and also serves as a key modulator between plant responses to environmental stresses and normal growth (Abeles et al. 1992). Physiological effects are detectable at ambient levels as low as 0.1 $\mu l.l^{-1}$. The first discovery of ethylene effects dates back to over a century ago when Neljubow (1901) showed that ethylene causes horizontal growth of pea seedlings, inhibition of elongation, and radial swelling. Ethylene controls ripening, abscission, senescence, and many aspects of vegetative growth. In fact, this hormone inhibits primary and lateral root elongation and induces root hair development (Vandenbussche et al. 2012). Furthermore, ethylene stimulates germination of dormant seeds and alleviates salt-induced inhibition of germination.

In plants, ethylene is synthesized through three enzymatic reaction steps: methionine is converted to *S*-adenosyl-methionine (*S*-AdoMet) by *S*-AdoMet synthetase; then the direct precursor of ethylene, aminocyclopropane-1-carboxylic acid (ACC), is synthesized from *S*-AdoMet by ACC synthase (ACS); finally, ethylene is produced through the oxidation of ACC by ACC oxidase (ACO) (Lin et al. 2009). The ethylene pathway positively or negatively affects salt stress tolerance (Kazan 2015). In fact, several key steps in ethylene biosynthesis could be affected by salinity. However, as a rate-limiting enzyme, ACS is the major target for regulation of ethylene production under stress (Tao et al. 2015) (Fig. 6.1). There are eight functional *ACS* genes in *Arabidopsis*. They play an important role in ACC biosynthesis (Tsuchisaka et al. 2009). The regulation of these genes under stress occurs at both transcriptional and post-transcriptional levels. Transcripts of *ACS2* and *ACS7* in *Arabidopsis* were increased dramatically following salt stress application (Achard et al. 2006). In tobacco, transcripts of *ACS1* were induced by salinity (Cao et al. 2006). Recent work in *Arabidopsis* found that four *ACSs* (*ACS2*, *ACS6*, *ACS7*, and *ACS8*) were induced by high salinity, while a low-salinity pretreatment alleviated this induction (Shen et al. 2014). This may mean that induction of ethylene production is necessary for plant response and adaptation to salinity stress (Shen et al. 2014). Additionally, some *ACSs* are also regulated post-transcriptionally under salinity, mainly through stress-activated MAPK cascades, which phosphorylate ACS protein to elevate ethylene production and then prevent 26S proteasome–mediated degradation (Liu and Zhang 2004).

In *Arabidopsis*, ethylene is perceived by a family of five receptors: two ethylene receptors (ETR1 and ETR2), two ethylene response sensors (ERS1 and ERS2), and Ethylene-Insensitive4 (EIN4). They are located at the Golgi and endoplasmic reticulum membranes. Ethylene binding is proposed to inhibit receptor function. The downstream component of the ethylene signaling pathway includes CTR1, EIN2, EIN3/EIL (EIN-Like), and ERF1 transcription factors (Lei et al. 2011). EIN2 is considered to play a central role in the ethylene signaling transduction pathway, which acts downstream from CTR1. Ethylene signaling is subsequently transduced into the nucleus to cause the accumulation of two master transcriptional activators—EIN3 and EIL1—which initiate transcriptional reprogramming in various ethylene responses.

Ethylene response factors (ERFs), a huge multigene family of transcription factors, regulate the expression of ethylene-dependent genes (Klay et al. 2014). ERFs are related to the salt stress response through ethylene signaling (Fig. 6.1). In *A. thaliana*, ERF1 was highly induced by high salinity and drought stress. ERF1-overexpressing lines (35S:ERF1) were more tolerant of drought and salt stress than wild-type plants (Cheng et al. 2013).

Currently, whether ethylene negatively affects salt tolerance or positively regulates salt response remains controversial. Kazan (2015) reported that ethylene negatively affects salt tolerance, because a correlation between increased ACC levels and reduced salt tolerance was found in *Arabidopsis*. In addition, the *acs7* mutant, with significantly reduced ethylene levels, shows increased salt tolerance during germination, suggesting that this gene negatively regulates salt tolerance. On the other hand, overexpression of the TαAOC1 gene, which encodes the wheat allene oxide cyclase (AOC) enzyme, increased salt sensitivity in *Arabidopsis* (Chen et al. 2014) (Table 6.1). However, ethylene could positively affect salt tolerance. For instance, Peng et al. (2014) reported that in *A. thaliana*, high salinity induced accumulation of EIN3/EIL1 proteins (two ethylene-activated transcription factors) and EBF1/2 protein degradation in an EIN2-independent manner. Moreover, EIN3 activation prevents excess ROS accumulation and increases salt tolerance. Lin et al. (2013) showed that in *A. thaliana*, the *ein3-1eil1-1* double mutant (lacking EIN3 and EIN3-Like1) and the *ein3-1* mutant (lacking EIN3) were hypersensitive to 150 mM NaCl when compared with wild-type plants; in contrast, an EIN3 overexpression mutant (EIN3ox) showed a high germination percentage under salinity when compared with wild-type plants. Furthermore, the two EIN3-deficient mutant seedlings accumulated high levels of hydrogen peroxide, which was thought to be an inhibitor of germination under salinity. It seems that EIN3 may function as a negative regulator of reactive oxygen species (ROS) metabolism in germinating seeds under salinity.

6.5 Cytokinins

Cytokinins (CKs) are adenine derivatives with either isoprenoids, which are widespread in nature, or aromatic side chains, such as N^6-(meta-hydroxybenzyl) adenine, which are found in plants at lower levels (Ha et al. 2012). Depending on hydroxylation and reduction of the side chain, isoprenoid CKs can be distinguished as isopentenyladenine (iPA)-, *trans*-zeatin (tZ)-, *cis*-zeatin (cZ)-, or dihydrozeatin-type derivatives (Ha et al. 2012). In plants, CK hormones are produced in the root tips and in developing seeds. They are translocated from the roots via the xylem to the shoot, where they regulate plant development and growth processes (Zahir et al. 2001). CKs are implicated in several plant growth and developmental processes, including cell division, chloroplast biogenesis, apical dominance, leaf senescence, vascular differentiation, nutrient mobilization, shoot differentiation, anthocyanin production, and photomorphogenic development (Fahad et al. 2014a, b). CKs are also known to alleviate the adverse effects of salinity on plant growth.

The key enzymes involved in CK metabolism are adenosine phosphate-isopentenyltransferases (IPTs) and CK oxidases/dehydrogenases (CKXs) (Werner and Schmülling 2009). *Arabidopsis* plants have two classes of IPTs acting on the adenine moiety, with seven genes for adenosine triphosphate/adenosine diphosphate (ATP/ADP) IPTs (IPT1, IPT3, IPT4, IPT5, IPT6, IPT7, and IPT8) and two genes for transfer RNA IPTs (IPT2 and IPT9). The ATP/ADP IPTs are required for the synthesis of iPA- and tZ-type CKs, and the transfer RNA IPTs are required for the synthesis of cZ-type CKs (Miyawaki et al. 2006). The iPA and tZ-type CKs are the major forms in dicotyledonous species, and cZ-type CKs are the major forms in monocotyledonous species (Sakakibara 2006). Formation of tZ by hydroxylation of iPA requires cytochrome P450 monooxygenases (Takei et al. 2004). CK catabolism is catalyzed by CKX enzymes. They are important players in regulating the CK pool and thereby influence plant growth and development. The CKX enzymes are encoded by a family of seven genes (CKX1–CKX7).

In plants, CKs are involved in responses to various environmental cues, including salinity (Žižková et al. 2015). For instance, seed priming with CKs can enhance salt tolerance in plants (Javid et al. 2011). In barley, salinity significantly decreased zeatin (Z), zeatin riboside (ZR), and iPA levels in shoots and roots. The addition of benzyl adenine suppressed the negative effects of salinity on growth and internal CK content (Kuiper et al. 1990). Cytokinin biosynthesis *IPT* genes can be upregulated by NaCl treatment (Nishiyama et al. 2011). Recent progress in genetic engineering of CKs has enabled control of plant CK content, increasing yield and improving plant adaptation to salt stress (Fig. 6.1) (Žižková et al. 2015; Sylva et al. 2017). Transcriptome analyses of IPT genes in *A. thaliana* under salt stress showed upregulation of AtIPT1, AtIPT2, and AtIPT8 and downregulation of AtIPT3, AtIPT5, AtIPT7, and AtIPT9 (Nishiyama et al. 2012). *Arabidopsis ipt*-deficient plants showed strong inhibition of shoot growth, elongation of primary and lateral roots, and reductions in tZ and iPA content (Miyawaki et al. 2006). Overexpression of IPT genes in plants resulted in faster shoot formation, shorter internodes, loss of apical dominance, delay of leaf senescence, higher photosynthetic rates, and accumulation of tZ and its riboside (Žižková et al. 2015). For instance, 35S::SlIPT3 tomato plants showed high salt tolerance, high accumulation of different CK metabolites, high accumulation of photosynthetic pigments, and a high K^+/Na^+ ratio (Table 6.1).

The components of cytokinin signaling also play diverse roles in responses to abiotic stresses including salinity. They include the type-B *Arabidopsis* response regulator (ARR) transcription factors and *Arabidopsis* histidine kinase receptors (AHKs). CKs are perceived at the plasma membrane by specific receptors, and the signal is transduced via type-B ARRs controlling transcription of type-A ARRs, which act as negative feedback regulators of CK signaling (Hwang et al. 2012). ARR transcription factors respond in different ways to salt stress and have been reported to regulate sodium accumulation in *Arabidopsis* (Mason et al. 2010). Salinity stress shows a strong impact on expression levels of CK receptors (Zalabák et al. 2013)—namely, AHK receptors. For instance, in *M. sativa*, downregulation of *AHK2* (histidine kinase 2 of cytokinin signaling) and *AHK4* and upregulation of *AHK3,* as well as its orthologue, were observed (Coba de la Peña et al. 2008; Argueso et al. 2009). *Arabidopsis* AHK1 acts as a positive regulator in responses to salt stress. A recent molecular study confirmed that in *Arabidopsis*, cytokinin signaling involves a set of three AHK receptors, which autophosphorylate upon binding of cytokinin. The phosphate is then shuttled from the AHK receptors via histidine-containing phospho-transfer proteins (AHPs) to ARRs. The type-B ARRs are transcription factors that, together with cytokinin response factors (CRFs) (Rashotte et al. 2006), regulate the transcription of cytokinin primary response genes, which include type-A ARRs (To and Kieber 2008).

6.6 Jasmonates

Jasmonates (JAs) are lipid-derived hormones and play multiple roles in regulation of plant development and plant responses to biotic and abiotic stresses (Wasternack and Strnad 2015). Jasmonic acid (JA) and its methyl ester, methyl jasmonate (MeJA) are ubiquitously found in the plant kingdom (Pirbalouti et al. 2014). MeJA and JA were isolated for the first time from jasmine (*Jasminum grandiflorum*) oil (Avanci et al. 2010) and culture of the fungus *Lasiodiplodia theobromae* (Tsukada et al. 2010), respectively. Other JA forms—particularly *cis*-jasmone, jasmonoyl ACC (JA-ACC), and jasmonoyl isoleucine (JA-Ile), with multiple biological functions—are present in plants (Ahmad et al. 2016). JAs influence developmental processes such as growth, lateral and adventitious root formation, seed germination, leaf senescence, glandular trichome formation, and embryo and pollen development (Wasternack and Strnad 2015).

JAs and MeJA are derivatives of fatty acid metabolism (Jalalpour et al. 2014). They are produced from α-linolenic acid (α-LeA) localized in chloroplast membranes (Wasternack and Kombrink 2010). The formation of α-LeA from lipids occurs through the action of phospholipases. An intermediate compound, 13-hydroperoxy-9,11,15-octadecatrienoic acid (13-HPOT), is formed by the addition of an oxygen molecule to α-LeA by the 13-lipoxygenase (LOX) enzyme. 13-HPOT is then oxidized to allene oxide (AO) by allene oxide synthase (AOS). AO is then converted to 12-oxo phytodienoic acid (12-OPDA) by the AOC enzyme (Wasternack and Hause 2013).

JA is an important hormone with versatile functions in development and in the response to environmental challenges, including drought and salinity (Valenzuela et al. 2016). In the two last decades, the involvement of JA in salt stress has been confirmed. The application of exogenous JAs improved salinity tolerance in soybean (Yoon et al. 2009) and rice (Kang et al. 2005). In rice, JA levels were higher in a salt-tolerant cultivar than in a salt-sensitive cultivar (Kang et al. 2005). In addition, the level of endogenous JAs increased under high salinity in rice roots (Moons et al. 1997) and in tomato (Pedranzani et al. 2007). Exogenous JA application effectively reduced sodium ion concentrations in rice plants under salinity (Kang et al. 2005), improved photosynthetic activity in several crops (Javid et al. 2011), and enhanced the activities of antioxidant enzymes such as SOD, POD, CAT, and APX in wheat (Qiu et al. 2014). Therefore, JA hormones could act as effective protectants against salt-mediated adverse effects in plants. At the molecular level, some JA biosynthesis genes (e.g., *AOC1*, *AOC2*, *AOS*, *LOX3*, and *OPR3*) are upregulated in roots under salt stress (Valenzuela et al. 2016). Overexpression of the wheat (*Triticum aestivum*) TaAOC1 gene (which encodes an AOC enzyme) in *Arabidopsis* resulted in an increase in JA levels and improved salt tolerance, suggesting that JAs positively regulate salt tolerance (Table 6.1). Thus, it seems that the JA signaling pathway is stimulated by salinity, which may induce physiological and growth changes in plants (Fig. 6.1). However, in a study of two grapevine cell lines with different salt tolerance levels, JA accumulation was greater in the salt-sensitive *Vitis riparia* than in the salt-tolerant *Vitis rupestris*. It seems that it is not the presence or absence of JA that decides salinity response but the right timing and control (Ismail et al. 2014). Recently the effects of salinity on two JA biosynthesis rice mutants (*cpm2* and *hebiba*) that were impaired in the function of ALLENE OXIDE CYCLASE (AOC), and their wild type, were studied. The two mutants accumulated significantly lower sodium ion concentrations in the shoots and showed higher chlorophyll content than the wild type. They also showed better scavenging of ROS under salt stress when compared with wild-type plants. Under stress conditions, the leaves of the wild type and JA mutants accumulated similar levels of ABA, and the levels of JA and its amino acid conjugate, JA-Ile, were not changed in the wild-type plants. However, the wild-type plants responded to salinity by strong induction of the JA precursor OPDA, in comparison with the two mutants. This was correlated with increased ROS-scavenging activity (such as glutathione *S*-transferase activity), lower level of

H_2O_2, and lower levels of malondialdehyde in the mutant plants (Hazman et al. 2015). OPDA is considered one of the highly reactive electrophile species (RES) responsible for signaling in chloroplasts (Farmer and Mueller 2013). OPDA can induce retrograde signaling when bound to its putative receptor, cyclophilin 20-3 (Park et al. 2013), leading to high sensitivity to salt stress.

6.7 Auxins

Auxins are plant hormones, with indole-3-acetic acid (IAA) being the predominant form. It regulates several aspects of plant growth and development, and is vital not only for plant growth and development but also for governing and/or coordinating plant growth under stress conditions (Wani et al. 2016). It regulates numerous complex plant processes such as apical dominance, lateral/adventitious root formation, tropisms, fruit set and development, and vascular differentiation, as well as embryogenesis.

Although IAA has been studied for over 100 years, its biosynthesis, transport, and signaling pathways are still not well understood (Kumar et al. 2016). Several interconnecting pathways have been proposed to synthesize IAA in plants, including tryptophan-dependent and tryptophan-independent pathways (Zhao 2010). Tryptophan-dependent IAA biosynthesis can proceed via four metabolic intermediates: indole-3-acetaldoxime (IAOx), indole-3-pyruvic acid (IPA), tryptamine (TAM), and indole-3-acetamide (IAM) (Zhao 2012). A major pathway in *A. thaliana* generates IAA in two reactions from tryptophan. Step 1 converts tryptophan to IPA—a reaction catalyzed by tryptophan aminotransferases—and step 2 converts IPA to IAA, catalyzed by members of the YUCCA family of flavin monooxygenases (FMOs) (Kim et al. 2013). Other proposed tryptophan-dependent pathways include cytochrome P450s, nitrilases, aldehyde oxidase, and IPA decarboxylase, but those pathways are not yet completely understood (Kim et al. 2013). IAA plays an important role in plant responses to biotic and abiotic stresses. IAA is involved in response to salt stress in plants (Fig. 6.1). It was reported that wheat seed germination declined with higher salinity levels, while this adverse effect was reversed by pretreatment of seeds with IAA (Ashraf and Foolad 2005). In corn plants, exogenous IAA application reduced some of the salt-induced adverse effects by enhancing essential inorganic nutrients, as well as by maintaining membrane permeability (Kaya et al. 2009). Overexpression of IAA biosynthesis–related YUCCA3 caused hypersensitivity to salt stress with an increase in IAA concentrations (Jung and Park 2011). In contrast, some studies have reported a significant reduction in IAA concentrations in crop plants such as rice and tomato (Kazan and Lyons 2014). Hence, salt-caused plant growth reduction could be a result of altered IAA biosynthesis, redistribution, and/or signaling. In *A. thaliana*, IAA does not affect the germination process in a salt-free medium. However, it acts as a negative regulator of seed germination under high salinity (Fig. 6.1). Recent work by Jung and Park (2011) showed that a membrane-bound NAC transcription factor, NTM2, mediates the signaling cross talk between IAA and salt stress via the *IAA30* gene during seed germination. In fact, in *A. thaliana*, germinated NTM2-deficient *ntm2-1* mutant seeds exhibited enhanced resistance to salt stress. However, salt resistance was reduced in *ntm2-1* mutants overexpressing the *IAA30* gene, which was induced by high salinity in an NTM2-dependent manner. In fact, in *A. thaliana* the transcript level of the *IAA30* gene was elevated under high salinity, and the salt induction of the *IAA30* gene largely disappeared in the *ntm2-1* mutant. These observations indicate that NTM2 is a molecular link that incorporates the IAA signal into salt stress signaling during seed germination, providing a role for IAA in modulating seed germination under high salinity, and the *IAA30* gene is a component of *NTM2*-mediated salt signaling.

6.8 Brassinosteroids

Brassinosteroids (BRs) are a newly discovered group of plant steroidal hormones, which are structurally similar to animal and insect steroids (Divi and Krishna 2009a). They are ubiquitous in the plant kingdom and regulate various aspects of plant growth and development, including cell elongation, photomorphogenesis, xylem differentiation, seed germination, and stress responses (Lopez-Gomez et al. 2016). More than 70 BRs have been isolated from plants. Brassinolide, 28-homobrassinolide, and 24-epibrassinolide are the most bioactive BRs. They are largely used in physiological studies (Wani et al. 2016). The first brassinosteroid to be isolated—brassinolide—was isolated in 1979 from *Brassica napus*, when Grove et al. (1979) showed that pollen from this species promoted stem elongation and cell division.

Brassinolide (BL), the most active BR, is synthesized from campesterol via several pathways. Recent studies have revealed the presence of two parallel pathways from campestanol to castasterone, called the early and late C-6 oxidation pathways (Divi and Krishna 2009a). Another two pathways—the early C-22 oxidation and C-23 oxidation pathways—have also been described (Divi and Krishna 2009a). The oxidation steps in BR biosynthesis are catalyzed by cytochrome P450 monooxygenases. The C-22 and C-23 hydroxylation reactions are mediated by the P450 DWF4 (Choe et al. 1998) and CPD genes (Szekeres et al. 1996), respectively. In *Arabidopsis* and tomato, brassinosteroid-6-oxidases catalyze multiple steps in BR biosynthesis: 6-deoxoteasterone to teasterone, 3-dehydro-6-deoxoteasterone to 3-dehydroteasterone, 6-deoxotyphasterol to typhasterol, and 6-deoxocastasterone to castasterone (Shimada et al. 2001). In the final step, castasterone is converted to brassinolide by lactonization of the B ring. The C-22 and C-23 hydroxylation and C-6 oxidation reactions are key regulatory steps in the BR biosynthesis pathway. Accordingly, the enzymes catalyzing these reactions are potential targets for engineering abiotic stress tolerance (Divi and Krishna 2009a).

Ameliorative effects of BRs under salt stress have been reported in various plant species, including *A. thaliana*, *B. napus*, *Brassica juncea*, *Solanum melongena*, *Capsicum annuum*, *Cucumis sativus*, *Phaseolus vulgaris*, and *Z. mays* (Ahammed et al. 2015). BR treatment improved salt stress tolerance in barley (Kulaeva et al. 1991), and improved germination rates in the presence of high salt were observed in *Eucalyptus camaldulensis* (Sasse et al. 1995) and *O. sativa* (Anuradha and Rao 2001). Treatment with 24-epibrassinolide (EBR) overcame salt stress–induced inhibition of seed germination in *A. thaliana* and *B. napus*. In *O. sativa*, 24-epibrassinolide (EBL) treatment enhanced growth, protein content, and proline content, reduced MDA content, and stimulated antioxidant enzyme activity under salinity (Sharma et al. 2013). Divi et al. (2010) showed that EBR abolished hypersensitivity to salt stress–induced inhibition of seed germination in ethylene-insensitive *ein2*. The positive effect of EBR was significantly greater in the ABA-deficient *aba1-1* mutant than in the wild type during seed germination under salt stress. This may indicate that ABA masks BR effects in plant salt stress responses (Fig. 6.1).

In recent decades, development of BR mutants in the model plant *Arabidopsis* has enabled researchers to carry out functional studies. Mutant plants with impaired BR biosynthetic capacity or impaired BR perception ability display severely abnormal phenotypes such as dwarfism, delays in flowering and senescence, low seed germination, decreased male fertility, and de-etiolation in the dark (Divi and Krishna 2009a). In contrast, overexpression of BR biosynthetic genes not only increases crop yield under normal conditions but also enhances stress tolerance under unfavorable conditions. For instance, overexpression of the AtHSD gene encoding a protein with homology to animal 11-b-hydroxysteroid dehydrogenase (HSD) from *A. thaliana* in *B. napus* increased growth and seed yield, as well as increasing tolerance of saline stress and reduced seed dormancy (Li et al. 2007) (Table 6.1). This confirmed that *AtHSD1* is linked to the BR biosynthesis pathway in plants. Hence, this class of hormones has the potential to enhance crop yield and confer tolerance of salinity, and their genetic manipulation presents a sound platform for producing high-yielding stress-tolerant crops (Kumar et al. 2016).

In plants, the major BR receptor BRI1 (brassinosteroid-insensitive 1) is a membrane-localized leucine-rich repeat receptor-like kinase (Li and Chory 1997). BR binding to BRI1 leads to dimerization of BRI1 with another receptor kinase, BAK1 (BRI1-associated receptor kinase 1), and kinase activation. This initiates a signaling cascade leading to gene expression. The downstream components of this cascade include the glycogen synthase kinase-3/SHAGGY-like kinase BIN2 (brassinosteroid-insensitive 2), the serine/threonine phosphatase BSU1 (bri1 suppressor 1), and the transcription factors BES1 (bri1-EMS-suppressor 1) and BZR1 (brassinazole-resistant 1) (Kagale et al. 2006). BIN2 negatively regulates BR signaling by phosphorylating BES1 and inhibiting its binding to BR target promoters, as well as its transcriptional activity (Vert and Chory 2006). Following BR binding, BES1 and BZR1 are rapidly dephosphorylated, probably by BSU1, leading to activation of their transcriptional activities (Divi and Krishna 2009b).

6.9 Conclusion

In recent decades, agriculture and food production have faced a large number of challenges, including increases in soil salinity. Biotechnological approaches are needed to overcome this problem. Plant hormones play vital roles in plant growth, development, and stress responses. They provide essential pathways adopted by plants for regulation of salt stress responses. During the past few years, the molecular mechanisms regulating hormonal synthesis, signaling, and action have been intensively investigated, and the roles of plant hormones in regulating responses to salt stress have been documented. However, despite the progress in the understanding of the role of plant hormones in plant responses to salt stress, there is still a large gap in our knowledge of hormone metabolism and action. More clarification is needed at the genetic level regarding the biosynthesis pathway of hormones such as IAA, the mechanism of upregulation of ABA biosynthesis genes by abiotic stress, and hormone homeostasis of GAs. The roles of cross talk signals between hormones and between hormones and secondary messengers (i.e., calcium or ROS) in regulating salt stress tolerance are not yet clear and also need to be investigated further.

References

Abeles S, Morgan PW, Saltveit ME (1992) Ethylene in plant biology. Academic Press, San Diego

Achard P, Genschik P (2009) Releasing the brakes of plant growth: how GAs shutdown DELLA proteins. J Exp Bot 60:1085–1092

Achard P, Cheng H, De Grauwe L et al (2006) Integration of plant responses to environmentally activated phytohormonal signals. Science 311:91–94

Ahammed GJ, Xia XJ, Li X et al (2015) Role of brassinosteroid in plant adaptation to abiotic stresses and its interplay with other hormones. Curr Protein Pept Sci 16:462–473

Ahmad P, Rasool S, Gul A et al (2016) Jasmonates: multifunctional roles in stress tolerance. Front Plant Sci 7:813. https://doi.org/10.3389/fpls.2016.00813

Ali-Rachedi S, Bouinot D, Wagner MH et al (2004) Changes in endogenous abscisic acid levels during dormancy release and maintenance of mature seeds: studies with the Cape Verde Islands ecotype, the dormant model of *Arabidopsis thaliana*. Planta 219:479–488

Anuradha S, Rao SSR (2001) Effect of brassinosteroids on salinity stress induced inhibition of germination and seedling growth of rice (*Oryza sativa* L.) Plant Growth Regul 33:151–153

Argueso CT, Ferreira JF, Kieber JJ (2009) Environmental perception avenues: the interaction of cytokinin and environmental response pathways. Plant Cell Environ 32:1147–1160

Ashraf M, Foolad MR (2005) Pre-sowing seed treatment—a shotgun approach to improving germination, plant growth and crop yield under saline and non-saline conditions. Adv Agron 88:223–271

Atia A, Debez A, Barhoumi Z et al (2009) ABA, GA3, and nitrate may control seed germination of *Crithmum maritimum* (Apiaceae) under saline conditions. C R Biol 332:704–710

Avanci NC, Luche DD, Goldman GH et al (2010) Jasmonates are plant hormones with multiple functions, including plant defense and reproduction. Genet Mol Res 9:484–505. https://doi.org/10.4238/vol9-1gmr754

Basu S, Rabara R (2017) Abscisic acid—an enigma in the abiotic stress tolerance of crop plants. Plant Gene. https://doi.org/10.1016/j.plgene.2017.04.008

Cao WH, Liu J, Zhou QY et al (2006) Expression of tobacco ethylene receptor NTHK1 alters plant responses to salt stress. Plant Cell Environ 29:1210–1219

Chen D, Ma X, Li C et al (2014) A wheat aminocyclopropane-1-carboxylate oxidase gene, TaACO1, negatively regulates salinity stress in *Arabidopsis thaliana*. Plant Cell Rep 33:1815–1827. https://doi.org/10.1007/s00299-014-1659-7

Cheng MC, Liao PM, Kuo WW et al (2013) The *Arabidopsis* ETHYLENE RESPONSE FACTOR1 regulates abiotic stress-responsive gene expression by binding to different *cis*-acting elements in response to different stress signals. Plant Physiol 162:1566–1582. https://doi.org/10.1104/pp.113.221911

Choe S, Dilkes BP, Fujioka S et al (1998) The DWF4 gene of *Arabidopsis* encodes a cytochrome P450 that mediates multiple 22alpha-hydroxylation steps in brassinosteroid biosynthesis. Plant Cell 10:231–243

Coba de la Peña T, Cárcamo CB, Almonacid L et al (2008) A salt stress-responsive cytokinin receptor homologue isolated from *Medicago sativa* nodules. Planta 227:769–779

Colebrook EH, Thomas SG, Phillips AL et al (2014) The role of gibberellin signalling in plant responses to abiotic stress. J Exp Biol 217:67–75

Cominelli E, Tonelli C (2010) Transgenic crops coping with water scarcity. New Biotechnol. https://doi.org/10.1016/j.nbt.2010.08.005

Cominelli E, Conti L, Tonelli C et al (2012) Challenges and perspectives to improve crop drought and salinity tolerance. New Biotechnol. https://doi.org/10.1016/j.nbt.2012.11.001

Davière JM, Achard P (2016) A pivotal role of DELLAs in regulating multiple hormone signals. Mol Plant 9:10–20. https://doi.org/10.1016/j.molp.2015.09.011

Divi, U.K. and Krishna, P. (2009) Brassinosteroids confer stress tolerance. In Plant Stress Biology (Hirt, H., ed.), Wiley–VCH Verlag GmbH & Co. KGaA, Weinheim pp. 119–135

Divi UK, Krishna P (2009a) Brassinosteroid: a biotechnological target for enhancing crop yield and stress tolerance. New Biotechnol 26:131–136

Divi UK, Krishna P (2009b) Brassinosteroids confer stress tolerance. In Plant Stress Biology (Hirt H, ed.), Wiley–VCH Verlag GmbH & Co. KGaA, Weinheim pp. 119–135

Divi UK, Rahman T, Krishna P (2010) Brassinosteroid-mediated stress tolerance in *Arabidopsis* shows interactions with abscisic acid, ethylene and salicylic acid pathways. BMC Plant Biol 10:151

Dong W, Wang M, Xu F et al (2013) Wheat oxophytodienoate reductase gene TaOPR1 confers salinity tolerance via enhancement of abscisic acid signaling and reactive oxygen species scavenging. Plant Physiol 161:1217–1228

Fahad S, Hussain S, Bano A et al (2014a) Potential role of plant hormones and plant growth–promoting rhizobacteria in abiotic stresses: consequences for changing environment. Environ Sci Pollut Res. https://doi.org/10.1007/s11356-014-3754-2

Fahad S, Hussain S, Matloob A et al (2014b) Plant hormones and plant responses to salinity stress: a review. Plant Growth Regul. https://doi.org/10.1007/s10725-014-0013-y

Farmer EE, Mueller MJ (2013) ROS-mediated lipid peroxidation and RES-activated signaling. Annu Rev Plant Biol 64:429–450

Finkelstein RR, Gampala SSL, Rock CD (2002) Abscisic acid signaling in seeds and seedlings. Plant Cell 14:S15–S45

Flowers TJ, Galal HK, Bromham L (2010) Evolution of halophytes: multiple origins of salt tolerance in land plants. Funct Plant Biol 37:604–612

Fukuda A, Tanaka Y (2006) Effects of ABA, auxin, and gibberellin on the expression of genes for vacuolar H^+-inorganic pyrophosphatase, H^+-ATPase subunit A, and Na^+/H^+ antiporter in barley. Plant Physiol Biochem 44(5–6):351–358

Grove MD, Spencer GF, Rohwedder WK et al (1979) Brassinolide, a plant growth promoting steroid isolated from *Brassica napus* pollen. Nature 281:216–217

Gupta B, Huang B (2014) Mechanism of salinity tolerance in plants: physiological, biochemical, and molecular characterization. Inter J Genomics. https://doi.org/10.1155/2014/701596

Gurmani AR, Bano A, Khan SU et al (2011) Alleviation of salt stress by seed treatment with abscisic acid (ABA), 6-benzylaminopurine (BA) and chlormequat chloride (CCC) optimizes ion and organic matter accumulation and increases yield of rice (*Oryza sativa* L.) Aus J Crop Sci 5(10):1278–1285

Ha S, Vankova R, Yamaguchi-Shinozaki K et al (2012) Cytokinins: metabolism and function in plant adaptation to environmental stresses. Trends Plant Sci. https://doi.org/10.1016/j.tplants.2011.12.005

Hazman M, Hause B, Eiche E et al (2015) Increased tolerance to salt stress in OPDA-deficient rice ALLENE OXIDE CYCLASE mutants is linked with an increased ROS-scavenging activity. J Exp Bot 66:3339–3352. https://doi.org/10.1093/jxb/erv142

Hwang I, Sheen J, Müller B (2012) Cytokinin signaling networks. Annu Rev Plant Biol 63:353–380

Ismail A, Seo M, Takebayashi Y et al (2014) Salt adaptation requires efficient fine-tuning of jasmonate signaling. Protoplasma 251:881–898. https://doi.org/10.1007/s00709-013-0591-y

Jalalpour Z, Shabani L, Khoraskani LA et al (2014) Stimulatory effect of methyl jasmonate and squalestatin on phenolic metabolism through induction of LOX activity in cell suspension culture of yew. Turk J Biol 38:76–82. https://doi.org/10.3906/biy-1306-91

Javid MG, Sorooshzadeh A, Moradi F et al (2011) The role of plant hormones in alleviating salt stress in crop plants. Aust J Crop Sci 5:726–734

Jewell MC, Bradley C, Campbell IDG (2010) Transgenic plants for abiotic stress resistance. In: Kole C et al (eds) Transgenic crop plants. https://doi.org/10.1007/978-3-642-04812-8_2

Jiang M, Zhang J (2002) Role of abscisic acid in water stress–induced antioxidant defence in leaves of maize seedlings. Free Radic Res 36:1001–1015. https://doi.org/10.1080/1071576021000006563

Jung JH, Park CM (2011) Auxin modulation of salt stress signaling in *Arabidopsis* seed germination. Plant Signal Behav 6:1198–1200

Kagale S, Divi UK, Krochko JE, Keller WA, Krishna P (2006) Brassinosteroid confers tolerance in *Arabidopsis thaliana* and Brassica napus to a range of abiotic stresses. Planta 225(2):353–364

Kang DJ, Seo YJ, Lee JD et al (2005) Jasmonic acid differentially affects growth, ion uptake and abscisic acid concentration in salt-tolerant and salt-sensitive rice cultivars. J Agron Crop Sci 191:273–282. https://doi.org/10.1111/j.1439-037X.2005.00153.x

Kaya C, Tuna AL, Yokas I (2009) The role of plant hormones in plants under salinity stress. In: Ashraf M, Ozturk M, Athar HR (eds) Salinity and water stress: improving crop efficiency. Springer, Berlin, pp 45–50

Kazan K, Lyons R (2014) Intervention of phytohormone pathways by pathogen effectors. Plant Cell 26(6):2285–2309

Kazan K (2015) Diverse roles of jasmonates and ethylene in abiotic stress tolerance. Trends Plant Sci. https://doi.org/10.1016/j.tplants.2015.02.001

Ke Q, Wang Z, Ji C et al (2015) Transgenic poplar expressing *Arabidopsis* YUCCA6 exhibits auxinoverproduction phenotypes and increased tolerance to abiotic stress. Plant Physiol Biochem 94:19–27. https://doi.org/10.1016/j.plaphy.2015.05.003

Kim DH, Yamaguchi S, Lim S et al (2008) SOMNUS, a CCCH-type zinc finger protein in *Arabidopsis*, negatively regulates light-dependent seed germination downstream of PIL5. Plant Cell 20:1260–1277

Kim JI, Baek D, Park HC et al (2013) Overexpression of *Arabidopsis* YUCCA6 in potato results in high-auxin developmental phenotypes and enhanced resistance to water deficit. Mol Plant 6(2):337–349

Klay I, Pirrello J, Riahi L et al (2014) Ethylene response factors Sl-ERFB3 is responsive to abiotic stresses and mediates salt and cold stress response regulation in tomato. Sci World J. https://doi.org/10.1155/2014/167681

Kuiper D, Schuit J, Kuiper PJC (1990) Actual cytokinin concentrations in plant tissue as an indicator for salt resistance in cereals. Plant Soil 123:243–250

Kulaeva ON, Burkhanova EA, Fedina AB et al (1991) Effect of brassinosteroids on protein synthesis and plant cell ultrastructure under stress conditions. In: Cutler HG, Yokota T, Adam G (eds) Brassinosteroids chemistry, bioactivity and application, ACS Symp Ser, vol 474. Am Chem Soc, Washington, pp 141–155

Kumar V, Sah SK, Khare T et al (2016) Engineering plant hormones for abiotic stress tolerance in crop plants. In: Ahammed GJ, Yu JQ (eds) Plant hormones under challenging environmental factors. https://doi.org/10.1007/978-94-017-7758-2_10

Kurotani K, Hayashi K, Hatanaka S et al (2015) Elevated levels of CYP94 family gene expression alleviate the jasmonate response and enhance salt tolerance in rice. Plant Cell Physiol 56:779–789

Lee JT, Prasad V, Yang PT et al (2003) Expression of *Arabidopsis* CBF1 regulated by an ABA/stress inducible promoter in transgenic tomato confers stress tolerance without affecting yield. Plant Cell Environ 26:1181–1190

Lei G, Shen M, Li ZG et al (2011) EIN2 regulates salt stress response and interacts with a MA3 domain-containing protein ECIP1 in *Arabidopsis*. Plant Cell Environ 34:1678–1692

Li J, Chory J (1997) A putative leucine-rich repeat receptor kinase involved in brassinosteroid signal transduction. Cell 90:929–938

Li F, Asami T, Wu X et al (2007) A putative hydroxysteroid dehydrogenase involved in regulating plant growth and development. Plant Physiol 145:87–97

Lim S, Park J, Lee N et al (2013) ABA-insensitive3, ABA-insensitive5, and DELLAs interact to activate the expression of SOMNUS and other high-temperature-inducible genes in imbibed seeds in *Arabidopsis*. Plant Cell 25:4863–4878

Lin Z, Zhong S, Grierson D (2009) Recent advances in ethylene research. J Exp Bot 60:3311–3336. https://doi.org/10.1093/jxb/erp204

Lin Y, Yang L, Chen D et al (2013) A role for Ethylene-Insensitive3 in the regulation of hydrogen peroxide production during seed germination under high salinity in *Arabidopsis*. Acta Physiol Plant 35:1701–1706. https://doi.org/10.1007/s11738-012-1176-7

Liu Y, Zhang S (2004) Phosphorylation of 1-aminocyclopropane-1-carboxylic acid synthase by MPK6, a stress responsive mitogen-activated protein kinase, induces ethylene biosynthesis in *Arabidopsis*. Plant Cell 16:3386–3399. https://doi.org/10.1105/tpc.104.026609

Lopez-Gomez M, Hidalgo-Castellanos J, Lluch C, Herrera-Cervera JA (2016) 24-Epibrassinolide ameliorates salt stress effects in the symbiosis *Medicago truncatula-Sinorhizobium meliloti* and regulates the nodulation in cross-talk with polyamines. Plant Physiol Biochem 108:212–221

Maggio A, Barbieri G, Raimondi G et al (2010) Contrasting effects of GA3 treatments on tomato plants exposed to increasing salinity. J Plant Growth Regul 29:63–72

Mao X, Zhang H, Tian S et al (2010) TaSnRK2.4, an SNF1-type serine/threonine protein kinase of wheat (*Triticum aestivum* L.), confers enhanced multistress tolerance in *Arabidopsis*. J Exp Bot 61:683–696

Mason MG, Jha D, Salt DE et al (2010) Type-B response regulators ARR1 and ARR12 regulate expression of AtHKT1;1 and accumulation of sodium in *Arabidopsis* shoots. Plant J 64:753–763

Miyawaki K, Tarkowski P, Matsumoto-Kitano M et al (2006) Roles of *Arabidopsis* ATP/ADP isopentenyltransferases and tRNA isopentenyltransferases in cytokinin biosynthesis. Proc Natl Acad Sci U S A 103(44):16598–16603

Moons A, Prinsen E, Bauw G et al (1997) Antagonistic effects of abscisic acid and jasmonates on salt stress-inducible transcripts in rice roots. Plant Cell 9:2243–2259. https://doi.org/10.1105/tpc.9.12.2243

Munns R, Tester M (2008) Mechanisms of salinity tolerance. Annu Rev Plant Biol 59:651–681

Neljubow D (1901) Ueber die horizontale Nutation der Stengel von *Pisum sativum* and einiger anderen Pflanzen. Beih Bot Zentralbl 10:128–139

Nishiyama R, Watanabe Y, Fujita Y et al (2011) Analysis of cytokinin mutants and regulation of cytokinin metabolic genes reveals important regulatory roles of cytokinins in drought, salt and abscisic acid responses, and abscisic acid biosynthesis. Plant Cell 23:2169–2183. https://doi.org/10.1105/tpc.111.087395

Nishiyama R, Le DT, Watanabe Y et al (2012) Transcriptome analyses of a salt-tolerantcytokinin-deficient mutant reveal differential regulation of salt stress response by cytokinin deficiency. PLoS One 7:e32124

Park HY, Seok HY, Park BK et al (2008) Overexpression of *Arabidopsis* ZEP enhances tolerance to osmotic stress. Biochem Biophys Res Commun 375:80–85

Park SW, Li W, Viehhauser A et al (2013) Cyclophilin 20-3 relays a 12-oxo-phytodienoic acid signal during stress responsive regulation of cellular redox homeostasis. Proc Natl Acad Sci U S A 110(23):9559–9564. https://doi.org/10.1073/pnas.1218872110

Pedranzani H, Sierra-de-Grado R, Vigliocco A et al (2007) Cold and water stresses produce changes in endogenous jasmonates in two population of *Pinus pinaster* Ait. Plant Growth Regul 52:111–116. https://doi.org/10.1007/s10725-007-9166-2

Peng J, Li Z, Wen X et al (2014) Salt-induced stabilization of EIN3/EIL1 confers salinity tolerance by deterring ROS accumulation in *Arabidopsis*. PLoS Genet. https://doi.org/10.1371/journal.pgen.1004664

Pirbalouti AG, Mirbagheri H, Hamedi B et al (2014) Antibacterial activity of the essential oils of myrtle leaves against *Erysipelothrix rhusiopathiae*. Asian Pac J Trop Biomed 4:505–509. https://doi.org/10.12980/APJTB.4.2014B1168

Pons R, Cornejo JM, Sanz A (2013) Is ABA involved in tolerance responses to salinity by affecting cytoplasm ion homeostasis in rice cell lines? Plant Physiol Biochem 62(2013):88–94

Qiu ZB, Guo JL, Zhu AJ et al (2014) Exogenous jasmonic acid can enhance tolerance of wheat seedlings to salt stress. Ecotoxicol Environ Saf 104:202–208

Raghavendra AS, Gonugunta VK, Christmann A et al (2010) ABA perception and signalling. Trends Plant Sci 15:395–401

Raja V, Majeed U, Kang H, Andrabi KI, John R (2017) Abiotic stress: interplay between ROS, hormones and MAPKs. Environ Exp Bot. https://doi.org/10.1016/j.envexpbot.2017.02.010

Rashotte AM, Mason MG, Hutchison CE et al (2006) A subset of *Arabidopsis* AP2 transcription factors mediates cytokinin responses in concert with a two-component pathway. Proc Natl Acad Sci U S A 103:11081–11085. https://doi.org/10.1073/pnas.0602038103

Rengasamy P (2006) World salinization with emphasis on Australia. J Exp Bot 57:1017–1023

Ruan CJ, Da Silva JAT, Mopper S et al (2010) Halophyte improvement for a salinized world. Crit Rev Plant Sci 29:329–359

Ruy H, Chao Y (2015) Plant hormones in salt stress tolerance. J Plant Biol 58:147–155. https://doi.org/10.1007/s12374-015-0103-z

Sakakibara H (2006) Cytokinins: activity, biosynthesis and translocation. Annu Rev Plant Biol 57:431–449

Sasse JM, Smith R, Hudson I (1995) Effects of 24-epibrassinolide on germination of seed of *Eucalyptus camaldulensis* in saline conditions. Proc Plant Growth Regul Soc Am 22:136–141

Sharma I, Ching E, Saini S et al (2013) Exogenous application of brassinosteroid offers tolerance to salinity by altering stress responses in rice variety Pusa Basmati-1. Plant Physiol Biochem 69:17–26

Shen X, Wang Z, Song X et al (2014) Transcriptomic profiling revealed an important role of cell wall remodeling and ethylene signaling pathway during salt acclimation in *Arabidopsis*. Plant Mol Biol 86:303–317. https://doi.org/10.1007/s11103-014-0230-9

Shimada Y, Fujioka S, Miyauchi N et al (2001) Brassinosteroid-6-oxidases from *Arabidopsis* and tomato catalyze multiple C-6 oxidations in brassinosteroid biosynthesis. Plant Physiol 126:770–779

Sylva P, Petre ID, Alena G, Petr H, Petr S, Vojtech K, Radomira V (2017) Hormonal dynamics during salt stress responses of salt-sensitive *Arabidopsis thaliana* and salt-tolerant *Thellungiella salsuginea*. Plant Sci. https://doi.org/10.1016/j.plantsci.2017.07.020

Szekeres M, Nemeth K, Koncz-Kalman Z et al (1996) Brassinosteroids rescue the deficiency of CYP90, a cytochrome P450, controlling cell elongation and de-etiolation in *Arabidopsis*. Cell 85:171–182

Takei K, Yamaya T, Sakakibara H (2004) *Arabidopsis* CYP735A1 and CYP735A2 encode cytokinin hydroxylases that catalyze the biosynthesis of *trans*-zeatin. J Biol Chem 279(40):41866–41872

Tao JJ, Chen HW, Ma B et al (2015) The role of ethylene in plants under salinity stress. Front Plant Sci 6:1059. https://doi.org/10.3389/fpls.2015.01059

To JPC, Kieber JJ (2008) Cytokinin signaling: two-components and more. Trends Plant Sci 13:85–92. https://doi.org/10.1016/j.tplants.2007.11.005

Tsuchisaka A, Yu G, Jin H et al (2009) A combinatorial interplay among the 1-aminocyclopropane-1-carboxylate isoforms regulates ethylene biosynthesis in *Arabidopsis thaliana*. Genetics 183:979–1003

Tsukada K, Takahashi K, Nabeta K (2010) Biosynthesis of jasmonic acid in a plant pathogenic fungus, *Lasiodiplodia theobromae*. Phytochemistry 71:2019–2023. https://doi.org/10.1016/j.phytochem.2010.09.013

Tuteja N (2007) Mechanisms of high salinity tolerance in plants. Methods Enzymol 428:419–438

Ueguchi-Tanaka M, Ashikari M, Nakajima M et al (2005) GIBBERELLIN INSENSITIVE DWARF1 encodes a soluble receptor for gibberellin. Nature 437:693–698

Valenzuela CE, Acevedo-Acevedo O, Miranda GS et al (2016) Salt stress response triggers activation of the jasmonate signaling pathway leading to inhibition of cell elongation in *Arabidopsis* primary root. J Exp Bot 67(14):4209–4220. https://doi.org/10.1093/jxb/erw202

Vandenbussche F, Vaseva I, Vissenberg K et al (2012) Ethylene in vegetative development: a tale with a riddle. New Phytol 194:895–909. https://doi.org/10.1111/j.1469-8137.2012.04100.x

Vert G, Chory J (2006) Downstream nuclear events in brassinosteroid signalling. Nature 441:96–100

Vishwakarma K, Upadhyay N, Kumar N et al (2017) Abscisic acid signaling and abiotic stress tolerance in plants: a review on current knowledge and future prospects. Front Plant Sci 8:161. https://doi.org/10.3389/fpls.2017.00161

Wang W, Liu B, Xu M et al (2015) ABA-induced CCCH tandem zinc finger protein OsC3H47 decreases ABA sensitivity and promotes drought tolerance in *Oryza sativa*. Biochem Biophys Res Commun 464:33–37

Wani SH, Kumar V, Shriram V et al (2016) Plant hormones and their metabolic engineering for abiotic stress tolerance in crop plants. Crops J 4:162–176

Wani SH, Dutta T, Neelapu RRN, Surekha C (2017) Transgenic approaches to enhance salt and drought tolerance in plants. Plant Gene. https://doi.org/10.1016/j.plgene.2017.05.006

Wasternack C, Hause B (2013) Jasmonates: biosynthesis, perception, signal transduction and action in plant stress response, growth and development. An update to the 2007 review. Ann Bot 111:1021–1058. https://doi.org/10.1093/aob/mct067

Wasternack C, Kombrink E (2010) Jasmonates: structural requirements for lipid-derived signals active in plant stress responses and development. ACS Chem Biol 5:63–77

Wasternack C, Strnad M (2015) Jasmonate signaling in plant stress responses and development—active and inactive compounds. New Biotechnol. https://doi.org/10.1016/j.nbt.2015.11.001

Werner T, Schmülling T (2009) Cytokinin action in plant development. Curr Opin Plant Biol 12:527–538

Yabuta T, Sumiki Y (1938) On the crystal of gibberellin, a substance to promote plant growth. J Agric Chem Soc Japan 14:1526

Yamaguchi S (2008) Gibberellin metabolism and its regulation. Annu Rev Plant Physiol 59:225–251

Yang R, Yang T, Zhang H et al (2014) Hormone profiling and transcription analysis reveal a major role of ABA in tomato salt tolerance. Plant Physiol Biochem 77:23–34

Yoon JY, Hamayun M, Lee S-K, Lee I-J (2009) Methyl jasmonate alleviated salinity stress in soybean. J Crop Sci Biotech 12(2):63–68

Zahir ZA, Asghar HN, Arshad M (2001) Cytokinin and its precursors for improving growth and yield of rice. Soil Biol Biochem 33:405–408

Zalabák D, Pospíšilová H, Šmehilová M et al (2013) Genetic engineering of cytokinin metabolism: prospective way to improve agricultural traits of crop plants. Biotechnol Adv 31:97–117

Zhang Y, Yang J, Lu S et al (2008) Overexpressing SgNCED1 in tobacco increases ABA level, antioxidant enzyme activities, and stress tolerance. J Plant Growth Regul 27:151–158. https://doi.org/10.1007/s00344-008-9041-z

Zhang W, Yang H, You S et al (2015) MhNCED3, a gene encoding 9-*cis*-epoxycarotenoid dioxygenase in *Malus hupehensis* Rehd., enhances plant tolerance to Cle stress by reducing Cl^- accumulation. Plant Physiol Biochem 89:85–91

Zhang J, Yu H, Zhang Y et al (2016) Increased abscisic acid levels in transgenic maize overexpressing AtLOS5 mediated root ion fluxes and leaf water status under salt stress. J Exp Bot 67(5):1339–1355

Zhao Y (2010) Auxin biosynthesis and its role in plant development. Annu Rev Plant Biol 61:49–64. https://doi.org/10.1146/annurev-arplant-042809-112308

Zhao Y (2012) Auxin biosynthesis: a simple two-step pathway converts tryptophan to indole-3-acetic acid in plants. Mol Plant 5:334–338

Zhao Y, Dong W, Zhang N et al (2014) A wheat allene oxide cyclase gene enhances salinity tolerance via jasmonate signaling. Plant Physiol 164:1068–1076

Žižková E, Dobrev PI, Muhovski Y et al (2015) Tomato (*Solanum lycopersicum* L.) SlIPT3 and SlIPT4 isopentenyltransferases mediate salt stress response in tomato. BMC Plant Biol 15:85. https://doi.org/10.1186/s12870-015-0415-7

Chapter 7
Transcription Factor-Based Genetic Engineering for Salinity Tolerance in Crops

Parinita Agarwal, Pradeep K. Agarwal, and Divya Gohil

Abstract Salinity is one of the major environmental factors limiting productivity of all important crops. With an alarming increase of the world's population, the sustainable agriculture produce is needed to feed the growing population. The decrease in agricultural land due to industrialization/urbanization and climate change poses a threat to agriculture. The genetic engineering of crops can serve as a promising technique and can significantly contribute to addressing food security and also add economic value to the world's farming systems. The deployment of stress-responsive transcription factors for providing salinity tolerance to crops is an interesting area at the forefront of research. In the present chapter, we discuss the mechanism of regulation of important transcription factors, namely, dehydration-responsive element-binding factors, myeloblastosis oncogene, myelocytomatosis oncogene, NACs and WRKYs, and also provide an update of the transgenics developed and characterized using these TFs for enhanced salinity tolerance.

Keywords Abiotic stress · DNA-binding domain · Genetic engineering · Ionic · Osmotic · Post-translational · Promoters · Reactive oxygen species · Salinity stress · Signalling · Transcription factors · Transgenics

Abbreviations

ABA	Abscisic acid
ABRE	ABA-responsive element
AP2/ERF	Apetala2/ethylene-responsive factor CDPK
bZIP	Basic leucine zipper

P. Agarwal (✉) · P. K. Agarwal · D. Gohil
Plant Omics Division, CSIR-Central Salt and Marine Chemicals Research Institute (CSIR-CSMCRI), Council of Scientific & Industrial Research (CSIR), Bhavnagar, Gujarat 364002, India
e-mail: parinitaa@csmcri.org

V. Kumar et al. (eds.), *Salinity Responses and Tolerance in Plants, Volume 1*,
https://doi.org/10.1007/978-3-319-75671-4_7

CaBD	Calmodulin-binding domain
CaMV	Cauliflower mosaic virus
CDPK	Calcium-dependent protein kinase
CK2	Casein kinase 2
DBD	DNA-binding domain
DRE	Dehydration-responsive element
DREB	Dehydration-responsive element-binding factor
ERD1	Early responsive to dehydration stress
ET	Ethylene
HTH	Helix–turn–helix
MAPK	Mitogen-activated protein kinase
MeJA	Methyl jasmonate
MYB	Myeloblastosis oncogene
MYC	Myelocytomatosis oncogene
NACRS	NAC recognition site
PKC	Protein kinase C
PP2Cs	Protein phosphatase 2C
ROS	Reactive oxygen species
SA	Salicylic acid
TF	Transcription factor
TRD	Transcription regulatory domain

7.1 Introduction

Plants with autotrophic life pattern serve as a support system of all life forms and play a significant role in establishing a stable environment on the planet Earth. Plants have intricate mechanisms to integrate a wide range of tissue, developmental and environmental signals to regulate complex patterns of gene expression established over a long period of evolution as sessile organisms (Wu et al. 2007). They regulate global weather, climate and surroundings in many ways at different levels such as molecular, cellular, organ, individual, community, regional ecosystem and largely at global ecosystems.

Among the different abiotic stresses (e.g., drought, salinity, cold, frost, heat, waterlogging and heavy metal toxicity), salinity is a key environmental factor limiting plant growth and productivity. Soil salinity poses a huge problem; globally 831.4 million (m) ha of land is affected by soil salinity (FAO 2014), and approximately 19.5% of the agricultural land is saline (FAO 2014). In India total geographical area is 329 m ha, of which cultivable land is 183 m ha. Salinity affects almost 9.38 m ha, out of which by good management practices an area of 6.75 m ha is now being used for cultivation (www.cssri.org, Database 2015). Salinity refers to the presence of a high concentration (>4 dS/m) of soluble salts in the soil. Soil salinity can be categorized as primary and secondary salinization based on the source of soil salinization. Primary or natural salinization results from naturally occurring processes such as weathering of minerals, soil from saline parent rocks, seawater incursions and climatic changes, whereas secondary salinization is caused by human activities like irrigation, deforestation, overgrazing and intensive cropping.

Saline soils consist of high concentrations of sodium (Na^+) and chloride (Cl^-) ions in the soil, which is toxic to the plants and leads to degraded soil structure. The high salt generates both osmotic stress (physical) at an early phase and ionic stress (chemical) at a later phase of plant growth which impairs plant growth, development and productivity (Munns and Tester 2008). The osmotic stress begins the accumulation of salt above the threshold level (approximately 40 mM NaCl or less for some sensitive plants like rice and *Arabidopsis*) in root vicinity leading to withdrawal of water from the cells (Shabala and Cuin 2008). During ionic stress plants are unable to take up K^+ and other vital micronutrients like calcium, nitrogen, phosphorus and magnesium (Shabala and Cuin 2008). The K^+ and Na^+ have similar ionic radius and ion hydration energies (Schachtman and Liu 1999) and thus compete for transport into the plant cells. However, living cells preferably use K^+ over Na^+ for various physiological functions like maintaining osmotic equilibrium and membrane potential and for cell growth, enzyme activity and protein synthesis. The multiplicity in such stress signals underlies the complexity of stress signalling. A single stress activates multiple signalling pathways differing in time, space and outputs; further, these pathways connect or interact with one another using shared components generating an intertwined network (Zhu 2001; Fujita et al. 2006; Shao et al. 2006).

The interaction of plant with the environment or vice versa involves a very complex mechanism involving interactions or crosstalk of multiple cellular signalling pathways, allowing the plants to respond with minimal and appropriate biological processes (Fraire-Velázquez et al. 2011). The adaptation and acclimation to stresses result from integration of various biochemical and physiological processes of plants. Transcriptome analysis using microarray technology (Bohnert et al. 2001; Seki et al. 2001; Zhu et al. 2001) has categorized the stress response into two groups based on their function: functional proteins that directly protect against environmental stress and regulatory proteins that regulate gene expression and function in signal transduction. The first group includes proteins that function by protecting cells through various osmoprotectant-related proteins, late embryogenesis abundant (LEA) proteins, heat shock proteins (HSP), cold-inducible (KIN) proteins, antifreeze proteins, carbohydrate metabolism-related proteins, water channel proteins, potassium transporters, proteases, senescence-related genes, protease inhibitors, ferritin, lipid transfer proteins, chaperones and detoxification enzymes. The second group includes the protein factors involved in the further regulation of signal transduction and gene expression which include transcription factors Apetala2/ethylene-responsive factor, myelocytomatosis oncogene, myeloblastosis oncogene (AP2/ERF, MYC, MYB), NAC, WRKY, basic leucine zipper (bZIP), protein kinases (MAPK, CDPK) and protein phosphatases, enzymes involved in phospholipid metabolism and other signalling molecules (Seki et al. 2002).

Responses to environmental/external cues occur by stimulus–response coupling; a stimulus is first perceived by the cell, and a signal is then generated, amplified and transmitted eliciting biochemical changes to alleviate stress. Plants respond and adapt to recurring biotic and abiotic stresses, causing an array of genes to be activated as part of the plant defence/stress response involving an intricate network of signalling pathways controlling perception of environmental signals, the generation of second messengers and signal transduction. The stress signal is perceived by the receptors (G-protein-coupled receptors, ion channel, receptor-like kinase or histidine kinase) at the membrane, followed by the generation of secondary signal molecules such as Ca^{2+}, inositol phosphates, ROS and abscisic acid (ABA). These secondary signalling molecules modulate intracellular Ca^{2+} levels, protein phosphorylation cascades that target proteins involved in cellular protection or transcription factors controlling specific sets of stress-responsive genes. Certain proteins may participate in the generation of regulatory molecules like the phytohormones abscisic acid (ABA), ethylene (ET) and salicylic acid (SA). These regulatory molecules further initiate another signalling cascade, which may crosstalk with different signalling component leading to improved salinity tolerance (Fig. 7.1).

In plants, as for most other eukaryotes, gene expression is largely controlled at the level of transcription, by a diverse group of proteins, collectively known as transcription factors (TFs). These are defined as proteins that recognize specific DNA sequences in the promoters of the genes they control. Through protein–protein interactions, TFs mediate the assembly of the basal transcription machinery resulting in the activation of RNA polymerase II and mRNA synthesis. TFs can be classified into >50 different families, on the basis of the presence of conserved folds in their DNA-binding domains (Riechmann and Ratcliffe 2000). These regions are responsible for binding to specific *cis*-regulatory DNA elements. The control of specific sets of genes is accomplished by the combinatorial interaction among TFs, between TFs and non-DNA-binding proteins and between TFs and *cis*-regulatory elements. Of all the protein encoding plant genes, 5–7% corresponds to TFs, and specific families of TFs have dramatically expanded in the plants due to significant complexity of plant metabolism, as compared to other kingdoms (Shiu et al. 2005). However, still the number of characterized TFs that participate in the regulation of plant metabolism during stress signalling remains small. The cellular processes controlled by many plant TFs are being established as part of a number of functional genomics initiatives in *Arabidopsis*, as well as in crops such as maize and rice (Alonso and Ecker 2006).

A number of abiotic stress-related TFs and regulatory sequences in plant promoters are characterized. The TFs interact with *cis*-elements in the promoter regions of abiotic stress-related genes and upregulate the expression of different responsive genes resulting in abiotic stress tolerance. The understanding of transcriptional control in plants is an exciting and important field of research. Rapid progress is being made on deciphering the transcriptional control, which should

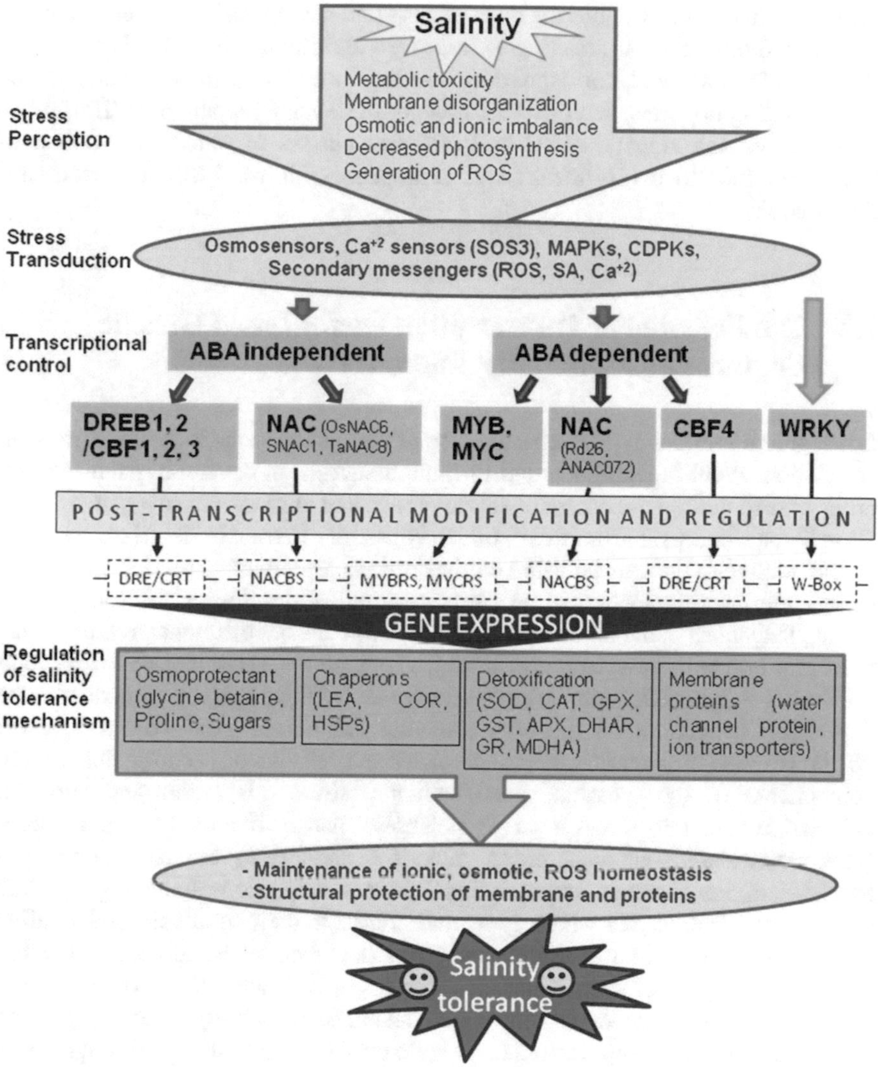

Fig. 7.1 Schematic representation of salinity tolerance mechanism by different ABA-dependent and ABA-independent transcriptional regulators. At the time of stress, different TFs like DREB, NAC, MYB, MYC, CBF4 and WRKY regulate the salinity tolerance by activating different stress-responsive genes. DREB/CBF dehydration-responsive element-binding protein/CRT element-binding factors, NAC NAM (no apical meristem) from petunia, ATAF1/2 and CUC2, MYB myeloblastoma oncogene, MYC myelocytomatosis oncogene, DRE/CRT dehydration-responsive element/C-repeat, NACBs NAC-binding sites, MYBRS/MYCRS MYB and MYC recognition sites, LEA late embryo abundant, COR cold-responsive, HSPs heat shock proteins, SOD superoxide dismutase, CAT catalase, GPX glutathione peroxidase, GST glutathione *S*-transferase, APX ascorbate peroxidase, DHAR dehydroascorbate reductase, GR glutathione reductase, and MDHAR monodehydroascorbate reductase

provide valuable insight into the mechanisms underlying various aspects of plant growth and development, leading to improved agriculture. In *Arabidopsis thaliana*, *cis*-elements and corresponding binding proteins, with distinct type of DNA-binding domains, such as AP2/ERF, basic leucine zipper, HD-ZIP (homeodomain leucine zipper), MYC and different classes of zinc-finger domains, have been identified (Shinozaki and Yamaguchi-Shinozaki 2000; Pastori and Foyer 2002).

7.2 The Potential of Transcription Factor-based Genetic Engineering for Salinity Tolerance

It is highly important to develop salt-tolerant productive crops to feed the growing population. Plant breeding has led to limited success in developing salt-tolerant crops, due to multigenic nature of salt tolerance and existence of low genetic variation in major crops. Furthermore, there is difficulty in transferring the gene from a wild relative to the domesticated cultivar, due to reproductive barrier. Therefore, genetic engineering appears to be a better option for developing salt tolerance in plants. The plants in the fields are subjected to not one but different stresses simultaneously and with varying magnitudes. Recent studies show that plants respond differently to individual and multiple stresses activating a specific cascade of gene expression depending on the environmental conditions (Atkinson and Urwin 2012). Transcription factors, kinase cascades and reactive oxygen species are key components of the crosstalk that facilitates stress adaptation and survival. Adaptation to salinity stress involves metabolic changes like production of osmolytes, antioxidative enzymes, photosynthetic machinery by the activation of ion transporters, ion channels, transcriptional factors and various signalling pathway components. Halophytes which grow and complete their life cycle under saline conditions are important candidates towards deciphering the molecular mechanism of salinity adaptation. Thus, research towards functional validation of the signalling components of halophytes in response to salinity stress is gaining momentum. Transcription factors respond to environmental stimuli through a signalling cascade and bind to specific cis-element sites on the promoter either by direct physical interaction or in combination with other proteins. The transcription factors along with their interacting proteins constitute a complex network that specifically regulates linear as well as crosstalks of the signalling pathways. Genetically engineering the expression of certain transcription factors can greatly influence plant stress tolerance. The transcription factor-based technologies are likely to be a prominent part of the next generation of successful biotechnology-derived crop (Century et al. 2008). The DREB, MYB, NAC and WRKY transcription factor family members have been characterized for their regulatory roles in the salinity stress.

7.3 Transcription Factors for Salinity Tolerance

7.3.1 DREB Transcription Factors

Dehydration-responsive element-binding (DREB) transcription factors are important members of the AP2/ERF family and unique to higher plants. However, DREBs have been reported from lower plants like moss *Physcomitrella patens*, a salt and osmotic stress-tolerant plant (Liu et al. 2007). These proteins share a conserved 58–59 amino acid domain (the ERF/AP2 domain) that binds two *cis*-elements, the GCC box, found in many PR (pathogen-related) gene promoters conferring ethylene responsiveness (Gu et al. 2000), and the C-repeat CRT/dehydration-responsive element (DRE) motif, involved in the expression of cold- and dehydration-responsive genes. DREB1A and DREB2A cDNAs for DRE-binding proteins have been isolated from *Arabidopsis* and many other plant species (Agarwal et al. 2006). DREB1A and its homologs were induced by low temperature, whereas DREB2A and its homolog show expression under dehydration and high-salt stress (Liu et al. 1998). The DREBs bind to the dehydration-responsive element/C-repeat (DRE/CRT) *cis*-element in the promoter region of many stress-responsive genes (Agarwal et al. 2007). The DREB1 genes are responsible for cold-regulated expression, whereas the DREB2 are involved in drought and salinity stress signalling, although significant overlap exists between the two pathways.

Amino acid alignment reports of DREB proteins shows high sequence similarity in the N-terminal nuclear localization signal and also some similarity in the C-terminal acidic domain. In the ERF/AP2 domain, the two amino acids, 14th valine and 19th glutamic acid, play a crucial role towards determining DNA-binding specificity (Liu et al. 1998; Cao et al. 2001; Sakuma et al. 2002). The conserved Ser-/Thr-rich region, close to the ERF/AP2 domain, is responsible for phosphorylation of DREB proteins (Liu et al. 1998). The DREB1/CBF1-type NLS consensus PKRPAGRTKFRETRHP separates these proteins from other ERF/AP2 proteins. The DSAW motif at the end of the ERF/AP2 domain and LWSY motif towards the C-terminal are conserved in most of the DREB1-type proteins.

The activity of DREB1/CBF is reported to be regulated mainly at the transcriptional level (Kidokoro et al. 2009). However, the mechanism of DREB2 expression is controlled at both transcriptional level and post-translational level like phosphorylation (Agarwal et al. 2007) and ubiquitination (Qin et al. 2008). In addition to these, alternative splicing is also reported for DREB2 TFs (Egawa et al. 2006; Qin et al. 2007; Matsukura et al. 2010; Vainonen et al. 2012).

DREB TFs have been isolated from many plants, and their overexpression has resulted in imparting enhanced abiotic stress tolerance in many plant species (Agarwal et al. 2017; Table 7.1). The constitutive overexpression of DREB2A using *CaMV 35S* promoter in transgenic *Arabidopsis* was not able to enhance tolerance of the transgenics. The domain analyses of *Arabidopsis* DREB2A gene revealed the presence of negative regulatory domain (136–165 aa); deletion of this region transforms DREB2A to a constitutive active form (DREB2A CA, Sakuma et al. 2006).

Table 7.1 Genetic engineering of DREB transcription factors for salinity tolerance

S. No.	Gene	Source	Transgenic plants	Performance of transgenic plants	References
DREB1/CBF-like genes					
1	*AaDREB1*	*Adonis amurensis*	*Arabidopsis thaliana*	Salt, drought and low-temperature tolerance	Zong et al. (2016)
2	*AhDREB1*	*Atriplex hortensis*	*Populus tomentosa*	Salt tolerance	Du et al. (2012b)
3	*AtDREB1A/ CBF3*	*Arabidopsis thaliana*	*Oryza sativa*	Salt, drought and low-temperature tolerance	Oh et al. (2005)
			Avena sativa	Salt tolerance	Orbay and Ahmad (2012)
4	*FaDREB1*	*Festuca arundinacea*	*Broussonetia papyrifera*	Salt and drought tolerance	Li et al. (2012)
5	*GmDREB1*	*Glycine max*	*Medicago sativa*	Salt tolerance	Jin et al. (2010)
6	*HsDREB1A*	*Hordeum spontaneum*	*Paspalum notatum*	Salt and dehydration tolerance	James et al. (2008)
7	*HvDREB1*	*Hordeum vulgare*	*Arabidopsis thaliana*	Salt tolerance	Xu et al. (2009)
8	*HvCBF4*	*Hordeum vulgare*	*Oryza sativa*	Salt, drought and low-temperature tolerance	Oh et al. (2007)
9	*LcDREB3a*	*Leymus chinensis*	*Arabidopsis thaliana*	Salt and dehydration tolerance	Xianjun et al. (2011)
10	*MtCBF4*	*Medicago truncatula*	*Medicago truncatula*	Salt tolerance	Li et al. (2011)
11	*OsDREB1A*	*Oryza sativa*	*Arabidopsis thaliana*	Salt and low-temperature tolerance	Dubouzet et al. (2003)
12	*OsDREB1D*	*Oryza sativa*	*Arabidopsis thaliana*	Salt tolerance	Zhang et al. (2009)
13	*OsDREB1F*	*Oryza sativa*	*Arabidopsis thaliana* *Oryza sativa*	Salt, dehydration and freezing tolerance	Wang et al. (2008)
DREB2-like genes					
14	*BpDREB2*	*Broussonetia papyrifera*	*Arabidopsis thaliana*	Salt and freezing tolerance	Sun et al. (2014)
15	*CkDREB*	*Caragana korshinskii*	*Nicotiana tabacum*	Salt and freezing tolerance	Wang et al. (2011)
16	*EsDREB2B*	*Eremosparton songoricum*	*Nicotiana tabacum*	Salt, dehydration, heat and freezing tolerance	Li et al. (2014b)

(continued)

Table 7.1 (continued)

S. No.	Gene	Source	Transgenic plants	Performance of transgenic plants	References
17	*OsDREB2A*	*Oryza sativa*	*Oryza sativa*	Salt and dehydration tolerance	Cui et al. (2011)
18	*OsDREB2A*	*Oryza sativa*	*Glycine max*	Salt tolerance	Zhang et al. (2013)
19	*PeDREB2*	*Populus euphratica*	*Nicotiana tabacum*	Salt tolerance	Chen et al. (2009a)
20	*PgDREB2A*	*Pennisetum glaucum*	*Nicotiana tabacum*	Salt and dehydration tolerance	Agarwal et al. (2010)
21	*SbDREB2A*	*Salicornia brachiata*	*Nicotiana tabacum*	Salt and dehydration tolerance	Gupta et al. (2014)
22	*SlDREB2*	*Solanum lycopersicum*	*Arabidopsis thaliana, Solanum lycopersicum*	Salt tolerance	Hichri et al. (2016)
23	*StDREB2*	*Solanum tuberosum*	*Solanum tuberosum*	Salt tolerance	Bouaziz et al. (2012)
24	*TaAIDFa*	*Triticum aestivum*	*Arabidopsis thaliana*	Salt and dehydration tolerance	Xu et al. (2008)
Other DREB genes with assigned subgroup					
25	*GmDREB2* A-5	*Glycine max*	*Arabidopsis thaliana*	Salt and dehydration tolerance	Chen et al. (2007)
26	*GmDREB3* A-5	*Glycine max*	*Arabidopsis thaliana*	Salt, dehydration and freezing tolerance	Chen et al. (2009b)
27	*PpDBF1* A-5	*Physcomitrella patens*	*Nicotiana tabacum*	Salt, dehydration and freezing tolerance	Liu et al. (2007)
28	*SsDREB* A-6	*Suaeda salsa*	*Nicotiana tabacum*	Salt and dehydration tolerance	Zhang et al. (2015)

This region possesses the PEST sequence (RSDASEVTSTSSQSEVCTVETPGCV) with many phosphorylation target sites for protein kinases such as PKC and CK2. The PEST sequence acts as signal peptide for protein degradation (Rogers et al. 1986). However, in some DREBs, the PEST sequence is found absent like SbDREB2A, ZmDREB2A and PgDREB2A (Qin et al. 2007; Agarwal et al. 2007; Gupta et al. 2014). Therefore, these TFs were overexpressed without deletion and resulted in enhanced stress tolerance in transgenics. DREB TF is one of the good choices of TFs which can be overexpressed in crop plants for stress tolerance.

7.3.2 MYB Transcription Factors

MYB (myeloblastoma) TFs are present in all eukaryotes, including plants. It is one of the largest TF families in plant that constitutes about 9% of total TF in *Arabidopsis* (Riechmann 2000). MYBs are the master regulator of various processes including regulation of primary and secondary metabolism, seed and floral development, cell fate and identity and abiotic and biotic stress tolerance (Dubos et al. 2010). The first plant MYB, c-MYB-like TF involved in anthocyanin biosynthesis, was isolated from maize (Paz-Ares et al. 1987); since then a number of MYB TFs have been isolated and characterized from different plants.

The MYB proteins are characterized by the presence of highly conserved MYB DNA-binding domain comprising of up to 4 imperfect repeats (R) of about 52 amino acids, each forming 3 α-helices. The second and third helices of each repeat build a helix–turn–helix (HTH) structure with three regularly spaced tryptophan (or hydrophobic) residues, forming a hydrophobic core in the 3D HTH structure (Ogata et al. 1996). The third helix of each repeat serves as 'recognition helix' facilitating intercalation of DNA in the major groove (Jia et al. 2004). The two MYB repeats are closely packed in the major groove, so that the two recognition helices bind cooperatively to the specific DNA sequence motif (Dubos et al. 2010). MYB proteins are classified into four different classes based on the number of repeats: 4R-MYB, 3R-MYB (R1R2R3), 2R-MYB (R2R3) and 1R-MYB. The 4R-MYB is the smallest group of the MYB family in plants which contains either four R1 or R2 repeats, and only one 4R-MYB gene is reported from *Arabidopsis* (Chen et al. 2006). The 3R-MYB also referred as R1R2R3-type MYB is evolutionary conserved group in plants, similar to the vertebrate MYB repeat (Ito 2005; Dubos et al. 2010; Li et al. 2014a). This group plays an important role in regulation of cell cycle at G2/M phase in tobacco (Haga et al. 2007). The 2R-MYB (R2R3-type) is the largest group of plant MYB family and is considered to have evolved either from 3R-MYB protein by loss of R1 or from an R1-MYB by duplication and is closely related to c-MYB from vertebrate slime moulds and ciliates (Feller et al. 2011). This group has >100 members in *Arabidopsis* (Yanhui et al. 2006), 157 in maize (Du et al. 2012a) and 90 in rice (Katiyar et al. 2012). The diversification and expansion of 2R-MYB proteins highlight their role in plant processes such as metabolism, development, cell fate and identity, hormone pathways and responses to stresses (Dubos et al. 2010; Li et al. 2014a). The 1R-MYB is the second largest group of the MYB family with 64 members in *Arabidopsis* and 70 members in rice (Dubos et al. 2010).

The MYB transcription factors participate in the ABA-dependent signal transduction pathway and upregulate a number of abiotic stress-responsive genes. Different MYB proteins bind to different *cis*-elements in their target gene's promoter. Several plant MYB proteins that bind to T/CAACG/TGA/C/TA/C/T (MBSI) will also bind to a second site, TAACTAAC (MBSII) (Romero et al. 1998). The AtMYB2 proteins bind to TGGTTAG *cis*-elements, of the rd22 promoter of *Arabidopsis*, and cooperatively activate this promoter (Abe et al. 1997).

Table 7.2 Genetic engineering of MYB transcription factors for salinity tolerance

S. No	Gene	Source	Transgenic plants	Performance of transgenic plants	References
1	*AmMYB1*	*Avicennia marina*	*Nicotiana tabacum*	Salinity tolerance	Ganesan et al. (2012)
2	*AtMYB2*	*Arabidopsis*	*Arabidopsis*	Salinity and drought tolerance	Abe et al. (2003)
3	*AtMYB15*	*Arabidopsis*	*Arabidopsis*	Salt and drought tolerance	Ding et al. (2009)
4	*AtMYB20*	*Arabidopsis*	*Arabidopsis*	Salinity tolerance	Cui et al. (2013)
5	*AtMYB44*	*Arabidopsis*	*Glycine max*	Salinity and drought tolerance	Seo et al. (2012)
6	*AtMYB88/ FLP*	*Arabidopsis*	*Arabidopsis*	Salinity and drought tolerance	Xie et al. (2010)
7	*AtMYB108/ BOS1*	*Arabidopsis*	*Arabidopsis (mutant analysis)*	Mutant shows impaired growth in salt	Mengiste et al. (2003)
8	*CmMYB2*	*Chrysanthemum*	*Arabidopsis*	Salinity and drought tolerance	Shan et al. (2012)
9	*GmMYB177*	*Glycine max*	*Arabidopsis*	Salinity tolerance	Liao et al. (2008)
10	*GmMYB76*	*Glycine max*	*Arabidopsis*	Salinity tolerance	
11	*GmMYB92*	*Glycine max*	*Arabidopsis*	Salinity tolerance	
12	*HPPBF-1*	*Arabidopsis*	*Arabidopsis*	Salinity tolerance	Nagaoka and Takano (2003)
13	*JAmyb*	Rice	*Arabidopsis*	Salinity tolerance	Yokotani et al. (2013)
14	*LcMYB1*	*Sheep grass*	*Arabidopsis*	Salinity tolerance	Cheng et al. (2013)
15	*MdoMYB121*	Apple	*Nicotiana tabacum*	Salinity, drought and cold tolerance	Cao et al. (2013)
16	*MdSIMYB1*	Apple	*Apple, Nicotiana tabacum*	Salinity, drought and cold tolerance	Wang et al. (2014)
17	*OsMPS/ MULTIPASS*	Rice	*Rice*	Salinity tolerance	Schmidt et al. (2013)
18	*OsMYB2*	Rice	Rice	Salinity, dehydration and cold tolerance	Yang et al. (2012)
19	*OsMYB3R-2*	Rice	*Arabidopsis*	Salinity, drought and freezing tolerance	Dai et al. (2007)
20	*OsMYB4*	Rice	*Arabidopsis*	Salinity, drought, freezing, chilling, ozone, UV and biotic stress tolerance	Vannini et al. (2004, 2006)
21	*OsMYB48-1*	Rice	*Rice*	Salinity and drought tolerance	Xiong et al. (2014)

(continued)

Table 7.2 (continued)

S. No	Gene	Source	Transgenic plants	Performance of transgenic plants	References
22	*TaMYB2A*	Wheat	*Arabidopsis*	Salinity, drought and cold tolerance	Mao et al. (2011a)
23	*TaMYB3R1*	Wheat	*Wheat*	Salinity, drought and cold tolerance	Cai et al. (2011)
24	*TaMYB33*	Wheat	*Arabidopsis*	Salinity and drought tolerance	Qin et al. (2012)
25	*TaMYB56-B*	Wheat	*Arabidopsis*	Salinity and freezing tolerance	Zhang et al. (2012)
26	*TaMYB73*	Wheat	*Arabidopsis*	Salinity tolerance	He et al. (2012)
27	*TaPIMP1*	Wheat	*Arabidopsis*	Salinity tolerance	Liu et al. (2011a)

The MYB transcription factors are induced by salinity stress, and its overexpression confers tolerance to salinity (Table 7.2) by increased ABA-induced expression of *rd22* and *AtADH1* (*AtMYB2*, Abe et al. 2003) or by enhanced transcript accumulation of ABA biosynthetic, signalling and responsive genes (*AtMYB15*, Ding et al. 2009) and also by regulating different TFs (*MdSIMYB1*, Wang et al. 2014). The MYB TFs also regulate stomatal opening in response to salinity; the *AtMYB44* confers drought and salinity tolerance in *Arabidopsis* by regulating ABA-induced stomatal closure by decreasing expression of serine/threonine protein phosphatase 2Cs (PP2Cs), which have been described as negative regulators of ABA signalling (Jung et al. 2008).

7.3.3 NAC Transcription Factors

NAC proteins are one of the largest families of plant-specific TFs which regulate both ABA-dependent and ABA-independent genes. The NAC is named from three earliest characterized proteins: NAM (no apical meristem) from petunia, ATAF1/2 and CUC2 (cup-shaped cotyledon) from *Arabidopsis* (Puranik et al. 2012). The first NAC gene isolated was *NAM* from petunia (Souer et al. 1996), which plays a critical role in determining meristem and primordia positions. However, since then, a number of NAC genes have been identified: 117 NAC in *Arabidopsis* (Tran et al. 2004), 151 in rice (Nuruzzaman et al. 2010), 79 in grape, 163 in poplar (Hu et al. 2010) and 152 each in soybean (Le et al. 2011) and tobacco (*Nicotiana tabacum,* Rushton et al. 2008). Apart from flowering plants, NAC TFs are also reported from land plants like gymnosperms, white spruce (36, Duval et al. 2014), maritime pine

(37, Pascual et al. 2015), also moss (5, Yao et al. 2012) and pteridophytes (Shen et al. 2009). The comparative genome analysis via computational modelling of the secondary cell wall-associated NAC genes (SND, NST and VND) from 19 higher plant species shows that only VND genes were found in *S. moellendorffii*. The NAC proteins possess a highly conserved N-terminus NAC DNA-binding domain (DBD), while the C-terminus is highly variable and contains a transcription regulatory domain (TRD, Nakashima et al. 2012). The NAC domain comprises of 150 amino acids in length and contains five conserved regions (A to E). Jensen et al. (2010) reported that biochemical and functional specificity of NAC TFs was associated with the structure of both the DBDs and the TRDs. The NAC TFs, responsive to drought, high salinity, ABA and MeJA (methyl jasmonate), interact with the putative MYC-like (CATGTG) NAC recognition site (NACRS), found in the ERD1 (early responsive to dehydration stress) promoter region (Tran et al. 2004). The NAC proteins could bind to NACRS even as multimers; heterodimerization might further regulate the transcriptional activity of the NAC proteins (Tran et al. 2004). They are expressed in different tissues at various developmental stages and are involved in many aspects of plant growth and development (Olsen et al. 2005) and furthermore also regulate abiotic and biotic stresses (Fujita et al. 2004; Tran et al. 2004, 2010; Nakashima et al. 2007). Distelfeld et al. (2012) showed that NAC transcription factor family has a dynamic nature since it has potential to evolve differently and perform novel functions. NAC gene from rice *OsNAC6*, a member of ATAF subfamily (Ooka et al. 2003), has been reported to mediate responses to cold, high salinity, drought, ABA and JA (Nakashima et al. 2007; Ohnishi et al. 2005). Many abiotic and biotic stress-responsive genes were upregulated in the *OsNAC6* transgenic plants, and the transgenics were tolerant to dehydration and high-salt stresses and even showed slightly improved tolerance to blast disease (Nakashima et al. 2007).

The activity of NAC TFs to perform a wide range of functions is tightly regulated at transcriptional, post-transcriptional and post-translational levels. The TFs such as ABREs (ABA-responsive elements) and DREs (dehydration-responsive elements) present upstream to transcriptionally regulate NACs, whereas certain micro-RNAs or alternative splicing participates in post-transcriptionally regulating NACs. The post-transnational modifications like ubiquitination, dimerization, phosphorylation or proteolysis also regulate NAC TF activity. Phosphorylation is essential for nuclear localization of OsNAC4 (Kaneda et al. 2009); similarly, ATAF1 requires phosphorylation for its subcellular localization, DNA-binding activity and protein interactions (Kleinow et al. 2009). Phosphorylation of ZmNAC84 by ZmCCaMK regulates the antioxidant defence by activation of downstream genes (Zhu et al. 2016). Miao et al. (2016) reported a tomato SINA ubiquitin ligase SINA3 that targets NAC1 for ubiquitination and degradation. It is the multi-mechanism regulation of NAC that facilitates its participation in a wide array of functions. The NAC candidate genes have been functionally characterized for enhancement of stress tolerance (Table 7.3).

Table 7.3 Genetic engineering of NAC transcription factors for salinity tolerance

S. No.	Gene	Source	Transgenic plants	Performance of transgenic plants	References
1	AhNAC2	*Arachis hypogea*	*Arabidopsis*	Salt and drought tolerance	Liu et al. (2011b)
2	AtNAC2	*Arabidopsis thaliana*	*Arabidopsis thaliana*	Salt tolerance	He et al. (2005)
3	ATAF1	*Arabidopsis thaliana*	*Oryza sativa*	Salt tolerance	Liu et al. (2011b)
4	DlNAC1	*Dendranthema lavandulifolium*	*Nicotiana*	Salinity and drought tolerance	Yang et al. (2016)
5	EcNAC1	*Eleusine coracana*	*Nicotiana tabacum*	Salt and drought tolerance	Ramegowda et al. (2012)
6	EcNAC67	*Eleusine coracana*	*Oryza sativa*	Salinity and drought tolerance	Rahman et al. (2016)
7	GhATAF1	*Gossypium hirsutum*	*Gossypium hirsutum*	Salt tolerance	He et al. (2016)
8	GhNAC2	*Gossypium hirsutum*	*Arabidopsis thaliana*	Salt and drought tolerance	Gunapati et al. (2016)
9	GmNAC11	*Glycine max*	*Arabidopsis thaliana*	Salt tolerance	Hao et al. (2011)
10	GmNAC20	*Glycine max*	*Arabidopsis thaliana*	Salt and freezing tolerance	
11	NAC042	*Musa*	*Banana*	Salinity and drought tolerance	Tak et al. (2016)
12	ONAC022	*Oryza sativa*	*Oryza sativa*	Salt and drought tolerance	Hong et al. (2016)
13	ONAC045	*Oryza sativa*	*Oryza sativa*	Salt and drought tolerance	Zheng et al. (2009)
14	ONAC063	*Oryza sativa*	*Arabidopsis*	Salinity and osmotic pressure tolerance	Yokotani et al. (2009)
15	OsNAC5	*Oryza sativa*	*Oryza sativa*	Salt tolerance	Takasaki et al. (2010)
16	OsNAC6	*Oryza sativa*	*Oryza sativa*	Salt, drought and freezing tolerance	Nakashima et al. (2007)
17	OsNAC10	*Oryza sativa*	*Oryza sativa*	Salinity, drought and low-temperature tolerance	Jeong et al. (2010)
18	OsNAP	*Oryza sativa*	*Oryza sativa*	Salt, drought and low-temperature tolerance	Chen et al. (2013)
19	SNAC1	*Oryza sativa*	*Oryza sativa*	Salt tolerance	Hu et al. (2006)
20	SNAC1	*Oryza sativa*	*Triticum aestivum*	Salt and drought tolerance	Saad et al. (2013)

(continued)

Table 7.3 (continued)

S. No.	Gene	Source	Transgenic plants	Performance of transgenic plants	References
21	SNAC2	*Oryza sativa*	*Oryza sativa*	Salt, drought and freezing tolerance	Hu et al. (2008)
22	TaNAC2	*Triticum aestivum*	*Arabidopsis thaliana*	Salt, drought and freezing tolerance	Mao et al. (2011b)
23	TaNAC29	*Triticum aestivum*	*Arabidopsis thaliana*	Salt and drought tolerance	Huang et al. (2015)
24	TaNAC29	*Triticum aestivum*	*Arabidopsis thaliana*	Salt stress	Xu et al. (2015)
25	TaNAC67	*Triticum aestivum*	*Arabidopsis thaliana*	Salt, drought and freezing tolerance	Mao et al. (2014)
26	VvNAC1	*Vitis vinifera*	*Arabidopsis*	Salt, drought, cold tolerance	Hénanff et al. (2013)
27	VaNAC26	*Vitis amurensis*	*Arabidopsis thaliana*	Salt and drought tolerance	Fang et al. (2016)

7.3.4 WRKY Transcription Factors

The WRKY transcription factors form the largest family of plant-specific TFs and are also found in protists, green algae such as *Chlamydomonas reinhardtii*, slime mould, fern and pine, indicating an ancestral origin of the family. The WRKY genes have shown evolution from unicellular to complex multicellular forms and are reported in large numbers in flowering plants, indicating important regulatory role in flowering plants. The first WRKY gene was identified in sweet potato, and since then, large number of WRKY proteins are reported from higher plants. The WRKY TFs have been reported from *Arabidopsis*, rice (74, 102 Baranwal et al. 2016), soybean (197, Schmutz et al. 2010), papaya, poplar, sorghum, *Physcomitrella patens* (66, 104, 68, 38, respectively, Pandey and Somssich 2009) and *Jatropha* (58, Xiong et al. 2013). The WRKY TFs get its name from its highly conserved 60 amino acid long WRKY domain, consisting of highly conserved heptapeptide WRKYGQK at amino (N)-terminus and a novel metal chelating zinc finger at the carboxy (C)-terminus. The WRKYGQK motif shows slight variation in *Arabidopsis*, rice, tobacco, barley and canola (Zhang and Wang 2005; Mangelsen et al. 2008; Xie et al. 2005; Yang et al. 2009). The WRKY TFs are classified into three groups depending on the number of WRKY motifs present and the constitution of their zinc-finger motif. Group I TFs contain two WRKY domains, while group II and group III possess only a single WRKY domain. Groups I and II have the C2H2-type zinc-finger motif (C-X4–5-C-X22–23-H-XH), whereas group III shows C2HC zinc-finger motif (C-X7-C-X23-H-X-C). Group II is further divided into subgroups a–e on the basis of additional amino acid motifs present outside the WRKY domain. Certain structures like nuclear localization signal (NLS), leucine zippers, Ser-/Thr-rich stretches, Gln- and Pro-rich stretches, kinase domains and pathogenesis-related TIR-NBS-LRR domains have also been reported (Jiang and Deyholos 2006).

The WRKY TFs show preferential binding to their cognate *cis*-acting element, the W-box (TTGACC/T, Ulker and Somssich 2004); the flanking sequence adjoining the W-box also plays a role in determining the binding of WRKY protein, thereby leading to distinct transcriptional outputs (Ciolkowski et al. 2008, Yamasaki et al. 2005). The N-terminal domain is not responsible for sequence-specific binding to DNA, it might participate in binding process by increasing affinity or specificity for their target sites, or it may provide an interface for protein–protein interactions, a known function of some zinc-finger-like domains (Mackay and Crossley 1998). In some group I WRKY members, SPF1, ZAP1 and PcWRKY1, the C-terminal WRKY domain is responsible for sequence-specific binding to DNA (Ishiguro and Nakamura 1994; de Pater et al. 1996; Eulgem et al. 1999).

The WRKY family has evolved in response to pressures imposed by environmental factors and phytopathogens, to facilitate distinct cellular, developmental and physiological roles in plants. The WRKY TFs play a broad-spectrum regulatory role as positive and negative regulators of plant defence regulation and abiotic stresses and are also involved in growth and development of plants such as embryogenesis, seed coat and trichome development, regulation of biosynthetic pathways and hormone signalling.

The WRKY family of TFs is one of the best-characterized classes of plant defence transcription factors and pathogen infection or pathogen elicitors or SA resulting in rapid induction of WRKY TF. The analysis of mutants shows that even closely related WRKY members display strong differences in their expression patterns and are most likely involved in distinct signalling pathways (Kalde et al. 2003). WRKY TFs regulate the plant stress response by interacting with other transcription factors and also by directly regulating some function genes. WRKY TFs exhibit autoregulation, cross regulation and also interaction with different proteins like MAP kinases, MAP kinase kinases, calmodulin, histone deacetylases, etc. for carrying out diverse plant functions (Rushton et al. 2010). Post-translational modification like phosphorylation also regulates the activity of WRKY transcription factors; some WRKY TFs are found to act downstream of MAPKs in regulating plant defence (Asai et al. 2002; Liu et al. 2004). It has been reported that the VQ proteins interact with WRKY TFs in yeast (Cheng et al. 2012). The VQ are activated by MAPK, and their interaction with WRKY TF alters its DNA-binding specificity (Chi et al. 2013), resulting in altered downstream response. The WRKY proteins also possess calmodulin (CaM)-binding domain (CaBD, Chi et al. 2013) and thus have tendency to interact with CaMs (Park et al. 2005). The WRKY TFs are regulated by hormones like SA, ABA and GA and through hormone crosstalk with different signalling pathways.

The genetic engineering by overexpressing WRKY TFs for enhanced salinity tolerance (Table 7.4) has often resulted in multiple stress tolerance. The WRKY25 and WRKY33 increased salt tolerance and ABA sensitivity independently of the SOS pathway when overexpressed in *A. thaliana* (Jiang and Deyholos 2009). The WRKY15-overexpressing plants showed endoplasmic reticulum-to-nucleus communication and a disrupted mitochondrial communication during salt stress (Vanderauwera et al. 2012). The overexpression of these TFs provides salinity tolerance by modulating ROS homeostasis through SA signalling (Agarwal et al. 2016).

Table 7.4 Genetic engineering of WRKY transcription factors for salinity tolerance

S. No.	Gene	Source	Transgenic plants	Performance of transgenic plants	References
1	AtWRKY25, 33	*Arabidopsis thaliana*	*Arabidopsis thaliana*	Salt tolerance	Jiang and Deyholos (2009)
2	BcWRKY46	*Brassica campestris*	*Nicotiana tabacum*	Salt, freezing, ABA and dehydration tolerance	Wang et al. (2012)
3	DgWRKY3	*Dendranthema grandiflorum*	*Nicotiana*	Salt tolerance	Liu et al. (2013)
4	GhWRKY41	*Gossypium hirsutum*	*Nicotiana benthamiana*	Salt and drought tolerance	Chu et al. (2015)
5	GhWRKY68	*Gossypium hirsutum*	*Nicotiana benthamiana*	Salt and drought tolerance	Jia et al. (2015)
6	GhWRKY25	*Gossypium hirsutum*	*Nicotiana benthamiana*	Salt and drought tolerance	Liu et al. (2016a)
7	GmWRKY54	*Glycine max*	*Arabidopsis thaliana*	Salt and drought tolerance	Zhou et al. (2008)
8	OsWRKY45	*Oryza sativa*	*Arabidopsis thaliana*	Salt and drought tolerance	Qiu and Yu (2009)
9	SpWRKY1	*Solanum pimpinellifolium*	*Nicotiana tabacum*	Salt and drought tolerance	Li et al. (2015)
10	TaWRKY2	*Triticum aestivum*	*Arabidopsis thaliana*	Salt and drought tolerance	Niu et al. (2012)
11	TaWRKY19	*Triticum aestivum*	*Arabidopsis thaliana*	Salt, drought and freezing tolerance	
12	TaWRKY10	*Triticum aestivum*	*Nicotiana tabacum*	Salt and drought tolerance	Wang et al. (2013)
13	TaWRKY44	*Triticum aestivum*	*Nicotiana*	Salt and drought tolerance	Wang et al. (2015)

7.4 Conclusion and Perspectives

The rapid progress being made on the isolation of important regulatory proteins, the development of in vitro transcription systems and the use of powerful genetic screening approaches for additional mutants using promoter/reporter gene fusions will facilitate further studies of transcriptional control in plants, which should provide valuable insight into the mechanisms underlying various aspects of plant growth and development and lead to agricultural benefits.

Acknowledgements The authors acknowledge the financial support of the Department of Science and Technology and Council of Scientific and Industrial Research, New Delhi.

References

Abe H, Yamaguchi-Shinozaki K, Urao T, Iwasaki T, Hosokawa D, Shinozaki K (1997) Role of Arabidopsis MYC and MYB homologs in drought- and abscisic acid-regulated gene expression. Plant Cell 9:1859–1868

Abe H, Urao T, Ito T, Seki M, Shinozaki K, Yamaguchi-Shinozaki K (2003) Arabidopsis AtMYC2 (bHLH) and AtMYB2 (MYB) function as transcriptional activators in abscisic acid signaling. Plant Cell 15:63–78

Agarwal PK, Agarwal P, Reddy MK, Sopory SK (2006) Role of DREB transcription factors in abiotic and biotic stress tolerance in plants. Plant Cell Rep 25:1263–1274

Agarwal P, Agarwal PK, Nair S, Sopory SK, Reddy MK (2007) Stress-inducible DREB2A transcription factor from *Pennisetum glaucum* is a phosphoprotein and its phosphorylation negatively regulates its DNA-binding activity. Mol Genet Genomics 277:189–198

Agarwal P, Agarwal PK, Joshi AJ, Reddy MK, Sopory SK (2010) Overexpression of PgDREB2A transcription factor enhances abiotic stress tolerance and activates downstream stress-responsive genes. Mol Biol Rep 37:1125–1135

Agarwal P, Dabi M, More P, Patel K, Jana K, Agarwal PK (2016) Improved shoot regeneration, salinity tolerance and reduced fungal susceptibility in transgenic tobacco constitutively expressing PR-10a gene. Front Plant Sci 7:217

Agarwal PK, Gupta K, Lopato S, Agarwal P (2017) Drought responsive element binding transcription factors and their applications for the engineering of stress tolerance. J Exp Bot 68:2135–2148

Alonso JM, Ecker JR (2006) Moving forward in reverse: genetic technologies to enable genome-wide phenomic screens in *Arabidopsis*. Nat Rev Genet 7:524–536

Asai T, Tena G, Plotnikova J, Willmann MR, Chiu WL, Gomez-Gomez L et al (2002) MAP kinase signalling cascade in Arabidopsis innate immunity. Nature 415:977–983

Atkinson NJ, Urwin PE (2012) The interaction of plant biotic and abiotic stresses: from genes to the field. J Exp Bot 63:3523–3543

Baranwal VK, Negi N, Khurana P (2016) Genome-wide identification and structural, functional and evolutionary analysis of WRKY components of mulberry. Sci Rep 6:30794

Bohnert HJ, Ayoubi P, Borchert C, Bressan RA, Burnap RL, Cushman JC, Cushman MA, Deyholos M, Fisher R, Galbraith DW, Hasegawa PM, Jenks M, Kawasaki S, Koiwa H, Kore-eda S, Lee B-H, Michalowski CB, Misawa E, Nomura M, Ozturk N, Postier B, Prade R, Song C-P, Tanaka Y, Wang H, Zhu JK (2001) A genomic approach towards salt stress tolerance. Plant Physiol Biochem 39:295–311

Bouaziz D, Pirrello J, Ben Amor H, Hammami A, Charfeddine M, Dhieb A, Bouzayen M, Gargouri-Bouzid R (2012) Ectopic expression of dehydration responsive element binding proteins (StDREB2) confers higher tolerance to salt stress in potato. Plant Physiol Biochem 60:98–108

Cai H, Tian S, Liu C, Dong H (2011) Identification of a MYB3R gene involved in drought, salt and cold stress in wheat (*Triticum aestivum* L.) Gene 485:146–152

Cao ZF, Li J, Chen F, Li YQ, Zhou HM, Liu Q (2001) Effect of two conserved amino-acid residues on DREB1A function. Biochemistry 66:623–627

Cao Z, Zhang S, Wang R, Zhang R, Hao Y (2013) Genome wide analysis of the apple MYB transcription factor family allows the identification of MdoMYB121 gene conferring abiotic stress tolerance in plants. PLoS One 8:e69955

Century K, Reuber TL, Ratcliffe OJ (2008) Regulating the regulators: the future prospects for transcription-factor based agricultural biotechnology products. Plant Physiol 147:20–29

Chen YH, Wu XM, Ling HQ, Yang WC (2006) Transgenic expression of DwMYB2 impairs iron transport from root to shoot in *Arabidopsis thaliana*. Cell Res 16:830–840

Chen M, Wang QY, Cheng XG, Xu ZS, Li LC, Ye XG, Xia LQ, Ma YZ (2007) GmDREB2, a soybean DRE-binding transcription factor, conferred drought and high-salt tolerance in transgenic plants. Biochem Biophys Res Commun 353:299–305

Chen J, Xia X, Yin W (2009a) Expression profiling and functional characterization of a DREB2-type gene from *Populus euphratica*. Biochem Biophys Res Commun 378:483–487

Chen M, Xu Z, Xia L, Li L, Cheng X, Dong J, Wang Q, Ma Y (2009b) Cold-induced modulation and functional analyses of the DRE-binding transcription factor gene GmDREB3 in soybean (*Glycine max* L.) J Exp Bot 60:121–135

Chen X, Wang Y, Lv B, Li J, Luo L, Lu S, Zhang X, Ma H, Ming F (2013) The NAC family transcription factor OsNAP confers abiotic stress response through the ABA pathway. Plant Cell Physiol 55:604–619

Cheng Y, Zhou Y, Yang Y, Chi YJ, Zhou J, Chen JY, Wang F, Fan B, Shi K, Zhou YH, Yu JQ, Chen Z (2012) Structural and functional analysis of VQ motif-containing proteins in *Arabidopsis* as interacting proteins of WRKY transcription factors. Plant Physiol 159:810–825

Cheng L, Li X, Huang X, Ma T, Liang Y, Ma X, Peng X, Jia J, Chen S, Chen Y, Deng B, Liu G (2013) Overexpression of sheep grass R1-MYB transcription factor LcMYB1 confers salt tolerance in transgenic *Arabidopsis*. Plant Physiol Biochem 70:252–260

Chi Y, Yang Y, Zhou Y, Zhou J, Fan B, Yu JQ, Chen Z (2013) Protein-protein interactions in the regulation of WRKY transcription factors. Mol Plant 6:287–300

Chu X, Wang C, Chen X, Lu W, Li H, Wang X et al (2015) The cotton WRKY gene GhWRKY41 positively regulates salt and drought stress tolerance in transgenic *Nicotiana benthamiana*. PLoS One 10:e0143022

Ciolkowski I, Wanke D, Birkenbihl RP, Somssich IE (2008) Studies on DNA-binding selectivity of WRKY transcription factors lend structural clues into WRKY-domain function. Plant Mol Biol 68:81–92

Cui M, Zhang W, Zhang Q, Xu Z, Zhu Z, Duan F, Wu R (2011) Induced overexpression of the transcription factor OsDREB2A improves drought tolerance in rice. Plant Physiol Biochem 49:1384–1391

Cui MH, Yoo KS, Hyoung S, Nguyen HTK, Kim YY, Kim HJ, Ok SH, Yoo SD, Shin JS (2013) An Arabidopsis R2R3-MYB transcription factor, AtMYB20, negatively regulates type 2C serine/threonine protein phosphatases to enhance salt tolerance. FEBS 587:1773–1778

Dai X, Xu Y, Ma Q, Xu W, Wang T, Xue Y, Chong K (2007) Overexpression of an R1R2R3 MYB gene OsMYB3R-2 increases tolerance to freezing, drought, and salt stress in transgenic *Arabidopsis*. Plant Physiol 143:1739–1751

De Pater S, Greco V, Pham K, Memelink J, Kijne J (1996) Characterization of a zinc-dependent transcriptional activator from *Arabidopsis*. Nucleic Acids Res 24:4624–4631

Ding Z, Li S, An X, Liu X, Qin H, Wang D (2009) Transgenic expression of MYB15 confers enhanced sensitivity to abscisic acid and improved drought tolerance in *Arabidopsis thaliana*. J Gen Genomics 36:17–29

Distelfeld A, Pearce SP, Avni R, Scherer B, Uauy C, Piston F, Slade A, Zhao R, Dubcovsky J (2012) Divergent functions of orthologous NAC transcription factors in wheat and rice. Plant Mol Biol 78:515–524

Du N, Liu X, Li Y, Chen S, Zhang J, Ha D, Deng W, Sun C, Zhang Y, Pijut PM (2012a) Genetic transformation of *Populus tomentosa* to improve salt tolerance. Plant Cell Tiss Org Cult 108:181–189

Du H, Yang SS, Liang Z, Feng BR, Liu L, Huang YB, Tang YX (2012b) Genome-wide analysis of the MYB transcription factor superfamily in soybean. BMC Plant Biol 12:106

Dubos C, Stracke R, Grotewold E, Weisshaar B, Martin C, Lepiniec L (2010) MYB transcription factors in *Arabidopsis*. Trends Plant Sci 15:573–581

Dubouzet JG, Sakuma Y, Ito Y, Kasuga M, Dubouzet EG, Miura S, Seki M, Shinozaki K, Yamaguchi-Shinozaki K (2003) OsDREB genes in rice *Oryza sativa* L. encode transcription activators that function in drought-, high-salt- and cold-responsive gene expression. Plant J 33:751–763

Duval I, Lachance D, Giguère I, Bomal C, Morency MJ, Pelletier G, Boyle B, MacKay JJ, Séguin A (2014) Large-scale screening of transcription factor-promoter interactions in spruce reveals a transcription. J Exp Bot 65:2319–2333

Egawa C, Kobayashi F, Ishibashi M, Nakamura T, Nakamura C, Takumi S (2006) Differential regulation of transcript accumulation and alternative splicing of a DREB2 homolog under abiotic stress conditions in common wheat. Genes Genet Syst 81:77–91

Eulgem T, Rushton PJ, Schmelzer E, Hahlbrock K, Somssich IE (1999) Early nuclear events in plant defence signalling: rapid gene activation by WRKY transcription factors. EMBO J 18:4689–4699

Fang L, Su L, Sun X, Li X, Sun M, Karungo SK, Fang S, Chu J, Li S, Xin H (2016) Expression of *Vitis amurensis* NAC26 in Arabidopsis enhances drought tolerance by modulating jasmonic acid synthesis. J Exp Bot 67:2829–2845

FAO (Food and Agriculture Organization of United Nations) (2014) http://www.fao.org/corp/statistics/en/

Feller A, Machemer K, Braun EL, Grotewold E (2011) Evolutionary and comparative analysis of MYB and bHLH plant transcription factors. Plant J 66:94–116

Fraire-Velázquez S, Rodríguez-Guerra R, Sánchez-Calderón L (2011) Abiotic and biotic stress response crosstalk in plants. In: Shanker A, Venkateswarlu B (eds) Abiotic stress response in plants: physiological, biochemical and genetic perspectives. Intech, Rijeka, pp 3–26

Fujita M, Fujita Y, Maruyama K, Seki M, Hiratsu K, Ohme-Takagi M, Tran LS, Yamaguchi-Shinozaki K, Shinozaki K (2004) A dehydration-induced NAC protein, RD26, is involved in a novel ABA-dependent stress-signaling pathway. Plant J 39:863–876

Fujita M, Fujita Y, Noutoshi Y, Takahashi F, Narusaka Y, Yamaguchi-Shinozaki K, Shinozaki K (2006) Crosstalk between abiotic and biotic stress responses: a current view from the points of convergence in the stress signaling networks. Curr Opin Plant Biol 9:436–442

Ganesan G, Sankararamasubramanian HM, Harikrishnan M, Ganpudi A, Parida A (2012) A MYB transcription factor from the grey mangrove is induced by stress and confers NaCl tolerance in tobacco. J Exp Bot 63:4549–4561

Gu YQ, Yang C, Thara VK, Zhou J, Martin GB (2000) Pti4 is induced by ethylene and salicylic acid, and its product is phosphorylated by the Pto kinase. Plant Cell 12:771–785

Gunapati S, Naresh R, Ranjan S, Nigam D, Hans A, Verma PC, Gadre R, Pathre UV, Sane AP, Sane VA (2016) Expression of GhNAC2 from *G. herbaceum*, improves root growth and imparts tolerance to drought in transgenic cotton and Arabidopsis. Sci Rep 6:24978

Gupta K, Jha B, Agarwal PK (2014) A dehydration-responsive element binding (DREB) transcription factor from the succulent halophyte *Salicornia brachiata* enhances abiotic stress tolerance in transgenic tobacco. Mar Biotechnol 16:657–667

Haga N, Kato K, Murase M, Araki S, Kubo M, Demura T, Suzuki K, Müller I, Voss U, Jürgens G, Ito M (2007) R1R2R3-Myb proteins positively regulate cytokinesis through activation of KNOLLE transcription in *Arabidopsis thaliana*. Development 134:1101–1110

Hao YJ, Wei W, Song QX, Chen HW, Zhang YQ, Wang F, Zou HF, Lei G, Tian AG, Zhang WK, Ma B, Zhang JS, Chen SY (2011) Soybean NAC transcription factors promote abiotic stress tolerance and lateral root formation in transgenic plants. Plant J 68:302–313

He XJ, Mu RL, Cao WH, Zhang ZG, Zhang JS, Chen SY (2005) AtNAC2, a transcription factor downstream of ethylene and auxin signaling pathways, is involved in salt stress response and lateral root development. Plant J 44:903–916

He Y, Li W, Lv J, Jia Y, Wang M, Xia G (2012) Ectopic expression of a wheat MYB transcription factor gene TaMYB73 improves salinity stress tolerance in *Arabidopsis thaliana*. J Exp Bot 63:1511–1522

He X, Zhu L, Xu L, Guo W, Zhang X (2016) GhATAF1, a NAC transcription factor, confers abiotic and biotic stress responses by regulating phytohormonal signaling networks. Plant Cell Rep 35:2167–2179

Hénanff GL, Profizi C, Courteaux B, Rabenoelina F, Gérard C, Clément C, Baillieul F, Cordelier S, Dhondt-Cordelier S (2013) Grapevine NAC1 transcription factor as a convergent node in developmental processes, abiotic stresses, and necrotrophic/biotrophic pathogen tolerance. J Exp Bot 64:4877–4893

Hichri I, Muhovski Y, Clippe A, Žižková E, Dobrev PI, Motyka V, Lutts S (2016) SlDREB2 a tomato dehydration-responsive element-binding 2 transcription factor mediates salt stress tolerance in tomato and Arabidopsis. Plant Cell Environ 39:62–79

Hong Y, Zhang H, Huang L, Liand D, Song F (2016) Overexpression of stress-responsive NAC transcription factor gene ONAC022 improves drought and salt tolerance in rice. Front Plant Sci 7:4

Hu H, Dai M, Yao J, Xiao B, Li X, Zhang Q, Xiong L (2006) Overexpressing a NAM, ATAF, and CUC (NAC) transcription factor enhances drought resistance and salt tolerance in rice. Proc Natl Acad Sci U S A 103:12987–12992

Hu H, You J, Fang Y, Zhu X, Qi Z, Xiong L (2008) Characterization of transcription factor gene SNAC2 conferring cold and salt tolerance in rice. Plant Mol Biol 67:169–181

Hu R, Qi G, Kong Y, Kong D, Gao Q, Zhou G (2010) Comprehensive analysis of NAC domain transcription factor gene family in *Populus trichocarpa*. BMC Plant Biol 10:145

Huang Q, Wang Y, Li B, Chang J, Chen M, Li K, Yang G, He G (2015) TaNAC29, a NAC transcription factor from wheat, enhances salt and drought tolerance in transgenic Arabidopsis. BMC Plant Biol 15:268

Ishiguro S, Nakamura K (1994) Characterization of a cDNA encoding a novel DNA-binding protein, SPF1, that recognizes SP8 sequences in the 50 upstream regions of genes coding for sporamin and b-amylase from sweet potato. Mol Gen Genet 244:563–571

Ito M (2005) Conservation and diversification of three repeat Myb transcription factors in plants. J Plant Res 118:61–69

James VA, Neibaur I, Altpeter F (2008) Stress inducible expression of the DREB1A transcription factor from xeric, *Hordeum spontaneum* L. in turf and forage grass (Paspalum notatum Flugge) enhances abiotic stress tolerance. Transgenic Res 17:93–104

Jensen MK, Kjaersgaard T, Nielsen MM, Galberg P, Petersen K, O'Shea C, Skriver K (2010) The *Arabidopsis thaliana* NAC transcription factor family: structure-function relationships and determinants of ANAC019 stress signalling. Biochem J 426:183–196

Jeong JS, Kim YS, Baek KH, Jung H, Ha SH, Choi YD, Kim M, Reuzeau C, Kim JK (2010) Root-specific expression of OsNAC10 improves drought tolerance and grain yield in rice under field drought conditions. Plant Physiol 153:185–197

Jia L, Clegg MT, Jiang T (2004) Evolutionary dynamics of the DNA-binding domains in putative R2R3-MYB genes identified from rice subspecies indica and japonica genomes. Plant Physiol 134:575–585

Jia H, Wang C, Wang F, Liu S, Li G, Guo X (2015) GhWRKY68 reduces resistance to salt and drought in transgenic *Nicotiana benthamiana*. PLoS One 10:e0120646

Jiang Y, Deyholos MK (2006) Comprehensive transcriptional profiling of NaCl-stressed Arabidopsis roots reveals novel classes of responsive genes. BMC Plant Biol 6:25

Jiang Y, Deyholos MK (2009) Functional characterization of Arabidopsis NaCl-inducible WRKY25 and WRKY33 transcription factors in abiotic stresses. Plant Mol Biol 69:91–105

Jin T, Chang Q, Li W, Yin D, Li Z, Wang D, Liu B, Liu L (2010) Stress-inducible expression of GmDREB1conferred salt tolerance in transgenic alfalfa. Plant Cell Tiss Org Cult 100:219–227

Jung C, Seo JS, Han SW, Koo YJ, Kim CH, Song SI, Nahm BH, Choi YD, Cheong JJ (2008) Overexpression of AtMYB44 enhances stomatal closure to confer abiotic stress tolerance in transgenic Arabidopsis. Plant Physiol 146:623–635

Kalde M, Barth M, Somssich IE, Lippok B (2003) Members of the Arabidopsis WRKY group III transcription factors are part of different plant defence signalling pathways. Mol Plant-Microbe Interact 16:295–305

Kaneda T, Taga Y, Takai R, Iwano M, Matsui H, Takayama S, Isogai A, Che FS (2009) The transcription factor OsNAC4 is a key positive regulator of plant hypersensitive cell death. EMBO J 4:740–742

Katiyar A, Smita S, Lenka SK, Rajwanshi R, Chinnusamy V, Bansal KC (2012) Genome wide classification and expression analysis of MYB transcription factor families in rice and Arabidopsis. BMC Genomics 13:544

Kidokoro S, Maruyama K, Nakashima K, Imura Y, Narusaka Y, Shinwari ZK, Osakabe Y, Fujita Y, Mizoi J, Shinozaki K, Yamaguchi-Shinozaki K (2009) The phytochrome-interacting factor PIF7 negatively regulates DREB1 expression under circadian control in Arabidopsis. Plant Physiol 151:2046–2057

Kleinow T, Himbert S, Krenz H, Jeske H, Koncz C (2009) NAC domain transcription factor ATAF1 interacts with SNF1-related kinases and silencing of its subfamily causes severe developmental defects in Arabidopsis. Plant Sci 177:360–370

Le DT, Nishiyama R, Watanabe Y, Mochida K, Yamaguchi-Shinozaki K, Shinozaki K, Tran LS (2011) Genome-wide survey and expression analysis of the plant-specific NAC transcription factor family in soybean during development and dehydration stress. DNA Res 18:263–276

Li D, Zhang Y, Hu X, Shen X, Ma L, Su Z, Wang T, Dong J (2011) Transcriptional profiling of *Medicago truncatula* under salt stress identified a novel CBF transcription factor MtCBF4 that plays an important role in abiotic stress responses. BMC Plant Biol 11:109

Li M, Li Y, Li H, Wu G (2012) Improvement of paper mulberry tolerance to abiotic stresses by ectopic expression of tall fescue FaDREB1. Tree Physiol 32:104–113

Li C, Ng CKY, Fan LM (2014a) MYB transcription factors active players in abiotic stress signaling. Environ Exp Bot 114:80–91

Li X, Zhang D, Li H, Wang Y, Zhang Y, Wood AJ (2014b) EsDREB2B a novel truncated DREB2-type transcription factor in the desert legume *Eremosparton songoricum*, enhances tolerance to multiple abiotic stresses in yeast and transgenic tobacco. BMC Plant Biol 14:44

Li JB, Luan YS, Liu Z (2015) Overexpression of SpWRKY1 promotes resistance to *Phytophthora nicotianae* and tolerance to salt and drought stress in transgenic tobacco. Physiol Plant 155:248–266

Liao Y, Zou HF, Wang HW, Zhang WK, Ma B, Zhang JS, Chen SY (2008) Soybean GmMYB76, GmMYB92, and GmMYB177 genes confer stress tolerance in transgenic Arabidopsis plants. Cell Res 18:1047–1060

Liu Q, Kasuga M, Sakuma Y, Abe H, Miura S, Yamaguchi-Shinozaki K, Shinozaki K (1998) Two transcription factors DREB1 and DREB2 with an EREBP/AP2 DNA binding domain separate two cellular signal transduction pathways in drought- and low temperature-responsive gene expression respectively in Arabidopsis. Plant Cell 10:1391–1406

Liu Y, Schiff M, Dinesh-Kumar SP (2004) Involvement of MEK1, MAPKK, NTF6, MAPK, WRKY/MYB transcription factors, COI1 and CTR1 in N-mediated resistance to tobacco mosaic virus. Plant J 38:800–809

Liu N, Zhong NQ, Wang GL, Li LJ, Liu XL, He YK (2007) Cloning and functional characterization of PpDBF1 gene encoding a DRE-binding transcription factor from *Physcomitrella patens*. Planta 226:827–838

Liu H, Zhou X, Dong N, Liu X, Zhang H, Zhang Z (2011a) Expression of a wheat MYB gene in transgenic tobacco enhances resistance to *Ralstonia solanacearum*, and to drought and salt stresses. Funct Integr Genomics 11:431–443

Liu X, Hong L, Li XY, Yao Y, Hu B, Li L (2011b) Improved drought and salt tolerance in transgenic Arabidopsis overexpressing a NAC transcriptional factor from *Arachis hypogaea*. Biosci Biotechnol Biochem 75:443–450

Liu Q, Zhong M, Li S, Pan Y, Jiang B, Jia Y, Zhang H (2013) Overexpression of a chrysanthemum transcription factor gene, DgWRKY3, in tobacco enhances tolerance to salt stress. Plant Physiol Biochem 69:27–33

Liu X, Song Y, Xing F, Wang N, Wen F, Zhu C (2016a) GhWRKY25, a group I WRKY gene from cotton, confers differential tolerance to abiotic and biotic stresses in transgenic *Nicotiana benthamiana*. Protoplasma 253:1265–1281

Liu Y, Sun J, Wu Y (2016b) Arabidopsis ATAF1 enhances the tolerance to salt stress and ABA in transgenic rice. J Plant Res 129:955–962

Mackay JP, Crossley M (1998) Zinc fingers are sticking together. Trends Biochem Sci 23:1–4

Mangelsen E, Kilian J, Berendzen KW, Kolukisaoglu ÜH, Harter K, Jansson C, Wanke D (2008) Phylogenetic and comparative gene expression analysis of barley (*Hordeum vulgare*) WRKY transcription factor family reveals putatively retained functions between monocots and dicots. BMC Genomics 9:194

Mao X, Jia D, Li A, Zhang H, Tian S, Zhang X, Jia J, Jing R (2011a) Transgenic expression of TaMYB2A confers enhanced tolerance to multiple abiotic stresses in Arabidopsis. Funct Integr Genomics 11:445–465

Mao X, Zhang H, Qian X, Li A, Zhao G, Jing R (2011b) TaNAC2, a NAC-type wheat transcription factor conferring enhanced multiple abiotic stress tolerances in Arabidopsis. J Exp Bot 63:2933–2946

Mao X, Chen S, Li A, Zhai C, Jing R (2014) Novel NAC transcription factor TaNAC67 confers enhanced multi-abiotic stress tolerances in Arabidopsis. PLoS One 9:e84359

Matsukura S, Mizoi J, Yoshida T, Todaka D, Ito Y, Maruyama K, Shinozaki K, Yamaguchi-Shinozaki K (2010) Comprehensive analysis of rice DREB2 type genes that encode transcription factors involved in the expression of abiotic stress- responsive genes. Mol Genet Genomics 283:185–196

Mengiste T, Chen X, Salmeron J, Dietrich R (2003) The BOTRYTIS SUSCEPTIBLE1 gene encodes an R2R3MYB transcription factor protein that is required for biotic and abiotic stress responses in Arabidopsis. Plant Cell 15:2551–2565

Miao M, Niu X, Kud J, Du X, Avila J, Devarenne TP, Kuhl JC, Liu Y, Xiao F (2016) The ubiquitin ligase SEVEN IN ABSENTIA (SINA) ubiquitinates a defense-related NAC transcription factor and is involved in defense signaling. New Phytol 211:138–148

Munns R, Tester M (2008) Mechanisms of salinity tolerance. Annu Rev Plant Biol 59:651–681

Nagaoka S, Takano T (2003) Salt tolerance related protein STO binds to a Myb transcription factor homologue and confers salt tolerance in Arabidopsis. J Exp Bot 54:2231–2237

Nakashima K, Tran LS, Van Nguyen D, Fujita M, Maruyama K, Todaka D, Ito Y, Hayashi N, Shinozaki K, Yamaguchi-Shinozaki K (2007) Functional analysis of a NAC-type transcription factor OsNAC6 involved in abiotic and biotic stress-responsive gene expression in rice. Plant J 51:617–630

Nakashima K, Takasaki H, Mizoi J, Shinozaki K, Yamaguchi-Shinozaki K (2012) NAC transcription factors in plant abiotic stress responses. Biochim Biophys Acta 1819:97–103

Niu CF, Wei W, Zhou QY, Tian AG, Hao YJ, Zhang WK, Ma B, Lin Q, Zhang ZB, Zhang JS, Chen SY (2012) Wheat WRKY genes TaWRKY2 and TaWRKY19 regulate abiotic stress tolerance in transgenic Arabidopsis plants. Plant Cell Environ 35:1156–1170

Nuruzzaman M, Manimekalai R, Sharoni AM, Satoh K, Kondoh H, Ooka H, Kikuchi S (2010) Genome-wide analysis of NAC transcription factor family in rice. Gene 465:30–44

Ogata K, Kanei-Ishii C, Sasaki M, Hatanaka H, Nagadoi A, Enari M, Nakamura H, Nishimura Y, Ishii S, Sarai A (1996) The cavity in the hydrophobic core of Myb DNA-binding domain is reserved for DNA recognition and trans-activation. Nat Struct Biol 3:178–187

Oh SJ, Song SI, Kim YS, Jang HJ, Kim SY, Kim M, Kim YK, Nahm BH, Kim JK (2005) Arabidopsis CBF3/DREB1A and ABF3 in transgenic rice increased tolerance to abiotic stress without stunting growth. Plant Physiol 138:341–351

Oh SJ, Kwon CW, Choi DW, Song SI, Kim JK (2007) Expression of barley HvCBF4 enhances tolerance to abiotic stress in transgenic rice. Plant Biotechnol J 5:646–656

Ohnishi T, Sugahara S, Yamada T, Kikuchi K, Yoshiba Y, Hirano HY, Tsutsumi N (2005) OsNAC6, a member of the NAC gene family, is induced by various stresses in rice. Genes Genet Syst 80:135–139

Olsen AN, Ernst HA, Leggio LL, Skriver K (2005) DNA-binding specificity and molecular functions of NAC transcription factors. Plant Sci 169:785–797

Ooka H, Ooka H, Satoh K, Nagata T, Otomo Y, Murakami K, Matsubara K, Osato N, Kawai J, Carninci P, Hayashizaki Y, Suzuki K, Kojima K, Takahara Y, Yamamoto K, Kikuchi S (2003) Comprehensive analysis of NAC family genes in Oryza sativa and Arabidopsis thaliana. DNA Res 247:239–247

Orbay H, Ahmad R (2012) Physiological and biochemical changes of CBF3 transgenic oat in response to salinity stress. Plant Sci 185:331–339

Pandey SP, Somssich IE (2009) The role of WRKY transcription factors in plant immunity. Plant Physiol 150:1648–1655

Park CY, Lee JH, Yoo JH, Moon BC, Choi MS, Kang YH, Lee SM, Kim HS, Kang KY, Chung WS, Lim CO, Cho MJ (2005) WRKY group II d transcription factors interact with calmodulin. FEBS Lett 579:1545–1550

Pascual MB, Cánovas FM, Ávila C (2015) The NAC transcription factor family in maritime pine (*Pinus Pinaster*): molecular regulation of two genes involved in stress responses. BMC Plant Biol 15:254

Pastori GM, Foyer CH (2002) Common components, networks and pathways of cross-tolerance to stress. The central role of 'redox' and abscisic-acid-mediated controls. Plant Physiol 129:460–468

Paz-Ares J, Ghosal D, Wlenand U, Peterson PA, Saedler H (1987) The regulatory c7 locus of *Zea mays* encodes a protein with homology to myb proto-oncogene products and with structural similarities to transcriptional activators. EMBO J 6:3553–3558

Puranik S, Sahu PP, Srivastava PS, Prasad M (2012) NAC proteins: regulation and role in stress tolerance. Trends Plant Sci 17:369–381

Qin F, Kakimoto M, Sakuma Y, Maruyama K, Osakabe Y, Tran LS, Shinozaki K, Yamaguchi-Shinozaki K (2007) Regulation and functional analysis of ZmDREB2A in response to drought and heat stresses in *Zea mays* L. Plant J 50:54–69

Qin F, Sakuma Y, Tran LS, Maruyama K, Kidokoro S, Fujita Y, Umezawa T, Sawano Y, Miyazono K, Tanokura M, Shinozaki K, Yamaguchi-Shinozaki K (2008) Arabidopsis DREB2A-interacting proteins function as RING E3 ligases and negatively regulate plant drought stress-responsive gene expression. Plant Cell 20:1693–1707

Qin Y, Wang M, Tian Y, He W, Han L, Xia G (2012) Over-expression of TaMYB33 encoding a novel wheat MYB transcription factor increases salt and drought tolerance in Arabidopsis. Mol Biol Rep 39:7183–7719

Qiu Y, Yu D (2009) Over-expression of the stress-induced OsWRKY45 enhances disease resistance and drought tolerance in Arabidopsis. Environ Exp Bot 65:35–47

Rahman H, Ramanathan V, Nallathambi J, Duraialagaraja S, Muthurajan R (2016) Over-expression of a NAC 67 transcription factor from finger millet (*Eleusine coracana* L.) confers tolerance against salinity and drought stress in rice. BMC Biotechnol 16:35

Ramegowda V, Kumar MS, Nataraja KN, Reddy MK, Mysore KS, Udayakumar M (2012) Expression of a finger millet transcription factor, EcNAC1, in tobacco confers abiotic stress-tolerance. PLoS One 7:e40397

Riechmann JL (2000) Arabidopsis transcription factors genome wide comparative analysis among eukaryotes. Science 290:2105–2110

Riechmann JL, Ratcliffe OJ (2000) A genomic perspective on plant transcription factors. Curr Opin Plant Biol 3:423–434

Rogers S, Wells R, Rechsteiner M (1986) Amino acid sequences common to rapidly degraded proteins: the PEST hypothesis. Science 234:364–368

Romero I, Fuertes A, Benito MJ, Malpica JM, Leyva A, Paz Ares J (1998) More than 80R2R3-MYB regulatory genes in the genome of *Arabidopsis thaliana*. Plant J 14:273–284

Rushton PJ, Bokowiec MT, Han S, Zhang H, Brannock JF, Chen X, Laudeman TW, Timko MP (2008) Tobacco transcription factors: novel insights into transcriptional regulation in the Solanaceae. Plant Physiol 147:280–295

Rushton PJ, Somssich IE, Ringler P, Shen QJ (2010) WRKY transcription factors. Trends Plant Sci 15:247–258

Saad ASI, Li X, Li HP, Huang T, Gao CS, Guo MW, Cheng W, Zhao GY, Liao YC (2013) A rice stress-responsive NAC gene enhances tolerance of transgenic wheat to drought and salt stresses. Plant Sci 35:1783–1798

Sakuma Y, Liu Q, Dubouzet JG, Abe H, Shinozaki K, Yamaguchi-Shinozaki K (2002) DNA-binding specificity of the ERF/AP2 domain of Arabidopsis DREBs transcription factors involved in dehydration and cold inducible gene expression. Biochem Biophys Res Commun 290:998–1009

Sakuma Y, Maruyama K, Osakabe Y, Qin F, Seki M, Shinozaki K, Yamaguchi-Shinozaki K (2006) Functional analysis of an Arabidopsis transcription factor DREB2A involved in drought responsive gene expression. Plant Cell 18:1292–1309

Schachtman DP, Liu WH (1999) Molecular pieces to the puzzle of the interaction between potassium and sodium uptake in plants. Trends Plant Sci 4:281–287

Schmidt R, Schippers JHM, Mieulet D, Obata T, Fernie AR, Guiderdoni E, Mueller Roeber B (2013) MULTIPASS a rice R2R3-type MYB transcription factor regulates adaptive growth by integrating multiple hormonal pathways. Plant J 76:258–273

Schmutz J, Cannon SB, Schlueter J, Ma J, Mitros T, Nelson W et al (2010) Genome sequence of the palaeopolyploid soybean. Nature 463:178–183

Seki M, Narusaka M, Abe H, Kasuga M, Yamaguchi-Shinozaki K, Carninci P, Hayashizaki Y, Shinozaki K (2001) Monitoring the expression pattern of 1300 Arabidopsis genes under drought and cold stresses by using a full-length cDNA microarray. Plant Cell 13:61–72

Seki M, Narusaka M, Ishida J, Nanjo T, Fujita M, Oono Y, Kamiya A, Nakajima M, Enju A, Sakurai T, Satou M, Akiyama K, Taji T, Yamaguchi-Shinozaki K, Carninci P, Kawai J, Hayashizaki Y, Shinozaki K (2002) Monitoring expression profile of 7000 Arabidopsis genes under drought, cold-and high-salinity stresses using a full-length cDNA microarray. Plant J 31:279–292

Seo JS, Sohn HB, Noh K, Jung C, An JH, Donovan CM, Somers DA, Kim DI, Jeong SC, Kim CG, Kim HM, Lee SH, Choi YD, Moon TW, Kim CH, Cheong JJ (2012) Expression of the Arabidopsis AtMYB44 gene confers drought/salt-stress tolerance in transgenic soybean. Mol Breed 29:601–608

Shabala S, Cuin TA (2008) Potassium transport and plant salt tolerance. Physiol Plant 133:651–669

Shan H, Chen S, Jiang J, Chen F, Chen Y, Gu C, Li P, Song A, Zhu X, Gao H, Zhou G, Li T, Yang X (2012) Heterologous expression of the Chrysanthemum R2R3-MYB transcription factor CmMYB2 enhances drought and salinity tolerance, increases hypersensitivity to ABA and delays flowering in Arabidopsis thaliana. Mol Biotechnol 51:160–173

Shao HB, Chu LY, Zhao CX, Guo QJ, Liu XA, Ribaut JM (2006) Plant gene regulatory network system under abiotic stress. Acta Biol Szegediensis 50:1–9

Shen H, Yin Y, Chen F, Xu Y, Dixon RA (2009) A bioinformatic analysis of NAC genes for plant cell wall development in relation to lignocellulosic bioenergy production. Bioenergy Res 2:217–223

Shinozaki K, Yamaguchi-Shinozaki K (2000) Molecular responses to dehydration and low temperature: differences and cross-talk between two stress signaling pathways. Curr Opin Plant Biol 3:217–223

Shiu SH, Shih MC, Li WH (2005) Transcription factor families have much higher expansion rates in plants than in animals. Plant Physiol 139:18–26

Souer E, Van Houwelingen A, Kloos D, Mol J, Koes R (1996) The no apical Meristem gene of petunia is required for pattern formation in embryos and flowers and is expressed at meristem and primordia boundaries. Cell 85:159–170

Sun J, Peng X, Fan W, Tang M, Liu J, Shen S (2014) Functional analysis of BpDREB2 gene involved in salt and drought response from a woody plant Broussonetia papyrifera. Gene 535:140–149

Tak H, Negi S, Ganapathi TR (2016) Banana NAC transcription factor MusaNAC042 is positively associated with drought and salinity tolerance. Protoplasma 254:803–816

Takasaki H, Maruyama K, Kidokoro S, Ito Y, Fujita Y, Shinozaki K, Yamaguchi-Shinozaki K, Nakashima K (2010) The abiotic stress-responsive NAC-type transcription factor OsNAC5 regulates stress-inducible genes and stress tolerance in rice. Mol Gen Genomics 284:173–183

Tran LP, Nakashima K, Sakuma Y, Simpson SD, Fujita Y, Maruyama K, Fujita M, Seki M, Shinozaki K, Yamaguchi-Shinozaki K (2004) Isolation and functional analysis of Arabidopsis Stress-inducible NAC transcription factors that bind to a drought-responsive cis-element in the early responsive to dehydration stress 1 Promoter. Plant Cell 16:2481–2498

Tran LS, Nishiyama R, Yamaguchi-Shinozaki K, Shinozaki K (2010) Potential utilization of NAC transcription factors to enhance abiotic stress tolerance in plants by biotechnological approach. GM Crops 1, 32–39. 10.4161/gmcr.1.1.10569

Ulker B, Somssich IE (2004) WRKY transcription factors from DNA binding towards biological function. Curr Opin Plant Biol 7:491–498

Vainonen JP, Jaspers P, Wrzaczek M, Lamminmaki A, Reddy RA, Vaahtera L, Brosche M, Kangasjarvi J (2012) RCD1-DREB2A interaction in leaf senescence and stress responses in *Arabidopsis thaliana*. Biochem J 442:573–581

Vanderauwera S, Vandenbroucke K, Inzé A, van de Cotte B, Mühlenbock P, De Rycke R, Naouar N, Van Gaever T, Van Montagu MC, Van Breusegem F (2012) AtWRKY15 perturbation abolishes the mitochondrial stress response that steers osmotic stress tolerance in Arabidopsis. Proc Natl Acad Sci U S A 109:20113–20118

Vannini C, Locatelli F, Bracale M, Magnani E, Marsoni M, Osnato M, Mattana M, Baldoni E, Coraggio I (2004) Overexpression of the rice Osmyb4 gene increases chilling and freezing tolerance of *Arabidopsis thaliana* plants. Plant J 37:115–127

Vannini C, Iriti M, Bracale M, Locatelli F, Faoro F, Croce P, Pirona R, Di Maro A, Coraggio I, Genga A (2006) The ectopic expression of the rice Osmyb4 gene in Arabidopsis increases tolerance to abiotic, environmental and biotic stresses. Physiol Mol Plant Pathol 69:26–42

Wang Q, Guan Y, Wu Y, Chen H, Chen F, Chu C (2008) Overexpression of a rice OsDREB1F gene increases salt, drought, and low temperature tolerance in both Arabidopsis and rice. Plant Mol Biol 67:589–602

Wang X, Chen X, Liu Y, Gao H, Wang Z, Sun G (2011) CkDREB gene in Caragana korshinskii is involved in the regulation of stress response to multiple abiotic stresses as an AP2/EREBP transcription factor. Mol Biol Rep 38:2801–2811

Wang F, Hou X, Tang J (2012) A novel cold-inducible gene from Pak-choi (Brassica campestris ssp. chinensis), BcWRKY46, enhances the cold, salt and dehydration stress tolerance in transgenic tobacco. Mol Biol Rep 39:4553–4564

Wang C, Deng P, Chen L, Wang X, Ma H, Hu W, Yao N, Feng Y, Chai R, Yang G, He G (2013) A wheat WRKY transcription factor TaWRKY10 confers tolerance to multiple abiotic stresses in transgenic tobacco. PLoS One 8:e65120

Wang RK, Cao ZH, Hao YJ (2014) Overexpression of a R2R3 MYB gene MdSIMYB1 increases tolerance to multiple stresses in transgenic tobacco and apples. Physiol Plant 150:76–87

Wang X, Zeng J, Li Y, Rong X, Sun J, Sun T, Li M (2015) Expression of TaWRKY44, a wheat WRKY gene, in transgenic tobacco confers multiple abiotic stress tolerances. Front Plant Sci 6:615

Wu G, Shao HB, Chu LY, Cai JW (2007) Insights into molecular mechanisms of mutual effect between plants and the environment. A review. Agron Sustain Dev 27:69–78

Xianjun P, Xingyong M, Weihong F, Man S, Liqin C, Alam I, Lee BH, Dongmei Q, Shihua S, Gongshe L (2011) Improved drought and salt tolerance of Arabidopsis thaliana by transgenic expression of a novel DREB gene from Leymus chinensis. Plant Cell Rep 30:1493–1502

Xie Z, Zhang ZL, Zou X, Huang J, Ruas P, Thompson D, Shen QJ (2005) Annotations and functional analyses of the rice WRKY gene superfamily reveal positive and negative regulators of abscisic acid signaling in aleurone cells. Plant Physiol 137:176–189

Xie Z, Li D, Wang L, Sack FD, Grotewold E (2010) Role of the stomatal development regulators FLP/MYB88 in abiotic stress responses. Plant J 64:731–739

Xiong W, Xu X, Zhang L, Wu P, Chen Y, Li M, Jiang H, Wu G (2013) Genome-wide analysis of the WRKY gene family in physic nut (Jatropha curcas L.) Gene 524:124–132

Xiong H, Li J, Liu P, Duan J, Zhao Y, Guo X, Li Y, Zhang H, Ali J, Li Z (2014) Overexpression of OsMYB48-1, a novel MYB-related transcription factor, enhances drought and salinity tolerance in rice. PLoS One 9:e92913

Xu ZS, Ni ZY, Liu L, Nie LN, Li LC, Chen M, Ma YZ (2008) Characterization of the TaAIDFa gene encoding a CRT/DRE binding factor responsive to drought, high-salt, and cold stress in wheat. Mol Genet Genomics 280:497–508

Xu ZS, Ni ZY, Li ZY, Li LC, Chen M, Gao DY, Yu XD, Liu P, Ma YZ (2009) Isolation and functional characterization of *HvDREB1-a* gene encoding a dehydration- responsive element binding protein in *Hordeum vulgare*. J Plant Res 122:121–130

Xu Z, Gongbuzhaxi, Wang C, Xue F, Zhang H, Ji W (2015) Wheat NAC transcription factor TaNAC29 is involved in response to salt stress. Plant Physiol Biochem 96:356–363

Yamasaki K, Kigawa T, Inoue M, Tateno M, Yamasaki T, Yabuki T, Aoki M, Seki E, Matsuda T, Tomo Y, Hayami N, Terada T, Shirouzu M, Tanaka A, Seki M, Shinozaki K (2005) Yokoyama S. Solution structure of an Arabidopsis WRKY DNA binding domain. Plant Cell 17:944–956

Yang B, Jiang Y, Rahman MH, Deyholos MK, Kav NN (2009) Identification and expression analysis of WRKY transcription factor genes in canola (*Brassica napus* L.) in response to fungal pathogens and hormone treatments. BMC Plant Biol 9(68):10–1186

Yang A, Dai X, Zhang W-H (2012) A R2R3-type MYB gene, OsMYB2, is involved in salt, cold, and dehydration tolerance in rice. J Exp Bot 63:2541–2556

Yang Y, Zhu K, Wu J, Liu L, Sun G, He Y, Chen F, Yu D (2016) Identification and characterization of a novel NAC-like gene in chrysanthemum (*Dendranthema lavandulifolium*). Plant Cell Rep 35:1783–1798

Yanhui C, Xiaoyuan Y, Kun H et al (2006) The MYB transcription factor superfamily of Arabidopsis expression analysis and phylogenetic comparison with the rice MYB family. Plant Mol Biol 60:107–124

Yao DX, Wei Q, Xu WY, Syrenne RD, Yuan JS, Su Z (2012) Comparative genomic analysis of NAC transcriptional factors to dissect the regulatory mechanisms for cell wall biosynthesis. BMC Bioinf 13:12

Yokotani N, Ichikawa T, Kondou Y, Matsui M, Hirochika H, Iwabuchi M, Oda K (2009) Tolerance to various environmental stresses conferred by the salt-responsive rice gene ONAC063 in transgenic Arabidopsis. Planta 229:1065–1075

Yokotani N, Ichikawa T, Kondou Y, Iwabuchi M, Matsui M, Hirochika H, Oda K (2013) Role of the rice transcription factor JAmyb in abiotic stress response. J Plant Res 126:131–139

Zhang Y, Wang L (2005) The WRKY transcription factor super family its origin in eukaryotes and expansion in plants. BMC Evol Biol 5:1–12

Zhang Y, Chen C, Jin XF, Xiong AS, Peng RH, Hong YH, Yao QH, Chen JM (2009) Expression of a rice DREB1 gene OsDREB1D enhances cold and high-salt tolerance in transgenic Arabidopsis. BMB Rep 42:486–492

Zhang L, Zhao G, Xia C, Jia J, Liu X, Kong X (2012) Overexpression of a wheat MYB transcription factor gene TaMYB56-B enhances tolerances to freezing and salt stresses in transgenic Arabidopsis. Gene 505:100–107

Zhang ZW, Feng LY, Cheng J, Tang H, Xu F, Zhu F, Zhao ZY, Yuan M, Chen YE, Wang JH, Yuan S, Lin HH (2013) The roles of two transcription factors ABI4 and CBFA, in ABA and plastid signalling and stress responses. Plant Mol Biol 83:445–458

Zhang X, Liu X, Wu L, Yu G, Wang X, Ma H (2015) The SsDREB transcription factor from the succulent halophyte Suaeda salsa enhances abiotic stress tolerance in transgenic tobacco. Int J Genomics 2015

Zheng X, Chen B, Lu G, Han B (2009) Overexpression of a NAC transcription factor enhances rice drought and salt tolerance. Biochem Biophy Res Comm 379:985–989

Zhou QY, Tian AG, Zou HF, Xie ZM, Lei G, Huang J, Wang CM, Wang HW, Zhang JS, Chen SY (2008) Soybean WRKY-type transcription factor genes, GmWRKY13, GmWRKY21, and GmWRKY54, confer differential tolerance to abiotic stresses in transgenic Arabidopsis plants. Plant Biotechnol J 6:486–503

Zhu JK (2001) Plant salt tolerance. Trends Plant Sci 6:66–71

Zhu T, Budworth P, Han B, Brown D, Chang HS, Zou G, Wang X (2001) Towards elucidating the global expression patterns of developing Arabidopsis: parallel analysis of 8300 genes by a high-density oligonucleotide probe array. Plant Physiol Biochem 39:221–242

Zhu Y, Yan J, Liu W, Liu L, Sheng Y, Sun Y, Li Y, Scheller HV, Jiang M, Hou X, Ni L, Zhang A (2016) Phosphorylation of a NAC transcription factor by ZmCCaMK regulates abscisic acid-induced antioxidant defense in maize. Plant Physiol 171:1651–1664

Zong JM, Li XW, Zhou YH, Wang FW, Wang N, Dong YY, Yuan YX, Chen H, Liu XM, Yao N, Li HY (2016) The AaDREB1 transcription factor from the cold-tolerant plant *Adonis amurensis* enhances abiotic stress tolerance in transgenic plant. Int J Mol Sci 17:E611

Chapter 8
Targeting Redox Regulatory Mechanisms for Salinity Stress Tolerance in Crops

Mohsin Tanveer and Sergey Shabala

Abstract Salinity stress is one of the major abiotic stresses that result in significant losses in agricultural crop production across the globe. Salinity stress results in osmotic stress, ionic stress, and oxidative stress; among these, oxidative stress is considered to be the most detrimental. Oxidative stress induces the production of different reactive oxygen species (ROS) at both intracellular and extracellular locations. Plants possess redox regulatory mechanisms by employing different enzymatic and nonenzymatic antioxidants to scavenge ROS. Different antioxidants have different tissue- and organelle-specific ROS-scavenging effects. However, the causal link between the amount of antioxidants and plant salinity stress tolerance is not as straightforward as one may assume, with controversial reports available in the literature. This chapter addresses those controversies and argues that there is a need for better understanding and development of tools for targeted regulation of plant redox systems in specific cellular compartments and tissues.

Keywords Antioxidant defense system · QTL · Redox regulation · ROS production · Salt stress · Tissue specific antioxidant activity

Abbreviations

$^1O_2$	Singlet oxygen
AKR1	NADPH-dependent aldo-ketoreductase
APX	Ascorbate peroxidase
ASC	Ascorbic acid
BADH	Betaine aldehyde dehydrogenase
CAT	Catalase
codA	Choline dehydrogenase gene

M. Tanveer · S. Shabala (✉)
School of Land and Food, University of Tasmania, Hobart, TAS, Australia
e-mail: Sergey.shabala@utas.edu.au

V. Kumar et al. (eds.), *Salinity Responses and Tolerance in Plants, Volume 1*,
https://doi.org/10.1007/978-3-319-75671-4_8

DHA	Dehydroascorbic acid
DHAR	Dehydroascorbate reductase
GPX	Glutathione peroxidase
GR	Glutathione reductase
GSH	Glutathione
GST	Glutathione S-transferase
L-ASC	L-Ascorbic acid
MAPK 1	Mitogen-activated protein kinase phosphatase
MDA	Malondialdehyde
MDHA	Monodehydroascorbic acid
MDHAR	Monodehydroascorbate reductase
MeOOH	Methyl hydrogen peroxide
NADP	Nicotinamide adenine dinucleotide phosphate
NADPH	Reduced NADP
O_2^-	Superoxide radical
OH^-	Hydroxyl radical
Orn-δ-OAT	Ornithine-δ-aminotransferase
P5CR	Pyrroline-5-carboxylate reductase
P5CS	Pyrroline-5-carboxylate synthase
POD	Peroxidase
PSI	Photosystem I
PSII	Photosystem II
PUFA	Polyunsaturated fatty acid
QTL	Quantitative trait locus
ROS	Reactive oxygen species
SOD	Superoxide dismutase

8.1 Introduction

Soil salinity is one of the most detrimental abiotic stresses that critically damage crops and cause major reductions in their yield (Munns et al. 2006; Tanveer and Shah 2017). Salinity is characterized by deleterious effects on plant growth, which are traditionally associated with reduced water availability under hyperosmotic saline conditions, and specific ion toxicity. In recent years, oxidative damage has been added to this list (Souza et al. 2012; Liu et al. 2017; López-Gómez et al. 2017). Like other aerobic organisms, higher plants require oxygen for efficient production of energy. During the reduction of O_2 to H_2O, reactive oxygen species (ROS)—namely, superoxide radicals (O_2^-), H_2O_2, and hydroxyl radicals (OH^-)—are formed (Demidchik 2015). Most cellular compartments in higher plants have the potential to become a source of ROS (Bose et al. 2014). Environmental stresses that limit CO_2

fixation, including salinity, reduce nicotinamide adenine dinucleotide phosphate ($NADP^+$) regeneration by the Calvin cycle. Consequently, the photosynthetic electron transport chain is over-reduced, producing O_2^- and singlet oxygen ($^1O_2$) in chloroplasts (Wu and Tang 2004; Shao and Chu 2005). To prevent overreduction of the electron transport chain under conditions that limit CO_2 fixation, higher plants have evolved the photorespiratory pathway to regenerate $NADP^+$ (Shao and Chu 2005). As a part of the photorespiratory pathway, H_2O_2 is produced in the peroxisomes, where it can also be formed during the catabolism of lipids as a by-product of β-oxidation of fatty acids (Foyer and Noctor 2005; Wu et al. 2007).

Because of the highly cytotoxic and reactive nature of ROS, their accumulation in plant tissues and intracellular compartments must be tightly controlled. Higher plant metabolism must be highly regulated in order to allow effective integration of a diverse spectrum of biosynthetic pathways that are reductive in nature (Crawford 2006; Grun et al. 2006). This requires the provision that the regulation does not completely avoid photodynamic or reductive activation of molecular oxygen to produce ROS, particularly superoxide, H_2O_2, and $^1O_2$ (Suzuki et al. 2012). However, in many cases, the production of ROS is genetically programmed, is induced during the course of development and by environmental fluctuations, and has complex downstream effects on both primary and secondary metabolism (Schurmann 2003; Link 2003; Rouhier et al. 2003). As a result, higher plants possess very efficient enzymatic and nonenzymatic antioxidant defense systems that allow scavenging of ROS and protection of plant cells from oxidative damage (Foyer and Noctor 2003; Anjum et al. 2016a, 2017).

For many years, the concept "the higher the antioxidant activity, the better the plant" has dominated the literature. However, in recent years it has become apparent that plants actively produce ROS as signaling molecules to control numerous physiological processes such as defense responses and cell death (Zhang et al. 2003), cross-tolerance (Bowler and Fluhr 2000), gravitropism (Mittler et al. 2004), stomatal aperture (Wang and Song 2008; Pei et al. 2000), cell expansion and polar growth (Joo et al. 2005; Pedreira et al. 2004), hormone action (Pei et al. 2000; Schopfer et al. 2002), and leaf and flower development (Sagi et al. 2004). In many cases, production of ROS is genetically programmed, and superoxide and H_2O_2 are used as second messengers (Foyer and Noctor 2005). A new concept of "oxidative signaling"—instead of "oxidative stress"—has been proposed (Foyer and Noctor 2005). The "positive" role of ROS has been reported at both the physiological level [e.g., regulation of ion channel activity (Foreman et al. 2003)] and the genetic level [e.g., control of gene expression (Shin and Schachtman 2004)]. This has prompted a need to rethink the above "the more, the better" concept and incorporate the signaling role of ROS and the redox state of the cell into breeding programs aimed at improving abiotic stress tolerance. Some aspects of this work are discussed in this chapter.

8.2 Antioxidant Defense System

The distinct subcellular localization and biochemical properties of antioxidant enzymes, their differential activation at the enzyme and gene expression level, and the plethora of nonenzymatic scavengers render the antioxidant system a very versatile and flexible unit that can control ROS accumulation temporally and spatially (Shao et al. 2007a, b; Anjum et al. 2015, 2016b). The sections below describe mechanisms involved in redox regulation for salinity stress tolerance in plants.

The redox state of a cell explains the ratio of the amount of oxidizing equivalents to the amount of reducing equivalents (Gabbita et al. 2000). The intracellular antioxidant systems form a powerful reducing buffer, which affects the ability of the cell to counteract the action of the prooxidant forces. The fine redox balance within a cell is thus governed by the levels of prooxidant molecules and antioxidant fluxes. Therefore, an appreciation of the different sources of oxidants and the counteracting antioxidant (or reducing) systems is necessary to understand what factors are involved in achieving a particular intracellular redox state. Salinity induces various ROS and, in response to that, plants develop complex antioxidant defense systems (Ismail et al. 2016) (Fig. 8.1). Antioxidants can be classified into three broad divisions: water-soluble compounds (reductants, ascorbate); lipid-soluble compounds (α-tocopherol, β-carotene); and enzymes (superoxide dismutase [SOD], catalase [CAT], peroxidase [POD], ascorbate peroxidase [APX], and glutathione reductase [GR]) (Chen and Dickman 2005; Gill and Tuteja 2010).

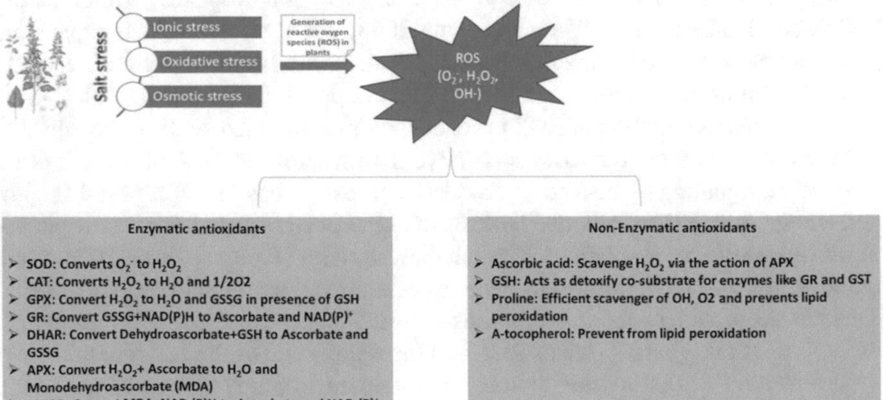

Fig. 8.1 Major plant antioxidants and their reactive oxygen species (ROS)–scavenging effects

8.2.1 Enzymatic Antioxidants

Among the different antioxidant enzymes, *SOD* is the most effective intracellular enzymatic antioxidant that is ubiquitous in all aerobic organisms. SOD can catalyze and reduce O_2^- to H_2O_2 because of its dismutation. SOD also indirectly reduces the risk of OH^- formation by using O_2^-. SOD catalyzes the first step of the enzymatic defense mechanism: the conversion of superoxide anions to hydrogen peroxide and water. If superoxide anions are not neutralized, oxidation occurs and hydroxyl radicals are formed. Hydroxyl radicals are extremely harmful because they are very reactive. Importantly, hydroxyl radicals cannot be scavenged by enzymatic means. Hydrogen peroxide can be decomposed by the activity of CATs and several classes of PODs, which act as important antioxidants.

CAT is another important enzymatic antioxidant; it is a tetrameric heme containing enzymes with the potential to directly dismutate H_2O_2 into H_2O and O_2, and it is indispensable for ROS detoxification during stressed conditions (Garg and Manchanda 2009). CAT has the highest turnover rate: one molecule of CAT can convert about 6 million molecules of H_2O_2 to H_2O and O_2 per minute. CAT enzymes remove H_2O_2 from peroxisomes by oxidases involved in β-oxidation of fatty acids, photorespiration, and purine catabolism. Different CAT isozymes have been reported in different plant species: two isozymes in barley (Azevedo et al. 1998), four in sunflower (Azpilicueta et al. 2007), and 12 in *Brassica* (Frugoli et al. 1996). Scandalias (1990) found three CAT isozymes in maize: CAT1 and CAT2 are localized in peroxisomes and the cytosol, whereas CAT3 is mitochondrial. It has also been reported that apart from reacting with H_2O_2, CAT also reacts with some hydroperoxides such as methyl hydrogen peroxide (MeOOH) (Ali and Alqurainy 2006). Overexpression of CAT encoded by the katE gene in rice conferred salt-induced oxidative stress tolerance (Nagamiya et al. 2007). Similarly, increased activity of CAT has been reported in chickpea following salt stress (Eyidogan and Oz 2005; Kukreja et al. 2005). Nonetheless, Srivastava et al. (2005) reported a decrease in CAT activity in *Anabaena doliolum* under NaCl and Cu^{2+} stress. Pan et al. (2006) studied the combined effect of salt and drought stress and found that it decreased CAT activity in *Glycyrrhiza uralensis* seedlings.

Thioredoxin and thiol-based glutathione are also important enzymes that play a crucial role in redox regulation. Thioredoxin includes a pleiotropic reduced NADP (NADPH)–dependent disulfide oxidoreductase, which catalyzes the reduction of exposed protein S–S bridges. Because of its dithiol-to-disulfide exchange activity, thioredoxin—acting as a hydrogen donor—determines the oxidation state of protein thiols (Lu and Holmgren 2014). This small 12 kDa protein contains a characteristic conserved catalytic Trp–Cys–Gly–Pro–Cys–Lys site. The two cysteine residues within this site can be oxidized reversibly to form a disulfide bridge. A specific selenoenzyme, thioredoxin reductase, is able to reduce the disulfide bond, utilizing NADPH as a hydrogen donor. The glutaredoxin system utilizes glutathione (a cysteine-containing tripeptide) in a manner similar to that used by the thioredoxin system. The antioxidant function of glutathione is implicated through two general mechanisms of reaction with ROS.

Glutathione peroxidase (GPX) belongs to a large family of diverse isozymes that use glutathione to reduce H_2O_2 and organic and lipid hydroperoxides, and therefore help plant cells with oxidative stress (Noctor et al. 2002). GPX uses glutathione to eliminate H_2O_2 and to decrease hydroperoxidation of lipids. Thus, these enzymes regulate intracellular levels of ROS such as H_2O_2 and O_2^- (Gabbita et al. 2000). Millar et al. (2003) identified a family of seven related proteins—named AtGPX1–AtGPX7—in the cytosol, chloroplast, mitochondria, and endoplasmic reticulum of *Arabidopsis*. Upregulation of the GPX gene was noted in response to $^1O_2$, showing the involvement of GPX in quenching $^1O_2$ (Leisinger et al. 2001). It was noted that GPX activity in transgenic cotton seedlings was 30–60% higher under normal conditions but no different from GPX activity in wild-type seedlings under salt stress conditions (Light et al. 2005).

8.2.2 Nonenzymatic Antioxidants

Ascorbic acid (ASC) is the most abundant, powerful, and water-soluble antioxidant that acts to prevent or minimize the damage caused by ROS in plants (Smirnoff 2005; Athar et al. 2008). ASC is considered a most powerful ROS scavenger because of its ability to donate electrons in a number of enzymatic and nonenzymatic reactions. It can provide protection to membranes by directly scavenging O_2^- and OH^-, and by regenerating α-tocopherol from the tocopheroxyl radical. In chloroplasts, ASC acts as a cofactor of violaxantin de-epoxidase, thus sustaining dissipation of excess excitation energy (Smirnoff 2005). In addition to the importance of ASC in the ascorbate–glutathione cycle, it also plays an important role in preserving the activity of enzymes that contain prosthetic transition metal ions (Noctor and Foyer 1998). The ASC redox system consists of L-ascorbic acid (L-ASC), monodehydroascorbic acid (MDHA), and dehydroascorbic acid (DHA). Both oxidized forms of ASC are relatively unstable in aqueous environments, while DHA can be chemically reduced by glutathione.

Tripeptide glutathione (glu–cys–gly) is one of the crucial metabolites in plants and is considered a most important intracellular component of defense against ROS-induced oxidative damage. It occurs abundantly in the reduced form in plant tissues and is localized in all cell compartments, including the cytosol, endoplasmic reticulum, vacuole, mitochondria, chloroplasts, and peroxisomes, as well as in the apoplast (Mittler and Zilinskas 1992; Jiménez et al. 1998). Glutathione provides a substrate for multiple cellular reactions that yield GSSG (i.e., two glutathione molecules linked by a disulfide bond). The balance between glutathione and GSSG is a central component in maintaining the cellular redox state (Foyer and Noctor 2005). Glutathione is necessary to maintain the normal reduced state of cells so as to counteract the inhibitory effects of ROS-induced oxidative stress

(Meyer 2008). It is a potent scavenger of $^1O_2$, H_2O_2, and most dangerous ROS such as OH^- (Op den Camp et al. 2003).

Proline is considered a potent antioxidant and a potential inhibitor of programmed cell death (Chen and Dickman 2005). Free proline has been proposed to also act as an osmoprotectant, a protein stabilizer, and a metal chelator, and helps in scavenging ROS (Ashraf and Foolad 2007; Trovato et al. 2008). In plants, there are two different precursors for proline. The first pathway is from glutamate, which is converted to proline by two successive reductions catalyzed by pyrroline-5-carboxylate synthase (P5CS) and pyrroline-5-carboxylate reductase (P5CR), respectively. P5CS is a bifunctional enzyme catalyzing first the activation of glutamate by phosphorylation and second the reduction of the labile intermediate c-glutamyl phosphate into glutamate semialdehyde, which is in equilibrium with the P5C form (Hu et al. 1992). An alternative precursor for proline biosynthesis is ornithine, which can be transaminated to P5C by ornithine-δ-aminotransferase (Orn-δ-OAT), a mitochondrially located enzyme. The glutamate pathway is the main pathway during osmotic stress. However, in young *Arabidopsis* plants, the ornithine pathway also seems to contribute and δ-OAT activity is enhanced (Roosens et al. 1998). Sorbitol, mannitol, myo-inositol, and proline have been tested for OH^--scavenging capacity, and it was found that proline appeared to be an effective scavenger of OH^- (Smirnoff and Cumbes 1989; Gill and Tuteja 2010). Therefore, proline is not only an important molecule in redox signaling but also an effective quencher of ROS formed under salt stress (Alia and Saradhi 1991; Tanveer and Shah 2017). Chen and Dickman (2005) showed that addition of proline to DARas mutant cells effectively quenched ROS and prevented cell death by inhibiting ROS-mediated apoptosis. Enhanced synthesis of proline under drought or salt stress has been implicated as a mechanism in alleviation of cytoplasmic acidosis and maintenance of the $NADPH:NADP^+$ ratio at a value compatible with metabolism (Hare and Cress 1997). Moreover, proline could maintain a low $NADPH:NADP^+$ ratio, decrease $^1O_2$ production from photosystem I (PSI), and lessen $^1O_2$ and OH^- damage to photosystem II (PSII) under stress (Szabados and Savouré 2010). Furthermore, proline has been reported to perform various antioxidant functions including (1) reductions in OH^-, H_2O_2, and $^1O_2$; (2) maintenance of a low $NADPH:NADP^+$ ratio to decrease $^1O_2$ production; (3) assistance in preventing programmed cell death; (4) reduction of detrimental effects of ROS on PSII; and (5) stabilization of mitochondrial respiration to protect the complex II phase of the electron transport chain in mitochondria (Szabados and Savouré 2010). Exogenously applied proline reduced the leaking of K^+ by reducing the production of hydroxyl radicals in *Arabidopsis* roots (Cuin and Shabala 2007).

Glycine betaine accumulates predominantly in chloroplasts and protects the photosynthetic apparatus during oxidative stress (Ashraf and Foolad 2007). In addition, glycine betaine stabilizes the structure and function of the oxygen-evolving complex of PSII and protects the photosynthetic apparatus under high salt stress (Papageorgiou and Murata 1995). The introduction of genes synthesizing glycine betaine into nonaccumulators of glycine betaine has been shown to be effective in increasing tolerance of various abiotic stresses (Sakamoto and Murata 2002).

In addition to the direct protective roles of glycine betaine, either through positive effects on enzyme and membrane integrity or as an osmoprotectant, glycine betaine may also protect cells from environmental stresses indirectly by participating in signal transduction pathways (Subbarao et al. 2000). Even in small concentrations, glycine betaine is very effective in stress amelioration. Exogenous application of glycine betaine (even in a submillimolar concentration) alleviates OH^--generated potassium ion leakage (Cuin and Shabala 2007), improving salinity stress tolerance. Consistent with this observation, halophytes (the most salt-tolerant species on the planet) may accumulate significant amounts (up to 90 μmol on a dry weight basis) of glycine betaine and thus can better withstand oxidative damage under saline conditions (Rhodes and Hanson 1993; Flowers and Colmer 2008). Transgenic crops overexpressing halophyte betaine aldehyde dehydrogenase (BADH)—a glycine betaine–synthesizing enzyme—were shown to possess enhanced salt and drought tolerance (Fitzgerald et al. 2009). Interestingly, overexpression of a plastid BADH in the salt-sensitive carrot resulted in remarkable salt tolerance (up to 400 mM NaCl), similar to that of halophytes (Kumar et al. 2004). The above findings could be attributed to both the osmotic role of glycine betaine and its ROS-scavenging ability.

Flavonoids are usually accumulated in the plant vacuole as glycosides, but they also occur as exudates on the surfaces of leaves and other aerial plant parts. Flavonoids can be classified into flavonols, flavones, isoflavones, and anthocyanins on the basis of their structure. Flavonoids are among the most bioactive plant secondary metabolites. Flavonoids serve as ROS scavengers by locating and neutralizing radicals before they damage the cell, which is important for plants to regulate redox potential under stressful conditions (Løvdal et al. 2010). Flavonoids function by virtue of the number and arrangement of their hydroxyl groups attached to ring structures. Their ability to act as antioxidants depends on the reduction potentials of their radicals and the accessibility of the radicals. Most flavonoids outperform well-known antioxidants, such as ASC and α-tocopherol (Hernandez et al. 2009).

Tocopherols are considered major antioxidants in biomembranes. The antioxidant ability of tocopherols against Fe^{2+} ascorbate–induced lipid peroxidation declined in the order of $\alpha > \beta \approx \gamma > \delta$, with each single molecule of these tocopherols protecting up to 220, 120, 100, and 30 molecules of polyunsaturated fatty acids, respectively, before being consumed (Fukuzawa et al. 1982). Tocopherols have been shown to prevent the chain propagation step in lipid auto-oxidation, which makes them an effective free-radical trap. α-Tocopherol (vitamin E) detoxifies $^1O_2$ and lipid peroxyl radicals, thus preventing lipid peroxidation under abiotic stress (Szarka et al. 2012). Among the different isoforms of tocopherols, α-tocopherol is the predominant form in plant green tissues. This isoform is synthesized in a plastid envelope and is stored in plastoglobuli of the chloroplast stroma and in thylakoid membranes, suggesting that α-tocopherol is pivotal to decreased production of ROS (primarily that of $^1O_2$) in the chloroplast during environmental stresses (Szarka et al. 2012). Recently, it has been found that oxidative stress activates the expression of genes responsible for the synthesis of tocopherols in higher plants (Wu et al. 2007). Increased levels of α-tocopherol and ASC have been found in tomato and in *A. doliolum*, helping to protect membranes from oxidative damage (Hsu and Kao 2007).

α-Tocopherol can scavenge $^1O_2$ in two ways. First, α-tocopherol quenches $^1O_2$ physically via resonance energy transfer (Fahrenholtz et al. 1974); following this, $^1O_2$ scavenging can happen through a chemical reaction (Falk and Munné-Bosch 2010). During the resonance energy transfer mechanism of $^1O_2$ scavenging, α-tocopherol content is not altered significantly and one molecule of α-tocopherol can deactivate 120 molecules of $^1O_2$, whereas the chemical scavenging mechanism implies involvement of an intermediate hydroperoxydienone that decomposes to form tocopherol quinone and tocopherol quinone epoxides, resulting in a significant decline in the α-tocopherol content (Munne-Bosch and Alegre 2002). A comparison between a halophyte (*Cakile maritima*) and a glycophyte (*Arabidopsis thaliana*) showed that *C. maritima* may detoxify salt-induced $^1O_2$ production through direct quenching, because the α-tocopherol level did not change significantly during salt stress. At the same time, *A. thaliana* achieved the same result through a chemical reaction, showing a reduction of 50% in the production of α-tocopherol (Ellouzi et al. 2011). α-Tocopherols also function as recyclable chain reaction terminators of polyunsaturated fatty acid (PUFA) radicals generated by lipid oxidation (Hare et al. 1998). α-Tocopherols scavenge lipid peroxy radicals and yield a tocopheroxyl radical that can be recycled back into the corresponding α-tocopherol by reacting with ascorbate or other antioxidants (Igamberdiev and Hill 2004).

8.3 Targeting Redox Regulation and Antioxidant Defense Systems in Breeding Programs

The number of papers linking oxidative stress with salinity has increased exponentially over the past two decades. There have also been reports of increased antioxidant activity in plants grown under saline conditions (Hernandez et al. 2000; Sairam and Srivastava 2002). It is hardly surprising that the idea of improving salinity stress tolerance by increased antioxidant production is gaining momentum (Table 8.1). However, many other reports have questioned the validity of this approach, reporting either no correlation or a negative correlation between the activity of antioxidant enzymes and plant salinity stress tolerance (Tables 8.2 and 8.3). The possible reasons for this discrepancy most likely lie in the fact that some ROS such as H_2O_2 play a very important signaling role in adaptive and developmental responses, and so tampering with them may result in pleiotropic effects (De Pinto et al. 2006). It is becoming increasingly evident that considerable variations exist in the production of both enzymatic and nonenzymatic antioxidants in response to salt stress in various plant tissues and at various time points. Hence, the interspecific or intraspecific aspects of ROS production and scavenging should be taken into account. Last, but not least, the diversity of known antioxidants should be accounted for. Some supporting arguments are given below.

Table 8.1 Selected examples of reported positive correlations between expression of antioxidant genes and salt stress tolerance in transgenic plant species

Plant species	Gene	Response of antioxidant system	References
Tobacco	AKR1	Increased production of SOD, APX, and GR; proline accumulation	Vemanna et al. (2017)
Sweet potato	CuZnSOD and APX	Increased production of SOD and APX	Yan et al. (2016)
Tobacco chloroplast	CuZnSOD and APX	Simultaneous expression of CuZnSOD and APX increased tolerance to oxidative stress as compared with individual expression of tolerance	Kwon et al. (2002)
Arabidopsis thaliana	Aldehyde dehydrogenase gene	Decreased accumulation of lipid peroxidation–derived reactive aldehydes	Sunkar et al. (2003)
Rice	codA	Increased accumulation of glycine betaine	Mohanty et al. (2002)
Arabidopsis thaliana	MAPK 1	Increased accumulation of SOD, POD, and CAT	Zaidi et al. (2016)
Arabidopsis thaliana	Mn-SOD	Increased expression of Mn-SOD (2-fold)	Wang et al. (2004)
Arabidopsis thaliana	DHAR1	Significant increase in DHAR and substantial improvement in salt tolerance	Ushimaru et al. (2006)
Rice	Mn-SOD	Increased SOD (1.7-fold) and APX (1.5-fold)	Tanaka et al. (1999)
Tobacco	GST and GPX	Transgenics had higher levels of glutathione and ascorbate than wild-type plants	Roxas et al. (2000)
Rice	Mn-SOD	Transgenic rice showed higher SOD activity and higher salt stress tolerance	Tanaka et al. (1999)

AKR1 NADPH-Dependent aldo-ketoreductase, *APX* ascorbate peroxidase, *CAT* catalase, *codA* choline dehydrogenase gene, *DHAR* dehydroascorbate reductase, *GPX* glutathione peroxidase, *GR* glutathione reductase, *GST* glutathione S-transferase, *MAPK 1* mitogen-activated protein kinase phosphatase, *NADPH* reduced nicotinamide adenine dinucleotide phosphate, *POD* peroxidase, *SOD* superoxide dismutase

8.3.1 Antioxidant Activity at the Tissue/Organ Level

Activation of the antioxidant defense system occurs via enzymatic and nonenzymatic mechanisms against salt stress. The antioxidant responses, though, are different in different organs and/or tissues (Turan and Tripathy 2013). Hamada et al. (2016) found that the total antioxidant capacity and polyphenol content in maize increased with increased salinity levels in roots and mature leaves but showed no changes in young leaves. They also showed that SOD, APX, glutathione (GSH), glutathione S-transferase (GST), and ASC content increased particularly in maize roots, while total tocopherol levels increased specifically in shoot tissues. Proline content was slightly decreased in young leaves in maize plants but did not show significant changes in maize roots and mature leaves under exposure to salinity stress (Hamada et al. 2016). Other studies have also reported tissue-specific

Table 8.2 Stress-induced changes in antioxidant activity of major enzymatic antioxidants in different plant species

Enzymatic antioxidant	Plant species	Antioxidant response	References
SOD	Rice	Increased	Lee et al. (2001)
	Brassica napus	Increased	Ashraf and Ali (2008)
	Wheat	Increased	Sairam et al. (2005)
	Pea	Increased	Noreen and Ashraf (2009)
	Alfalfa	Increased	Wang et al. (2009)
	Wheat	Decreased	Mandhania et al. (2006)
	Rice	Decreased	Khan and Panda (2008)
	Maize	Decreased	de Azevedo Neto et al. (2006)
	Foxtail millet	Decreased	Sreenivasulu et al. (2000)
CAT	Chickpea	Increased	Eyidogan and Oz (2005), Kukreja et al. (2005)
	Rice	Increased	Khan and Panda (2008)
	Pea	Increased	Hernandez et al. (2000), Noreen and Ashraf (2009)
	Brassica napus	Increased	Ashraf and Ali (2008)
	Wheat	Increased	Sairam et al. (2005), Mandhania et al. (2006)
	Alfalfa	Increased	Wang et al. (2009)
	Rice	Decreased	Khan and Panda (2008)
	Glycyrrhiza uralensis	Decreased	Pan et al. (2006)
	Pea	Decreased	Noreen and Ashraf (2009)
	Maize	Decreased	de Azevedo Neto et al. (2006)
APX	Rice	Increased	Khan and Panda (2008)
	Wheat	Increased	Mandhania et al. (2006)
	Alfalfa	Increased	Wang et al. (2009)
	Cotton	Increased	Desingh and Kanagaraj (2007)
	Foxtail millet	Increased	Sreenivasulu et al. (2000)
	Maize	Decreased	de Azevedo Neto et al. (2006)
	Pea	No change	Hernandez et al. (2000)
	Foxtail millet	Decreased	Sreenivasulu et al. (2000)
	Rice	No change	Demiral and Türkan (2005)
GPX	Rice	Increased	Vaidyanathan et al. (2003)
	Tobacco	Increased	Roxas et al. (2000)
	Tomato	Increased	Wang et al. (2005)
	Pea	No change	Hernandez et al. (2000)

APX Ascorbate peroxidase, *CAT* catalase, *GPX* glutathione peroxidase, *SOD* superoxide dismutase

responses of some other antioxidants. For example, ASC and tocopherol content has been shown to increase in the leaves of tomato to protect them against oxidative stress under high salinity (Salama et al. 1994; Tuna 2014); however, the levels of ASC and tocopherols declined in rice leaves under salinity stress (Turan and

Table 8.3 Stress-induced changes in antioxidant activity of major nonenzymatic antioxidants in different plant species

Enzymatic antioxidant	Plant species	Antioxidant response	References
ASC	Rice	Increased	Vaidyanathan et al. (2003)
	Wheat	Increased	Ahanger and Agarwal (2017)
	Catharanthus roseus	Increased	Jaleel et al. (2007a)
	Cassia angustifolia	Decreased	Agarwal and Pandey (2004)
	Wheat	Decreased	Sairam et al. (2005), Sairam and Srivastava (2002)
Proline	Wheat (*Triticum aestivum*)	Increased	Sairam et al. (2002)
	Sugar beet	Increased	Ghoulam et al. (2002)
	Sesame	Increased	Koca et al. (2007)
	Rice	Increased	Lutts et al. (1996)
	Green gram	Increased	Misra and Gupta (2005)
	Wheat (*Triticum durum*)	Increased	Demiral and Türkan (2005)
	Cassia angustifolia	Decreased	Agarwal and Pandey (2004)
	Rice	Decreased	Lutts et al. (1996)
	Wheat (*Triticum durum*)	Decreased	Demiral and Türkan (2005)
	Prosopis alba	No change	Meloni et al. (2004)
Glycine betaine	Wheat	Increased	Sairam et al. (2002)
	Catharanthus roseus	Increased	Jaleel et al. (2007b)
	Spinach	Increased	Di Martino et al. (2003)
	Prosopis alba	No change	Meloni et al. (2004)
	Wheat	Decreased	Sairam et al. (2002)
Tocopherols	Cotton	Increased	Gossett et al. (1994)
	Pea	Increased	Noreen and Ashraf (2009)
	Pea	Decreased	Noreen and Ashraf (2009)
	Rice	Decreased	Turan and Tripathy (2013)
Flavonoids	Rice	Increased	Chutipaijit et al. (2009)
	Maize	Increased	Hichem and Mounir (2009)
	Potato	Increased	Daneshmand et al. (2010)
	Safflower	Increased	Gengmao et al. (2015)
	Thellungiella callus	Decreased	Zhao et al. (2009)
	Arabidopsis thaliana	Decreased	Zhao et al. (2009)

ASC Ascorbic acid

Tripathy 2013). There was no significant change in proline content in *Sorghum bicolor* leaves under salinity stress, but proline content in *Sorghum sudanense* decreased slightly with salinity (De Oliveira et al. 2013). Redox changes estimated by the ratios of redox couples (ASC:total ascorbate and GSH:total glutathione) showed significant decreases in maize roots (Hamada et al. 2016). Tocopherol, on the other hand, appears to be a more shoot-specific antioxidant in maize seedlings

(Hamada et al. 2016). In lentil, root tissues were less affected by salt stress and had higher activity of Cu/ZnSOD, APX, and GR in roots than in shoots (Bandeoğlu et al. 2004). In common bean (*Phaseolus vulgaris*), salinity stress led to reductions in the activity of SOD and APX in nodules but not in the bulk of the roots (Jebara et al. 2005). These conflicting results could be due to plant/tissue specificity. Moreover, plant age–related differences in antioxidant responses in plant cells, tissues, and organs could also explain the reason behind the interspecific or intraspecific aspects of ROS production and antioxidant activity. For instance, mature maize leaf cells (in the distal leaf parts) are more sensitive to high salinity than younger cells (from actively expanding leaf parts), which have higher antioxidant activity (Kravchik and Bernstein 2013). These studies have indicated possible roles of ROS in the systemic signaling from roots to leaves and activation of antioxidants for better protection against oxidative stress and/or salt stress.

8.3.2 Antioxidant Activity at the Organelle Level

Besides tissue-specific and/or organ-specific responses, organelle-specific responses of antioxidants under salt stress have been observed. The activity of different antioxidants within different compartments of a cell, tissue, or organ plays a role in protection against oxidative stress caused by salt stress. Hernandez et al. (2000) showed that salt stress tolerance in pea was associated with high SOD activity in the apoplast and high activity of dehydroascorbate reductase (DHAR), GR, and monodehydroascorbate reductase (MDHAR) in the symplast. Moreover, Mittova et al. (2004) showed that the mitochondria and peroxisomes of salt-treated roots of wild tomato had increased levels of lipid peroxidation and H_2O_2, coupled with decreased activity of SOD, POD, ASC, and GSH, suggesting that improved endogenous production of antioxidants in the mitochondria and/or peroxisome could improve salt stress tolerance. In another study, higher activity of SOD, APX, and GR was noted in the chloroplastic fraction as compared with the mitochondrial fraction and cytosolic fraction (Sairam and Srivastava 2002).

8.3.3 Antioxidant Activity at Different Time Points

Antioxidant activity also shows a pronounced time dependence and thus may be different at various time points. Mhadhbi et al. (2011) reported increased activity of CAT, SOD, and POD in salt-treated *Medicago truncatula* roots at 24 h; however, this activity was lost after 48 h of salt stress. In pea, SOD activity increased after 48 h of salt stress, while GR showed an increase after 24 h of salt stress Hernandez et al. (2000). These authors also found no change in APX during salt stress at any time point. Shalata et al. (2001) showed that under long-term salt stress conditions, the activity of SOD, APX, CAT, and MDHAR was increased and reached a maximum 16 days after the beginning of salinization, and then decreased, while GR activity decreased from the start of salt stress. All of these studies reported highly varied responses and production of antioxidants at various time points under salt stress, and suggested that the variations could have been due to differences in experimental conditions, data collection, and plant species.

8.3.4 Targeting Quantitative Trait Loci for Antioxidant Activity

The practical aspect of targeting antioxidant activity in breeding programs should also be considered. Jiang et al. (2013) conducted quantitative trait locus (QTL) analysis of the activity of antioxidant enzymes and malondialdehyde content in wheat seeds during germination. These authors discovered 22 unconditional QTLs on 1A, 1B, 1D, 2B, 2D, 3A, 4A, 6B, 7A, and 7B, scattered on nine chromosomes. Eight of these QTLs were for SOD activity, five for POD, and six for CAT. In rice, two QTLs for malondialdehyde (MDA) content in rice leaves were detected on chromosome 1, with additive effects from maternal and paternal parents accounting for 4.33% and 4.62% of phenotype variations, respectively. Rousseaux et al. (2005) found 20 QTLs in tomato; of these, five were for total antioxidant activity, five for ASC, and nine for total phenolics. Frary et al. (2010) carried out QTL analysis and found nine QTLs for phenolic compounds and 14 for flavonoids in tomato. Given these numbers, it will be very difficult—if it is at all possible—to make a valid recommendation to breeders as to which of these QTLs play bigger role(s) and thus should be transferred into high-yielding varieties to increase their stress tolerance.

It has also been argued (Bose et al. 2014) that truly salt-tolerant plants do not allow formation of harmful ROS in the first instance and, as such, require no higher antioxidant activity. This can be achieved by increasing the rate of sodium exclusion from the cytosol, either into vacuoles (via mechanisms including NHX tonoplast Na^+/H^+ exchange) (Blumwald 2000) or into the apoplast (via SOS1-mediated Na^+ exclusion from the cell) (Shi et al. 2000). Halophytes show high levels of salt stress tolerance and possess high levels of antioxidant production at an intrinsic level. As described in Sects. 8.1 and 8.2.1, SOD rapidly converts O_2^- to H_2O_2 at the initial level of salt stress and the latter plays an important role as a second messenger, triggering cascades of different adaptive responses at the genetic and physiological levels; thus, rapid conversion of O_2^- to H_2O_2 could be essential for early defense in halophytes. Nonetheless, it still remains to be established how the stress-induced increase in H_2O_2 production is ultimately converted into plant adaptive responses (Miller et al. 2010). Because of similarities between Ca^{2+}- and H_2O_2-induced signatures, the roles of other enzymatic antioxidants may be attributed to the need to decrease the basal levels of H_2O_2 once the signaling has been processed. In this context, the roles of CAT and APX in shaping the H_2O_2 signature may be similar to those in Ca^{2+} efflux systems (Bose et al. 2011).

8.4 Concluding Remarks

ROS are produced as a result of a large number of metabolic processes, occurring at both intracellular and extracellular locations. Plants possess different enzymatic and nonenzymatic antioxidants to scavenge ROS, regarded as redox regulatory mechanisms in plants. Different antioxidants have different ROS-scavenging effects (Fig. 8.1), and their effects are highly tissue specific and organelle specific. A significant variability exists among different plant species and/or among different genotypes of the same plant species in terms of the kinetics of antioxidant production and activity. Different reports have described controversial results regarding increases or decreases in the activity of different antioxidants in response to salinity. In the light of this, it appears not to be highly fruitful to try and improve plant stress tolerance by increasing the activity of some specific antioxidants via either genetic engineering or a MAS-based approach, without accounting for the tissue and time dependence of this process. There is a need for better understanding of the role of ROS as signaling components of plant adaptive mechanisms, and development of tools for targeted regulation of plant redox systems in specific cellular compartments and tissues.

Acknowledgements This work was supported by the Australian Research Council and Qatar National Science Foundation (NPRP-8-126-1-024) grants to Sergey Shabala.

References

Agarwal S, Pandey V (2004) Antioxidant enzyme responses to NaCl stress in *Cassia angustifolia*. Biol Plant 48:555–560

Ahanger MA, Agarwal RM (2017) Salinity stress induced alterations in antioxidant metabolism and nitrogen assimilation in wheat (*Triticum aestivum* L.) as influenced by potassium supplementation. Plant Physiol Biochem 115:449–460

Ali AA, Alqurainy F (2006) Activities of antioxidants in plants under environmental stress. In: Motohasci A (ed) The lutein-prevention and treatment for diseases. Transworld Research Network, New Delhi, pp 187–256

Alia P, Saradhi P (1991) Proline accumulation under heavy metal stress. J Plant Physiol 138:554–558

Anjum SA, Tanveer M et al (2015) Cadmium toxicity in maize: consequences on antioxidative systems, reactive oxygen species and cadmium accumulation. Environ Sci Pollut Res 22:17022–17030

Anjum SA, Tanveer M, Hussain S et al (2016a) Osmoregulation and antioxidant production in maize under combined cadmium and arsenic stress. Environ Sci Pollut Res 23:11864–11875

Anjum SA, Tanveer M et al (2016b) Effect of progressive drought stress on growth, leaf gas exchange, and antioxidant production in two maize cultivars. Environ Sci Pollut Res 23:17132–17141

Anjum SA, Ashraf U, Tanveer M, Khan I, Hussain S, Shahzad B, Zohaib A, Abbas F, Saleem MF, Ali I, Wang LC (2017) Drought induced changes in growth, osmolyte accumulation and antioxidant metabolism of three maize hybrids. Front Plant Sci 8:69. https://doi.org/10.3389/fpls.2017.00069

Ashraf M, Ali Q (2008) Relative membrane permeability and activities of some antioxidant enzymes as the key determinants of salt tolerance in canola (*Brassica napus* L.) Environ Exp Bot 63:266–273

Ashraf M, Foolad MR (2007) Roles of glycine betaine and proline in improving plant abiotic stress resistance. Environ Exp Bot 59:206–216

Athar HR, Khan A, Ashraf M (2008) Exogenously applied ascorbic acid alleviates salt-induced oxidative stress in wheat. Environ Exp Bot 63:224–231

Azevedo RA, Alas RM, Smith RJ, Lea PA (1998) Response of antioxidant enzymes to transfer from elevated carbon dioxide to air and ozone fumigation, in leaves and roots of wild-type and catalase-deficient mutant of barley. Physiol Plant 104:280–292

Azpilicueta CE, Benavides MP, Tomaro ML, Gallego SM (2007) Mechanism of CATA3 induction by cadmium in sunflower leaves. Plant Physiol Biochem 45:589–595

Bandeoğlu E, Eyidoğan F, Yücel M, Öktem HA (2004) Antioxidant responses of shoots and roots of lentil to NaCl-salinity stress. Plant Growth Regul 42:69–77

Blumwald E (2000) Sodium transport and salt tolerance in plants. Curr Opin Cell Biol 12:431–434

Bose J, Pottosin I, Shabala SS, Palmgren MG, Shabala S (2011) Calcium efflux systems in stress signaling and adaptation in plants. Front Plant Sci 2:1–17

Bose J, Rodrigo-Moreno A, Shabala S (2014) ROS homeostasis in halophytes in the context of salinity stress tolerance. J Exp Bot 65:1241–1257

Bowler C, Fluhr R (2000) The role of calcium and activated oxygens as signals for controlling cross-tolerance. Trends Plant Sci 5:241–246

Chen C, Dickman MB (2005) Proline suppresses apoptosis in the fungal pathogen *Colletotrichum trifolii*. PNAS 102:3459–3464

Chutipaijit S, Cha-Um S, Sompornpailin K (2009) Differential accumulations of proline and flavonoids in Indica rice varieties against salinity. Pak J Bot 41:2497–2506

Crawford NM (2006) Mechanisms for nitric oxide synthesis in plants. J Exp Bot 57:471–478

Cuin TA, Shabala S (2007) Compatible solutes reduce ROS-induced potassium efflux in *Arabidopsis* roots. Plant Cell Environ 30:875–885

Daneshmand F, Arvin MJ, Kalantari KM (2010) Physiological responses to NaCl stress in three wild species of potato in vitro. Acta Physiol Plant 32:91–101

de Azevedo Neto AD, Prisco JT, Enéas-Filho J, de Abreu CEB, Gomes-Filho E (2006) Effect of salt stress on antioxidative enzymes and lipid peroxidation in leaves and roots of salt-tolerant and salt-sensitive maize genotypes. Environ Exp Bot 56:87–94

de Oliveira VP, Marques EC, de Lacerda CF, Prisco JT, Gomes Filho E (2013) Physiological and biochemical characteristics of *Sorghum bicolor* and *Sorghum sudanense* subjected to salt stress in two stages of development. Afr J Agric Res 8:660–670

De Pinto MC, Paradiso A, Leonetti P, De Gara L (2006) Hydrogen peroxide, nitric oxide and cytosolic ascorbate peroxidase at the crossroad between defence and cell death. Plant J 48:784–795

Demidchik V (2015) Mechanisms of oxidative stress in plants: from classical chemistry to cell biology. Environ Exp Bot 109:212–228

Demiral T, Türkan I (2005) Comparative lipid peroxidation, antioxidant defense systems and proline content in roots of two rice cultivars differing in salt tolerance. Environ Exp Bot 53:247–257

Desingh R, Kanagaraj G (2007) Influence of salinity stress on photosynthesis and antioxidative systems in two cotton varieties. Gen Appl Plant Physiol 33:221–234

Di Martino C, Delfine S, Pizzuto R, Loreto F, Fuggi A (2003) Free amino acids and glycine betaine in leaf osmoregulation of spinach responding to increasing salt stress. New Phytol 158:455–463

Ellouzi H, Ben Hamed K, Cela J, Munne-Bosch S, Abdelly C (2011) Early effects of salt stress on the physiological and oxidative status of *Cakile maritima* (halophyte) and *Arabidopsis thaliana* (glycophyte). Physiol Plant 142:128–143

Eyidogan F, Oz MT (2005) Effect of salinity on antioxidant responses of chickpea seedlings. Acta Physiol Plant 29:485–493

Fahrenholtz S, Doleiden F, Trozzolo A, Lamola A (1974) On the quenching of singlet oxygen by α-tocopherol. Photochem Photobiol 20:505–509

Falk J, Munné-Bosch S (2010) Tocochromanol functions in plants: antioxidation and beyond. J Exp Bot 61:1549–1566
Fitzgerald TL, Waters DL, Henry RJ (2009) Betaine aldehyde dehydrogenase in plants. Plant Biol 11(2):119–30
Flowers TJ, Colmer TD (2008) Salinity tolerance in halophytes. New Phytol 179:945–963
Foreman J, Demidchik V, Bothwell JHF, Mylona P, Miedema H, Torres MA, Linstead P, Costa S, Brownlee C, Jones JDG, Davies JM, Dolan L (2003) Reactive oxygen species produced by NADPH oxidase regulate plant cell growth. Nature 422(6930):442–446
Foyer CH, Noctor G (2003) Redox sensing and signaling associated with reactive oxygen in chloroplasts, peroxisomes and mitochondria. Physiol Plant 119:355–364
Foyer CH, Noctor G (2005) Redox homeostasis and antioxidant signaling: a metabolic interface between stress perception and physiological responses. Plant Cell 17:1866–1875
Frary A, Göl D, Keleş D, Ökmen B, Pınar H et al (2010) Salt tolerance in *Solanum pennellii*: antioxidant response and related QTL. BMC Plant Biol 10:58–73
Frugoli JA, Zhong HH, Nuccio ML, McCourt P, McPeek MZ, Thomas TL, McClung CR (1996) Catalase is encoded by a multigene family in *Arabidopsis thaliana* (L.) Plant Physiol 112:327–336
Fukuzawa K, Tokumura A, Ouchi S, Tsukatani H (1982) Antioxidant activities of tocopherols on Fe^{2+}-ascorbate-induced lipid peroxidation in lecithin liposomes. Lipids 17:511–513
Gabbita SP, Robinson KA, Stewart CA, Floyd RA, Hensley K (2000) Redox regulatory mechanisms of cellular signal transduction. Arch Biochem Biophys 376:1–13
Garg N, Manchanda G (2009) ROS generation in plants: boon or bane? Plant Biosyst 143:8–96
Gengmao Z, Yu H, Xing S, Shihui L, Quanmei S, Changhai W (2015) Salinity stress increases secondary metabolites and enzyme activity in safflower. Ind Crop Prod 64:175–181
Ghoulam C, Foursy A, Fares K (2002) Effects of salt stress on growth, inorganic ions and proline accumulation in relation to osmotic adjustment in five sugar beet cultivars. Environ Exp Bot 47:39–50
Gill SS, Tuteja N (2010) Reactive oxygen species and antioxidant machinery in abiotic stress tolerance in crop plants. Plant Physiol Biochem 48:909–930
Gossett DR, Millhollon EP, Lucas M (1994) Antioxidant response to NaCl stress in salt-tolerant and salt-sensitive cultivars of cotton. Crop Sci 34:706–714
Grun S, Lindermayr C, Sell S (2006) Nitric oxide and gene regulation in plants. J Exp Bot 57:507–516
Hamada AbdElgawad GZ, Hegab MM, Pandey R, Asard H, Abuelsoud W (2016) High salinity induces different oxidative stress and antioxidant responses in maize seedlings organs. Front Plant Sci 7:276–287
Hare PD, Cress WA (1997) Metabolic implications of stress-induced proline accumulation in plants. Plant Growth Regul 21:79–102
Hare PD, Cress WA, Van Staden J (1998) Dissecting the roles of osmolyte accumulation during stress. Plant Cell Environ 21:535–553
Hernandez JA, Jimerez A, Mullineaux P, Sevilla F (2000) Tolerance of pea (*Pisum sativum*) to long term salt stress is associated with induction of antioxidant defences. Plant Cell Biol 23:853–862
Hernandez I, Chacón O, Rodriguez R, Portieles R, López Y, Pujol M, Borrás-Hidalgo O (2009) Black shank resistant tobacco by silencing of glutathione S-transferase. Biochem Biophys Res Commun 387:300–304
Hichem H, Mounir D (2009) Differential responses of two maize (*Zea mays* L.) varieties to salt stress: changes on polyphenols composition of foliage and oxidative damages. Ind Crop Prod 30:144–151
Hsu YT, Kao CH (2007) Heat shock-mediated H_2O_2 accumulation and protection against Cd toxicity in rice seedlings. Plant Soil 300:137–147
Hu CA, Delauney AJ, Verma DPS (1992) A bifunctional D1-enzymepyrroline-5-carboxylate synthetase catalyzes the first two steps in proline biosynthesis in plants. PNAS 89:9354–9358

Igamberdiev AU, Hill RD (2004) Nitrate, NO and haemoglobin in plant adaptation to hypoxia: an alternative to classic fermentation pathways. J Exp Bot 55:2473–2482

Ismail H, Maksimović JD, Maksimović V, Shabala L, Živanović BD, Tian Y, Jacobsen S, Shabala S (2016) Rutin, a flavonoid with antioxidant activity, improves plant salinity tolerance by regulating K^+ retention and Na^+ exclusion from leaf mesophyll in quinoa and broad beans. Funct Plant Biol 43(1):75–86

Jaleel CA, Gopi R, Manivannan P, Panneerselvam R (2007a) Responses of antioxidant defense system of *Catharanthus roseus* (L.) G. Don. to paclobutrazol treatment under salinity. Acta Physiol Plant 29:205–209

Jaleel CA, Gopi R, Sankar B, Manivannan P, Kishorekumar A, Sridharan R, Panneerselvam R (2007b) Studies on germination, seedling vigour, lipid peroxidation and proline metabolism in *Catharanthus roseus* seedlings under salt stress. S Afr J Bot 73:190–195

Jebara S, Jebara M, Limam F, Aouani ME (2005) Changes in ascorbate peroxidase, catalase, guaiacol peroxidase and superoxide dismutase activities in common bean (*Phaseolus vulgaris*) nodules under salt stress. J Plant Physiol 162:929–936

Jiang P, Wan Z, Wang Z, Li S, Sun Q (2013) Dynamic QTL analysis for activity of antioxidant enzymes and malondialdehyde content in wheat seed during germination. Euphytica 190:75–85

Jiménez A, Hernández JA, Pastori G, del Rıo LA, Sevilla F (1998) Role of the ascorbate–glutathione cycle of mitochondria and peroxisomes in the senescence of pea leaves. Plant Physiol 118:1327–1335

Joo JH, Yoo HJ, Hwang I, Lee JS, Nam KH, Bae YS (2005) Auxin-induced reactive oxygen species production requires the activation of phosphatidylinositol 3-kinase. FEBS Lett 579:1243–1248

Khan MH, Panda SK (2008) Alterations in root lipid peroxidation and antioxidative responses in two rice cultivars under NaCl-salinity stress. Acta Physiol Plant 30:81–89

Koca H, Bor M, Özdemir F, Türkan I (2007) The effect of salt stress on lipid peroxidation, antioxidative enzymes and proline content of sesame cultivars. Environ Exp Bot 60:344–351

Kravchik M, Bernstein N (2013) Effects of salinity on the transcriptome of growing maize leaf cells point at cell-age specificity in the involvement of the antioxidative response in cell growth restriction. BMC Genomics 14:24. https://doi.org/10.1186/1471-2164-14-24

Kukreja S, Nandval AS, Kumar N, Sharma SK, Sharma SK, Unvi V, Sharma PK (2005) Plant water status, H_2O_2 scavenging enzymes, ethylene evolution and membrane integrity of *Cicer arietinum* roots as affected by salinity. Biol Plant 49:305–308

Kumar S, Dhingra A, Daniell H (2004) Plastid-expressed betaine aldehyde dehydrogenase gene in carrot cultured cells, roots, and leaves confers enhanced salt tolerance. Plant Physiol 136:2843–2854

Kwon SY, Jeong YJ, Lee HS, Kim JS, Cho KY, Allen RD, Kwak SS (2002) Enhanced tolerances of transgenic tobacco plants expressing both superoxide dismutase and ascorbate peroxidase in chloroplasts against methyl viologen mediated oxidative stress. Plant Cell Environ 25:873–882

Lee DH, Kim YS, Lee CB (2001) The inductive responses of the antioxidant enzymes by salt stress in the rice (*Oryza sativa* L.) J Plant Physiol 158:737–745

Leisinger U, Rüfenacht K, Fischer B, Pesaro M, Spengler A, Zehnder AJB, Eggen RIL (2001) The glutathione peroxidase homologous gene from *Chlamydomonas reinhardtii* is transcriptionally up-regulated by singlet oxygen. Plant Mol Biol 46:395–408

Light GG, Mahan JR, Roxas VP, Allen RD (2005) Transgenic cotton (*Gossypium hirsutum* L.) seedlings expressing a tobacco glutathione S-transferase fail to provide improved stress tolerance. Planta 222:346–354

Link G (2003) Redox regulation of chloroplast transcription. Antioxid Redox Signal 5:79–87

Liu CG, Wang QW, Jin YQ, Pan KW, Wang YJ (2017) Photoprotective and antioxidative mechanisms against oxidative damage in *Fargesia rufa* subjected to drought and salinity. Funct Plant Biol 44:302–311

López-Gómez M, Hidalgo-Castellanos J, Muñoz-Sánchez JR, Marín-Peña AJ, Lluch C, Herrera-Cervera JA (2017) Polyamines contribute to salinity tolerance in the symbiosis *Medicago truncatula–Sinorhizobium meliloti* by preventing oxidative damage. Plant Physiol Biochem 116:9–17

Løvdal T, Olsen KM, Slimestad R, Verheul M, Lillo C (2010) Synergetic effects of nitrogen depletion, temperature, and light on the content of phenolic compounds and gene expression in leaves of tomato. Phytochemistry 71:605–613

Lu J, Holmgren A (2014) The thioredoxin antioxidant system. Free Radic Biol Med 66:75–87

Lutts S, Kinet JM, Bouharmont J (1996) Effects of salt stress on growth, mineral nutrition and proline accumulation in relation to osmotic adjustment in rice (*Oryza sativa* L.) cultivars differing in salinity resistance. Plant Growth Regul 19:207–218

Mandhania S, Madan S, Sawhney V (2006) Antioxidant defense mechanism under salt stress in wheat seedlings. Biol Plant 50:227–231

Meloni DA, Gulotta MR, Martínez CA, Oliva MA (2004) The effects of salt stress on growth, nitrate reduction and proline and glycine betaine accumulation in *Prosopis alba*. Braz J Plant Physiol 16:39–46

Meyer AJ (2008) The integration of glutathione homeostasis and redox signaling. J Plant Physiol 165:1390–1403

Mhadhbi H, Fotopoulos V, Mylona PV, Jebara M, Elarbi Aouani M, Polidoros AN (2011) Antioxidant gene–enzyme responses in *Medicago truncatula* genotypes with different degree of sensitivity to salinity. Physiol Plant 141:201–214

Millar AH, Mittova V, Kiddle G, Heazlewood JL, Bartoli CG, Theodoulou FL, Foyer CH (2003) Control of ascorbate synthesis by respiration and its implication for stress responses. Plant Physiol 133:443–447

Miller GAD, Suzuki N, Ciftci-Yilmaz S, Mittler RON (2010) Reactive oxygen species homeostasis and signalling during drought and salinity stresses. Plant Cell Environ 33:453–467

Misra N, Gupta AK (2005) Effect of salt stress on proline metabolism in two high yielding genotypes of green gram. Plant Sci 169:331–339

Mittler R, Zilinskas BA (1992) Molecular cloning and characterization of a gene encoding pea cytosolic ascorbate peroxidase. J Biol Chem 267:21802–21807

Mittler R, Vanderauwera S, Gollery M, Van Breusegem F (2004) Reactive oxygen gene network of plants. Trends Plant Sci 9:490–498

Mittova V, Guy M, Tal M, Volokita M (2004) Salinity up-regulates the antioxidative system in root mitochondria and peroxisomes of the wild salt-tolerant tomato species *Lycopersicon pennellii*. J Exp Bot 55:1105–1113

Mohanty A, Kathuria H, Ferjani A, Sakamoto A, Mohanty P, Murata N, Tyagi AK (2002) Transgenics of an elite Indica rice variety Pusa Basmati 1 harbouring the codA gene are highly tolerant to salt stress. Theor Appl Genet 106:51–57

Munne-Bosch S, Alegre L (2002) The function of tocopherols and tocotrienols in plants. Crit Rev Plant Sci 21:31–57

Munns R, James RA, Launchli A (2006) Approaches to increasing the salt tolerance of wheat and other cereals. J Exp Bot 57:1025–1043

Nagamiya K, Motohasci T, Nakao K, Prodhan SH, Hattori E, Hirose S, Ozawa K, Ohkawa Y, Takabe T, Takabe T, Komamine A (2007) Enhancement of salt tolerance in transgenic rice expressing an *Escherichia coli* catalase gene, katE. Plant Biotechnol Rep 1:49–55

Noctor G, Foyer CH (1998) A re-evaluation of the ATP:NADPH budget during C3 photosynthesis. A contribution from nitrate assimilation and its associated respiratory activity? J Exp Bot 49:1895–1908

Noctor G, Gomez L, Vanacker H, Foyer CH (2002) Interactions between biosynthesis, compartmentation, and transport in the control of glutathione homeostasis and signaling. J Exp Bot 53:1283–1304

Noreen Z, Ashraf M (2009) Assessment of variation in antioxidative defense system in salt-treated pea (*Pisum sativum*) cultivars and its putative use as salinity tolerance markers. J Plant Physiol 166:1764–1774 op den Camp RG, Przybyla D, Ochsenbein C, Laloi C, Kim C, Danon A, Wagner D, Hideg E, Gobel C, Feussner I, Nater M, Apel K (2003) Rapid induction of distinct stress responses after the release of singlet oxygen in *Arabidopsis*. Plant Cell 15:2320–2332

Pan Y, Wu LJ, Yu ZL (2006) Effect of salt and drought stress on antioxidant enzymes activities and SOD isoenzymes of liquorice (*Glycyrrhiza uralensis* Fisch). Plant Growth Regul 49:157–165

Papageorgiou GC, Murata N (1995) The unusually strong stabilizing effects of glycine betaine on the structure and function of the oxygen-evolving photosystem-II complex. Photosynth Res 44:243–252

Pedreira J, Sanz N, Pena MJ, Sanchez M, Queijeiro E, Revilla G, Zarra I (2004) Role of apoplastic ascorbate and hydrogen peroxide in the control of cell growth in pine hypocotyls. Plant Cell Physiol 45:530–534

Pei ZM, Murata Y, Benning G, Thomine S, Klüsener B, Allen GJ, Grill E, Schroeder JI (2000) Hydrogen peroxide-activated Ca^{2+} channels mediate guard cell abscisic acid signaling. Nature 406:731–734

Rhodes D, Hanson AD (1993) Quaternary ammonium and tertiary sulfonium compounds in higher plants. Annu Rev Plant Physiol Plant Mol Biol 44:357–384

Roosens NH, Thu TT, Iskandar HM, Jacobs M (1998) Isolation of the ornithine-δ-aminotransferase cDNA and effect of salt stress on its expression in *Arabidopsis thaliana*. Plant Physiol 117:263–271

Rouhier N, Vlamis-Gardicas A, Lilling CH (2003) Characerization of redox properties of poplar glutaredoxin. Antioxid Redox Signal 5:15–22

Rousseaux MC, Jones CM, Adams D, Chetelat R, Bennett A, Powell A (2005) QTL analysis of fruit antioxidants in tomato using *Lycopersicon pennellii* introgression lines. Theor Appl Genet 111:1396–1408

Roxas VP, Lodhi SA, Garrett DK, Mahan JR, Allen RD (2000) Stress tolerance in transgenic tobacco seedlings that overexpress glutathione S-transferase/glutathione peroxidase. Plant Cell Physiol 41:1229–1234

Sagi M, Davydov O, Orazova S, Yesbergenova Z, Ophir R, Stratmann JW, Fluhr R (2004) Plant respiratory burst oxidase homologs impinge on wound responsiveness and development in *Lycopersicon esculentum*. Plant Cell 16:616–628

Sairam RK, Srivastava GC (2002) Changes in antioxidant activity in sub-cellular fractions of tolerant and susceptible wheat genotypes in response to long term salt stress. Plant Sci 162:897–904

Sairam RK, Rao KV, Srivastava GC (2002) Differential response of wheat genotypes to long term salinity stress in relation to oxidative stress, antioxidant activity and osmolyte concentration. Plant Sci 163:1037–1046

Sairam RK, Srivastava GC, Agarwal S, Meena RC (2005) Differences in antioxidant activity in response to salinity stress in tolerant and susceptible wheat genotypes. Biol Plant 49:85–91

Sakamoto A, Murata N (2002) The role of glycine betaine in the protection of plants from stress: clues from transgenic plants. Plant Cell Environ 25:163–171

Salama S, Trivedi S, Busheva M, Arafa AA, Garab G, Erdei L (1994) Effects of NaCl salinity on growth, cation accumulation, chloroplast structure and function in wheat cultivars differing in salt tolerance. J Plant Physiol 144:241–247

Scandalias JG (1990) Response of plant antioxidant defense genes to environmental stress. Adv Genet 28:1–41

Schopfer P, Liszkay A, Bechtold M, Frahry G, Wagner A (2002) Evidence that hydroxyl radicals mediate auxin-induced extension growth. Planta 214:821–828

Schurmann P (2003) Redox signaling in the chloroplast: the ferredoxin/thioredoxin system. Antioxid Redox Signal 5:69–78

Shalata A, Mittova V, Volokita M, Guy M, Tal M (2001) Response of the cultivated tomato and its wild salt-tolerant relative *Lycopersicon pennellii* to salt-dependent oxidative stress: the root antioxidative system. Physiol Plant 112:487–494

Shao HB, Chu LY (2005) Plant molecular biology in China: opportunities and challenges. Plant Mol Biol Report 23:345–358

Shao HB, Jiang SY, Li FM, Chu LY, Zhao CX, Shao MA, Zhao XN, Li F (2007a) Some advances in plant stress physiology and their implications in the systems biology era. Biointerphases 54:33–36

Shao HB, Guo QJ, Chu LY et al (2007b) Understanding molecular mechanism of higher plant plasticity under abiotic stress. Biointerphases 54:37–45

Shi HZ, Ishitani M, Kim CS, Zhu JK (2000) The *Arabidopsis thaliana* salt tolerance gene SOS1 encodes a putative Na^+/H^+ antiporter. PNAS 97:6896–6901

Shin R, Schachtman DP (2004) Hydrogen peroxide mediates plant root cell response to nutrient deprivation. PNAS 101:8827–8832

Smirnoff N, Cumbes QJ (1989) Hydroxyl radical scavenging activity of compatible solutes. Phytochemistry 28(4):1057–1060

Smirnoff N (2005) Ascorbate, tocopherol and carotenoids: metabolism, pathway engineering and functions. In: Smirnoff N (ed) Antioxidants and reactive oxygen species in plants. Blackwell, Oxford, pp 53–86

Souza ER, Freire MBGS, Cunha KPV, Nascimento CWA, Ruiz HA et al (2012) Biomass, anatomical changes and osmotic potential in *Atriplex nummularia* Lindl. cultivated in sodic saline soil under water stress. Environ Exp Bot 82:20–27

Sreenivasulu N, Grimm B, Wobus U, Weschke W (2000) Differential response of antioxidant compounds to salinity stress in salt-tolerant and salt-sensitive seedlings of foxtail millet (*Setaria italica*). Physiol Plant 109:435–442

Srivastava AK, Bhargava P, Rai LC (2005) Salinity and copper-induced oxidative damage and changes in antioxidative defense system of *Anabaena doliolum*. Microb Biotechnol 22:1291–1298

Subbarao GV, Nam NH, Chauhan YS, Johansen C (2000) Osmotic adjustment, water relations and carbohydrate remobilization in pigeon pea under water deficits. J Plant Physiol 157:651–659

Sunkar R, Bartels D, Kirch HH (2003) Overexpression of a stress inducible aldehyde dehydrogenase gene from *Arabidopsis thaliana* in transgenic plants improves stress tolerance. Plant J 35:452–464

Suzuki N, Koussevitzky S, Mittler RON, Miller GAD (2012) ROS and redox signaling in the response of plants to abiotic stress. Plant Cell Environ 35:259–270

Szabados L, Savouré A (2010) Proline: a multifunctional amino acid. Trends Plant Sci 15:89–97

Szarka A, Tomasskovics B, Bánhegyi G (2012) The ascorbate–glutathione–α-tocopherol triad in abiotic stress response. Int J Mol Sci 13:4458–4483

Tanaka Y, Hibin T, HayASCi Y, Tanaka A, Kishitani S, Takabe T, Yokota S, Takabe T (1999) Salt tolerance of transgenic rice overexpressing yeast mitochondrial Mn-SOD in chloroplasts. Plant Sci 148:131–138

Tanveer M, Shah AN (2017) An insight into salt stress tolerance mechanisms of *Chenopodium album*. Environ Sci Pollut Res 24:16531–16535

Trovato M, Mattioli R, Costantino P (2008) Multiple roles of proline in plant stress tolerance and development. Rend Lincei 19:325–346

Tuna AL (2014) Influence of foliarly applied different triazole compounds on growth, nutrition, and antioxidant enzyme activities in tomato ('*Solanum lycopersicum*' L.) under salt stress. Aust J Crop Sci 8:71–79

Turan S, Tripathy BC (2013) Salt and genotype impact on antioxidative enzymes and lipid peroxidation in two rice cultivars during de-etiolation. Protoplasma 250:209–222

Ushimaru T, Nakagawa T, Fujioka Y, Daicho K, Naito M, Yamauchi Y, Nonaka H, Amako K, Yamawaki K, Murata N (2006) Transgenic *Arabidopsis* plants expressing the rice dehydroascorbate reductase gene are resistant to salt stress. J Plant Physiol 163:1179–1184

Vaidyanathan H, Sivakumar P, Chakrabarty R, Thomas G (2003) Scavenging of reactive oxygen species in NaCl-stressed rice (*Oryza sativa* L.)—differential response in salt-tolerant and sensitive varieties. Plant Sci 165:1411–1418

Vemanna RS, Babitha KC, Solanki JK, Reddy VA, Sarangi SK, Udayakumar M (2017) Aldo-keto reductase-1 (AKR1) protect cellular enzymes from salt stress by detoxifying reactive cytotoxic compounds. Plant Physiol Biochem 113:177–186

Wang P, Song CP (2008) Guard-cell signalling for hydrogen peroxide and abscisic acid. New Phytol 178:703–718

Wang Y, Ying Y, Chen J, Wang X (2004) Transgenic *Arabidopsis* overexpressing Mn-SOD enhanced salt-tolerance. Plant Sci 167:671–677

Wang Y, Wisniewski M, Meilan R, Cui M, Webb R, Fuchigami L (2005) Overexpression of cytosolic ascorbate peroxidase in tomato confers tolerance to chilling and salt stress. J Am Soc Hortic Sci 130:167–173

Wang WB, Kim YH, Lee HS, Kim KY, Deng XP, Kwak SS (2009) Analysis of antioxidant enzyme activity during germination of alfalfa under salt and drought stresses. Plant Physiol Biochem 47:570–577

Wu YS, Tang KX (2004) MAP kinase cascades responding to environmental stress in plants. Acta Bot Sin 46:127–136

Wu G, Wei ZK, Shao HB (2007) The mutual responses of higher plants to environment: physiological and microbiological aspects. Biointerphases 59:113–119

Yan H, Li Q, Park SC, Wang X, Liu YJ, Zhang YG et al (2016) Overexpression of CuZnSOD and APX enhance salt stress tolerance in sweet potato. Plant Physiol Biochem 109:20–27

Zaidi I, Ebel C, Belgaroui N, Ghorbel M, Amara I, Hanin M (2016) The wheat MAP kinase phosphatase 1 alleviates salt stress and increases antioxidant activities in *Arabidopsis*. J Plant Physiol 193:12–21

Zhang W, Wang C, Qin C, Wood T, Olafsdottir G, Welti R, Wang X (2003) The oleate-stimulated phospholipase D, PLDδ, and phosphatidic acid decrease H_2O_2-induced cell death in *Arabidopsis*. Plant Cell 15:2285–2295

Zhao X, Tan HJ, Liu YB, Li XR, Chen GX (2009) Effect of salt stress on growth and osmotic regulation in Thellungiella and Arabidopsis callus. Plant Cell Tissue Org Cult 98:97–103

Chapter 9
Manipulating Metabolic Pathways for Development of Salt-Tolerant Crops

Melike Bor and Filiz Özdemir

Abstract Engineering plants for salt stress tolerance is a complex process due to the multiple-sided characteristics of stress coping mechanisms. The common approach was first identification of the components of signalling and regulatory pathways for salinity tolerance, then transformation of plants with one of those genes and phenotyping the transgenic plant subjected to salt stress at controlled conditions. Plant biology literature is full of research papers on the success of such plants to acclimate and survive under salinity; however, to date none of them was able to become a commercial variety having improved performance at field conditions. Disturbing or interfering with complex networks and pathways can result in unexpected effects on plant growth and development. Furthermore, tolerance against one stress would not be efficient to cope with different environmental stress factors which field-grown plants encounter during a single growth season.

Instead of targeting signalling or regulatory networks, manipulating metabolic routes for higher osmotic and ionic stress tolerance would be more realistic to mitigate negative impact of salt stress on crop plants. At field conditions coping well with osmotic and ionic stresses will double the chance of crop plants to overcome other challenges such as drought, nutrient and high temperature. Absolutely manipulating plant metabolism is also a complex task, but it is worth to put an effort since modifying common tolerance routes such as osmoregulation, antioxidant capacity and ion transport mechanisms will be more promising at field conditions for the whole plant life cycle. In this chapter we tried to collect and point out the recent information on metabolism in relation to salt stress tolerance and focused on more feasible efforts for the achievement of this purpose in crop plants at field conditions.

Keywords Primary metabolism · Cellular homeostasis · Metabolite · Abiotic stress · Salinity tolerance · Osmotic adjustment · Reactive oxygen species · Trade-offs · Polyamine metabolism · Sugar metabolism · Halophytes · Glycophytes

---

M. Bor (✉) · F. Özdemir
University of Ege, Department of Biology, İzmir, Turkey
e-mail: melike.bor@ege.edu.tr

V. Kumar et al. (eds.), *Salinity Responses and Tolerance in Plants, Volume 1*,
https://doi.org/10.1007/978-3-319-75671-4_9

Abbreviations

ABA	Abscisic acid
ADC	Arginine decarboxylase
GABA	Gamma amino butyric acid
GC-MS	Gas chromatography-mass spectrometry
H_2O_2	Hydrogen peroxide
HKT	High-affinity potassium transporter
LEA	Late embryogenesis abundant
LTP	Lipid transfer protein
MDH	Malate dehydrogenase
MTMM	Multi-trait mixed model
NaCl	Sodium chloride
NATA1	N-acetyltransferase activity 1
ODC	Ornithine decarboxylase
PA	Polyamine
PLC1	Phospholipase C1
PLD	Phospholipase D
Put	Putrescine
ROS	Reactive oxygen species
SAMDC	S-adenosylmethionine synthetase
SA	Salicylic acid
SFR	Sensitive to freezing
SKC1	Sodium ion transmembrane transporter activity 1
SnRK1	Sucrose nonfermenting-related kinase1
SOS1	Salt overly sensitive 1
Spd	Spermidine
Spm	Spermine
TCA	Tricarboxylic acid
T6P	Trehalose-6-phosphate
UDPGlc	Uridine-diphosphate-glucose

9.1 Introduction

Salt stress is one of the major environmental constraints that affect crop performance and yield in relation to its osmotic and ionic effects. High levels of Na^+ and Cl^- in the plant tissues have negative impact on growth and development by disturbing metabolic homeostasis. Since the demand for food production significantly increases worldwide, the severity of salinity problem in agricultural lands becomes more and more striking. Determination of functionally conserved traits among different crop species which enable them to cope with abiotic stresses including salinity might be a promising strategy to improve performance and yield (Mickelbart et al. 2015).

Nevertheless, crop lines engineered for the traits which are efficient for stress tolerance are still successful at laboratory or greenhouse scale. Field trials of these lines usually do not provide extensive solutions to encounter efficient tolerance and productivity at stress conditions.

Harvested energy from photosynthesis is used for biomass production, synthesis of essential macromolecules and uptake and transfer of nutrients depending on the developmental stage of plants. At salinity conditions, photosynthetic efficiency declines because of the alterations in leaf biochemistry and low levels of available CO_2. In relation to this, imbalance between energy gain and lost takes place leading to the inhibition of plant growth (Chaves et al. 2009; Munns and Gilliham 2015). In salt-tolerant plant species whether by the synthesis of organic solutes (sugar alcohols, glycine betaine, sucrose, proline, etc.) or by the use of ions, osmotic adjustment is among the most efficient and common mechanism for the maintenance of growth under salinity conditions (Flowers et al. 2015). Both of these processes are energy demanding; therefore, directing the available energy to cope with salt stress instead of investing into biomass production is a big burden for crop plants. As indicated by Munns and Gilliham (2015), targeting the energy use efficiency of crop plants in the metabolic processes has a potential to mitigate this burden and lead the diversion of energy for better yield and defence at certain periods of growth and during exposure to stress. Improving biomass at stress conditions is of great importance and can also be targeted by engineering approaches. In a recent study, a cytochrome P450 encoding gene, PLASTOCHRON1 (PLA1), was introduced to maize which increased the proliferative and undifferentiated state of the dividing cells during leaf elongation (Sun et al. 2017). Extended growth period was advantageous for plants to overcome the growth reduction under mild drought and cold stress (Sun et al. 2017).

Plant metabolism is a very complex network of signalling, synthesis and degradation processes which are all governed by strict developmental and diurnal programme. Metabolism is composed of two types of events: anabolism (energy requiring biosynthesis of cellular components) and catabolism (degradation of nutrients and complex molecules in order to acquire energy) (Jones et al. 2013). These two processes contribute to metabolic turnover and continuous flux in which metabolites enter or exit pools of metabolic precursors or intermediates. Accumulation and depletion of some distinct metabolites are good clues for us to understand the dynamics of plant growth and development within its life cycle (Jones et al. 2013). Even at normal conditions, the maintenance of this balance needs well-organized and integrated events at different regulatory levels; hence, when plants are exposed to different types of environmental stresses, both of these reactions can easily get out of hand. Growth is the key process for a plant within its lifetime, and it is important for the overall performance in agricultural practices. Plant growth is maintained by the conversion of primary metabolites into cells as building blocks depending on the balance between resources and sinks (White et al. 2015). Therefore any redirection of energy from primary metabolism to defence should inhibit growth and reproduction which means that defensive traits have a cost that results in trade-offs in plant growth and reproduction (Züst and Agrawal 2017).

Trade-offs in growth defence usually result from the nutrient allocation decisions in order to respond to variable environment conditions (Züst and Agrawal 2017). Reallocation of nutrients between source and sink tissues leads to the adaption of plants to different environmental stresses (Roitsch et al. 2003; Roitsch and González 2004; Albacete et al. 2014). For example, high root-to-shoot ratio in stress-treated plants can be an adaptive mechanism to improve the uptake of water and nutrient from the roots (Sharp 2002; Albacete et al. 2008; Albacete et al. 2014). Fusari et al. (2017) studied enzyme activities, metabolites, structural components and biomass in *A. thaliana* by genome-wide association approach. They found that a multi-trait quantitative trait loci (QTL) on chromosome 4 affected enzyme activities, metabolite and protein levels and biomass (Fusari et al. 2017). Variation in the alleles of ACCELERATED CELL DEATH6 was necessary for the modification of these traits which pointed out a trade-off between defence and metabolism (Fusari et al. 2017). In this perspective, defining metabolic processes and alleles by means of such trade-offs would also improve our understanding of stress coping processes in plants.

The same principle applies to plants as to human beings; when we get sick, the first thing we consider is to take a blood test which will evaluate the levels of some distinct metabolites which are critical in a certain type of health condition. Likewise, comparing the metabolite profiles of normal- and stress-treated plants would provide important hints and clues for the overall status of plants. Moreover, metabolite profiling of glycophytes and comparing this data with their halophyte relatives would be helpful to figure out salt stress coping mechanisms which would either be enhanced by transgenic approach or conventional breeding. The osmoprotectant molecule "glycine betaine" is a good example of this foresight. Glycine betaine is accumulated to high levels in most of the halophytes during salt stress. Crop plants which are transformed by betaine biosynthetic genes isolated from halophytes have better protection against drought and salt stresses. Also crossing wheat with its glycine betaine accumulating halophytic relatives showed increased glycine betaine and improved salt tolerance (Rhodes and Hanson 1993; Flowers and Colmer 2008; Fitzgerald et al. 2009). The mechanism of this gained salt tolerance is still unknown. Is it due to the osmotic adjustment process or ROS scavenging capacity of glycine betaine or any another unknown mechanism? The accumulated glycine betaine levels are usually low for facilitating osmotic adjustment (Fitzgerald et al. 2009; Bose et al. 2014).

Effective and regular primary metabolism (amino acids, lipids, sugars, sugar alcohols and TCA intermediates) is required to overcome the adverse effects of salinity; only by this way biomass build-up can be maintained (Arbona et al. 2013). Similar tendencies but different alteration rates were usually observed at the levels of these metabolites between halophytes and glycophytes under salinity conditions (Arbona et al. 2013).On the other hand, sometimes a stress-specific change in the level of a distinct metabolite might be the result of the inhibition or activation of that stress-specific metabolic pathway (Obata and Fernie 2012). Accordingly, understanding the regulatory mechanisms acting on such pathways will also be helpful for us to distinguish good candidates for the manipulation of metabolism under stress conditions. With this perspective, evaluation and comparison of the metabolite profiling data derived from the salt sensitive and tolerant genotypes of the same species and halophytes will be helpful to discover novel metabolites responsible for salinity tolerance. Definitely it would be hard to generalize and address these molecules to all crop plants. Recent studies on cereals showed the importance of allelic variations for Na^+ and K^+ transporters to be important for salinity tolerance (Mickelbart et al. 2015). Same approach can be valid for the genes encoding enzymes necessary in the biosynthetic pathways of some distinct metabolites in sensitive and tolerant genotypes. One has to keep in mind the importance and difficulty of choosing the best candidate of salt-tolerant model for crop species. As far as we know, the salt-tolerant relatives of most of the crop species such as wheat, barley and rice were poorly studied and characterized in this context (Slama et al. 2015).

A large number of studies have been carried out to examine the responses and tolerance mechanisms of plants against salts stress. Nowadays the most realistic and emerging approach is to integrate genomics, transcriptomics, proteomics and metabolomics data on salinity research in order to provide solutions which would best fit to crop plants. Plant metabolism under salinity conditions is one of the hot topics of the plant stress research area, and it is progressing very rapidly. Knowledge accumulated in the salt stress research field is no longer the tip of the ice, now we are far from that; step by step we are getting more and more close to the solution. Here in this chapter, we tried to summarize the recent findings and focus on the best candidates of metabolism for the manipulation to achieve salinity tolerance in crop plants.

9.2 Plants Are Troubled by Salt Stress: How Do They Manage?

Early effects of salinity are known to be similar to that of drought stress, but, in the long term as a consequence of accumulated Na^+ and Cl^- in the leaves, hyper-osmotic and hyper-ionic stresses double the severity and impact of this stress (Munns 2002; Chaves et al. 2009). Salt stress responses and tolerance mechanisms of plants are dominated by complex networks and multiple signalling pathways (Maron et al. 2016). Salt-tolerant plants share common traits such as efficient osmotic adjustment, vacuolar ion sequestration and reduced ion uptake from the roots (Munns and Tester 2008). On the other hand, there is an important discrepancy between plants by means of salt tolerance; survival is substantial for plants growing at natural salt habitats, while for crop plants yield performance is noteworthy at field conditions. Scientific literature of salt stress responses and tolerance mechanisms in plants is full of detailed information on how plants are affected by salinity. In order to avoid repetitions, we will briefly go through these data and summarize what we know so far about the effects of salinity on plants.

Stomatal closure, increased leaf temperature and inhibition of shoot elongation are among the rapid emerging effects of salt stress. These events are defined as the "shoot salt accumulation independent effects" since they occur before Na^+ and Cl^- accumulation have reached to high levels in the shoots (Roy et al. 2014). The toxic effects of accumulated Na^+ and Cl^- take place as the second phase of salinity which is remarkable by the senescence of older leaves (Munns and Tester 2008; Roy et al. 2014). Generally, salt stress tolerance mechanisms are categorized according to the phase of the salinity. Usually stress coping mechanisms of the osmotic phase are separated from the ionic phase strategies. However, this distinction is a little bit difficult to achieve due to two reasons. First, to date it is not clear whether one type of strategy serves the best for one phase or not. Second, the sequence and onset of these stress coping mechanisms are not fully understood. Roy et al. (2014) categorized salt tolerance mechanisms in three groups: osmotic tolerance, tissue tolerance and tolerance by ion exclusion. Osmotic tolerance depends on long-distance signalling to avoid growth reduction, while tissue and ion exclusion involve ion transporters, pumps and compartmentation mechanisms for exclusion and/or sequestration of Na^+ and Cl^- at both plasma membrane and tonoplast (Roy et al. 2014).

Ion exclusion capacity of plants can be altered by manipulating high-affinity potassium transporter (HKT) gene family in wheat, rice and tomato (Roy et al. 2014; Mickelbart et al. 2015). The expressions of these genes are usually dependent on cell type and induced by characterized or uncharacterized stress-responsive molecules. Therefore, their potential to improve salinity tolerance in crop plants is promising but limited (Roy et al. 2014). However, marker-assisted selection approach for the introgression of *HKT*1;5-A from *T. monococcum* into durum wheat resulted in improved grain yield by increasing Na^+ exclusion capacity at saline fields (Munns et al. 2012; Roy et al. 2014). At salinity conditions, maintenance of Na^+ and Cl^- in the roots is one of the common avoiding strategies of crop plants (maize, barley, wheat, etc.), but this capacity is limited to short term (Peng et al. 2016). Salt tolerance along with high-yield performance needs more sophisticated mechanisms such as formation of salt-secreting structures for exclusion of salt crystals from the leaves. Salt-tolerant cotton is a good example among crop plants for its leaf glandular trichomes which can secrete vacuole sequestered Na^+ from the leaves (Peng et al. 2016). These kind of structures are also present in cereal plants, and their composition can be enhanced by salinity; however, their sizes are usually not sufficient for efficient Na^+ exclusion (Ramadan and Flowers 2004; Shabala 2013). Manipulation of trichomes or glandular formations by means of size, shape and density would be an important tool for better Na^+ sequestration and/or exclusion in crop plants (Shabala 2013).

Accordingly keeping low Na^+ but high K^+ levels within the plant cell is an important trait for salinity tolerance. K^+ loss is induced by the channels activated by reactive oxygen species (ROS) during salt stress, and K^+ loss can also be used as a measure of salt tolerance threshold in plants (Shabala and Cuin 2008; Shabala 2009; Jayakannan et al. 2013, 2015). Tissue tolerance can be improved by altering the synthesis of compatible solutes, by increasing antioxidative capacity and by efficient Na^+ sequestration into vacuoles (Roy et al. 2014). Biosynthesis of compatible

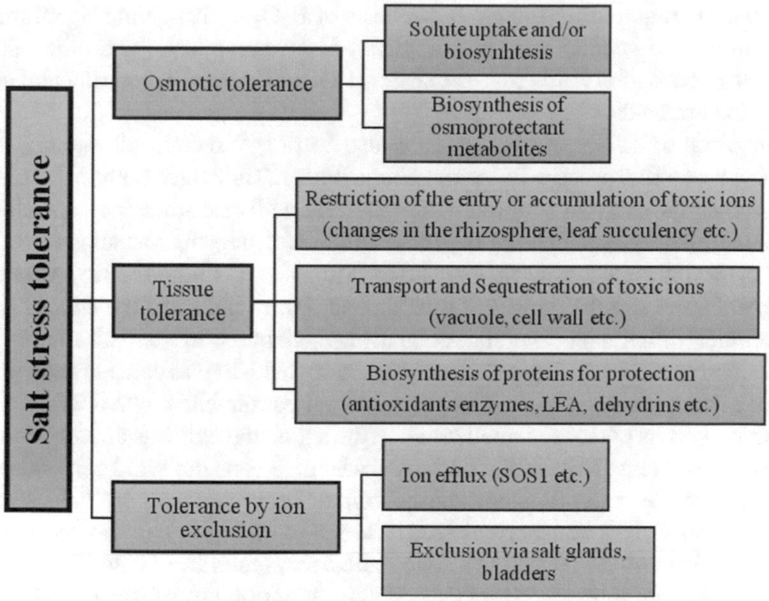

Fig. 9.1 Overview of the mechanisms for salt stress tolerance in plants

organic molecules is an important salt stress defence mechanism. Small carbohydrates, polyols, amino acids, methylamines, methylsulphonium solutes and urea are among those molecules with high osmoprotective capacity (Yancey 2005; Hoang et al. 2016) (Fig. 9.1).

ROS levels significantly increase due to the impairment of photosynthesis and respiration under salinity (Bor et al. 2003; Mittova et al. 2003; Ozgur et al. 2013; Yolcu et al. 2016). In relation to their signalling roles, sometimes propagation and transduction of ROS are favoured for the activation of defence, but, excessive accumulation of these molecules has to be avoided due to their interference with metabolism (Foyer and Noctor 2005; Xia et al. 2015). The balance between ROS production and efficiency of antioxidant metabolites is critical for cellular homeostasis. As it is well known that synthesis of enzymatic antioxidants such as superoxide dismutase, ascorbate peroxidase, catalase and non-enzymatic antioxidants, ascorbate, glutathione, tocopherol and carotenoids is the basic step for the regulation of ROS levels within plant cells (Mittler et al. 2004; Sekmen et al. 2014; Xia et al. 2015; Yolcu et al. 2016). Salt tolerance has been proven to be tightly linked to an efficient antioxidative capacity, and even the constitutive levels of antioxidants were reported to be high in extreme salt-tolerant plants “halophytes” (Bor et al. 2003; Hamed et al. 2013; Ozgur et al. 2013; Sekmen et al. 2014). On the other hand, although the production and scavenging of ROS have significant importance in salt stress tolerance, “a true salt-tolerant plant” was proposed to be dependent on efficient Na^+ exclusion along with high antioxidative protection (Bose et al. 2014; Shabala and Munns 2017). Bose et al. (2014) also suggested that high superoxide dismutase activity in halophytes is required for the rapid synthesis of H_2O_2 for triggering signalling networks to initiate genetic and physiological adaptive responses, and other antioxidants are needed afterwards for decreasing H_2O_2 and preventing detrimental effects of hydroxyl radicals.

Sequestration of Na^+ within the cell is also a strategy to cope with salinity. Some halophytes use this process for osmotic adjustment. They have higher Na^+ concentrations than the external solution within the cell, while salt-tolerant non-halophyte plants (e.g. barley) prefer to compartmentalize Na^+ into the vacuoles in order to avoid cytosol accumulation (Shabala 2013; Munns and Gilliham 2015). One of the conserved traits in relation to salt tolerance is tightly linked to the control of Na^+ intracellular influx and vacuolar compartmentalization in root and shoot cells (Mickelbart et al. 2015). Peng et al. (2016) found that Na^+ was concentrated in roots and then transported to the shoots in salt-tolerant cotton. Na^+ content was higher in the leaves, and its compartmentalization in the shoot and leaf was an important trait for tolerance (Peng et al. 2016). However, when we consider yield performance of crop plants under salinity, there can be contradictory opinions by means of Na^+ accumulation. A Na^+ transporter (SKC1) which regulates K^+/Na^+ homeostasis has been identified and introduced to rice to increase salt tolerance; however, yield performance was not improved (Ren et al. 2005). According to Albacete et al. (2014), yield performance at salinity conditions is governed by traits which regulate flower/grain sink activity rather than leaf Na^+ accumulation which indicates also the importance of the regulation of metabolic processes.

As indicated above accumulation of compatible solutes for osmotic adjustment is a common response in salt-tolerant plants (Munns 2005). *Arabidopsis* and tobacco overexpressing proline biosynthetic enzyme, Δ-pyrroline-5-carboxylate synthase, had high proline levels and enhanced salinity tolerance (Liu and Zhu 1997). Overexpression of glycine betaine synthesizing enzyme choline oxidase in rice resulted in salt stress resistance (Sakamoto and Murata 1998). Transgenic wheat for mannitol-1-phosphate dehydrogenase increased tolerance against drought and salt stresses (Abebe et al. 2003). Manipulation of compatible solute biosynthesis has high potential for salinity tolerance at field conditions for crop plants such as wheat, tomato, rice and tobacco. However, besides some successful results at green house trials, these plants tend to perform not so well at field conditions. Most of the compatible solutes are N-containing molecules; therefore, manipulation of nitrogen metabolism becomes crucial for salt tolerance. Wang et al. (2012) found a great discrepancy between nitrogen metabolism of old and young rice leaves subjected to salinity, which leads to a better protection in the latter one. In the old leaves, high accumulation of NO_3^-, Na^+ and Cl^- were reported in relation to the upregulated expression of transporter encoding genes with different transmission mechanisms (Wang et al. 2012). Manipulating nitrogen metabolism to cope with salinity has a high potential, but the key homeostatic importance and complexity of this process has to be taken and treated with precaution. Likewise, we are still far from understanding the mechanism(s) of how plants can switch nitrogen metabolism on different leaves (or organs) under stress conditions at the same time.

Metabolic responses of plants to salinity are governed by interconnected signalling pathways and reprogramming of whole plant metabolism in order to achieve energetic and developmental homeostasis (Chaves et al. 2009). Acclimation and survival under salt stress can be possible by the continuation of metabolic activity with the modulation of important metabolites. Salt-acclimated plants tend to have altered levels of some distinct metabolites belonging to amino acid, carbohydrate and lipid metabolism. Fluctuations and adjustments in the levels of these metabolites at cellular level contribute to ameliorate the effects of salinity. For instance, keeping low levels of organic acids is an important mechanism for the compensation of ionic imbalance at salinity conditions (Sanchez et al. 2007). According to a metabolite profiling study, levels of 45 metabolites were changed in the leaves of *Arabidopsis* subjected to different abiotic stresses including salinity. Among these metabolites the most significant differences were seen in the levels of proline, sucrose and raffinose at salt stress conditions (Obata and Fernie 2012).

9.3 Manipulating Metabolism for Salinity Tolerance: Aren't We Still There?

Plant responses to abiotic stresses are arranged by diverse metabolic and signalling networks. Different metabolic routes can take in action for a specific type of stress, or one metabolic route can be activated/inactivated for variety of stresses. Metabolic engineering approaches for some certain type of stress-specific metabolites were effective for increasing the tolerance of some crop plants; however, manipulating plant metabolism is still a difficult task due to its complexity. For instance, the balance between amino acid anabolism and catabolism is important not only in stress-related reactions but also in the conversion of storage proteins into carbohydrates during germination and in the recycling of energy-rich compounds during senescence (Hildebrandt et al. 2015). Amino acid and protein metabolism are known to be highly altered by abiotic stress conditions. In stressed cells the rate of amino acid and protein biosynthesis declines, and protein degradation along with the accumulation of some certain amino acids (e.g. proline) is highly induced (Szabados and Savouré 2010; Hildebrandt et al. 2015). Therefore, targeting amino acid or protein biosynthesis by means of metabolic engineering would not be rationale since homeostasis of these two processes are critical both at normal and stress conditions.

Lipid biosynthesis and metabolism are also involved in stress sensitivity and/or tolerance as reported by several studies. In maize subjected to salt stress, not only glycolysis and amino acid biosynthesis but also fatty acid biosynthesis was highly induced; hence, palmitic and oleic acid levels in the shoots were increased (Guo et al. 2017). Tomato sensitive to freezing (*SFR2*) is important for the integrity of chloroplast membranes at freezing temperature. *SFR2*was found to be effective in alleviating the negative impact of salinity by increasing trigalactosyl diacylglycerol levels that leads to lipid remodelling of membranes (Wang et al. 2016a). On the other hand, Pitzschke et al. (2014) reported that overexpression of lipid transfer protein (LTP) mediated salinity tolerance in *A. thaliana* and wheat LTP homolog was expressed in salt stress-resistant wheat cultivars during the osmotic phase (Takahashi et al. 2015). Participation of lipids to salt stress tolerance both as a signalling and/or a structural component has been proposed and proved (DeWald et al. 2001; Darwish et al. 2009; Munnik 2014; Li et al. 2017). Bargmann et al. (2009) pointed out the contribution of phospholipase D (PLD) to salt stress tolerance in tomato and *A. thaliana*. Although their work has defined the signalling role of PLD via formation of phosphatidic acid, they also emphasized that PLD activity might be important for stress-induced membrane remodelling and rearrangements (Gigon et al. 2004; Bargmann et al. 2009). Moreover, the recruitment of phospholipase C1 (PLC1) from cytoplasm to plasma membrane was shown to be induced by salt stress in *Oryza sativa*, and hydrolyzation of phosphatidylinositol-4-phosphate by this enzyme-elicited stress-induced Ca^{2+} signals which regulated Na^{+} accumulation in the leaves in order to maintain whole plant salt tolerance (Li et al. 2017). In addition to these, deposition of hydrocarbons in the form of cuticle or wax for avoiding water loss from the leaves is an advantage at osmotic stress. Comparisons of the lipid profiles of desiccation susceptible and tolerant plants showed alterations in the levels of monogalactosyl diacylglycerol, galactolipids, phosphatidylinositol and phosphatidic acid (Gasulla et al. 2013; McLoughlin et al. 2013; Okazaki and Saito 2014). Nevertheless, interfering with such complex and critical metabolic networks for the enhancement of stress tolerance can lead to the emergence of undesired pleiotropic effects such as growth and developmental alterations (Cabello et al. 2014).

Metabolic modelling of the acclimation of plants to the environmental stresses is a handy tool for a better understanding of plant metabolism (Baghalian et al. 2014). Genetic studies targeting the manipulation of a single enzyme in a complex metabolic pathway usually fail to be successful since such pathways are controlled by multi-reaction systems (Morandini 2009, 2013; Baghalian et al. 2014). For instance, in *Triticium aestivum* different metabolic shifts were reported in the leaves and roots under salinity (Guo et al. 2015). The authors have identified 75 metabolites by GC-MS analysis which indicated enhancement of gluconeogenesis, glycolysis and amino acid synthesis at salt stress conditions. In wheat leaves glucose, fructose, trehalose, leucine, isoleucine, valin and proline levels were increased, while maltose, malic acid, shikimic acid and fumaric acid levels were decreased. Roots exhibited a different trend in which glucose, glucose 6-phosphate, GABA, sorbitol and valine were decreased as proline, galactose, sucrose and myoinositol were increased. Such results are usually difficult to interpret since without tagging analysis it would be hard to estimate the source of amino acids or carbohydrates. They can either be degradation products or their biosynthesis is activated. Sometimes it is possible to figure out this kind of results as it was indicated by Guo et al. (2015). High levels of proline by the decreased levels of GABA and glutamate indicated the conversion of glutamate to proline via Δ-pyrroline-5-carboxylate synthetase in salt-stressed maize (Guo et al. 2015).

When one considers the overall information about plants metabolism and salt stress tolerance, it looks like "sky will be the limit." However, as it was indicated before, for crop plants not only coping with different environmental constraints but also efficient biomass production is important. Enhanced agricultural performance and productivity under salinity depend on the ability of a plant to develop biomass at high Na^+, low available water and CO_2 and high-energy demanding defence processes such as biosynthesis of antioxidants, hormones and osmolytes (Shabala 2013; Munns and Gilliham 2015). Generally, what we experience in halophytes could not be applicable to crop plants since their survival strategies take place in the expense of growth (biomass). Halophytes are the best salt-tolerant or salt-adapted plants; their stress coping mechanisms also involve cross tolerance, anticipation, stress memory and predictive abilities (Hamed et al. 2013; Slama et al. 2015). These concepts are still so unknown and unfamiliar for us to apply to crop plants due to their complex and restricted nature, but it is still worth to work on and characterize.

All of the metabolic routes mentioned so far all serve whether to acclimation or survival at salt stress conditions. Here we want to focus more on polyamine and sugar metabolisms which would be more advantageous to manipulate by means of tolerance in crop plants at field conditions. Concentrating on polyamines for being the common stress-protecting metabolites and sugars for their signalling, energetic and structural properties will be more helpful since they also have the ability to alleviate both osmotic and ionic effects of salinity. Moreover, these two metabolic routes are the ones in which we could find the exact counterparts of the information gathered from the model plants and/or halophytes in crop plants.

9.3.1 Polyamine Metabolism

Polyamines are involved in a broad range of processes in plants including growth, development and stress responses (Moschou and Roubelakis-Angelakis 2013). The presence of two pathways in plants (except *Arabidopsis*) is also an indicator of the importance of polyamines for plant metabolism with or without stress. Chloroplast-localized arginine decarboxylase (ADC) and mitochondria- and nucleus-localized ornithine decarboxylase (ODC) pathways are the two important biosynthetic routes for polyamines (Moschou and Roubelakis-Angelakis 2013; Do et al. 2014). There is a consensus on the importance of polyamines in stress responses, and usually polyamine accumulation is attributed to stress tolerance (Takahashi and Kakehi 2010; Bose et al. 2014). Nevertheless, under stress conditions the exact role of polyamines is still undefined although a vast amount of information is available about their contribution to the maintenance of ion balance, free radical scavenging capacity, prevention of senescence, protein phosphorylation, regulation of vacuolar and plasma membrane channels and stabilization of membranes (Do et al. 2014).

Integration or overexpression of polyamine biosynthesis-related genes usually resulted in better performance under salinity, while lack or inefficient expression of these genes decreases tolerance. Rice plants overexpressing oat ADC and *Tritodermum* S-adenosylmethionine synthetase (SAMDC) were resistant to salt stress with high levels of polyamines (Roy and Wu 2001, 2002). On the other hand, loss-of-function *Arabidopsis* mutants in spermine (Spm) and thermospermine biosynthesis suffered from high Na^+ accumulation (Alet et al. 2012). Salt sensitivity of barley genotypes is genetically variable in relation to polyamine- and ROS-induced alterations in K^+ homeostasis (Velarde-Buendía et al. 2012). Apoplastic polyamines have been shown to potentiate hydroxyl radical-induced K^+ efflux which is well correlated with salinity tolerance in different barley genotypes (Zepeda-Jazo et al. 2011; Velarde-Buendía et al. 2012; Bose et al. 2014). Do et al. (2014) did an extensive study with 18 rice cultivars which differed in salinity tolerance. They reported remarkable discrepancies between putrescine (Put), spermidine (Spd) and Spm levels in sensitive and tolerant cultivars subjected to 50 and 100 mM NaCl. High Put levels were kept constant in tolerant cultivars along with slightly increased Spm, while in the sensitive cultivars decreased Put but increased Spm levels were detected with a possible time course difference (quick or prolonged response) in accordance with the transcription profiles (Do et al. 2014).

Several of salinity tolerance traits including K^+ homeostasis, Na^+ sequestration and osmotic adjustment are related to and/or controlled by polyamine metabolism (Shabala and Munns 2012; Pottosin and Shabala 2014); therefore, the manipulation of polyamine biosynthesis has high potential for improved crop performance under salt stress. On the other hand, as it was indicated by Pottosin and Shabala (2014), “the more PA the better for stress tolerance” concept is not valid for most of the salt stress-tolerant species. Since, there can be fluctuations in the biosynthesis and conversion of polyamines under salinity, manipulating the timing and localization of these processes is seemed to be the key component for improving crop performance. Maintenance of high K^+ low Na^+ levels in plant cells at salinity conditions can be facilitated by the biosynthesis and regulation of proteins responsible for K^+ transport and the efficient vacuolar sequestration of Na^+ via slow and fast vacuolar channels. These processes are all known to be regulated by polyamines (Pottosin and Shabala 2014). Accordingly the limitation of Na^+ entry into cytosol via the inhibition of non-selective cation channels by polyamines is also an intriguing opportunity for improving salt tolerance in crop plants (Demidchik and Maathuis 2007; Pottosin and Shabala 2014).

At the end of this part, we want to emphasize two recent studies about the importance of polyamine conversion in stress responses of *A. thaliana*. These new findings can be integrated and used in crop plants for improving salt tolerance. Since, polyamine metabolism is an evolutionary conserved route not only for plant development but also for stress responses, it would be rationale to consider that same trends may be applicable to the agricultural plants. Therefore, deciphering polyamine conversion processes in more detail would provide an excellent opportunity for improving salt stress resistance or tolerance in crop plants. Zarza et al. (2017) studied the contribution of polyamine back-conversion process to salt tolerance in *A. thaliana* (Zarza et al. 2017). This reaction reverses the polyamine biosynthesis; first spermidine and, then, putrescine are synthesized from spermine and thermospermine by the activity of polyamine oxidases (Tavladoraki et al. 2006; Takahashi et al. 2010; Kim et al. 2014; Zarza et al. 2017). Loss-of-function *atpao5* mutants exhibited enhanced salt stress tolerance in relation to high thermospermine levels. Besides that, alanine, arabitol, erythritol, fumarate, citrate, glucose, sucrose, rhamnose and ethanolamine metabolites were highly elicited in these mutants according to the metabolic analysis. Each of these metabolites either directly or indirectly contributes to salt stress tolerance mechanisms. The authors concluded that metabolic and transcriptional reprogramming which was driven by thermospermine improved salt stress tolerance in *A. thaliana* (Zarza et al. 2017). Another intriguing process related to polyamine conversion is the acetylation of polyamines which is catalysed by N-acetyltransferase activity 1 (NATA1) enzyme in *A. thaliana* (Lou et al. 2016). Acetylated polyamines have been reported to be present in a diverse group of plants, but, still, not much is known about their exact role and how they function. Their role in defence against *Pseudomonas syringae* was studied, and it was found that coronatine-regulated putrescine acetylation attenuated antimicrobial defence in a way by competing with polyamine oxidases (Lou et al. 2016). This process is yet to be elucidated at abiotic stress conditions, which is most likely expected to be important for overall polyamine metabolism in plants.

9.3.2 Sugar Metabolism

Sugars are involved not only in the maintenance of growth but also in regulation of the expression of genes responsible for several biochemical processes; therefore, the effects of alterations in sugar levels become more remarkable and dramatic at stress conditions (Hare et al. 1998; Slama et al. 2015). Targeting sugar metabolism for enhancing salt stress tolerance in plants will be beneficial since it provides precursors of hormones and signalling molecules and has ROS scavenging activity (Cabello et al. 2014). There are several reports indicating the relevance of sugar metabolism both in the short and long terms of salinity. In different plants, sucrose, glucose and fructose levels were induced by salinity, while starch content was decreased. Accumulation of simple sugars is reported to be necessary to maintain cell turgor, stabilization of membranes and prevention of protein degradation (Arbona et al. 2013). Metabolome analysis of halophyte *Thellungiella* and its comparison with *Arabidopsis* has revealed the involvement of sugar metabolism for efficient salt tolerance. Under salinity *Thellungiella* had higher levels of fructose, sucrose and complex sugars which contributed not only to the osmoprotectant capacity but also to ROS scavenging capacity. On the other hand, even in non-stressed conditions, the levels of these metabolites were higher in *Thellungiella* than that of *Arabidopsis*. Likewise, prestress proline levels were highly induced in *Thellungiella* as reported by Taji et al. (2004). Halophytes are constitutively well-prepared for the negative impact of salinity as compared to glycophytes (Hamed et al. 2013).

In a recent publication, maize root and shoot metabolite profiles were compared in response to neutral and alkaline salt stresses (Guo et al. 2017). Among the sugars fructose, sucrose, talose and myoinositol in the roots and raffinose and galactinol in the shoots were increased; however, glucose levels were decreased in both. The authors concluded that gluconeogenesis was enhanced, and maintenance of carbon source and osmotic adjustment were possible by the degradation of polysaccharides under salinity in maize (Guo et al. 2017). Wu et al. (2013) reported the differences in the metabolite profiles of salt stress-sensitive and salt stress-tolerant barley genotypes. In the tolerant genotype, proline and raffinose were the common osmolytes both in roots and shoots. Sucrose, trehalose, mannitol and inositol were root-specific, while, asparagine, glycine, isoleucine and serine amino acids were the leaf-specific ones in the long term of the salinity. According to their findings, induced levels of proline and sugars served not only in osmotic adjustment and stabilization of the membranes but also as antioxidants since ascorbic acid and other antioxidants levels were not increased (Wu et al. 2013). Vulnerability of the germination and early seedling stage to salinity is found out to be related to the lower mobilization of sugars from seed storage reserves (Rosa et al. 2009). Pandey and Penna (2017) proposed that measurement of total soluble sugar content within the first 24 h of germination would provide an important clue for the tolerance level to salt stress. They found that at 150 mM NaCl treatment, there was a rapid increase in the soluble sugar content of mildly salt-tolerant *Brassica juncea* which facilitated the radicle emergence (Pandey and Penna 2017). Moreover, transcription of genes related to source and sink activity was altered by salinity. Genes involved in photosynthesis, photo assimilate export and nutrient mobilization were downregulated as the transcription of genes involved in carbohydrate degradation, and lipid storage, and protein and polysaccharide synthesis were upregulated (Stitt et al. 2007; Chaves et al. 2009).

Trehalose as an osmoprotectant and stabilization agent contributes to tolerance strategies of plants when they are exposed to different abiotic stresses (Bianchi et al. 1993; Drennan et al. 1993; Ghaffari et al. 2016). Trehalose-6-phosphate (T6P) is synthesized from UDPGlc and glucose-6 phosphate by trehalose-6-phosphate synthase, and it is dephosphorylated to trehalose by trehalose-6-phosphate phosphatase enzyme in plants (Paul et al. 2008). High levels of trehalose were found in rice plants which had better growth performance under salt stress (Garcia et al. 1997; Hoang et al. 2016). Henry et al. (2015) reported elevated levels of trehalose-6-phosphate in leaf, cob and kernels of maize plants subjected to salt stress. According to their overall findings, they proposed that T6P and SUCROSE NONFERMENTING-RELATED KINASE1 (SnRK1) may have different roles in source and sink tissues, which has a potential to target for improving maize performance under salt stress conditions (Henry et al. 2015). Accordingly overexpression of a rice trehalose-6-phosphate phosphatase gene (driven by an OsMads6 promoter) improved harvest index and yield of maize plants subjected to water deficit at field conditions. Data from different field trials showed that crop performance was increased by the modified T6P/sucrose ratio at mild and severe drought conditions (Nuccio et al. 2015). In a recent study, metabolic and transcriptional profiling of central metabolic components in barley colonized with endophytic fungus Piriformospora indica showed a positive correlation with trehalose metabolism and increased salt stress tolerance (Ghaffari et al. 2016).

Although there are some success stories of improved salt stress tolerance by manipulating sugar metabolism, some studies emphasized the undesirable traits and growth penalties in relation to modified sugar homeostasis. For example, Cortina and Culiáñez-Macià (2005) reported thick shoot, dark-green leaves and aberrant root development in tomato plants overexpressing the yeast trehalose-6-phosphate synthase (TPS1). Stunted growth was found in potato and alfalfa transgenic plants overexpressing *E. coli* trehalose biosynthetic genes (Jang et al. 2003; Suárez et al. 2009). In such kind of transgenic approaches, the decision on the use of appropriate promoter was found to be important in order to overcome the undesired effects. When trehalose genes were expressed under the light-regulated promoter of the Rubisco small subunit or an ABA-inducible promoter, besides higher trehalose accumulation better vegetative growth, higher photosynthetic capacity, low Na^+ accumulation in the leaves and delayed senescence were reported (Garg et al. 2002; Albacete et al. 2014).

9.4 New Possibilities for Metabolic Manipulation

Integration of bioinformatics and experimental data provides a wide range of possibilities for us to better understand plant metabolism under salt stress. Thoen et al. (2017) used a novel approach – multi-trait mixed model (MTMM) – to analyse the genetic architecture of complex traits against environmental stresses by genome-wide association in *A. thaliana* that can be used as a model strategy for producing next-generation crops starting with *Brassica* species. They compared 30 traits under different stress conditions including salt stress and selected candidate genes for multiple stress-responsive traits in 350 *Arabidopsis* accessions (Thoen et al. 2017). Among these several genes were found to be both abiotic and biotic responsive with contrasting effects such as *TCH4* (encoding a cell wall modifying enzyme), *AtCCR2* (lignin biosynthesis related), *ASG1* (ABA and salt responsive) and *RMG1* (encoding a disease resistance protein with a conserved pathogen recognition pattern). The last one is of great interest as its expression is upregulated by salt stress in roots, and it was previously reported as an activator of salicylic acid (SA) pathway (Yu et al. 2013). A possible integration of SA signalling pathway to salt stress tolerance can be proposed. Jayakannan et al. (2015) reported that salt and oxidative stress tolerance is controlled by non-expresser of pathogenesis-related gene 1 (*NPR1*) dependent SA signalling in *A. thaliana*. They found that *NPR1*has an important function in the regulation of Na^+ entry into the roots and its delivery to the shoots. The interaction between salt stress and SA signalling is still unclear and looks deeper than it seems. Wang et al. (2016b) concluded that enhanced salinity tolerance in malate dehydrogenase (MDH) overexpressing apple was correlated to SA signalling since free and total SA levels were high in this plant. They also emphasized that in order to provide appropriate SA concentration, both biosynthetic and degradation routes were induced. MDH activity is critical for malate valves involved in central metabolism and redox homeostasis between different organelles (Wang et al. 2016b).

In the light of MTMM results, cell wall-modifying enzyme encoding *TCH4* reminded a recent finding by Endler et al. (2015) about the mechanism of sustained cellulose synthesis during salt stress. Plant cell walls are mainly composed of cellulose that is synthesized by the activity of plasma membrane-localized and microtubule-guided cellulose synthase enzyme complexes (Endler et al. 2015). Growth and development of plants depend on regular and proportional organization of primary and secondary cell walls. Their behaviours have remarkable effects on the plant biomass which would be more significant under stress conditions. When plant cells are subjected to salt stress, the integrity of cellulose synthase enzyme complexes is affected, and these complexes are disassembled from the plasma membrane. Endler et al. (2015) recently reported that two plant-specific proteins named as companion of cellulose synthase 1 and 2 proteins enable microtubule dynamics and cellulose synthase enzyme complex activity to support biomass production under salinity in *A. thaliana*. Biomass development is improved by the alterations in metabolism-mediated gene expression in salt-tolerant wheat genotypes which is of great importance for higher-yield performance under salinity conditions (Takahashi et al. 2015). Especially in long-term salinity, maintenance of high cellulose synthesizing capacity might have an important impact on crop performance and yield. Reports providing evidence for the importance of cell wall in salt responses are rapidly increasing. In a recent article, Xiao et al. (2017) found that the mRNA level of *Arabidopsis* glucuronokinase (AtGlcAK) was highly induced after 8 h of drought and salt treatments. This enzyme contributes to the phosphorylation of glucuronic acid and its conversion into UDP-glucuronic acid, which is an important molecule of sugar interconversion and cell wall biosynthesis (Pieslinger et al. 2010; Garlock et al. 2012; Xiao et al. 2017).

Two birds with one stone approach will make it possible to breed crop plants resistant to several environmental stresses. For sure, this kind of interesting links and connections resulted from the outcomes of MTMM analysis would widen our salt stress tolerance perspective; however, these promising analyses are still restricted to model plants. Moreover, integrated transcriptomic and metabolomics data put forward very clearly that the regulation of these processes took place not only at transcriptional level but also at post-transcriptional, translational and post-translational levels.

9.5 Conclusion

Extensive research has been carried out to examine salt stress responses and tolerance mechanisms of plants for decades. In most of the cases, field applications of these inputs are still far from the desired high-yield performance for crop plants. There are several reasons for this failure such as the unexpected negative effects on growth and development due to the interference to signalling and regulatory pathways, combination of different stressors at field conditions at the same time, the discrepancies between the phenotypes of tolerant plants which are grown at green house and field, etc. Alterations in different metabolic routes and pathways via synthesis or degradation of some certain metabolites have been reported to be necessary for acclimation and survival under salinity. Generally plants tend to manipulate the levels of amino acids, lipids, sugars, sugar alcohols and TCA intermediates to cope with osmotic and ionic phases of salt stress, but the efficiency of these processes differs according to the plant species. Manipulating plant metabolism for higher levels of glycine betaine and phenolic compounds or enabling them to accumulate and use Na^+ for osmotic adjustment or engineering plant leaves for salt glands just like halophytes will be the best solutions for high salt tolerance in crop plants. However, the feasibility of such modifications is very low and there are still so many unknowns about the mode of action. On the other hand, these mechanisms are usually extreme features which are unique to salt-tolerant plants and mostly to halophytes. The osmotic and ionic characteristics of salt stress challenge plants in two different but simultaneous ways; therefore, manipulating metabolites which could be effective in both phases will be more beneficial. Although it is a complicated task, manipulating metabolic routes for higher osmotic and ionic stress tolerance will be more realistic for increasing the chance of crop plants to overcome salinity and other stresses. Among the metabolites studied so far, polyamines and sugars are the most promising ones in relation to their contribution not only to protection of membranes, proteins, leaf water status and cellular homeostasis but also for efficient biomass production under salinity conditions. On the other hand, comparative metabolic analysis provided information that most of the time variations between levels of polyamines and sugars in the roots and leaves of crop plants are good indicators of the degree of salt sensitivity or tolerance. These two metabolic routes are the best candidates for acquiring resistance against salinity; therefore, the more we learn about these processes, the closer we get to the final goal of manipulating crops for increased yield performance at field conditions.

References

Abebe T, Guenzi AC, Martin B et al (2003) Tolerance of mannitol accumulating transgenic wheat to water stress and salinity. Plant Physiol 131:1748–1755

Albacete A, Ghanem ME, Martínez-Andújar C et al (2008) Hormonal changes in relation to biomass partitioning and shoot growth impairment in salinized tomato (*Solanum lycopersicum* L.) plants. J Exp Bot 59:4119–4131

Albacete AA, Martínez-Andújar C, Pérez-Alfocea F (2014) Hormonal and metabolic regulation of source–sink relations under salinity and drought: from plant survival to crop yield stability. Biotechnol Adv 32:12–30

Alet AI, Sánchez DH, Cuevas JC et al (2012) New insights into the role of spermine in *Arabidopsis thaliana* under long-term salt stress. Plant Sci 182:94–100

Arbona V, Manzi M, Ollas C et al (2013) Metabolomics as a tool to investigate abiotic stress tolerance in plants. Int J Mol Sci 14:4885–4911

Baghalian K, Hajirezaei RM, Schreiber F (2014) Plant metabolic modelling: achieving new insight into metabolism and metabolic engineering. Plant Cell 26:3847–3866

Bargmann BOR, Laxalt AM, Riet B et al (2009) Multiple PLDs required for high salinity and water deficit tolerance in plants. Plant Cell Physiol 50(1):78–89

Bianchi G, Gamba A, Limiroli R et al (1993) The unusual sugar composition in leaves of the resurrection plant *Myrothamnus flabellifolia*. Physiol Plant 87:223–226

Bor M, Ozdemir F, Turkan I (2003) The effect of salt stress on lipid peroxidation and antioxidants in leaves of sugar beet *Beta vulgaris* L. and wild beet *Beta maritima* L. Plant Sci 164:77–84

Bose J, Rodrigo-Moreno A, Shabala S (2014) ROS homeostasis in halophytes in the context of salinity stress tolerance. J Exp Bot 65:1241–1257

Cabello JV, Lodeyro AF, Zurbriggen MD (2014) Novel perspectives for the engineering of abiotic stress tolerance in plants. Curr Opin Biotechnol 26:62–70

Chaves MM, Flexas J, Pinheiro C (2009) Photosynthesis under drought and salt stress: regulation mechanisms from whole plant to cell. Ann Bot 103:551–560

Cortina C, Culianez-Macia FA (2005) Tomato abiotic stress enhanced tolerance by trehalose biosynthesis. Plant Sci 169:75–82

Darwish E, Testerink C, Khalil M et al (2009) Phospholipid signalling responses in salt-stressed rice leaves. Plant Cell Physiol 50:986–997

Demidchik V, Maathuis FJM (2007) Physiological roles of non-selective cation channels in plants: from salt stress to signalling and development. New Phytol 175:387–404

DeWald DB, Torabinejad J, Jones CA et al (2001) Rapid accumulation of phosphatidylinositol 4,5-bisphosphate and inositol 1,4,5-trisphosphate correlates with calcium mobilization in salt-stressed Arabidopsis. Plant Physiol 126:759–769

Do TP, Drechsel O, Heyer GA et al (2014) Changes in free polyamine levels, expression of polyamine biosynthesis genes, and performance of rice cultivars under salt stress: a comparison with responses to drought. Front Plant Sci 5(182):1–16

Drennan P, Smith M, Goldsworthy D et al (1993) The occurrence of trehalose in the leaves of the desiccation-tolerant angiosperm *Myrothamnus flabellifolius* welw. J Plant Physiol 142:493–496

Endler A, Kesten C, Schneider R et al (2015) A mechanism for sustained cellulose synthesis during salt stress. Cell 162:1353–1364

Fitzgerald TL, Waters DLE, Henry RJ (2009) Betaine aldehyde dehydrogenase in plants. Plant Biol 11:119–130

Flowers TJ, Colmer TD (2008) Salinity tolerance in halophytes. New Phytol 179:945–963

Flowers T, Munns R, Colmer TD (2015) Sodium chloride toxicity and the cellular basis of salt tolerance in halophytes. Ann Bot 115(3):419–431

Foyer CH, Noctor G (2005) Oxidant and antioxidant signalling in plants: a re-evaluation of the concept of oxidative stress in a physiological context. Plant Cell Environ 28:1056–1071

Fusari CM, Kooke R, Lauxmann MA et al (2017) Genome-wide association mapping reveals that specific and pleiotropic regulatory mechanisms fine-tune central metabolism and growth in Arabidopsis. Plant Cell. https://doi.org/10.1105/tpc.17.00232

Garcia AB, Engler J, Iyer S et al (1997) Effects of osmoprotectants upon NaCl stress in rice. Plant Physiol 115:159–169

Garg AK, Kim JK, Owens TG et al (2002) Trehalose accumulation in rice plants confers high tolerance levels to different abiotic stresses. PNAS 99:15898–15903

Garlock RJ, Wong YS, Balan V et al (2012) AFEX pretreatment and enzymatic conversion of black locust (*Robinia pseudoacacia* L.) to soluble sugars. Bioenergy Res 5(2):306–318

Gasulla F, Vom Dorp K, Dombrink I et al (2013) The role of lipid metabolism in the acquisition of desiccation tolerance in *Craterostigma plantagineum*: a comparative approach. Plant J 75:726–741

Ghaffari MR, Ghabooli M, Khatabi B et al (2016) Metabolic and transcriptional response of central metabolism affected by root endophytic fungus *Piriformospora indica* under salinity in barley. Plant Mol Biol 90:699–717

Gigon A, Matos AR, Laffray D et al (2004) Effect of drought stress on lipid metabolism in the leaves of *Arabidopsis thaliana* (ecotype Columbia). Ann Bot 94:345–351

Guo R, Yang Z, Li F et al (2015) Comparative metabolic responses and adaptive strategies of wheat (*Triticum aestivum*) to salt and alkali stress. BMC Plant Biol 15:170

Guo R, Shi XL, Yan C et al (2017) Ionomic and metabolic responses to neutral salt or alkaline salt stresses in maize (*Zea mays* L.) seedlings. BMC Plant Biol 17(41):1–13

Hamed KB, Ellouzi H, Talbi OZ et al (2013) Physiological response of halophytes to multiple stresses. Funct Plant Biol 40:883–896

Hare PD, Cress WA, Van Staden J (1998) Dissecting the roles of osmolyte accumulation during stress. Plant Cell Environ 21:535–553

Henry C, Samuel Bledsoe SW, Griffiths CA et al (2015) Differential role for trehalose metabolism in salt-stressed maize. Plant Physiol 169:1072–1089

Hildebrandt TM, Nunes Nesi A, Araujo WL et al (2015) Amino acid catabolism in plants. Mol Plant 8:1563–1579

Hoang TML, Tran NT, Nguyen TKT et al (2016) Improvement of salinity stress tolerance in rice: challenges and opportunities. Agronomy 6(54):1–23

Jang IC, Oh SJ, Seo JS et al (2003) Expression of a bifunctional fusion of the Escherichia coli genes for trehalose-6-phosphate synthase and trehalose-6-phosphate phosphatase in transgenic rice plants increases trehalose accumulation and abiotic stress tolerance without stunting growth. Plant Physiol 131:516–524

Jayakannan M, Bose J, Babourina O et al (2013) Salicylic acid improves salinity tolerance in Arabidopsis by restoring membrane potential and preventing salt-induced K+ loss via a GORK channel. J Exp Bot 64:2255–2268

Jayakannan M, Bose J, Babourina O et al (2015) The NPR1-dependent salicylic acid signalling pathway is pivotal for enhanced salt and oxidative stress tolerance in Arabidopsis. J Exp Bot 66(7):1865–1875

Jones R, Ougham H, Howard T et al (2013) The molecular life of plants, 1st edn. Wiley Hoboken (New Jersey) USA

Kim DW, Watanabe K, Murayama C et al (2014) Polyamine oxidase5 regulates Arabidopsis growth through thermospermine oxidase activity. Plant Physiol 165:1575–1590

Li L, Wang F, Yan P et al (2017) A phosphoinositide-specific phospholipase C pathway elicits stress-induced Ca^{2+} signals and confers salt tolerance to rice. New Phytol 214:1172–1187

Liu JP, Zhu JK (1997) Proline accumulation and salt-stress induced gene expression in a salt-hypersensitive mutant of Arabidopsis. Plant Physiol 114:591–596

Lou YR, Bor M, Yan J et al (2016) Arabidopsis NATA1 acetylates putrescine and decreases defence-related hydrogen peroxide accumulation. Plant Physiol 171(2):1443–1455

Maron GL, Pineros AM, Kochian VL et al (2016) Redefining 'stress resistance genes', and why it matters. J Exp Bot 67(19):5588–5591

McLoughlin F, Arisz SA, Dekker HL et al (2013) Identification of novel candidate phosphatidic acid-binding proteins involved in the salt-stress response of Arabidopsis thaliana roots. Biochem J 450:573–581

Mickelbart MV, Hasegawa PM, Bailey-Serres J (2015) Genetic mechanisms of abiotic stress tolerance that translate to crop yield stability. Nat Rev Genet 16:237–251

Mittler R, Vanderauwera S, Gollery M et al (2004) Reactive oxygen gene network of plants. Trends Plant Sci 9:490–498

Mittova V, Tal M, Volokita M et al (2003) Up-regulation of the leaf mitochondrial and peroxisomal antioxidative systems in response to salt-induced oxidative stress in the wild salt-tolerant tomato species *Lycopersicon pennellii*. Plant Cell Environ 26:845–856

Morandini P (2009) Rethinking metabolic control. Plant Sci 176:441–451

Morandini P (2013) Control limits for accumulation of plant metabolites: brute force is no substitute for understanding. Plant Biotechnol J 11:253–267

Moschou PN, Roubelakis-Angelakis KA (2013) Polyamines and programmed cell death. J Exp Bot 65:1285–1296

Munnik T (2014) PI-PLC: phosphoinositide-phospholipase C in plant signalling. In: Wang X (ed) Phospholipases in plant signalling. Springer, Berlin, pp 27–54

Munns R (2002) Comparative physiology of salt and water stress. Plant Cell Environ 25:239–250

Munns R (2005) Genes and salt tolerance: bringing them together. New Phytol 167:645–663

Munns R, Gilliham M (2015) Salinity tolerance of crops – what is the cost? New Phytol 208:668–673

Munns R, Tester M (2008) Mechanisms of salinity tolerance. Annu Rev Plant Biol 59:651–681

Munns R, James RA, Xu B et al (2012) Wheat grain yield on saline soils is improved by an ancestral Na+ transporter gene. Nat Biotechnol 30:360–364

Nuccio ML, Wu J, Mowers R et al (2015) Expression of trehalose-6-phosphate phosphatase in maize ears improves yield in well-watered and drought conditions. Nat Biotechnol 33(8):862–874

Obata T, Fernie AR (2012) The use of metabolomics to dissect plant responses to abiotic stresses. Cell Mol Life Sci 69:3225–3243

Okazaki Y, Saito K (2014) Roles of lipids as signaling molecules and mitigators during stress response in plants. Plant J 79:584–596

Ozgur R, Uzilday B, Sekmen AH et al (2013) Reactive oxygen species regulation and antioxidant defence in halophytes. Funct Plant Biol 40:832–847

Pandey M, Penna S (2017) Time course of physiological, biochemical, and gene expression changes under short-term salt stress in *Brassica juncea* L. Crop J 5(3):219–230

Paul MJ, Primavesi LF, Jhurreea D et al (2008) Trehalose metabolism and signalling. Annu Rev Plant Biol 59:417–441

Peng Z, He S, Sun J et al (2016) Na+ compartmentalization related to salinity stress tolerance in upland cotton (*Gossypium hirsutum*) seedlings. Sci Rep 6:34548

Pieslinger AM, Hoepflinger MC, Tenhaken R (2010) Cloning of glucuronokinase from Arabidopsis thaliana, the last missing enzyme of the myo-inositol oxygenase pathway to nucleotide sugars. J Biol Chem 285:2902–2910

Pitzschke A, Datta S, Persak H (2014) Salt stress in Arabidopsis: lipid transfer protein AZI1 and its control by mitogen-activated protein kinase MPK3. Mol Plant 7(4):722–738

Pottosin I, Shabala S (2014) Polyamines: control of cation transport across plant membranes: implications for ion homeostasis and abiotic stress signalling. Front Plant Sci 5(154):1–16

Ramadan T, Flowers TJ (2004) Effects of salinity and benzyl adenine on development and function of microhairs of *Zea mays* L. Planta 219:639–648

Ren ZH, Gao JP, Li LG et al (2005) A rice quantitative trait locus for salt tolerance encodes a sodium transporter. Nat Genet 37:1141–1146

Rhodes D, Hanson AD (1993) Quaternary ammonium and tertiary sulfonium compounds in higher plants. Annu Rev Plant Physiol Plant Mol Biol 44:357–384

Roitsch T, González MC (2004) Function and regulation of plant invertases: sweet sensations. Trends Plant Sci 9:606–613

Roitsch T, Balibrea ME, Hofmann M et al (2003) Extracellular invertase: key metabolic enzyme and PR protein. J Exp Bot 54:513–524

Rosa M, Prado C, Podazza G et al (2009) Soluble sugars—metabolism, sensing and abiotic stress: a complex network in the life of plants. Plant Signal Behav 4:388–393

Roy M, Wu R (2001) Arginine decarboxylase transgene expression and analysis of environmental stress tolerance in transgenic rice. Plant Sci 160:869–875

Roy M, Wu R (2002) Overexpression of S-adenosylmethionine decarboxylase gene in rice increases polyamine level and enhances sodium chloride stress tolerance. Plant Sci 163:987–992

Roy SJ, Negrao S, Tester M (2014) Salt resistant crop plants. Curr Opin Biotechnol 26:115–124

Sakamoto A, Murata A (1998) Metabolic engineering of rice leading to biosynthesis of glycinebetaine and tolerance to salt and cold. Plant Mol Biol 38:1011–1019

Sanchez DH, Siahpoosh MR, Roessner U et al (2007) Plant metabolomics reveals conserved and divergent metabolic responses to salinity. Physiol Plant 132:209–219

Sekmen AH, Bor M, Ozdemir F et al (2014) Current concepts about salinity and salinity tolerance in plants. In: Tuteja N, Gill SS (eds) Climate change and plant abiotic stress tolerance. Wiley-VCH Verlag GmbH Co. KGaA, Weinheim, pp 163–188

Shabala S (2009) Salinity and programmed cell death: unravelling mechanisms for ion specific signalling. J Exp Bot 60:709–711
Shabala S (2013) Learning from halophytes: physiological basis and strategies to improve abiotic stress tolerance in crops. Ann Bot 112:1209–1221
Shabala S, Cuin TA (2008) Potassium transport and plant salt tolerance. Physiol Plant 133:651–669
Shabala S, Munns R (2012) Salinity stress: physiological constraints and adaptive mechanisms. In: Shabala S (ed) Plant stress physiology, 1st edn. CABI, Wallingford, pp 59–93
Shabala S, Munns R (2017) Salinity stress: physiological constraints and adaptive mechanisms. In: Shabala S (ed) Plant stress physiology, 2nd edn. CABI, Wallingford, pp 24–63
Sharp RE (2002) Interaction with ethylene: changing views on the role of abscisic acid in root and shoot growth responses to water stress. Plant Cell Environ 25:211–222
Slama I, Abdelly C, Bouchereau A et al (2015) Diversity, distribution and roles of osmoprotective compounds accumulated in halophytes under abiotic stress. Ann Bot 115:433–447
Stitt M, Gibon Y, Lunn JE et al (2007) Multilevel genomics analysis of carbon signalling during low carbon availability: coordinating the supply and utilisation of carbon in a fluctuating environment. Funct Plant Biol 34:526
Suárez R, Calderón C, Iturriaga G (2009) Enhanced tolerance to multiple abiotic stresses in transgenic alfalfa accumulating trehalose. Crop Sci 49:1791–1799
Sun X, Cahill J, Van Hautegem T et al (2017) Altered expression of maize PLASTOCHRON1 enhances biomass and seed yield by extending cell division duration. Nat Commun 8(14752):1–11
Szabados L, Savouré A (2010) Proline: a multifunctional amino acid. Trends Plant Sci 15:89–97
Taji T, Seki M, Satou M et al (2004) Comparative genomics in salt tolerance between Arabidopsis and Arabidopsis-related halophyte salt cress using Arabidopsis microarray. Plant Physiol 135:1697–1709
Takahashi T, Kakehi JI (2010) Polyamines: ubiquitous polycations with unique roles in growth and stress responses. Ann Bot 105:1–6
Takahashi Y, Cong R, Sagor GHM et al (2010) Characterization of five polyamine oxidase isoforms in Arabidopsis thaliana. Plant Cell Rep 29:955–965
Takahashi F, Tilbrook J, Trittermann C et al (2015) Comparison of leaf sheath transcriptome profiles with physiological traits of bread wheat cultivars under salinity stress. PLoS One 10(8):1–23
Tavladoraki P, Rossi MN, Saccuti G et al (2006) Heterologous expression and biochemical characterization of a polyamine oxidase from Arabidopsis involved in polyamine back conversion. Plant Physiol 141:1519–1532
Thoen MPM, Davila Olivas NH, Kloth KJ et al (2017) Genetic architecture of plant stress resistance: multi-trait genome-wide association mapping. New Phytol 213:1346–1362
Velarde-Buendía AM, Shabala S, Cvikrova M et al (2012) Salt-sensitive and salt-tolerant barley varieties differ in the extent of potentiation of the ROS-induced K+ efflux by polyamines. Plant Physiol Biochem 61:18–23
Wang H, Zhang M, Guo R et al (2012) Effects of salt stress on ion balance and nitrogen metabolism of old and young leaves in rice (*Oryza sativa* L.) BMC Plant Biol 12(194):1–11
Wang K, Hersh LH, Benning C (2016a) Sensitive to freezing2 aids in resilience to salt and drought in freezing-sensitive tomato. Plant Physiol 172:1432–1442
Wang JQ, Sun H, Dong LQ et al (2016b) The enhancement of tolerance to salt and cold stresses by modifying the redox state and salicylic acid content via the cytosolic malate dehydrogenase gene in transgenic apple plants. Plant Biotechnol J 14:1986–1997
White AC, Rogers A, Rees M et al (2015) How can we make plants grow faster? A source–sink perspective on growth rate. J Exp Bot 67:31–45
Wu D, Cai S, Chen M et al (2013) Tissue metabolic responses to salt stress in wild and cultivated barley. PLoS One 8(1):1–11
Xia JX, Zhou YH, Shi K et al (2015) Interplay between reactive oxygen species and hormones in the control of plant development and stress tolerance. J Exp Bot 66(10):2839–2856

Xiao W, Hu S, Zhou X et al (2017) A glucuronokinase gene in Arabidopsis, AtGlcAK, is involved in drought tolerance by modulating sugar metabolism. Plant Mol Biol Report 35:298–311

Yancey PH (2005) Organic osmolytes as compatible, metabolic and counteracting cytoprotectants in high osmolarity and other stresses. J Exp Biol 208:2819–2830

Yolcu S, Özdemir F, Güler A et al (2016) Histone acetylation influences the transcriptional activation of POX in *Beta vulgaris* L. and *Beta maritima* L. under salt stress. Plant Physiol Biochem 100:37–46

Yu A, Lepere G, Jay F et al (2013) Dynamics and biological relevance of DNA demethylation in Arabidopsis antibacterial defense. PNAS USA 110:2389–2394

Zarza X, Atanasov KE, Marco F et al (2017) Polyamine oxidase 5 loss-of-function mutations in Arabidopsis thaliana trigger metabolic and transcriptional reprogramming and promote salt stress tolerance. Plant Cell Environ 40:527–542

Zepeda-Jazo I, Velarde-Buendía AM, Enríquez-Figueroa R et al (2011) Polyamines interact with hydroxyl radicals in activating Ca^{2+} and K^{+} transport across the root epidermal plasma membranes. Plant Physiol 157:2167–2180

Züst T, Agrawal AA (2017) Trade-offs between plant growth and defense against insect herbivory: an emerging mechanistic synthesis. Annu Rev Plant Biol 68:513–534

Chapter 10
The Glyoxalase System: A Possible Target for Production of Salinity-Tolerant Crop Plants

Tahsina Sharmin Hoque, David J. Burritt, and Mohammad Anwar Hossain

Abstract Among the various abiotic stressors, soil salinity is one of the most detrimental, restricting the growth and productivity of major agricultural crops worldwide. Apart from ionic, osmotic, and oxidative stress, one of the most important biochemical impacts of salt stress on plants is overaccumulation of methylglyoxal (MG), a cytotoxic compound that can cause degradation of proteins, lipids, and nucleic acids, inactivation of antioxidant systems and, finally, the death of plants. However, plants possess a complex network of enzymatic and nonenzymatic scavenging and detoxification systems to defend against MG-induced glycation and oxidative stress. Among the various defense mechanisms employed by plants, the glyoxalase system (composed mainly of two enzymes—glyoxalase I and glyoxalase II) is the most important, playing a crucial role in detoxifying MG, as well as regulating glutathione homeostasis and reactive oxygen species metabolism. Apart from its deleterious effects on plant growth and development, MG also has important signaling roles associated with stress tolerance. Recent genetic engineering studies have shown that overexpression of glyoxalase genes confers tolerance of various abiotic stresses, including salinity stress. This chapter summarizes the current knowledge and understanding of MG and the glyoxalase pathway, with respect to salinity stress tolerance and the potential for use of genetic engineering of glyoxalase genes into crop plants to improve crop yields under salt stress.

T. S. Hoque
Department of Soil Science, Bangladesh Agricultural University, Mymensingh, Bangladesh

D. J. Burritt
Department of Botany, University of Otago, Dunedin, New Zealand

M. A. Hossain (✉)
Department of Genetics and Plant Breeding, Bangladesh Agricultural University, Mymensingh, Bangladesh

Laboratory of Plant Nutrition and Fertilizers, Graduate School of Agricultural and Life Sciences, University of Tokyo, Tokyo, Japan
e-mail: anwargpb@bau.edu.bd

V. Kumar et al. (eds.), *Salinity Responses and Tolerance in Plants, Volume 1*,
https://doi.org/10.1007/978-3-319-75671-4_10

Keywords Salinity stress · Ion homeostasis · Methylglyoxal · Glyoxalase system · Oxidative stress · Antioxidant defense · Transgenic plants · Stress tolerance · Osmoprotectants · Hormones · Methylglyoxal signaling

Abbreviations

$^1O_2$	Singlet oxygen
ABA	Abscisic acid
APX	Ascorbate peroxidase
AsA	Ascorbate
BR	Brassinosteroid
CAT	Catalase
CK	Cytokinin
DHAP	Dihydroxyacetone phosphate
DHAR	Dehydroascorbate reductase
ET	Ethylene
GA	Gibberellin
GAP	Glyceraldehyde-3-phosphate
GB	Glycine betaine
Gly	Glyoxalase
GPX	Glutathione peroxidase
GR	Glutathione reductase
GSH	Reduced glutathione
GSSG	Oxidized glutathione
GST	Glutathione-*S*-transferase
HKT	High-affinity potassium transporter
IAA	Indole-3-acetic acid
JA	Jasmonate
MDA	Malondialdehyde
MDHAR	Monodehydroascorbate reductase
MG	Methylglyoxal
NAC	N-acetyl-L-cysteine
NHX	Na^+/H^+ exchanger
$O_2^{\bullet-}$	Superoxide
$OH^{\bullet}$	Hydroxyl radical
PCD	Programmed cell death
POD	Peroxidase
Pro	Proline
PSII	Photosystem II
ROS	Reactive oxygen species
SA	Salicylic acid
SLG	*S*-D-lactoylglutathione
SOD	Superoxide dismutase
SOS	Salt Overly Sensitive

10.1 Introduction

Plants are often exposed to abiotic and biotic stressors, which can reduce their growth and productivity. Abiotic stressors have a major impact on global agriculture and can reduce growth by more than 50% in most plant species (Wang et al. 2003; Rodríguez et al. 2005; Qin et al. 2011). Changing climatic conditions, combined with the demand for increased global food production because of population increases, has resulted in a demand for development of stress-tolerant crops (Takeda and Matsuoka 2008; Newton et al. 2011). Increased knowledge regarding the underlying mechanisms by which plants respond to stressors is crucial if broad-spectrum stress-tolerant crop varieties are to be developed.

Salinity stress due to poor agricultural practices or climatic changes is one of the most important abiotic stresses limiting global crop productivity, affecting large areas of land worldwide (Jagadish et al. 2012). It has been predicted that increased soil salinity in agricultural lands will have devastating effects globally by the middle of the twenty-first century, resulting in loss of up to 50% of cultivable land (Mahajan and Tuteja 2005). In plants, high soil salinity primarily causes osmotic stress and ion toxicity (Roy et al. 2014). In addition, secondary effects such as lower assimilate production, reduced cell expansion, membrane dysfunction, altered metabolism, and production of reactive compounds such as methylglyoxal (MG; CH_3COCHO) and reactive oxygen species (ROS) (Hossain et al. 2009; El-Shabrawi et al. 2010; Upadhyaya et al. 2011; Kumar 2013; Mostofa et al. 2015a; Acosta-Motos et al. 2017; Gupta et al. 2017) also occur. Excessive production of MG and ROS in plants causes cellular damage due to the ability of these compounds to react with proteins, lipids, and nucleic acids, leading to disruption of cellular homeostasis. It is well established that excessive production of MG and ROS is a general response of plants to stressors, including salinity (Veena et al. 1999; Yadav et al. 2005a, b; Singla-Pareek et al. 2006; Hossain and Fujita 2009, 2010; Banu et al. 2010; El-Shabrawi et al. 2010; Hossain et al. 2009, 2010, 2011a, 2014a, b; Gupta et al. 2017).

In order to avoid cellular damage from excessive MG and ROS production, plants must upregulate their detoxification and scavenging processes in order to survive. Among various scavenging mechanisms employed by plants, the glyoxalase system is the most efficient for detoxifying MG and helping to control ROS levels. The importance of glyoxalases has been well established, as many recent studies have demonstrated that increasing the activity of the glyoxalase pathway increases plants' tolerance of various abiotic stressors. For example, overexpression of glyoxalase pathway genes in transgenic plants confers tolerance of salinity and other abiotic or biotic stresses (Yadav et al. 2005a, b; Singla-Pareek et al. 2006, 2008; Lin et al. 2010; Saxena et al. 2011; Wani and Gosal 2011; Tuomainen et al. 2011; Alvarez Viveros et al. 2013; Wu et al. 2013; Mustafiz et al. 2014; Kaur et al. 2014a, b, c; Ghosh et al. 2014, 2016; Alvarez-Gerding et al. 2015, Rajwanshi et al. 2016; Zeng et al. 2016; Gupta et al. 2017). This chapter discusses the effects of salinity stress on plants and plant salinity tolerance mechanisms, focusing on the roles of the glyoxalases. Furthermore, it summarizes the findings of recent studies that have used genetic engineering to modify the glyoxalase pathway to develop salinity-tolerant plants that also show broad-spectrum abiotic stress tolerance.

10.2 Effects of Salt Stress in Plants

It has been estimated that more than 6% of the world's land and 30% of cultivated and irrigated agricultural land are salt affected (Pitman and Läuchli 2002; Chaves et al. 2009). Depending upon the severity and duration of exposure, salinity causes alterations in many physiological and metabolic processes in plants, leading to growth inhibition and reduced productivity or even death of plants (Munns 2005; Rozema and Flowers 2008; Rahnama et al. 2010; James et al. 2011; Munns and Gilliham 2015). In vegetative plants, salt stress causes reduced cell turgor and slows the rates of root and shoot elongation, and leaf expansion (Werner and Finkelstein 1995; Fricke et al. 2006), because of the negative impact of salinity on water uptake. High soil salinities cause both hyperosmotic stress and hyperionic stress, with plant growth initially being suppressed by osmotic stress; this is then followed by ion toxicity (Munns 2005; James et al. 2011; Rahnama et al. 2010), which can ultimately lead to programmed cell death (Zhu 2007). High soil salinity stress reduces the water absorption capacity of roots, and leaf water loss is enhanced; this negatively impacts plant metabolism, disrupts cellular homeostasis, and can uncouple major physiological and biochemical processes (Munns 2005). The negative impacts on plants of hyperionic stress resulting from high soil salinity are due to excessive accumulation of Na^+ and Cl^- ions in plant cells. Entry of excessive Na^+ and Cl^- ions into plant cells causes a severe ion imbalance resulting in disruption of plant metabolism and physiological processes (Gupta and Huang 2014). High cellular Na^+ levels disrupt K^+ and Ca^{2+} nutrition and interfere with stomatal regulation, which results in reduced photosynthesis and plant growth. High cellular concentrations of Cl^- disturb cellular integrity and affect photosynthetic processes through both membrane damage and enzyme inhibition (Tavakkoli et al. 2010). For example, high Cl^- concentrations reduce the photosynthetic capacity and quantum yield of photosystem II (PSII) because of chlorophyll degradation (Tavakkoli et al. 2010). Furthermore, salinity stress rapidly induces production and accumulation of reactive carbonyl compounds such as MG (Yadav et al. 2005a, b; Hossain et al. 2009; Upadhyaya et al. 2011; Gupta et al. 2017), which can react with proteins, lipids, and nucleic acids (DNA and RNA) to form adducts (Martins et al. 2001; Hoque et al. 2012a) that can disrupt cellular functions and eventually lead to cell death. Studies have also shown that salt stress significantly enhances the production of ROS, including singlet oxygen ($^1O_2$), superoxide ($O_2^{\bullet -}$), the hydroxyl radical ($OH^{\bullet}$), and H_2O_2 (Mahajan and Tuteja 2005; Tanou et al. 2009; Ahmad et al. 2010, 2012; Ahmad and Umar 2011; Ahmad and Prasad 2012a, b; Hossain et al. 2013a, b; Adem et al. 2014; Bose et al. 2014; You and Chan 2015). Hence, salinity-induced ROS formation can also lead to oxidative damage to proteins, lipids, and nucleic acids, which in turn disrupts key cellular functions of plants (El-Shabrawi et al. 2010; Bose et al. 2014).

In general, high salinity causes membrane dysfunction, nutrient imbalances, accumulation of MG and ROS and impairment of their detoxification, inhibition of cell expansion, inactivation of enzymes (including those involved in antioxidant

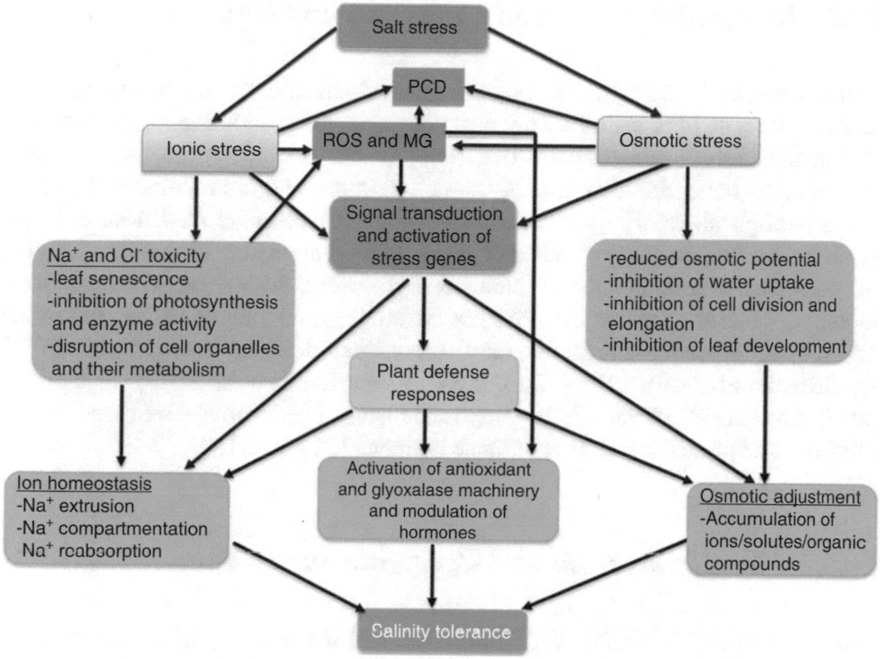

Fig. 10.1 Detrimental effects of high salinity on plants and the corresponding defense responses used to overcome salt stress. Salt stress induces activation of complex regulatory networks, which lead to establishment of the defense responses required for tolerance. High salinity causes osmotic and ionic stress, resulting in reduced water uptake, ion toxicity, excessive production of methylglyoxal (MG) and reactive oxygen species (ROS), programmed cell death (PCD) and, ultimately, inhibition of plant growth and development. MG and ROS can also act as signal molecules which, through signal transduction pathways, modulate expression of stress-responsive genes. This subsequently regulates various cellular processes, including those associated with ion homeostasis, osmotic adjustment, activation of antioxidant and glyoxalase defenses, and modulation of hormone levels, which ultimately lead to salinity tolerance (Modified from Horie et al. 2012, with permission)

metabolism), altered levels of plant hormones, decreased stomatal opening, and lower photosynthetic activity, which can result in a range of physiological and metabolic disorders (Fig. 10.1; Mahajan and Tuteja 2005; Munns and Tester 2008; Rahnama et al. 2010; Hasanuzzaman et al. 2012b; Hossain et al. 2011b, 2013a, b; Bose et al. 2014; Roy et al. 2014; Nahar et al. 2015a; Mostofa et al. 2015a; Akram et al. 2017; Gupta et al. 2017). In addition, high soil salinities can cause reduced plant fertility, premature plant senescence (Krasensky and Jonak 2012; Allu et al. 2014; Hussain et al. 2016), and limited nodulation, reducing the efficiency of nitrogen fixation and hence resulting in lower leguminous crop yields (Dong et al. 2013; Do et al. 2016). While high salinity can inhibit plant growth, development, and yields, it is important to note that the severity depends on the salt concentration, the time of exposure, the plant genotype, and interactions with various other environmental factors.

10.3 Mechanisms of Salinity Tolerance in Plants

Plants have developed numerous physiological, biochemical, and molecular mechanisms to tolerate high salinity. The mechanisms include ion homeostasis and compartmentalization, biosynthesis of osmoprotectants and compatible solutes, activation of the antioxidant and glyoxalase systems, and modulation of hormone levels (Abogadallah 2010; El-Shabrawi et al. 2010; Hossain et al. 2013a, b; Gupta et al. 2017). Salinity tolerance is a complex trait involving coordinated expression of many genes, which encode proteins that perform a diverse range of functions, including control of water loss through the stomata, ion sequestration, metabolic adjustment, osmotic adjustment, and antioxidant defense (Abogadallah 2010; El-Shabrawi et al. 2010; Roy et al. 2014; Acosta-Motos et al. 2017; Gupta et al. 2017). An overview of the effects of high salinity and the resulting defense responses of plants that confer salt stress tolerance is shown in Fig. 10.1.

10.3.1 Ion Homeostasis and Compartmentalization

Under salt stress, high Na^+ and Cl^- ion concentrations in the cytoplasm are harmful for both glycophytes and halophytes; thus, maintaining ion homeostasis and compartmentalization is important for plant growth and development (Serrano et al. 1997; Hasegawa 2013; Gupta and Huang 2014). Glycophytes restrict salt uptake and adjust their cellular osmotic potentials by synthesizing compatible solutes; in contrast, halophytes compartmentalize accumulated ions within vacuoles, and this helps to maintain a high cellular cytosolic K^+/Na^+ ratio. Thus, protection from salinity stress is achieved by transportation of excess ions into vacuoles, with high levels of ions often being sequestered in older tissues/organs—for example, leaves, which will eventually be lost from the plant (Zhu 2003). Membranes, along with their associated components, play a vital role in regulating ion uptake and transport, which is essential for maintenance of ion homeostasis during salt stress (Sairam and Tyagi 2004; Osakabe et al. 2014). Ion transport is carried out by a range of carrier proteins, channel proteins, antiporters, and symporters. A large number of genes and proteins that encode K^+ transporters and channels have been identified and the genes cloned in various plant species (Gupta and Huang 2014). Two examples are the plasma membrane transporter HKT (high-affinity potassium transporter), which regulates Na^+ and K^+ ion transport, and intracellular NHX (Na^+/H^+ exchanger), which maintains K^+ homeostasis as well as regulating endosomal pH, thereby playing an essential role in salt tolerance. Overexpression of genes encoding proteins involved in the SOS (Salt Overly Sensitive) stress signaling pathway, which plays a pivotal role in ion homeostasis in plants and involves three key proteins (SOS1, SOS2, and SOS3), confers salinity tolerance (Sanders 2000; Hasegawa et al. 2000; Ishitani et al. 2000; Shi et al. 2002; Yang et al. 2009; Chakraborty et al. 2012; Ji et al. 2013; Zhou et al. 2014; Yang et al. 2015; Quan et al. 2017).

10.3.2 Biosynthesis of Osmoprotectants and Compatible Solutes

Accumulation of a diverse range of osmolytes or compatible solutes is a well-known adaptive feature of plants growing under saline conditions. Plants produce a wide range of organic solutes or osmoprotectants, including ammonium compounds (e.g., glycine betaine [GB]), amino acids (e.g., proline [Pro]), carbohydrates (e.g., sugars—namely, glucose, fructose, fructans, and trehalose—and starch), and polyols or sugar alcohols (e.g., mannitol and pinitol) (Rhodes and Hanson 1993; Koyro et al. 2012). These osmolytes are synthesized and accumulate to various extents in many plant species, with cellular concentrations of osmolytes maintained either by irreversible synthesis of the compounds or by a combination of synthesis and degradation. The accumulation of osmoprotectants is often proportional to the external osmolarity. Osmolytes play a key role in protecting cellular structures, maintaining ion homeostasis and the activities of enzymes, and maintaining the cellular osmotic balance, by decreasing cellular water potentials and hence facilitating water uptake in plants grown in saline soils (Wang et al. 2003; Attipali et al. 2004; Ashraf and Foolad 2007; Kusano et al. 2007; Hayat et al. 2012; Ranganayakulu et al. 2013). Osmolytes can also act as low molecular weight chaperones because of their hydrophilic nature, and directly scavenge ROS (Gupta and Huang 2014; Hossain et al. 2014b). A number of studies have also shown that osmoprotectants—including proline, glycine betaine, some sugars and sugar alcohols, and polyamines—can promote increases in the activities of antioxidant enzymes and hence reduce oxidative stress in plants grown in saline soils (Ashraf and Foolad 2007; Kubiś 2008; Hayat et al. 2012). For example, treatment of plants with mannitol, sorbitol and trehalose, polyamines, proline, and glycine betaine improves the performance of various plant species under saline conditions, including rice (*Oryza sativa*), maize (*Zea mays*), sugarcane (*Saccharum* spp.), tobacco (*Nicotiana tabacum*), ryegrass (*Lolium perenne*), olive (*Olea europaea*), soybean (*Glycine max*), and pistachio (*Pistacia vera*) (Theerakulpisut and Gunnula 2012; Nounjan et al. 2012; Sobahan et al. 2012; Mostofa et al. 2015a; Kaya et al. 2013; Patade et al. 2014; Hoque et al. 2008; Hu et al. 2012; Malekzadeh 2015; Kamiab et al. 2014).

10.3.3 Activation of the Antioxidant Defense System and Glyoxalase System

High salinity induces formation of ROS and MG and, in response to severe oxidative and MG stress, plants upregulate both antioxidant defenses and the glyoxalase system to minimize stress-induced damage (Bose et al. 2014; Gupta et al. 2017). Both enzymatic and nonenzymatic antioxidants play critical roles in detoxifying the ROS induced by salinity stress. In plants, salinity tolerance is positively correlated with the activities of antioxidant enzymes—such as superoxide dismutase (SOD), catalase (CAT), peroxidase (POD), glutathione peroxidase (GPX), ascorbate peroxidase (APX), glutathione-*S*-transferase (GST), monodehydroascorbate reductase (MDHAR), dehydroascorbate reductase (DHAR), and glutathione reductase (GR)—and with the accumulation of the nonenzymatic antioxidants glutathione (GSH), ascorbate (AsA), tocopherol, carotenoids, and flavonoids (e.g., anthocyanins), which scavenge excess ROS (Gill and Tuteja 2010; Hossain and Fujita 2010; Kadioglu et al. 2011; Vardharajula et al. 2011; Kaya et al. 2013; Van Oosten et al. 2013; Kubiś et al. 2014). Recently, Vardharajula et al. (2011) and Wu et al. (2012) reported increased activities of a diverse range of antioxidant enzymes—including SOD, POD, CAT, and APX—in several plant species grown at high salinity. In addition, increased levels of antioxidants have been observed in salt-tolerant cultivars of various crop plants—including tomato (*Lycopersicon pennellii*), pea (*Pisum sativum*), broccoli (*Brassica oleracea*), marigold (*Calendula officinalis*), and Barbados nut (*Jatropha curcas*) (Mittova et al. 2004; Hernández et al. 2000, 2010; Chaparzadeh et al. 2004; Gao et al. 2008)—suggesting an important role for antioxidants in alleviating salt stress–induced oxidative damage. Also, several studies have confirmed that exogenous application of nonenzymatic antioxidants such as GSH, AsA, and α-tocopherol, which interact with multiple stress-responsive pathways, provides plants with increased tolerance of salinity (Azzedine et al. 2011; Dehghan et al. 2011; Salama and Al-Mutawa 2009; Farouk 2011; Ejaz et al. 2012; Wang et al. 2014; Nahar et al. 2015a; Akram et al. 2017). Hence, antioxidant levels could potentially be selection criteria for improving plant salinity tolerance.

10.3.4 Plant Hormones and Salinity Tolerance

Plant hormones are well known as key signaling molecules involved in regulating physiological and molecular responses (Fahad et al. 2015), plant development, and tolerance of or susceptibility to a diverse array of stresses, including high salinity (Ryu and Cho 2015). The major plant hormones abscisic acid (ABA), Indole-3-acetic acid (IAA), cytokinins (CKs), brassinosteroids (BRs), jasmonates (JAs), salicylic acid (SA), gibberellins (GAs), and ethylene (ET) are all known to play roles, to various degrees, in tolerance of high salinity (Thomas et al. 1992; Vaidyanathan et al. 1999; Mahajan and Tuteja 2005; Zhang et al. 2006; Shakirova et al. 2010; Yang et al. 2013; Amjad et al. 2014; Kissoudis et al. 2016). For example, high salinity causes osmotic stress and water deficit in plants, resulting in increased production of ABA in both shoots and roots (Popova et al. 1995; Cramer and Quarrie 2002; Kang et al. 2005; Cabot et al. 2009). Accumulation of ABA can to some extent mitigate the negative impacts of salinity on photosynthesis, growth, and translocation of assimilates (Popova et al. 1995; Jeschke et al. 1997). BRs and SA have also been shown to regulate biochemical and physiological processes that lead to improved salt tolerance in plants (Ashraf et al. 2010). With respect to both the antioxidant and glyoxalase systems, recent studies have shown that exogenous application of hormones can increase the activities of both systems and promote increased salinity tolerance (Yoon et al. 2009; Gurmani et al. 2011; Torabian 2011; Sharma et al. 2013; Bastam et al. 2013; Sripinyowanich et al. 2013; Fariduddin et al. 2014; Liu et al. 2016).

10.4 Production of Methylglyoxal in Plants in Response to High Salinity

Methylglyoxal is an oxygenated short aldehyde and a toxic metabolite, which accumulates in plants under abiotic stresses, including high salinity (Yadav et al. 2005a, b; Hossain et al. 2009; Upadhyaya et al. 2011; Gómez Ojeda et al. 2013; Li et al. 2017a). In plants, MG is formed spontaneously as a by-product of several metabolic pathways, including glycolysis, lipid peroxidation, and oxidative degradation of glucose and glycated proteins (Shangari and O'Brien 2004; Kaur et al. 2014a). MG is produced in plants mainly by a nonenzymatic route from glyceraldehyde-3-phosphate (GAP)—which is an intermediate of glycolysis and photosynthesis—and from dihydroxyacetone phosphate (DHAP) (Richard 1993). Formation of MG from triose phosphates occurs through elimination of the phosphoryl group from the 1,2-enediolate of trioses, with a nonenzymatic formation rate of approximately 0.1 mM day^{-1} (Richard 1993). In plants exposed to high salinity or to other abiotic or biotic stressors, increased carbohydrate, protein, and lipid metabolism results in excessive production of aldehydes such as MG (Fig. 10.2; Mano et al. 2009; Ahuja et al. 2012; Kaur et al. 2014a, b, c; Mano 2012). The processes of metabolism of aminoacetone, acetone, and ketones are considered minor routes for the production of MG in plants (Lyles and Chalmers 1992; Casazza et al. 1984; Koop and Casazza 1985; Kalapos 1999; Kaur et al. 2014c). Under normal metabolic conditions, MG concentrations in plants range from 30 to 90 μM (Yadav et al. 2005a; Hossain et al. 2009), with a two- to sixfold increase in MG levels being observed in response to various abiotic stressors (Yadav et al. 2005a; Hossain et al. 2009; Mostofa et al. 2015a, b; Gupta et al. 2017). For example, MG levels were found to increase by 77%, 67%, and 50% in pumpkin (*Cucurbita maxima* L.), tobacco (*N. tabacum* L., cv. BY-2 and cv. Petit Havana), and potato (*Solanum tuberosum* L. cv. Taedong Valley) plants, respectively, under high salinity, in comparison with control plants (Hossain et al. 2009; Banu et al. 2010; Ghosh et al. 2014; Upadhyaya et al. 2011). Recently, Gupta et al. (2017) showed a 70–95% increase in MG levels in rice plants in response to high salinity and drought stress. Hence, MG accumulation is closely associated with stress responses (Yadav et al. 2005a, b). In addition to directly causing damage, high MG levels in plant cells cause increased oxidative stress due to enhanced ROS production (Maeta et al. 2005; Kalapos 2008); inhibit photosynthesis (Saito et al. 2011); generate advanced glycation end products (Thornalley 2003); modify proteins, nucleic acids, and basic phospholipids (Thornalley 2008; Thornalley et al. 2010); disable antioxidant defense systems (Hoque et al. 2010; Hoque et al. 2012a); and interfere with cell division (Ray et al. 1994).

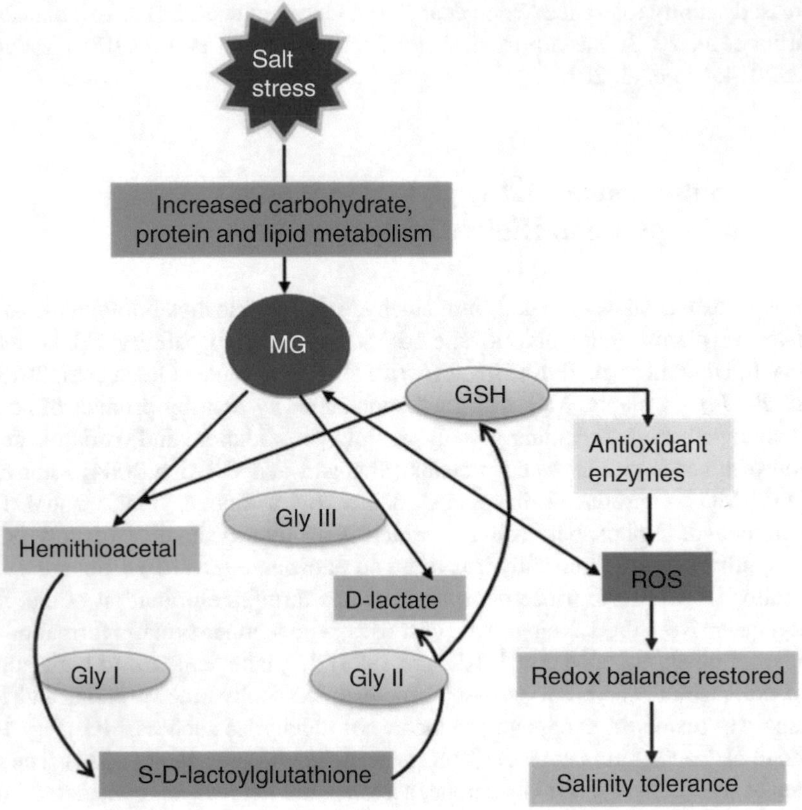

Fig. 10.2 Role of the glyoxalase (Gly) system in plant tolerance of high salinity. Under salinity, methylglyoxal (MG) production is enhanced by increased carbohydrate, protein, and lipid metabolism, and higher MG levels induce increased reactive oxygen species (ROS) generation. The enzymes Gly I and Gly II efficiently detoxify MG by converting it to D-lactate in a reduced glutathione (GSH)–dependent manner. In addition, MG is degraded to D-lactate in a GSH-independent manner by the activity of the enzyme Gly III. Under high salinity the Gly pathway helps to maintain proper functioning of the antioxidant system, reducing ROS and maintaining cellular redox homeostasis by breaking down MG and regenerating GSH (Modified from Hoque et al. 2015, with permission)

10.5 The Glyoxalase Pathway as the Main Methylglyoxal Detoxification System in Plants

In the course of their evolution, plants have developed the glyoxalase system to protect against excessive cellular MG levels induced by exposure to multiple abiotic stressors. The glyoxalase system is an essential and ubiquitous mechanism, which plants use to detoxify cytotoxic MG that would otherwise rise to lethal concentrations under stressful conditions. The glyoxalase system consists of two major enzymes—glyoxalase I (Gly I) and glyoxalase II (Gly II)—which work together to catalyze GSH-dependent detoxification of MG and other reactive aldehydes

(Fig. 10.2; Yadav et al. 2008; Hossain et al. 2011a; Hoque et al. 2017). Gly I (*S*-D-lactoylglutathione lyase; EC 4.4.1.5) is a metalloenzyme, which requires divalent metal ions for activation, binding two metal ions, one per active site (Ridderström et al. 1998). Gly II (*S*-2-hydroxyacylglutathione hydrolase; EC 3.1.2.6) belongs to the superfamily of metallo-β-lactamases (Ghosh et al. 2014). MG detoxification is accomplished by sequential action of Gly I and Gly II and involves two irreversible reactions. GSH and MG spontaneously form hemithioacetal, and the isomerization of hemithioacetal adducts to *S*-D-lactoylglutathione (SLG) is catalyzed by Gly I. SLG is then converted to D-lactate, with GSH being recycled back into the system by the action of Gly II (Thornalley 1993; Kaur et al. 2014a; Li 2016). For effective MG detoxification the availability and cellular concentration of GSH are very important, as low levels of GSH can lead to MG accumulation. In addition to Gly I and Gly II, a third glyoxalase—glyoxalase III (Gly III; containing a DJ-1/PfpI domain)—has recently been identified in plants. This enzyme directly converts MG to D-lactate in a single GSH-independent step (Ghosh et al. 2016).

In addition to the glyoxalase system, enzymes such as aldo-keto reductases and aldehyde dehydrogenases play minor roles in detoxifying MG (Simpson et al. 2009; Narawongsanont et al. 2012; Kirch et al. 2005). However, the glyoxalase system is by far the most efficient MG detoxification system in plants under both normal and stressful environmental conditions (Singla-Pareek et al. 2006; Alvarez Viveros et al. 2013; Ghosh et al. 2016). For example, a significant increase in Gly I activity was observed after 24 h in the presence of 25 mM MG (Hossain et al. 2009) suggesting that an increased MG level causes a rapid increase in Gly I activity. Furthermore, Hoque et al. (2017) showed rapid increases in the activities of Gly I and Gly II in response to higher cellular MG levels. Glyoxalases are known to be differentially regulated in plants under abiotic stress, and recent studies have demonstrated the involvement of glyoxalases in several plant stress responses. Glyoxalases not only are involved in MG detoxification but also are important for maintaining the ratio of cellular GSH to oxidized glutathione (GSSG) and for proper functioning of antioxidant enzymes that directly or indirectly utilize and regenerate GSH—e.g., GR, GST, GPX, and DHAR (Yadav et al. 2005b). Hence the glyoxalase system plays a crucial role in cellular defense against glycation and oxidative stress (Fig. 10.2); thus, glyoxalases are considered potential markers associated with plant stress susceptibility or tolerance (Thornalley 1993; Kalapos 2008; Kaur et al. 2014c; Gupta et al. 2017).

10.6 Interactions Between the Glyoxalase and Antioxidant Systems for High Salinity Tolerance

As mentioned in Section 10.5, in plants both the glyoxalase and antioxidant systems are largely dependent on GSH (Fig. 10.3) (Hossain et al. 2011a). GSH is a multifunctional molecule with diverse roles in plant stress tolerance and stress signaling (Noctor et al. 2012). Recently the use of stress-tolerant or stress-sensitive transgenic

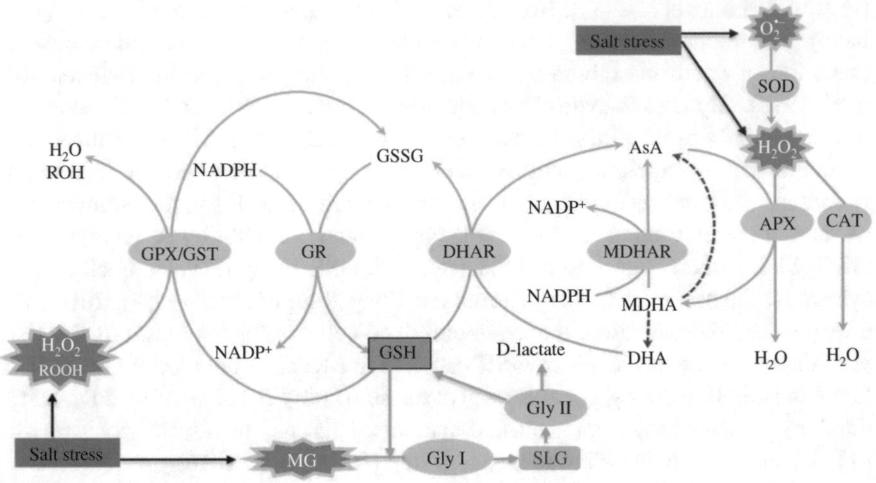

Fig. 10.3 Mechanistic interactions of the antioxidant defense and glyoxalase systems for regulation of reactive oxygen species (ROS) levels, methylglyoxal (MG) levels, and signaling functions in plants under salinity or abiotic stress. For further discussion, please see the text (Modified from Hossain et al. 2011a, b with permission)

plants has helped to advance our understanding of abiotic stress/tolerance response networks in plants. Coordinated induction of both the glyoxalase and antioxidant systems is required to render plants more tolerant of abiotic stress, as synergistic reductions in MG and ROS levels and efficient regulation of GSH levels, as well as GSH-utilizing and -regenerating enzymes, are required for high levels of tolerance (Yadav et al. 2005b; El-Shabrawi et al. 2010; Hossain et al. 2011b; Upadhyaya et al. 2011). The interaction between ROS and MG detoxification systems was first described by Yadav et al. (2005b) in transgenic tobacco (*N. tabacum* var. Petit Havana) plants overexpressing both Gly I and Gly II genes. Transgenic plants maintained higher glyoxalase enzyme activities and showed reduced ROS- and MG-induced damage under conditions of high salinity. Surprisingly, the transgenic plants maintained higher APX, GR, GST, and GPX activities. Later, Upadhyaya et al. (2011) reported that transgenic potato (*S. tuberosum* L. cv. Taedong Valley) plants overexpressing AsA biosynthetic genes also showed the importance of an interaction between both ROS and MG detoxification systems with respect to high salinity tolerance. Importantly, transgenic potato plants maintained a higher GSH/GSSG ratio; higher APX, DHAR, GR, GST, GPX, Gly I, and Gly II activities; and lower MG levels than wild-type plants. El-Shabrawi et al. (2010) showed that the salt-tolerant rice genotype Pokkali maintained higher GSH levels; a higher GSH/GSSG ratio; and higher activity of Gly I and Gly II, SOD, CAT, POD, and GPX than the salt-sensitive genotype IR64. The tolerant genotype also showed lower ROS accumulation and less ROS-induced DNA damage. These findings support the importance of GSH for high salinity tolerance and demonstrate that coordinated induction of GSH biosynthesis and GSH-metabolizing enzymes is correlated with high salinity tolerance in plants. Studies on heat- or cold-priming-induced salinity and drought stress tolerance have also demonstrated the relationship between the glyoxalase and antioxidant systems with respect to abiotic stress tolerance (Hossain et al. 2013a, b). Recently several studies treating plants with plant growth regulators, osmoprotectants, signaling molecules, or other chemicals have also confirmed the importance of the relationship between the glyoxalase and antioxidant systems for abiotic stress tolerance (Yadav et al. 2005b; Hoque et al. 2008; Kumar and Yadav 2009; Hossain et al. 2010, 2011b; Hasanuzzaman et al. 2011a, b; Hasanuzzaman et al. 2012a, b; Mostofa et al. 2014a, b; Hossain et al. 2014a, b; Mostofa et al. 2015a, b; Nahar et al. 2015a, b; Jin et al. 2015; Gupta et al. 2017).

10.7 Signaling Roles of Methylglyoxal and Exogenous Methylglyoxal–Mediated High Salinity Tolerance in Plants

Though MG is most often regarded as a toxic molecule, inhibiting growth and development processes in plants, at low levels it acts as an important signaling molecule and is involved in regulating a diverse range of processes, including cell proliferation and survival, reproductive growth and development, control of the redox status of cells, and many aspects of general metabolism and cellular homeostasis (reviewed by Hoque et al. 2016; Sankaranarayanan et al. 2017). MG can regulate plant stress responses by controlling stomatal opening and closure, production of ROS, cytosolic Ca^{2+} oscillations, activation of inward-rectifying potassium channels, accumulation of anthocyanins, the activities of glyoxalases, and expression of many stress-responsive genes (Hoque et al. 2012b, c, d; Kaur et al. 2014b; Hoque et al. 2015, 2016, 2017). It has been suggested that MG plays important roles in plant signal transduction as it transmits and amplifies cellular signals and functions promoting plant adaptation to salinity and other adverse environmental conditions. Recent studies have shown that treatment of plants with low concentrations of MG improves abiotic stress tolerance. Li et al. (2017a) demonstrated that application of MG to wheat (*Triticum aestivum* L. cv. Yunmai 41) plants increased salinity tolerance. In controls the imposition of salt stress inhibited seed germination and root and shoot growth, whereas application of exogenous MG reduced the inhibitory effects of salinity. The activities of Gly I and Gly II increased, whereas the endogenous MG level decreased, in MG-treated seedlings. Importantly, MG treatment increased the activities of SOD, CAT, APX, and GR; increased GSH and AsA levels; and reduced the levels of ROS and malondialdehyde (MDA). MG treatment also altered betaine, proline, and sugar content under salt stress, whereas the positive effects of MG were neutralized when seedlings were treated with the MG scavenger N-acetyl-L-cysteine (NAC). Li and coworkers also demonstrated that treatment with MG improves cadmium tolerance in wheat (Li et al. 2017b).

10.8 Genetic Engineering of Glyoxalase Pathway Enzymes to Improve Salinity Tolerance in Plants

Modification of the glyoxalase system has the potential to enhance abiotic stress tolerance in genetically modified crop plants (Kaur et al. 2014a, c; Gupta et al. 2017). The levels of glyoxalase enzymes can be used as phenomic biomarkers to indicate the degree of stress tolerance a plant processes, and plants with high glyoxalase enzyme levels have been shown to be potentially tolerant of a range of abiotic stressors (Kaur et al. 2014c; Gupta et al. 2017). In recent years, our understanding of the roles of glyoxalases has been broadened by various transcriptomic, proteomic, and metabolomic analyses of various plant species (Singla-Pareek et al. 2003, 2006; Hossain et al. 2009; Lin et al. 2010; Mustafiz et al. 2011; Gupta et al. 2017), and glyoxalase genes (*Gly I* and *Gly II*) from different plant species have been cloned and characterized, with promising results obtained in genetically engineered plants that overexpress *Gly I* or *Gly II* or both, with respect to increased abiotic and biotic stress tolerance, including salinity (Table 10.1) (Veena et al. 1999; Reddy and Sopory 1999; Singla-Pareek et al. 2003, 2006, 2008; Yadav et al. 2005a, b; Verma et al. 2005; Bhomkar et al. 2008; Roy et al. 2008; Lin et al. 2010; Saxena et al. 2011; Wani and Gosal 2011; Tuomainen et al. 2011; Alvarez Viveros et al. 2013; Wu et al. 2013; Mustafiz et al. 2014; Kaur et al. 2014a, b, c; Ghosh et al. 2014, 2016; Alvarez-Gerding et al. 2015, Rajwanshi et al. 2016; Zeng et al. 2016; Gupta et al. 2017). Generally, transgenic plants that overexpress glyoxalase pathway genes have better glutathione homeostasis and retain higher antioxidant and glyoxalase enzyme activities that help to maintain optimum cellular level of MG and ROS, ensuring signaling functions but restricting ROS- and MG-induced cellular damage. The transgenic plants also show higher photosynthetic rates, higher yields and yield-attributing traits, and increased yields under stressful conditions (Gupta et al. 2017). Thus, genetic manipulation of the glyoxalase system in plants has the potential to contribute to production of crop plants with resistance to high salinity and multiple abiotic and biotic stresses.

10.9 Conclusions and Future Perspectives

Salinity-induced overaccumulation of damage causing MG and ROS is a universal response in plants and affects plant growth, development, and survival. To survive under saline conditions, plants need to sense and respond appropriately to the level of stress. Activation of the glyoxalase system and maintenance of optimum levels of MG and ROS are important for salinity tolerance. Over the past few years, significant progress has been made in understanding the roles of the glyoxalase system in regulating MG, as well as ROS levels, in plants exposed to salinity or other environmental stressors, but the signaling roles of MG still require further investigation. In addition, a large number of transcriptome, proteomic, metabolomic, and genetic

Table 10.1 Genetic modification of glyoxalase genes to confer abiotic stress tolerance

Gene	Plant species	Response phenotype	References
Gly I	Tobacco (*Nicotiana tabacum*)	Improved salt stress tolerance	Veena et al. (1999)
Gly I	Tobacco (*Nicotiana tabacum*)	Improved tolerance of MG and salt stress	Reddy and Sopory (1999)
Gly I	Black gram (*Vigna mungo*)	Improved salt stress tolerance	Bhomkar et al. (2008)
Gly I	*Arabidopsis thaliana*	Improved salt stress tolerance	Roy et al. (2008)
Gly I	Rice (*Oryza sativa*)	Improved salt stress tolerance	Verma et al. (2005)
Gly I	Tobacco (*Nicotiana tabacum*)	Improved salt stress tolerance	Yadav et al. (2005a)
Gly I	Tobacco (*Nicotiana tabacum*)	Improved Zn tolerance	Lin et al. (2010)
Gly I	Tobacco (*Nicotiana tabacum*)	Improved tolerance of Zn, Cd, and Pb	Tuomainen et al. (2011)
Gly I	Tobacco (*Nicotiana tabacum*)	Improved tolerance of MG, salt stress, mannitol, and H_2O_2	Wu et al. (2013)
Gly I	Tobacco (*Nicotiana tabacum*)	Improved tolerance of MG and salt stress	Mustafiz et al. (2014)
Gly I	Rice (*Oryza sativa*)	Improved tolerance of salt stress, heavy metal, and mannitol	Zeng et al. (2016)
Gly I	Mustard (*Brassica juncea*)	Improved tolerance of salt, heavy metal, and mannitol	Rajwanshi et al. (2016)
Gly II	Rice (*Oryza sativa*)	Improved salt stress tolerance	Singla-Pareek et al. (2008)
Gly II	Mustard (*Brassica juncea*)	Improved salt stress tolerance	Saxena et al. (2011)
Gly II	Rice (*Oryza sativa*)	Improved salt stress tolerance	Wani and Gosal (2011)
Gly II	*Arabidopsis thaliana*	Improved salt and anoxic stress tolerance	Devanathan et al. (2014)
Gly II	Tobacco (*Nicotiana tabacum*)	Improved salt stress tolerance	Ghosh et al. (2014)
Gly I + *Gly II*	Tobacco (*Nicotiana tabacum*)	Improved salinity tolerance and setting of viable seeds in Zn-spiked soils	Singla-Pareek et al. (2003, 2006) and Yadav et al. (2005b)
Gly I + *Gly II*	Tomato (*Solanum lycopersicum*)	Improved salt stress tolerance	Alvarez Viveros et al. (2013)
Gly I + *Gly II*	Carrizo citrange (*Citrus sinensis* × *Poncirus trifoliata*)	Improved salt stress tolerance	Alvarez-Gerding et al. (2015)
Gly I + *Gly II*	Rice (*Oryza sativa*)	Improved salinity, drought, heat, and sheath blight tolerance	Gupta et al. (2017)
Gly III	Rice (*Oryza sativa*)	Improved tolerance of MG	Ghosh et al. (2016)

Gly glyoxalase, *MG* methylglyoxal

engineering studies have revealed linkages between the glyoxalase and ROS detoxification systems with respect to regulating salinity tolerance. However, further studies investigating the importance of the glyoxalase system in abiotic stress tolerance should be conducted in different cellular organelles, at various stages of plant growth, in different crops. In-depth investigation of the physiological, biochemical, and metabolic changes observed in glyoxalase-overexpressing transgenic crop plants is also needed for better understanding of the regulatory roles of glyoxalases in plants, if genetic engineering is to be used to produce salinity-tolerant and multiple stressor–tolerant crop plants.

Acknowledgements Mohammad Anwar Hossain thankfully acknowledges his postdoctoral fellowship from the Japan Society for the Promotion of Science (JSPS).

References

Abogadallah GM (2010) Antioxidative defense under salt stress. Plant Signal Behav 5:369–374

Acosta-Motos JR, Ortuño MF, Bernal-Vicente A et al (2017) Plant responses to salt stress: adaptive mechanisms. Agronomy 7:18

Adem GD, Roy SJ, Zhou M et al (2014) Evaluating contribution of ionic, osmotic and oxidative stress components towards salinity tolerance in barley. BMC Plant Biol 14:113

Ahmad P, Prasad MNV (2012a) Abiotic stress responses in plants: metabolism, productivity and sustainability. Springer, New York

Ahmad P, Prasad MNV (2012b) Environmental adaptations and stress tolerance in plants in the era of climate change. Springer Science + Business Media, New York

Ahmad P, Umar S (2011) Oxidative stress: role of antioxidants in plants. Studium Press, New Delhi

Ahmad P, Jaleel CA, Salem MA et al (2010) Roles of enzymatic and non-enzymatic antioxidants in plants during abiotic stress. Crit Rev Biotechnol 30:161–175

Ahmad P, Hakeem KR, Kumar A et al (2012) Salt-induced changes in photosynthetic activity and oxidative defense system of three cultivars of mustard (*Brassica juncea* L.) Afr J Biotechnol 11:2694–2703

Ahuja I, Kissen R, Bones AM (2012) Phytoalexins in defense against pathogens. Trends Plant Sci 17:73–90

Akram S, Siddiqui MN, Hussain BMN et al (2017) Exogenous glutathione modulates salinity tolerance of soybean [*Glycine max* (L.) Merrill] at reproductive stage. J Plant Growth Regul 36:877–888. https://doi.org/10.1007/s00344-017-9691-9

Allu AD, Soja AM, Wu A et al (2014) Salt stress and senescence: identification of cross-talk regulatory components. J Exp Bot 65:3993–4008

Alvarez Viveros MF, Inostroza-Blancheteau C, Timmermann T et al (2013) Overexpression of *GlyI* and *GlyII* genes in transgenic tomato (*Solanum lycopersicum* mill.) plants confers salt tolerance by decreasing oxidative stress. Mol Biol Rep 40:3281–3290

Alvarez-Gerding X, Cortés-Bullemore R, Medina C et al (2015) Improved salinity tolerance in Carrizo Citrange rootstock through overexpression of glyoxalase system genes. Biomed Res Int 2015:827951

Amjad M, Akhtar J, Anwar-ul-Haq M et al (2014) Integrating role of ethylene and ABA in tomato plants adaptation to salt stress. Sci Hort 172:109–116

Ashraf M, Foolad MR (2007) Roles of glycine betaine and proline in improving plant abiotic stress resistance. Environ Exp Bot 59:206–216

Ashraf M, Akram NA, Arteca RN et al (2010) The physiological, biochemical and molecular roles of brassinosteroids and salicylic acid in plant processes and salt tolerance. Crit Rev Plant Sci 29:162–190

Attipali RR, Kolluru VC, Munusamy V (2004) Drought induced responses of photosynthesis and antioxidant metabolism in higher plants. J Plant Physiol 161:1189–1202

Azzedine F, Gherroucha H, Baka M (2011) Improvement of salt tolerance in durum wheat by ascorbic acid application. J Stress Physiol Biochem 7:27–37

Banu MNA, Hoque MA, Watanabe-Sugimoto M et al (2010) Proline and glycinebetaine ameliorated NaCl stress via scavenging of hydrogen peroxide and methylglyoxal but not superoxide or nitric oxide in tobacco cultured cells. Biosci Biotechnol Biochem 74:2043–2049

Bastam N, Baninasab B, Ghobadi C (2013) Improving salt tolerance by exogenous application of salicylic acid in seedlings of pistachio. Plant Growth Regul 69:275–284

Bhomkar P, Upadhyay CP, Saxena M et al (2008) Salt stress alleviation in transgenic *Vigna mungo* L. Hepper (blackgram) by overexpression of the *glyoxalase I* gene using a novel *Cestrum* yellow leaf curling virus (CmYLCV) promoter. Mol Breed 22:169–181

Bose J, Rodrigo-Moreno A, Shabala S (2014) ROS homeostasis in halophytes in the context of salinity stress tolerance. J Exp Bot 65:1241–1257

Cabot C, Sibole JV, Barcelo J et al (2009) Abscisic acid decreases leaf Na^+ exclusion in salt-treated *Phaseolus vulgaris* L. J Plant Growth Regul 28:187–192

Casazza JP, Felver ME, Veech RL (1984) The metabolism of acetone in rat. J Biol Chem 259:231–236

Chakraborty K, Sairam RK, Bhattacharya R (2012) Differential expression of Salt Overly Sensitive pathway genes determines salinity stress tolerance in *Brassica* genotypes. Plant Physiol Biochem 51:90–101

Chaparzadeh N, D'Amico ML, Khavari-Nejad RA et al (2004) Antioxidative responses of *Calendula officinalis* under salinity conditions. Plant Physiol Biochem 42:695–701

Chaves MM, Flexas J, Pinheiro C (2009) Photosynthesis under drought and salt stress: regulation mechanisms from whole plant to cell. Ann Bot 103:551–560

Cramer GR, Quarrie SA (2002) Abscisic acid is correlated with the leaf growth inhibition of four genotypes of maize differing in their response to salinity. Funct Plant Biol 29:111–115

Dehghan G, Rezazadeh L, Habibi G (2011) Exogenous ascorbate improves antioxidant defense system and induces salinity tolerance in soybean seedlings. Acta Biol Szeged 55:261–264

Devanathan S, Erban A, Rodolfo PJ et al (2014) *Arabidopsis thaliana* glyoxalase 2-1 is required during abiotic stress but is not essential under normal plant growth. PLoS One 9:e9597

Do TD, Chen H, Hien VTT et al (2016) *Ncl* synchronously regulates Na^+, K^+, and Cl^- in soybean and greatly increases the grain yield in saline field conditions. Sci Rep 6:19147

Dong Z, Shi L, Wang Y et al (2013) Identification and dynamic regulation of microRNAs involved in salt stress responses in functional soybean nodules by high throughput sequencing. Int J Mol Sci 14:2717–2738

Ejaz B, Sajid ZA, Aftab F (2012) Effect of exogenous application of ascorbic acid on antioxidant enzyme activities, proline contents, and growth parameters of *Saccharum* spp. hybrid cv. HSF-240 under salt stress. Turk J Biol 35:630–640

El-Shabrawi H, Kumar B, Kaul T et al (2010) Redox homeostasis, antioxidant defense, and methylglyoxal detoxification as markers for salt tolerance in Pokkali rice. Protoplasma 24:85–96

Fahad S, Hussain S, Bano A et al (2015) Potential role of phytohormones and plant growth–promoting rhizobacteria in abiotic stresses: consequences for changing environment. Environ Sci Pollut Res 22:4907–4921

Fariduddin Q, Mir BA, Yusuf M et al (2014) 24-Epibrassinolide and/or putrescine trigger physiological and biochemical responses for the salt stress mitigation in *Cucumis sativus* L. Photosynthetica 52:464–474

Farouk S (2011) Ascorbic acid and α-tocopherol minimize salt-induced wheat leaf senescence. J Stress Physiol Biochem 7:58–79

Fricke W, Akhiyarova G, Wei WX et al (2006) The short-term growth response to salt of the developing barley leaf. J Exp Bot 57:1079–1095
Gao S, Ouyang C, Wang S et al (2008) Effects of salt stress on growth, antioxidant enzyme and phenylalanine ammonia-lyase activities in *Jatropha curcas* L. seedlings. Plant Soil Environ 54:374–381
Ghosh A, Pareek A, Sopory SK et al (2014) A glutathione responsive rice glyoxalase II, OsGLYII-2, functions in salinity adaptation by maintaining better photosynthesis efficiency and anti-oxidant pool. Plant J 80:93–105
Ghosh A, Kushwaha HR, Hasan MR et al (2016) Presence of unique glyoxalase III proteins in plants indicates the existence of shorter route for methylglyoxal detoxification. Sci Rep 6:18358
Gill SS, Tuteja N (2010) Reactive oxygen species and antioxidant machinery in abiotic stress tolerance in crop plants. Plant Physiol Biochem 48:909–930
Gómez Ojeda A, Corrales Escobosa AR, Wrobel K et al (2013) Effect of cd(II) and se(IV) exposure on cellular distribution of both elements and concentration levels of glyoxal and methylglyoxal in *Lepidium sativum*. Metallomics 5:1254–1261
Gupta B, Huang B (2014) Mechanisms of salinity tolerance in plants: physiological, biochemical, and molecular characterization. Int J Genom 2014:1–18
Gupta BK, Sahoo KK, Ghosh A et al (2017) Manipulation of glyoxalase pathway confers tolerance to multiple stresses in rice. Plant Cell Environ. https://doi.org/10.1111/pce.12968
Gurmani AR, Bano A, Khan SU et al (2011) Alleviation of salt stress by seed treatment with abscisic acid (ABA), 6-benzylaminopurine (BA) and chlormequat chloride (CCC) optimizes ion and organic matter accumulation and increases yield of rice (*Oryza sativa* L.) Aust J Crop Sci 5:1278–1285
Hasanuzzaman M, Hossain MA, Fujita M (2011a) Nitric oxide modulates antioxidant defense and the methylglyoxal detoxification system and reduces salinity-induced damage of wheat seedlings. Plant Biotechnol Rep 5:353–365
Hasanuzzaman M, Hossain MA, Fujita M (2011b) Selenium-induced up-regulation of the antioxidant defense and methylglyoxal detoxification system reduces salinity-induced damage in rapeseed seedlings. Biol Trace Elem Res 143:1704–1721
Hasanuzzaman M, Hossain MA, Fujita M (2012a) Exogenous selenium pretreatment protects rapeseed seedlings from cadmium-induced oxidative stress by up-regulating antioxidant defense and methylglyoxal detoxification systems. Biol Trace Elem Res 149:248–261
Hasanuzzaman M, Hossain MA, Teixeira da Silva JA et al (2012b) Plant responses and tolerance to abiotic oxidative stress: antioxidant defenses is a key factors. In: Bandi V, Shanker AK, Shanker C, Mandapaka M (eds) Crop stress and its management: perspectives and strategies. Springer, Berlin, pp 261–316
Hasegawa PM (2013) Sodium (Na^+) homeostasis and salt tolerance of plants. Environ Exp Bot 92:19–31
Hasegawa PM, Bressan RA, Zhu JK et al (2000) Plant cellular and molecular responses to high salinity. Annu Rev Plant Physiol Plant Mol Biol 51:463–499
Hayat S, Hayat Q, Alyemeni MN et al (2012) Role of proline under changing environment. Plant Signal Behav 7:1456–1466
Hernández JA, Jiménez A, Mullineaux P et al (2000) Tolerance of pea (*Pisum sativum* L.) to long-term salt stress is associated with induction of antioxidant defences. Plant Cell Environ 23:853–862
Hernández M, Fernandez-Garcia N, Diaz-Vivancos P et al (2010) A different role for hydrogen peroxide and the antioxidative system under short and long salt stress in *Brassica oleracea* roots. J Exp Bot 61:521–535
Hoque MA, Banu MNA, Nakamura Y et al (2008) Proline and glycine betaine enhance antioxidant defense and methylglyoxal detoxification systems and reduce NaCl-induced damage in cultured tobacco cells. J Plant Physiol 165:813–824
Hoque MA, Uraji M, Banu MNA et al (2010) The effects of methylglyoxal on glutathione *S*-transferase from *Nicotiana tabacum*. Biosci Biotechnol Biochem 74:2124–2126
Hoque MA, Uraji M, Banu MNA et al (2012a) Methylglyoxal inhibition of cytosolic ascorbate peroxidase from *Nicotiana tabacum*. J Biochem Mol Toxicol 26:315–321

Hoque TS, Okuma E, Uraji M et al (2012b) Inhibitory effects of methylglyoxal on light-induced stomatal opening and inward K^+ channel activity in *Arabidopsis*. Biosci Biotechnol Biochem 76:617–619

Hoque TS, Uraji M, Tuya A et al (2012c) Methylglyoxal inhibits seed germination and root elongation and up-regulates transcription of stress-responsive genes in ABA-dependent pathway in *Arabidopsis*. Plant Biol 14:854–858

Hoque TS, Uraji M, Ye W et al (2012d) Methylglyoxal-induced stomatal closure accompanied by peroxidase-mediated ROS production in *Arabidopsis*. J Plant Physiol 169:979–986

Hoque TS, Hossain MA, Mostofa MG et al (2015) Signalling roles of methylglyoxal and the involvement of the glyoxalase system in plant abiotic stress responses and tolerance. In: Azooz MM, Ahmad P (eds) Plant–environment interaction: responses and approaches to mitigate stress. Wiley, Chichester, pp 311–326

Hoque TS, Hossain MA, Mostofa MG et al (2016) Methylglyoxal: an emerging signaling molecule in plant abiotic stress responses and tolerance. Front Plant Sci 7:1341

Hoque TS, Uraji M, Hoque MA et al (2017) Methylglyoxal induces inhibition of growth, accumulation of anthocyanin, and activation of glyoxalase I and II in *Arabidopsis thaliana*. J Biochem Mol Toxicol 2017:e21901

Horie T, Karahara I, Katsuhara M (2012) Salinity tolerance mechanisms in glycophytes: an overview with the central focus on rice plants. Rice 5:11

Hossain MA, Fujita M (2009) Purification of glyoxalase I from onion bulbs and molecular cloning of its cDNA. Biosci Biotechnol Biochem 73:2007–2013

Hossain MA, Fujita M (2010) Evidence for a role of exogenous glycinebetaine and proline in antioxidant defense and methylglyoxal detoxification systems in mung bean seedlings under salt stress. Physiol Mol Biol Plants 16:19–29

Hossain MA, Hossain MZ, Fujita M (2009) Stress-induced changes of methylglyoxal level and glyoxalase I activity in pumpkin seedlings and cDNA cloning of glyoxalase I gene. Aust J Crop Sci 3:53–64

Hossain MA, Hasanuzzaman M, Fujita M (2010) Up-regulation of antioxidant and glyoxalase systems by exogenous glycinebetaine and proline in mung bean confer tolerance to cadmium stress. Physiol Mol Biol Plants 16:259–272

Hossain MA, Teixeira da Silva JA, Fujita M (2011a) Glyoxalase system and reactive oxygen species detoxification system in plant abiotic stress response and tolerance: an intimate relationship. In: Shanker A, Venkateswarlu B (eds) Abiotic stress in plants—mechanisms and adaptations. INTECH-Open Access Publisher, Rijeka, pp 235–266

Hossain MA, Hasanuzzaman M, Fujita M (2011b) Coordinate induction of antioxidant defense and glyoxalase system by exogenous proline and glycinebetaine is correlated with salt tolerance in mung bean. Front Agric China 5:1–14

Hossain MA, Mostofa MG, Fujita M (2013a) Cross protection by cold-shock to salinity and drought stress–induced oxidative stress in mustard (*Brassica campestris* L.) seedlings. Mol Plant Breed 4:50–70

Hossain MA, Mostofa MG, Fujita M (2013b) Heat-shock positively modulates oxidative protection of salt and drought-stressed mustard (*Brassica campestris* L.) seedlings. J Plant Sci Mol Breed 2:1–14

Hossain MA, Mostofa MG, Burritt DJ et al (2014a) Modulation of reactive oxygen species and methylglyoxal detoxification systems by exogenous glycinebetaine and proline improves drought tolerance in mustard (*Brassica juncea* L.) Int J Plant Biol Res 2(2):2014

Hossain MA, Hoque MA, Burritt DJ et al (2014b) Proline protects plants against abiotic oxidative stress: biochemical and molecular mechanisms. In: Ahmad P (ed) Oxidative damage to plants. Elsevier, USA, pp 477–522

Hu L, Hu T, Zhang X et al (2012) Exogenous glycine betaine ameliorates the adverse effect of salt stress on perennial ryegrass (*Lolium perenne* L.) J Amer Soc Hort Sci 137:38–46

Hussain BMN, Akram S, Raffi SA et al (2016) Exogenous glutathione improves salinity stress tolerance in rice (*Oryza sativa* L.) Plant Gene Trait 8:1–17

Ishitani M, Liu J, Halfter U et al (2000) SOS3 function in plant salt tolerance requires N-myristoylation and calcium binding. Plant Cell 12:1667–1677

Jagadish S, Septiningsih E, Kohli A et al (2012) Genetic advances in adapting rice to a rapidly changing climate. J Agron Crop Sci 198:360–373

James RA, Blake C, Byrt CS et al (2011) Major genes for Na^+ exclusion, Nax1 and Nax2 (wheat HKT1;4 and HKT1;5), decrease Na^+ accumulation in bread wheat leaves under saline and waterlogged conditions. J Exp Bot 62:2939–2947

Jeschke WD, Peuke AD, Pate JS et al (1997) Transport, synthesis and catabolism of abscisic acid (ABA) in intact plants of castor bean (*Ricinus communis* L.) under phosphate deficiency and moderate salinity. J Exp Bot 48:1737–1747

Ji H, Pardo JM, Batelli G et al (2013) The Salt Overly Sensitive (SOS) pathway: established and emerging roles. Mol Plant 6:275–286

Jin SH, Li XQ, Wang GG et al (2015) Brassinosteroids alleviate high-temperature injury in *Ficus concinna* seedlings via maintaining higher antioxidant defence and glyoxalase systems. AoB Plants 7:plv009

Kadioglu A, Saruhan N, Sağlam A et al (2011) Exogenous salicylic acid alleviates effects of long term drought stress and delays leaf rolling by inducing antioxidant system. Plant Growth Regul 64:27–37

Kalapos MP (1999) Methylglyoxal in living organisms: chemistry, biochemistry, toxicology biological implications. Toxicol Lett 110:145–175

Kalapos MP (2008) The tandem of free radicals and methylglyoxal. Chem Biol Interact 171:251–271

Kamiab F, Talaie A, Khexri M et al (2014) Exogenous application of free polyamines enhance salt tolerance of pistachio (*Pistacia vera* L.) seedlings. Plant Growth Regul 72:257–268

Kang DJ, Seo YJ, Lee JD et al (2005) Jasmonic acid differentially affects growth, ion uptake and abscisic acid concentration in salt-tolerant and salt-sensitive rice cultivars. J Agron Crop Sci 191:273–282

Kaur C, Ghosh A, Pareek A et al (2014a) Glyoxalases and stress tolerance in plants. Biochem Soc Trans 42:485–490

Kaur C, Mustafiz A, Sarkar A et al (2014b) Expression of abiotic stress inducible ETHE1-like protein from rice is higher in roots and is regulated by calcium. Physiol Plant 152:1–16

Kaur C, Singla-Pareek SL, Sopory SK (2014c) Glyoxalase and methylglyoxal as biomarkers for plant stress tolerance. Crit Rev Plant Sci 33:429–456

Kaya C, Sonmez O, Aydemir S et al (2013) Exogenous application of mannitol and thiourea regulates plant growth and oxidative stress responses in salt-stressed maize (*Zea mays* L.) J Plant Interact 3:234–241

Kirch HH, Schlingensiepen S, Kotchoni S et al (2005) Detailed expression analysis of selected genes of the aldehyde dehydrogenase (ALDH) gene superfamily in *Arabidopsis thaliana*. Plant Mol Biol 57:315–332

Kissoudis C, Seifi A, Yan Z et al (2016) Ethylene and abscisic acid signaling pathways differentially influence tomato resistance to combined powdery mildew and salt stress. Front Plant Sci 7:2009

Koop DR, Casazza JP (1985) Identification of ethanol-inducible P-450 isozyme 3a as the acetone and acetol monooxygenase of rabbit microsomes. J Biol Chem 260:13607–13612

Koyro HW, Ahmad P, Geissler N (2012) Abiotic stress responses in plants: an overview. In: Ahmad P, Prasad MNV (eds) Environmental adaptations and stress tolerance of plants in the era of climate change. Springer, New York, pp 1–28

Krasensky J, Jonak C (2012) Drought, salt, and temperature stress induced metabolic rearrangements and regulatory networks. J Exp Bot 63:1593–1608

Kubiś J (2008) Exogenous spermidine alters in different ways activities of some scavenging system enzymes, H_2O_2 and superoxide radical levels in water stressed cucumber leaves. J Plant Physiol 165:397–406

Kubiś J, Floryszak-Wieczorek J, Arasimowicz-Jelonek M (2014) Polyamines induce adaptive responses in water deficit stressed cucumber roots. J Plant Res 127:151–158
Kumar V, Yadav SK (2009) Proline and betaine provide protection to antioxidant and methylglyoxal detoxification systems during cold stress in *Camellia sinensis* (L.) O. Kuntze. Acta Physiol Plant 31:261–269
Kumar M (2013) Crop plants and abiotic stresses. J Biomol Res Ther 3:e125
Kusano T, Yamaguchi K, Berberich T et al (2007) The polyamine spermine rescues *Arabidopsis* from salinity and drought. Plant Signal Behav 2:251–252
Li ZG (2016) Methylglyoxal and glyoxalase system in plants: old players, new concepts. Bot Rev 82:183–203
Li ZG, Duan XQ, Min X (2017a) Methylglyoxal as a novel signal molecule induces the salt tolerance of wheat by regulating the glyoxalase system, the antioxidant system, and osmolytes. Protoplasma 254(5):1995–2006. https://doi.org/10.1007/s00709-017-1094-z
Li ZG, Duan XQ, Xia YM et al (2017b) Methylglyoxal alleviates cadmium toxicity in wheat (*Triticum aestivum* L.) Plant Cell Rep 36:367–370
Lin F, Xu J, Shi J et al (2010) Molecular cloning and characterization of a novel glyoxalase I gene *TaGly I* in wheat (*Triticum aestivum* L.) Mol Biol Rep 37:729–735
Liu W, Zhang Y, Yuan X et al (2016) Exogenous salicylic acid improves salinity tolerance of *Nitraria tangutorum*. Russ J Plant Physiol 63:132–142
Lyles GA, Chalmers J (1992) The metabolism of aminoacetone to methylglyoxal by semicarbazide-sensitive amine oxidase in human umbilical artery. Biochem Pharmacol 43:1409–1414
Maeta K, Izawa S, Inoue Y (2005) Methylgyoxal, a metabolite derived from glycolysis, functions as a signal initiator of the high osmolarity glycerol-mitogen-activated protein kinase cascade and calcineurin/Crz1-mediated pathway in *Saccharomyces cerevisiae*. J Biol Chem 280:253–260
Mahajan S, Tuteja N (2005) Cold, salinity and drought stresses: an overview. Arch Biochem Biophys 444:139–158
Malekzadeh P (2015) Influence of exogenous application of glycinebetaine on antioxidative system and growth of salt-stressed soybean seedlings (*Glycine max* L.) Physiol Mol Biol Plants 21:225–232
Mano J (2012) Reactive carbonyl species: their production from lipid peroxides, action in environmental stress, and the detoxification mechanism. Plant Physiol Biochem 59:90–97
Mano J, Miyatake F, Hiraoka E et al (2009) Evaluation of the toxicity of stress-related aldehydes to photosynthesis in chloroplasts. Planta 230:639–648
Martins AMTBS, Coedeiro CAA, Freire AMJP (2001) In situ analysis of methylglyoxal metabolism in *Saccharomyces cerevisiae*. FEBS Lett 499:41–44
Mittova V, Guy M, Tal M et al (2004) Salinity upregulates the antioxidative system in root mitochondria and peroxisomes of the wild salt-tolerant tomato species *Lycopersicon pennellii*. J Exp Bot 55:1105–1113
Mostofa MG, Seraj ZI, Fujita M (2014a) Exogenous sodium nitroprusside and glutathione alleviate copper toxicity by reducing copper uptake and oxidative damage in rice (*Oryza sativa* L.) seedlings. Protoplasma 251:1373–1386
Mostofa MG, Yoshida N, Fujita M (2014b) Spermidine pretreatment enhances heat tolerance in rice seedlings through modulating antioxidative and glyoxalase systems. Plant Growth Regul 73:31–44
Mostofa MG, Hossain MA, Fujita M (2015a) Trehalose pretreatment induces salt tolerance in rice seedlings: oxidative damage and co-induction of antioxidant defense and glyoxalase systems. PRO 252:461–475
Mostofa MG, Hossain MA, Fujita M et al (2015b) Physiological and biochemical mechanism associated with trehalose-induced copper-stress tolerance in rice. Sci Rep 5:11433
Munns R (2005) Genes and salt tolerance: bringing them together. New Phytol 167:645–663
Munns R, Gilliham M (2015) Salinity tolerance of crops—what is the cost? New Phytol 208:668–673

Munns R, Tester M (2008) Mechanisms of salinity tolerance. Annu Rev Plant Biol 59:651–681
Mustafiz A, Singh AK, Pareek A et al (2011) Genome-wide analysis of rice and *Arabidopsis* identifies two glyoxalase genes that are highly expressed in abiotic stresses. Funct Integr Genomics 11:293–305
Mustafiz A, Ghosh A, Tripathi AK et al (2014) A unique Ni^{2+}-dependent and methylglyoxal-inducible rice glyoxalase I possesses a single active site and functions in abiotic stress response. Plant J 78:951–963
Nahar K, Hasanuzzaman M, Alam MM et al (2015a) Roles of exogenous glutathione in antioxidant defense system and methylglyoxal detoxification during salt stress in mungbean. Biol Plant 59:745–756
Nahar K, Hasanuzzaman M, Alam MM et al (2015b) Exogenous glutathione confers high temperature stress tolerance in mung bean (*Vigna radiata* L.) by modulating antioxidant defense and methylglyoxal detoxification system. Environ Exp Bot 112:44–54
Narawongsanont R, Kabinpong S, Auiyawong B et al (2012) Cloning and characterization of AKR4C14, a rice aldo-ketoreductase, from Thai jasmine rice. Protein J 31:35–42
Newton AC, Johnson SN, Gregory PJ (2011) Implications of climate change for diseases, crop yields and food security. Euphytica 179:3–18
Noctor G, Mhamdi A, Chaouch S et al (2012) Glutathione in plants: an integrated overview. Plant Cell Environ 35:454–484
Nounjan N, Nghia PT, Theerakulpisut P (2012) Exogenous proline and trehalose promote recovery of rice seedlings from salt stress and differentially modulate antioxidant enzymes and expression of related genes. J Plant Physiol 169:596–604
Osakabe Y, Osakabe K, Shinozaki K et al (2014) Response of plants to water stress. Front Plant Sci 5:86
Patade VY, Lokhande V, Suprasanna P (2014) Exogenous application of proline alleviates salt induced oxidative stress more efficiently than glycine betaine in sugarcane cultured cells. Sugar Tech 16:22–29
Pitman MG, Läuchli A (2002) Global impact of salinity and agricultural ecosystems. In: Läuchli A, Lüttge U (eds) Salinity: environment–plants–molecules. Kluwer, Dordrecht, pp 3–20
Popova LP, Stoinova ZG, Maslenkova LT (1995) Involvement of abscisic acid in photosynthetic process in *Hordeum vulgare* L. during salinity stress. J Plant Growth Reg 14:211–218
Qin F, Shinozaki K, Yamaguchi-Shinozaki K (2011) Achievements and challenges in understanding plant abiotic stress responses and tolerance. Plant Cell Physiol 52:1569–1582
Quan R, Wang J, Yang D et al (2017) EIN3 and SOS2 synergistically modulate plant salt tolerance. Sci Rep 7:44637
Rahnama A, James RA, Poustini K et al (2010) Stomatal conductance as a screen for osmotic stress tolerance in durum wheat growing in saline soil. Funct Plant Biol 37:255–263
Rajwanshi R, Kumar D, Yusuf M et al (2016) Stress-inducible overexpression of glyoxalase I is preferable to its constitutive overexpression for abiotic stress tolerance in transgenic *Brassica juncea*. Mol Breed 36:1–15
Ranganayakulu GS, Veeranagamallaiah G, Sudhakar C (2013) Effect of salt stress on osmolyte accumulation in two groundnut cultivars (*Arachis hypogaea* L.) with contrasting salt tolerance. African J Plant Sci 12:586–592
Ray S, Dutta S, Halder J et al (1994) Inhibition of electron flow through complex I of the mitochondrial respiratory chain of Ehrlich ascites carcinoma cells by methylglyoxal. Biochem J 303:69–72
Reddy VS, Sopory SK (1999) Glyoxalase I from *Brassica juncea*: molecular cloning, regulation and its over-expression confer tolerance in transgenic tobacco under stress. Plant J 17:385–395
Rhodes D, Hanson AD (1993) Quaternary ammonium and tertiary sulfonium compounds in higher plants. Annu Rev Plant Physiol Plant Mol Biol 44:357–384
Richard JP (1993) Mechanism for the formation of methylglyoxal from triosephosphates. Biochem Soc Trans 21:549–553
Ridderström M, Cameron AD, Jones TA et al (1998) Involvement of an active-site Zn^{2+} ligand in the catalytic mechanism of human glyoxalase I. J Biol Chem 273:21623–21628

Rodríguez M, Canales E, Borrás-Hidalgo O (2005) Molecular aspects of abiotic stress in plants. Biotechnol Appl 22:1–10

Roy SD, Saxena M, Bhomkar PS et al (2008) Generation of marker free salt tolerant transgenic plants of *Arabidopsis thaliana* using the *gly I* gene and *cre* gene under inducible promoter. Plant Cell Tissue Organ Cult 95:1–11

Roy SJ, Negrão S, Tester M (2014) Salt resistant crop plants. Curr Opin Biotechnol 26:115–124

Rozema J, Flowers T (2008) Ecology: crops for a salinized world. Science 322:1478–1380

Ryu H, Cho YG (2015) Plant hormones in salt stress tolerance. J Plant Biol 58:147–155

Sairam RK, Tyagi A (2004) Physiology and molecular biology of salinity stress tolerance in plants. Curr Sci 86:407–421

Saito R, Yamamoto H, Makino A, Sugimoto T, Miyake C (2011) Methylglyoxal functions as Hill oxidant and stimulates the photoreduction of O_2 at photosystem I: a symptom of plant diabetes. Plant Cell Environ 34:1454–1464

Salama KHA, Al-Mutawa MM (2009) Glutathione-triggered mitigation in salt-induced alterations in plasmalemma of onion epidermal cells. Int J Agric Biol 11:639–642

Sanders D (2000) Plant biology: the salty tale of *Arabidopsis*. Curr Biol 10:486–488

Sankaranarayanan S, Jamshed M, Kumar A et al (2017) Glyoxalase goes green: the expanding roles of glyoxalase in plants. Int J Mol Sci 18:898

Saxena M, Roy SD, Singla-Pareek SL et al (2011) Overexpression of the glyoxalase II gene leads to enhanced salinity tolerance in *Brassica juncea*. Open Plant Sci J 5:23–28

Serrano R, Márquez JA, Rios G (1997) Crucial factors in salt stress tolerance. In: Hohmann S, Mager WH (eds) Yeast stress responses. RG Landes Company, Austin, pp 147–169

Shakirova FM, Avalbaev AM, Bezrukova MV et al (2010) Role of endogenous hormonal system in the realization of the antistress action of plant growth regulators on plants. Plant Stress 4:32–38

Shangari N, O'Brien PJ (2004) The cytotoxic mechanism of glyoxal involves oxidative stress. Biochem Pharmacol 68:1433–1442

Sharma I, Ching E, Saini S (2013) Exogenous application of brassinosteroid offers tolerance to salinity by altering stress responses in rice variety Pusa Basmati-1. Plant Physiol Biochem 69:17–26

Shi H, Quintero FJ, Prado JM et al (2002) The putative plasma membrane Na^+–H^+ antiporter SOS1 controls long-distance Na^+ transport in plants. Plant Cell 14:465–477

Simpson PJ, Tantitadapitak C, Reed AM et al (2009) Characterization of two novel aldo-keto reductases from Arabidopsis: expression patterns, broad substrate specificity, and an open active-site structure suggest a role in toxicant metabolism following stress. J Mol Biol 392:465–480

Singla-Pareek SL, Reddy MK, Sopory SK (2003) Genetic engineering of the glyoxalase pathway in tobacco leads to enhanced salinity tolerance. Proc Natl Acad Sci U S A 100:14672–14677

Singla-Pareek SL, Yadav SK, Pareek A et al (2006) Transgenic tobacco overexpressing glyoxalase pathway enzymes grow and set viable seeds in zinc-spiked soils. Plant Physiol 140:613–623

Singla-Pareek SL, Yadav SK, Pareek A et al (2008) Enhancing salt tolerance in a crop plant by overexpression of glyoxalase II. Transgenic Res 17:171–180

Sobahan MA, Akter N, Ohno M et al (2012) Effects of exogenous proline and glycinebetaine on the salt tolerance of rice cultivars. Biosci Biotechnol Biochem 76:1568–1570

Sripinyowanich S, Klomsakul P, Boonburapong B et al (2013) Exogenous ABA induces salt tolerance in indica rice (*Oryza sativa* L.): the role of OsP5CS1 and OsP5CR gene expression during salt stress. Environ Exp Bot 86:94–105

Takeda S, Matsuoka M (2008) Genetic approaches to crop improvement: responding to environmental and population changes. Nat Rev Genet 9:444–457

Tanou G, Molassiotis A, Diamantidis G (2009) Induction of reactive oxygen species and necrotic death–like destruction in strawberry leaves by salinity. Environ Exp Bot 65:270–281

Tavakkoli E, Rengasamy P, McDonald GK (2010) High concentrations of Na^+ and Cl^- ions in soil solution have simultaneous detrimental effects on growth of faba bean under salinity stress. J Exp Bot 61:4449–4459

Theerakulpisut P, Gunnula W (2012) Exogenous sorbitol and trehalose mitigated salt stress damage in salt-sensitive but not salt-tolerant rice seedlings. Asian J Crop Sci 4:165–170

Thomas JC, McElwain EF, Bohnert HJ (1992) Convergent induction of osmotic stress-responses: abscisic acid, cytokinin, and the effects of NaCl. Plant Physiol 100:416–423

Thornalley PJ (1993) The glyoxalase system in health and disease. Mol Asp Med 14:287–371

Thornalley PJ (2003) Glyoxalase I—structure, function and a critical role in the enzymatic defence against glycation. Biochem Soc Trans 31:1343–1348

Thornalley PJ (2008) Protein and nucleotide damage by glyoxal and methylglyoxal in physiological systems—role in ageing and disease. Drug Metabol Drug Interact 23:125–150

Thornalley PJ, Waris S, Fleming T et al (2010) Imidazopurinones are markers of physiological genomic damage linked to DNA instability and glyoxalase 1–associated tumour multidrug resistance. Nucleic Acids Res 38:5432–5442

Torabian AR (2011) Effect of salicylic acid on germination and growth of alfalfa (*Medicago sativa* L.) seedlings under water potential loss at salinity stress. Plant Ecophysiol 2:151–155

Tuomainen M, Ahonen V, Kärenlampi SO et al (2011) Characterization of the glyoxalase 1 gene TcGLX1 in the metal hyperaccumulator plant *Thlaspi caerulescens*. Planta 233:1173–1184

Upadhyaya CP, Venkatesh J, Gururani MA et al (2011) Transgenic potato overproducing L-ascorbic acid resisted an increase in methylglyoxal under salinity stress via maintaining higher reduced glutathione level and glyoxalase enzyme activity. Biotechnol Lett 33:2297–2307

Vaidyanathan R, Kuruvilla S, Thomas G (1999) Characterization and expression pattern of an abscisic acid and osmotic stress responsive gene from rice. Plant Sci 140:21–30

Van Oosten MJ, Sharkhuu A, Batelli G et al (2013) The *Arabidopsis thaliana* mutant air1 implicates SOS3 in the regulation of anthocyanins under salt stress. Plant Mol Biol 83:405–415

Vardharajula S, Ali SZ, Grover M et al (2011) Drought-tolerant plant growth promoting *Bacillus* spp.: effect on growth, osmolytes, and antioxidant status of maize under drought stress. J Plant Interact 1:1–14

Veena, Reddy VS, Sopory SK (1999) Glyoxalase I from *Brassica juncea*: molecular cloning, regulation and its over-expression confer tolerance in transgenic tobacco under stress. Plant J 17:385–395

Verma M, Verma D, Jain RK et al (2005) Overexpression of glyoxalase I gene confers salinity tolerance in transgenic *japonica* and *indica* rice plants. Rice Genet Newslett 22:58–62

Wang WX, Vinocur B, Altman A (2003) Plant responses to drought, salinity and extreme temperatures: towards genetic engineering for stress tolerance. Planta 218:1–14

Wang R, Liu S, Zhou F et al (2014) Exogenous ascorbic acid and glutathione alleviate oxidative stress induced by salt stress in the chloroplasts of (*Oryza sativa* L.) J Biosci 69:226–236

Wani SH, Gosal SS (2011) Introduction of OsglyII gene into Oryza sativa for increasing salinity tolerance. Biol Plantarum 55:536–540

Werner JE, Finkelstein RR (1995) *Arabidopsis* mutants with reduced response to NaCl and osmotic stress. Physiol Plant 93:659–666

Wu X, Zhu Z, Li X et al (2012) Effects of cytokinin on photosynthetic gas exchange, chlorophyll fluorescence parameters and antioxidative system in seedlings of eggplant (*Solanum melongena* L.) under salinity stress. Acta Physiol Plant 34:2105–2114

Wu C, Ma C, Pan Y et al (2013) Sugar beet M14 glyoxalase I gene can enhance plant tolerance to abiotic stresses. J Plant Res 126:415–425

Yadav SK, Singla-Pareek SL, Ray M et al (2005a) Methylglyoxal levels in plants under salinity stress are dependent on glyoxalase I and glutathione. Biochem Biophys Res Commun 337:61–67

Yadav SK, Singla-Pareek SL, Ray M et al (2005b) Transgenic tobacco plants overexpressing glyoxalase enzymes resist an increase in methylglyoxal and maintain higher reduced glutathione levels under salinity stress. FEBS Lett 579:6265–6271

Yadav SK, Singla-Pareek SL, Sopory SK (2008). An overview on the role of methylglyoxal and glyoxalases in plants. Drug Metabol Drug Interact 23:51–68

Yang Q, Chen ZZ, Zhou XF et al (2009) Overexpression of *SOS* (*Salt Overly Sensitive*) genes increases salt tolerance in transgenic *Arabidopsis*. Mol Plant 2:22–31

Yang L, Zu YG, Tang Z (2013) Ethylene improves *Arabidopsis* salt tolerance mainly via K^+ in shoots and roots rather than decreasing tissue Na^+ content. Environ Exp Bot 86:60–69

Yang Y, Tang RJ, Jiang CM et al (2015) Overexpression of the *PtSOS2* gene improves tolerance to salt stress in transgenic poplar plants. Plant Biotechnol J 13:962–973

Yoon JY, Hamayun M, Lee S-K, Lee I-J (2009) Methyl jasmonate alleviated salinity stress in soybean. J Crop Sci Biotechnol 12:63–68

You J, Chan Z (2015) ROS regulation during abiotic stress responses in crop plants. Front Plant Sci 6:1092

Zeng Z, Xiong F, Yu X et al (2016) Overexpression of a glyoxalase gene, OsGly I, improves abiotic stress tolerance and grain yield in rice (*Oryza sativa* L.) Plant Physiol Biochem 109:62–71

Zhang J, Jia W, Yang Y et al (2006) Role of ABA in integrating plant responses to drought and salt stresses. Field Crops Res 97:111–119

Zhou J, Wang J, Bi Y et al (2014) Overexpression of *PtSOS2* enhances salt tolerance in transgenic poplars. Plant Mol Biol Rep 32:185–197

Zhu JK (2003) Regulation of ion homeostasis under salt stress. Curr Opin Plant Biol 6:441–445

Zhu JK (2007) Plant salt stress. Encyclopaedia of life sciences. Wiley, Chichester, pp 1–3

Chapter 11
Cross-Protection by Oxidative Stress: Improving Tolerance to Abiotic Stresses Including Salinity

Vokkaliga T. Harshavardhan, Geetha Govind, Rajesh Kalladan, Nese Sreenivasulu, and Chwan-Yang Hong

Abstract Abiotic stresses severely limit crop productivity. Plants being sessile, they are continuously exposed to a broad range of environmental stresses. Hence, multiple stress situations are more likely to occur in field conditions. Nevertheless, plants have evolved strategies to sense their environment to modulate its growth. However, its prime aim is to survive under adverse conditions and complete its life cycle. It is with the idea to increase or sustain productivity under adverse conditions that we are interested in. The response of plants to adverse environmental condition is sensed by changes in ROS leading to oxidative stress. Hence, it can be speculated that plants that are tolerant to oxidative stress would also be tolerant to multiple abiotic stress (abiotic stress-induced oxidative stress). In other words, cross-protection to multiple abiotic stresses can be achieved by developing plants tolerant to oxidative stress. Cross-protection can be enhanced by developing inherent tolerance by using conventional breeding or genetic engineering techniques or induced tolerance by priming. Here we try to compile the opinion of using oxidative stress tolerance as first line of defense against multiple abiotic stresses leading to cross-protection in field conditions.

V. T. Harshavardhan (✉) · C.-Y. Hong (✉)
Department of Agricultural Chemistry, College of Bio-resources and Agriculture, National Taiwan University, Taipei, Taiwan
e-mail: cyhong@ntu.edu.tw

G. Govind
Department of Agricultural Chemistry, College of Bio-resources and Agriculture, National Taiwan University, Taipei, Taiwan

Department of Crop Physiology, College of Agriculture, Hassan, Karnataka, India

R. Kalladan
Institute of Plant and Microbial Biology, Academia Sinica, Taipei, Taiwan

N. Sreenivasulu
Grain Quality and Nutrition Center, Plant Breeding Division, International Rice Research Institute, Los Baños, Philippines

V. Kumar et al. (eds.), *Salinity Responses and Tolerance in Plants, Volume 1*,
https://doi.org/10.1007/978-3-319-75671-4_11

Keywords Cross-protection · Broad-spectrum stress tolerance · Redox homeostasis · Energy balance · Inherent tolerance · Genetic engineering · Priming · Oxidative stress · Drought · Salinity · Extreme temperatures · Heavy metal stress

Abbreviations

$^{\bullet}OH$	Hydroxyl radical
ABA	Abscisic acid
APX	Ascorbate peroxidase
AsA	Ascorbate
CAT	Catalase
DHAR	Dehydroascorbate reductase
GPX	Glutathione peroxidase
GR	Glutathione reductase
GSH	Glutathione
GST	glutathione S-transferase
H_2O_2	Hydrogen peroxide
MDHAR	Monodehydroascorbate reductase
MV	Methyl viologen
NA	Not available
NO	Nitric oxide
$O^{2\bullet-}$	Superoxide radical
OE	Overexpression
PEG	Polyethylene glycol
POX	Peroxidase
PSI	Photosystem I
PSII	Photosystem II
RONSS	Reactive oxygen, nitrogen, and sulfur species
ROS	Reactive oxygen species
SA	Salicylic acid
SOD	Superoxide dismutase

11.1 Introduction

In field, plants are exposed to multiple environmental factors (abiotic) that may or may not be suboptimum for their growth and development. However, plants have developed various strategies to respond to its ever-changing environmental conditions, by monitoring its surroundings and adjusting its metabolic systems accordingly to maintain homeostasis and survive stressful conditions (Mittler 2006; Mittler and Blumwald 2010; Sreenivasulu et al. 2007, 2012). Nevertheless, with the severity of environmental stress, the plant's genetic background and its memory of stress exposure determine its daily survival or death. These factors determine the performance of the individual. Therefore the genome-environment interaction is essential in elucidating the nature of the phenotypic variation leading to the plant's successful performance in response to environmental cues (Pastori and Foyer 2002). Most of the crop varieties currently used in agriculture have been selected for their high productivity to cope with food security with increasing population demand and not for their stress tolerance traits (Gilliham et al. 2017; Taiz 2013).

Abiotic stress is an important feature in agriculture that has a huge impact on reduction in growth and productivity with additional challenge on plants by biotic stress leading to substantial yield loss. Stress combination studies instead of individual stresses have now been recognized and gained importance (Rizhsky et al. 2004; Mittler 2006; Kissoudis et al. 2014; Suzuki et al. 2014; Ramegowda and Senthil-Kumar 2015; Zandalinas et al. 2017; Pandey et al. 2017). Lab studies have demonstrated that plant's response to a combination of stresses is different from plant's response to individual stresses (Mittler 2006; Mittler and Blumwald 2010; Atkinson and Urwin 2012). However, outcome with experiments with plants under combination of stresses depends on the plant age, its developmental stage, severity of stress factors, and its inherent capacity of stress tolerance. The responses observed are inclined to the dominant stress factor. In addition, research on multiples stresses has been to simulate controlled stress to mimic natural conditions in the field, but nevertheless field conditions are not controlled. In nature, the environment of the plant is highly variable with abiotic stress factors that can occur in combinations or in successions. The loss in agricultural productivity can be minimized by reduction in yield caused by abiotic stress if the plants have broad-spectrum stress tolerance or cross-protection or tolerance.

Cross-protection, or tolerance, is the phenomena in which plants displaying tolerance to one kind of stress also display tolerance to other one or more stress factors (Pastori and Foyer 2002; Mittler 2006). Targeting responses that are shared across various kinds of stress could be the best components to be targeted to impart cross-protection or broad-spectrum stress tolerance (Fig. 11.1). Plants adapt to its variable environment mainly owing to the plasticity reflected by the cells to adjust the dynamics of cellular reduction/oxidation (redox) processes. In all aerobic organisms, metabolic processes such as respiration and photosynthesis unavoidably lead to the production of ROS in the mitochondria, chloroplast, and peroxisome (Apel and Hirt 2004; Gill and Tuteja 2010). Although under optimal growth conditions, ROS are produced at a low level in organelles, and they get elevated during stress. Till date, all kinds of abiotic and biotic stresses trigger a generalized stress response called oxidative stress caused by the accumulation of activated oxygen molecules called reactive oxygen species (ROS). Under abiotic stress condition, stomatal closure limits the uptake of CO_2 which results in the overproduction of ROS due to overreduction of photosynthetic electron transport chain (Apel and Hirt 2004; Noctor et al. 2014).

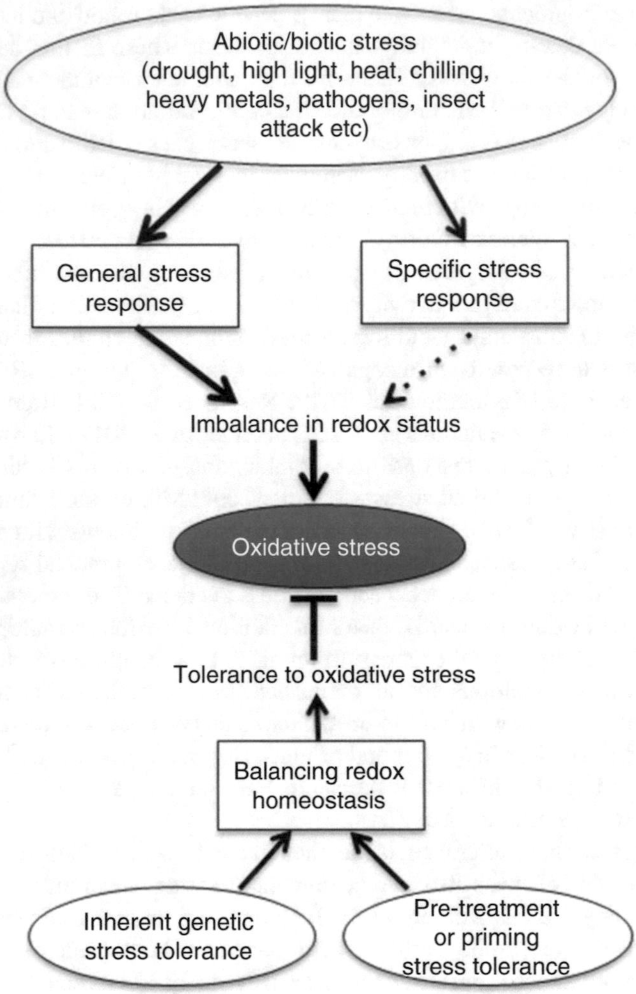

Fig. 11.1 Oxidative stress is a general stress response, and tolerance to oxidative stress imparts cross-protection or multiple stress tolerance

11.2 Oxidative Stress: A Generalized Response

All kinds of abiotic and biotic stresses trigger a generalized stress response called oxidative stress caused by the accumulation of reactive oxygen species (ROS). In general, oxidative stress can be defined as a physiological state wherein oxidation exceeds reduction leading to oxidative damage to cell compounds. Hence, it is an imbalance of the reduction/oxidation (redox) state of the cell caused by lack of electrons resulting in imbalance between ROS generation and its detoxification.

Interestingly, it is believed that O_2 derivatives are one of the oldest stresses on the planet (Dowling and Simmons 2009). Therefore, plants have evolved the ability to employ oxidative stress – imbalance in ROS – not only for signaling needs and sensing other stresses but for regulation of growth, polarity, and death (Demidchik et al. 2003, 2010; Foreman et al. 2003), sensing hormones and regulatory agents (Murata et al. 2001; Demidchik et al. 2004, 2009; Krishnamurthy and Rathinasabapathi 2013) and many other processes that are not primarily related to stress or oxidation. Environmental stresses like salinity, extreme temperature, or drought result in a marked increase in ROS level leading to oxidative damage in plant cells (Robinson and Bunce 2000).

One of the most important consequence of abiotic stresses is the overaccumulation of ROS like $O_2^{\bullet -}$, H_2O_2, and $^{\bullet}OH$ in plant cells (Mittler et al. 2004). Plants have several strategies to withstand an oxidative stress, mainly by use of antioxidants – enzymes and nonenzymatic substances that scavenge ROS and free radicals (reviewed by Alscher et al. 2002; Dietz et al. 2006; Dietz 2003; Mittler et al. 2004; Gill and Tuteja 2010; Wani et al. 2016; Upadhyaya and Hossain 2016). It has now become quite evident that changes in cellular redox environment play a pivotal role in integrating external stimuli and stress signaling network in plants (Fujita et al. 2006; Spoel and Loake 2011; Suzuki et al. 2012; Scheibe and Dietz 2012). Therefore, the cell must tightly regulate its ROS levels to avoid cellular damage.

Many reviews have focused on ROS scavenging involved in plant defense against oxidative stress (Smirnoff 2005; Dietz et al. 2006; Pitzschke et al. 2006; Vieira Dos Santos and Rey 2006; Moller et al. 2007; Foyer and Noctor 2009, 2011; Gill and Tuteja 2010; Asensi-Fabado and Munne-Bosch 2010; Farmer and Mueller 2013; Zagorchev et al. 2013). Detailed reviews have focused on ROS describing its metabolism (Apel and Hirt 2004; Noctor et al. 2014), signaling networks (Miller et al. 2010; Suzuki et al. 2012; Baxter et al. 2014), and cross talk with other signaling molecules involved in developmental and stress response processes (Suzuki et al. 2012; Noctor et al. 2014).

Most of the injury to plants due to abiotic stresses is related to ROS-initiated oxidative damage. Hence, enhancing tolerance to several environmental stresses could be achieved through the modulation of gene expression to different ROS-scavenging enzymes ultimately reducing ROS in cells. Nevertheless, to improve tolerance to oxidative stress, plants have been overexpressed for genes encoding either ROS-scavenging enzymes or enzymes modulating the cellular antioxidant capacity, which have been found to be successful (Liu et al. 2013; Zhai et al. 2013; Lee et al. 2007; Diaz-Vivancos et al. 2013; Xu et al. 2014).

In plants, the main sites of ROS production are localized in different cellular compartments: chloroplast, peroxisomes, mitochondria, plasma membrane, and apoplast. In light, chloroplast and peroxisomes are the major source of ROS production, while in the dark, it is the mitochondria (Foyer and Noctor 2003; Moller 2001; Rhoads et al. 2006). This undesirable by-product of cellular metabolism, ROS, is known to play a crucial role in molecular signaling.

The major ROS-producing sites during abiotic stress are the chloroplast, mitochondria, peroxisome, and apoplast (Dietz et al. 2016; Gilroy et al. 2016; Huang et al. 2016; Kerchev et al. 2016; Rodriguez-Serrano et al. 2016; Takagi et al. 2016). The major contributors of ROS are the chloroplast and mitochondria performing normal physiological processes like photosynthesis and respiration. Their rate of ROS production steeply increases under abiotic and biotic stress resulting in oxidative stress (Hossain and Fujita 2013; Hossain et al. 2010, 2013a, b, 2015; Mittler 2002).

11.3 Photosynthesis: Sensor of Environmental Stress

Chloroplast and mitochondrial redox state and ROS metabolism serve as the source of retrograde signals that regulate nuclear gene expression and modulates the acclimation of plants to environmental stimuli (Rhoads and Subbaiah 2007; Pogson et al. 2008; Woodson and Chory 2008; Shapiguzov et al. 2012; Szechyńska-Hebda and Karpiński 2013). Chloroplast retrograde signaling plays a major role in networking signals received from different organelles and ROS, which might be an important mediator role. As observed, chloroplastic ROS production and photosynthetic functions are also regulated by cues perceived by cell wall or apoplastic spaces (Padmanabhan and Dinesh-Kumar 2010). Mitochondrial ROS production is much lower when compared to chloroplast ROS production. However, ROS generated in mitochondria also regulate a number of cellular processes, including stress adaptations and programmed cell death (Hossain et al. 2015; Robson and Vanlerberghe 2002).

Chloroplast is the vital source of ROS, generated during light reaction in PSII and PSI, which is increased during stress also due to limited CO_2 availability and impaired ATP synthesis (Takahashi and Murata 2008; Yamamoto et al. 2008; Nishiyama and Murata 2014; Noctor et al. 2014) leading to oxidative stress (Foyer and Shigeoka 2011). Reduced rate of photosynthetic carbon fixation is commonly observed under various abiotic stresses including salinity, drought, extreme temperature, and heavy metals (Abogadallah 2011; Cruz de Carvalho 2008; Kaushal et al. 2011; Kim and Portis 2004; Sanda et al. 2011; Wise 1995).

Plant stress adaptive mechanism involves the reestablishment of cellular energy balance. Hence, photosynthesis that plays a major role in modulating energy signaling and balance has significant contribution in energy homeostasis of the whole plant. The chloroplast involves a series of photochemical and biochemical reactions that are interconnected with several redox components.

ROS generation in chloroplast by abiotic stresses (drought, high or low temperature, heavy metal toxicity, high light, salinity) results in photoinhibition (Allakhverdiev et al. 2002; Allakhverdiev and Murata 2004; Ohnishi and Murata 2006; Allakhverdiev et al. 2008; Takahashi et al. 2009; Foyer and Noctor 2005). Chloroplast response to variations in environmental stress has been widely studied and reviewed (Biswal et al. 2011; Li et al. 2009; Sage and Kubien 2007; Allakhverdiev et al. 2008; Lawlor 2009; Lawlor and Tezara 2009; Pfannschmidt et al. 2009). Alterations in pigment composition, structural organization, primary photochemistry, and CO_2 fixations all act as stress sensors with PSII and RuBisCO, the major stress sensors (Biswal and Raval 2003; Biswal 2005). The imbalance in energy between source and sink helps plants to regulate stress adaptation (Ensminger et al. 2006; Wilson et al. 2006). Under stress, the energy balance and redox homeostasis between source and sink is perturbed, and the altered redox homeostasis signals readjust to tolerate stress. However, the mechanism of energy sensing has not yet clearly been understood. The plants exhibit dynamic stress adaptation behavior and as much as possible try to maintain optimal photosynthetic efficiency (Biswal et al. 2008).

11.4 Oxidative Stress-Tolerant Plants Are Cross-Protected to Other Abiotic Stresses

Oxidative stress is a highly regulated process, wherein the fate of the plant is determined by the equilibrium between ROS and antioxidative capacity of the plant. Several reports confirmed that enhanced antioxidant defense combats oxidative stress induced by abiotic stressors like salinity (Hasanuzzaman et al. 2011; Hossain et al. 2011), drought (Selote and Khanna-Chopra 2010; Hasanuzzaman and Fujita 2011), heat (Chakraborty and Pradhan 2011; Rani et al. 2013), cold (Zhao et al. 2009; Yang et al. 2011), flooding (Li et al. 2011), heavy metal toxicity (Hossain et al. 2010; Gill et al. 2011), UV radiation (Kumari et al. 2010; Li et al. 2010; Ravindran et al. 2010), and ozone (Yan et al. 2010). Developing plants with higher antioxidative potential provides an opportunity to develop plants with enhanced tolerance to abiotic stresses.

To overcome and withstand oxidative stress, plants have an effective antioxidative system that includes both enzymatic and nonenzymatic compounds to regulate redox homeostasis. The enzymatic antioxidant defense system includes mainly superoxide dismutase (SOD), ascorbate peroxidase (APX), monodehydroascorbate reductase (MDHAR), dehydroascorbate reductase (DHAR), glutathione reductase (GR), catalase (CAT), glutathione peroxidase (GPX), glutathione S-transferase (GST), and peroxidase (POX). The nonenzymatic antioxidant defense system mainly includes ascorbate (AsA), glutathione (GSH), tocopherol, carotenoids, and flavonoids (Apel and Hirt 2004; Hossain and Fujita 2013).

Plants genetically modified to reduce oxidative damage under stress by manipulation of ROS-detoxifying enzymes or nonenzymatic oxidants are reported to be tolerant to a range of abiotic stresses, providing cross-protection. Crop plants which showed improved tolerance to abiotic stresses, which are manipulated for ROS detoxifying and nonenzymatic oxidants, are listed in Table 11.1.

Table 11.1 Crop plants altered with ROS-scavenging antioxidants by transgenic approach showing improved or decreased tolerance to abiotic stresses

Transgenic crop	Gene source	Gene	Approach	Promoter	Improved stress tolerance to	Mechanism of tolerance	Reference
Rice	*Suaeda*	*GST and CAT*	OE	CaMV35s	Oxidative and salt stresses	Increased SOD and CAT activity	Zhao and Zhang (2006)
	Mangrove	*Cu/ZnSOD*	OE	Ubiquitin	Oxidative, salt, and drought stresses	NA	Prashanth et al. (2008)
	Yeast	*MnSOD*	OE	CaMV35s	Oxidative and salt stresses	Higher activity of antioxidant enzymes	Tanaka et al. (1999)
	Suaeda	*GST and CAT*	OE	CaMV35s	Cd, high temperature, and oxidative stress	Higher activity of antioxidant enzymes	Zhao et al. (2009)
	Mustard	*GR*	OE	CaMV35s	Photooxidative stress and heat tolerance in presence of low conc of MV	Protection against photobleaching of chlorophyll	Kouřil et al. (2003)
	Pea	*MnSOD*	OE	Oxidative stress inducible	Oxidative and drought stress	Increased SOD activity	Wang et al. (2005)
Maize	*Arabidopsis*	*FeSOD*	OE	CaMV35s	Oxidative and chilling stresses	Enhanced SOD activity	Van Breusegem et al. (1999a)
	Tobacco	*MnSOD*	OE	CaMV35s	Oxidative and chilling stresses	Higher activity of antioxidant enzymes	Van Breusegem et al. (1999b)

Tomato	Tomato	*MnSOD*	OE	CaMV35s	Oxidative and salt stresses	Higher activity of SOD and APX	Wang et al. (2007)
	Potato	*DHAR*	OE	CaMV35s	Oxidative and salt stresses	Increased ascorbic acid	Li et al. (2012)
	Bacteria	*CAT/KatE*	OE	Rubisco	Oxidative damage caused by drought and chilling stress at high light intensity	Increased catalase activity	Mohamed et al. (2003)
	Tomato	*Cat*	Antisense	CaMV35s	Sensitivity to oxidative and chilling stresses	Decreased catalase activity	Kerdnaimongkol and Woodson (1999)
	Pea	*APX*	OE	CaMV35s	Chilling and salt stress	Increased APX activity	Wang-Yueju et al. (2005)
	Pea	*APX*	OE	CaMV35s	Heat, UV-B, and drought stress	Increased APX activity	Wang et al. (2006)
Potato	Potato	*Cu/ZnSOD and APX*	OE	Oxidative stress inducible	Oxidative and high temperature stresses	Scavenging of ROS at their site of generation and prevent formation of toxic hydroxyl radicals	Tang et al. (2006)
	Tomato	*Cu/ZnSOD*	OE	CaMV35s	Oxidative stress	Enhanced SOD activity	Perl et al. (1993)
Potato	*Arabidopsis*	*AtDHAR1*	OE	CaMV35s	Herbicide, drought, and salt stress	Increased DHAR activity	Eltayeb et al. (2011)
	Rat	*GLOase*	OE	CaMV35s	Salt, mannitol, and methyl viologen	Increased ascorbic acid	Hemavathi et al. (2010)
	Potato	*Cu/ZnSOD*, *APX*, and *NDPK2*	OE	Oxidative stress inducible	Oxidative and high temperature stresses	Higher activity of antioxidant enzymes	Kim et al. (2010)

(continued)

Table 11.1 (continued)

Transgenic crop	Gene source	Gene	Approach	Promoter	Improved stress tolerance to	Mechanism of tolerance	Reference
Sweet potato	Sweet potato and pea	*Cu/ZnSOD* and *APX*	OE	Oxidative stress inducible	Oxidative and chilling stresses	Increased SOD and APX activity	Lim et al. (2007)
Cabbage	NA	*Cu/ZnSOD* and *CAT*	OE	Rubisco	Salt and oxidative stress	Increased APX and GR activity	Tseng et al. (2007)
Sugar beet	Tomato	*Cu/ZnSOD*	OE	CaMV35s	Oxidative agents and fungus	NA	Tertivanidis et al. (2004)
Cassava	Cassava	*Cu/ZnSOD* and *APX*	OE	Vascular specific and CaMV35s	Oxidative and chilling stresses	Higher activity of antioxidant enzymes in transgenic under stress	Xu et al. (2014)
Pearl millet	Pearl millet	*GPx*	OE	CaMV35s	Salinity and drought stresses	Higher activity of antioxidant enzymes	Islam et al. (2015)
Alfalfa	Tobacco	*Mn SOD*	OE	CaMV35s	Freezing and herbicide stresses	Increased SOD activity	McKersie et al. (1993)

11.5 Genetic Engineering for Abiotic Oxidative Stress Tolerance

Transgenic tobacco, rice, wheat, and *Medicago* plants overexpressing isoform of *SOD* were found to be tolerant to multiple abiotic stresses. Transgenic tobacco plants displayed enhanced tolerance to salt, drought, and PEG-induced drought stress (Badawi et al. 2004; Faize et al. 2011). Transgenic rice was tolerant to MV-induced oxidative stress, salt, and drought stress (Prashanth et al. 2008; Wang et al. 2005). Similarly, transgenic wheat and *Medicago* also displayed oxidative stress tolerance along with photooxidative stress tolerance caused by other abiotic stresses (Melchiorre et al. 2008; Rubio et al. 2002). In tomato, overexpression of *SOD* leads to enhanced tolerance to MV-induced oxidative stress and salinity (Wang et al. 2007). Transgenic *Arabidopsis* and cotton overexpressing *SOD* were tolerant to drought (Liu et al. 2013; Zhang et al. 2014).

Overexpression of *APX* in tobacco resulted in enhanced tolerance of plants to salinity, drought, PEG-induced water deficit, osmotic stress, and MV- and paraquat-induced oxidative stress (Fotopoulos et al. 2008; Badawi et al. 2004; Sun et al. 2010; Singh et al. 2014). Similarly, transgenic tomato expressing *APX* displayed enhanced tolerance to drought, UV-B, heat, and chilling stress (Wang-Yueju et al. 2005; Wang et al. 2006). Overexpression of *APX* in *Arabidopsis* and rice resulted in enhanced tolerance to salinity and cold stress, respectively (Sato et al. 2011). Moreover, constitutively expressed cytosolic ascorbate peroxidase 1 (*APX1*) in *Arabidopsis* increased tolerance to heat and drought combined stress (Koussevitzky et al. 2008). Transgenic plants overexpressing *CAT* displayed enhanced tolerance to oxidative stress caused by drought, cold, high light, and salinity (Mohamed et al. 2003; Matsumura et al. 2002). Transgenic studies in tobacco, tomato, rice, and potato overexpressing either *MDHAR* or *DHAR* showed enhanced tolerance to salt, drought, PEG-induced water deficit, ozone, and MV-induced oxidative stress (Eltayeb et al. 2006, 2007, 2011; Kavitha et al. 2010; Li et al. 2010, 2012; Sultana et al. 2012). Likewise, improved oxidative stress tolerance and multiple stress tolerance toward salinity, drought, MV-induced oxidative stress, and cold were also observed in transgenic plants overexpressing other enzymatic antioxidant defense system *GR*, *GPX* or *GST* (Ding et al. 2009; Le Martret et al. 2011; Kouřil et al. 2003; Yoshimura et al. 2004; Gaber et al. 2006; Qi et al. 2010; Ji et al. 2010; Jha et al. 2011; Liu et al. 2013).

Pyramiding two or more genes is an alternative approach to increase tolerance to a desired level (Ahmad et al. 2010; Lee et al. 2007). Transgenic plants co-expressing multiple antioxidant genes (various combinations of two or more genes *SOD*, *APX*, *CAT*, *GST*, or *DHAR*) displayed better oxidative stress tolerance compared to single gene transformants (Ahmad et al. 2010; Faize et al. 2011; Lee et al. 2007; Wang et al. 2011; Zhao and Zhang 2006; Zhao et al. 2009).

Apart from the enzymatic components of the antioxidant system of ascorbate-glutathione (AsA-GSH) pathway, it also includes the nonenzymatic components – AsA and GSH. The response of cells to environmental conditions depends on the altered redox ratios of AsA/DHA or GSH/GSSG (Tausz et al. 2004). Under field conditions, the ratio of AsA/DHA is relatively lower compared to the ratio GSH/GSSG (Noctor and Foyer 1998). Studies have highlighted the importance of high ratio of GSH/GSSG and/or AsA/DHA in abiotic stress tolerance that is achieved by either increased GSH and AsA or diminution of GSSG and DHA (Szalai et al. 2009). Transgenic plants overexpressing the enzymes involved in the biosynthesis of AsA or GSH have been reported to have improved abiotic oxidative stress tolerance (Hemavathi et al. 2010; Lim et al. 2012; Lisko et al. 2013; Liu et al. 2013; Upadhyaya et al. 2011; Zhang et al. 2011).

11.6 Induced Cross-Protection by Chemicals

Chemical priming for crop stress management is a promising and emerging field (Antoniou et al. 2016). Use of chemical agents overcomes the drawback of implementing inherent cross-protection by methodologies that are time-consuming (conventional breeding) and unacceptable by many countries (transgenic approach).

A number of agrochemicals are available that are targeted as defense inducers. The principle action of these chemicals is an initial disruption of ROS homeostasis (induce accumulation of H_2O_2) in cells on application of inducer, followed by triggering enhanced detoxification capacity of the cells. Direct application of ROS by oxidative agents has shown to induce cross-protection against abiotic stresses. Best known among them is the application of H_2O_2 and O_3. Reactive oxygen (H_2O_2), nitrogen (NO), and sulfur (H_2S) species (RONSS) have shown to play an important role in stress acclimation (Savvides et al. 2016). When these are applied at low concentrations, they act as priming agents and enhance abiotic stress tolerance. For example, strawberry, tomato, and soybean plants pretreated with chemical primers displayed enhanced abiotic stress tolerance (Christou et al. 2013, 2014; İşeri et al. 2013; Radhakrishnan and Lee 2013). H_2O_2, a known activator of antioxidant defense, is used for wheat seed pretreatment to sow in an environment that experienced drought and/or salt stress (Wahid et al. 2007; He and Gao 2009). The priming increases the antioxidant capacity that is correlated with increased transcripts of enzymatic antioxidants and or AsA and GSH biosynthesis (Christou et al. 2013, 2014).

Priming with exposure to stress or chemical compounds can also modulate abiotic stress tolerance (Filippou et al. 2013; Hossain and Fujita 2013; Mostofa and Fujita 2013; Borges et al. 2014; Mostofa et al. 2015; Nahar et al. 2015; Wang et al. 2014). Recent studies provide evidence that initial exposure to chemical priming agents (such as H_2O_2, ABA, NO, SA, etc.) renders plants more tolerant to abiotic stresses (Hasanuzzaman et al. 2011; Mostofa and Fujita 2013; Mostofa et al. 2014; Sathiyaraj et al. 2014; Teng et al. 2014). Exogenous application of H_2O_2 induced tolerance to salinity, drought, chilling, and high temperatures and heavy metal stress, all of which cause elevated H_2O_2 production (Gong et al. 2001; Uchida et al. 2002; de Azevedo Neto et al. 2005; Chao et al. 2009; Liu et al. 2010; Wang et al. 2010, 2014; Ishibashi et al. 2011; Gondim et al. 2012, 2013; Hossain and Fujita 2013).

Although these chemical agents are effective at low concentrations, they can be deleterious at higher concentration (Sagor et al. 2013; Sathiyaraj et al. 2014; Bajwa et al. 2014). In addition, exogenous priming agents may have physiological effect on plants. The effects depend on the type of priming agent used, concentration, and frequency of application and method of application, environmental condition, and plant species. Hence, there is a need for development of technologies for appropriate application of appropriate concentration that is plant specific. In addition, their impact on environment (beneficial microorganisms, pollinators, aquatic habitat, etc.) must be studied. Chemical priming agents, as seed treatment is most desirable as it reduces cost of later priming treatment in field conditions.

11.7 Conclusion

Oxidative stress that results due to ROS accumulation accompanies almost all abiotic stresses. Combinations of abiotic stresses result in an even more severe oxidative stress condition. Hence, regulation and control of cellular ROS homeostasis play a primary defense mechanism for enhanced multiple stress tolerance or cross-protection against abiotic stresses. Plants with cross-protection against multiple abiotic stresses provide us with improved agronomic and genetic tool for sustainable agriculture that is climate resilient. For example, the antioxidant enzyme ascorbate peroxidase 1 has been proved to be crucial for drought and heat combined stress tolerance. Genetic engineering of transcription factor provides another opportunity to master regulate many stress-responsive genes. In addition, hormonal cross talk modulates plant responses to multiple stresses, which also opens new vistas for engineering plants tolerant to stress combination.

Acknowledgments The project was supported by the Ministry of Science and Technology (MOST) of Taiwan (Grant no. MOST 104-2313-B-002-013-MY3 and 106-2628-B-002-036-MY3) and National Taiwan University (Grant no. NTU-CDP-106R7721) to C.-Y. Hong.

References

Abogadallah GM (2011) Differential regulation of photorespiratory gene expression by moderate and severe salt and drought stress in relation to oxidative stress. Plant Sci 180(3):540–547

Ahmad R, Kim YH, Kim MD, Kwon SY, Cho K, Lee HS, Kwak SS (2010) Simultaneous expression of choline oxidase, superoxide dismutase and ascorbate peroxidase in potato plant chloroplasts provides synergistically enhanced protection against various abiotic stresses. Physiol Plant 138(4):520–533

Allakhverdiev SI, Murata N (2004) Environmental stress inhibits the synthesis de novo of proteins involved in the photodamage–repair cycle of photosystem II in Synechocystis sp. PCC 6803. Biochimica et Biophysica Acta (BBA)-Bioenergetics 1657(1):23–32

Allakhverdiev SI, Nishiyama Y, Miyairi S, Yamamoto H, Inagaki N, Kanesaki Y, Murata N (2002) Salt stress inhibits the repair of photodamaged photosystem II by suppressing the transcription and translation of psbAGenes in Synechocystis. Plant Physiol 130(3):1443–1453

Allakhverdiev SI, Kreslavski VD, Klimov VV, Los DA, Carpentier R, Mohanty P (2008) Heat stress: an overview of molecular responses in photosynthesis. Photosynth Res 98(1–3):541

Alscher RG, Erturk N, Heath LS (2002) Role of superoxide dismutases (SODs) in controlling oxidative stress in plants. J Exp Bot 53:372.1331

Antoniou C, Savvides A, Christou A, Fotopoulos V (2016) Unravelling chemical priming machinery in plants: the role of reactive oxygen–nitrogen–sulfur species in abiotic stress tolerance enhancement. Curr Opin Plant Biol 33:101–107

Apel K, Hirt H (2004) Reactive oxygen species: metabolism, oxidative stress, and signal transduction. Annu Rev Plant Biol 55:373–399

Asensi-Fabado MA, Munne-Bosch S (2010) Vitamins in plants: occurrence, biosynthesis and antioxidant function. Trends Plant Sci 15(10):582–592

Atkinson NJ, Urwin PE (2012) The interaction of plant biotic and abiotic stresses: from genes to the field. J Exp Bot 63(10):3523–3543

de Azevedo Neto AD, Prisco JT, Enéas-Filho J, Medeiros J-VR, Gomes-Filho E (2005) Hydrogen peroxide pre-treatment induces salt-stress acclimation in maize plants. J Plant Physiol 162(10):1114–1122

Badawi GH, Yamauchi Y, Shimada E, Sasaki R, Kawano N, Tanaka K, Tanaka K (2004) Enhanced tolerance to salt stress and water deficit by overexpressing superoxide dismutase in tobacco (*Nicotiana tabacum*) chloroplasts. Plant Sci 166(4):919–928

Bajwa VS, Shukla MR, Sherif SM, Murch SJ, Saxena PK (2014) Role of melatonin in alleviating cold stress in *Arabidopsis thaliana*. J Pineal Res 56(3):238–245

Baxter A, Mittler R, Suzuki N (2014) ROS as key players in plant stress signalling. J Exp Bot 65(5):1229–1240

Biswal B (2005) Photosynthetic response of green plants to environmental stress. In: Handbook of Photosynthesis, 2nd edn. CRC Press, Boca Raton

Biswal UC, Raval MK (2003) Chloroplast biogenesis: from proplastid to gerontoplast. Springer Science & Business Media

Biswal B, Raval MK, Biswal UC, Joshi P (2008) Response of photosynthetic organelles to abiotic stress: modulation by Sulfur metabolism. In: Khan NA, Singh S, Umar S (eds) Sulfur assimilation and abiotic stress in plants. Springer, Berlin Heidelberg, pp 167–191

Biswal B, Joshi P, Raval M, Biswal U (2011) Photosynthesis, a global sensor of environmental stress in green plants: stress signaling and adaptation. Curr Sci 101(1):47–56

Borges AA, Jiménez-Arias D, Expósito-Rodríguez M, Sandalio LM, Pérez JA (2014) Priming crops against biotic and abiotic stresses: MSB as a tool for studying mechanisms. Front Plant Sci 5:642

Chakraborty U, Pradhan D (2011) High temperature-induced oxidative stress in *Lens culinaris*, role of antioxidants and amelioration of stress by chemical pre-treatments. J Plant Interact 6(1):43–52

Chao Y-Y, Hsu Y, Kao C (2009) Involvement of glutathione in heat shock and hydrogen peroxide induced cadmium tolerance of rice (Oryza sativa L.) seedlings. Plant Soil 318(1–2):37

Christou A, Manganaris GA, Papadopoulos I, Fotopoulos V (2013) Hydrogen sulfide induces systemic tolerance to salinity and non-ionic osmotic stress in strawberry plants through modification of reactive species biosynthesis and transcriptional regulation of multiple defence pathways. J Exp Bot 64(7):1953–1966

Christou A, Filippou P, Manganaris GA, Fotopoulos V (2014) Sodium hydrosulfide induces systemic thermo tolerance to strawberry plants through transcriptional regulation of heat shock proteins and aquaporin. BMC Plant Biol 14(1):42

Cruz de Carvalho MH (2008) Drought stress and reactive oxygen species: production, scavenging and signaling. Plant Signal Behav 3(3):156–165

Demidchik V, Shabala SN, Coutts KB, Tester MA, Davies JM (2003) Free oxygen radicals regulate plasma membrane Ca^{2+} and K^+ permeable channels in plant root cells. J Cell Sci 116(1):81–88

Demidchik V, Essah PA, Tester M (2004) Glutamate activates cation currents in the plasma membrane of Arabidopsis root cells. Planta 219(1):167–175

Demidchik V, Shang Z, Shin R, Thompson E, Rubio L, Laohavisit A, Mortimer JC, Chivasa S, Slabas AR, Glover BJ, Schachtman DP, Shabala SN, Davies JM (2009) Plant extracellular

ATP signalling by plasma membrane NADPH oxidase and Ca^{2+} channels. Plant J: for cell and molecular biology 58(6):903–913

Demidchik V, Cuin TA, Svistunenko D, Smith SJ, Miller AJ, Shabala S, Sokolik A, Yurin V (2010) Arabidopsis root K^+ efflux conductance activated by hydroxyl radicals: single-channel properties, genetic basis and involvement in stress-induced cell death. J Cell Sci 123(9):1468–1479

Diaz-Vivancos P, Faize M, Barba-Espin G, Faize L, Petri C, Hernández JA, Burgos L (2013) Ectopic expression of cytosolic superoxide dismutase and ascorbate peroxidase leads to salt stress tolerance in transgenic plums. Plant Biotechnol J 11(8):976–985

Dietz KJ (2003) Redox control, redox signaling, and redox homeostasis in plant cells. Int Rev Cytol 228:141–193

Dietz KJ, Jacob S, Oelze ML, Laxa M, Tognetti V, de Miranda SM, Baier M, Finkemeier I (2006) The function of peroxiredoxins in plant organelle redox metabolism. J Exp Bot 57(8):1697–1709

Dietz KJ, Turkan I, Krieger-Liszkay A (2016) Redox- and reactive oxygen species-dependent Signaling into and out of the photosynthesizing chloroplast. Plant Physiol 171(3):1541–1550

Ding S, Lu Q, Zhang Y, Yang Z, Wen X, Zhang L, Lu C (2009) Enhanced sensitivity to oxidative stress in transgenic tobacco plants with decreased glutathione reductase activity leads to a decrease in ascorbate pool and ascorbate redox state. Plant Mol Biol 69(5):577–592

Dowling DK, Simmons LW (2009) Reactive oxygen species as universal constraints in life-history evolution. Proc R Soc Lond B Biol Sci 276(1663):1737–1745

Eltayeb AE, Kawano N, Badawi GH, Kaminaka H, Sanekata T, Morishima I, Shibahara T, Inanaga S, Tanaka K (2006) Enhanced tolerance to ozone and drought stresses in transgenic tobacco overexpressing dehydroascorbate reductase in cytosol. Physiol Plant 127(1):57–65

Eltayeb AE, Kawano N, Badawi GH, Kaminaka H, Sanekata T, Shibahara T, Inanaga S, Tanaka K (2007) Overexpression of monodehydroascorbate reductase in transgenic tobacco confers enhanced tolerance to ozone, salt and polyethylene glycol stresses. Planta 225(5):1255–1264

Eltayeb AE, Yamamoto S, Habora MEE, Yin L, Tsujimoto H, Tanaka K (2011) Transgenic potato overexpressing Arabidopsis cytosolic *AtDHAR1* showed higher tolerance to herbicide, drought and salt stresses. Breed Sci 61(1):3–10

Ensminger I, Busch F, Huner N (2006) Photostasis and cold acclimation: sensing low temperature through photosynthesis. Physiol Plant 126(1):28–44

Faize M, Burqos L, Piqueras A, Nicolas E, Barba-Espin G, Clement-Moreno MJ, Alcobendas R, Artlip T, Hernandez JA (2011) Involvement of cytosolic ascorbate peroxidase and Cu/Zn-superoxide dismutase for improved tolerance against drought stress. J Exp Bot 62(8):2599–2613

Farmer EE, Mueller MJ (2013) ROS-mediated lipid peroxidation and RES-activated signaling. Annu Rev Plant Biol 64:429–450

Filippou P, Tanou G, Molassiotis A, Fotopoulos V (2013) Plant acclimation to environmental stress using priming agents. In: Plant Acclimation to Environmental Stress. Springer, New York, NY, pp. 1–27

Foreman J, Demidchik V, Bothwell JH, Mylona P, Miedema H, Torres MA, Linstead P, Costa S, Brownlee C, Jones JD (2003) Reactive oxygen species produced by NADPH oxidase regulate plant cell growth. Nature 422(6930):442–446

Fotopoulos V, De Tullio MC, Barnes J, Kanellis AK (2008) Altered stomatal dynamics in ascorbate oxidase over-expressing tobacco plants suggest a role for dehydroascorbate signaling. J Exp Bot 59(4):729–737

Foyer CH, Noctor G (2003) Redox sensing and signaling associated with reactive oxygen in chloroplasts, peroxisomes and mitochondria. Physiol Plant 119(3):355–364

Foyer CH, Noctor G (2005) Oxidant and antioxidant signaling in plants: a re-evaluation of the concept of oxidative stress in a physiological context. Plant Cell Environ 28(8):1056–1071

Foyer CH, Noctor G (2009) Redox regulation in photosynthetic organisms: signaling, acclimation, and practical implications. Antioxid Redox Signal 11(4):861–905

Foyer CH, Noctor G (2011) Ascorbate and glutathione: the heart of the redox hub. Plant Physiol 155(1):2–18

Foyer CH, Shigeoka S (2011) Understanding oxidative stress and antioxidant functions to enhance photosynthesis. Plant Physiol 155(1):93–100

Fujita M, Fujita Y, Noutoshi Y, Takahashi F, Narusaka Y, Yamaguchi-Shinozaki K, Shinozaki K (2006) Crosstalk between abiotic and biotic stress responses: a current view from the points of convergence in the stress signaling networks. Curr Opin Plant Biol 9(4):436–442

Gaber A, Yoshimura K, Yamamoto T, Yabuta Y, Takeda T, Miyasaka H, Nakano Y, Shigeoka S (2006) Glutathione peroxidase-like protein of Synechocystis PCC 6803 confers tolerance to oxidative and environmental stresses in transgenic Arabidopsis. Physiol Plant 128(2):251–262

Gill SS, Tuteja N (2010) Reactive oxygen species and antioxidant machinery in abiotic stress tolerance in crop plants. Plant Physiol Biochem: PPB 48(12):909–930

Gill SS, Khan NA, Tuteja N (2011) Differential cadmium stress tolerance in five Indian mustard (*Brassica juncea* L.) cultivars: an evaluation of the role of antioxidant machinery. Plant Signal Behav 6(2):293–300

Gilliham M, Able JA, Roy SJ (2017) Translating knowledge about abiotic stress tolerance to breeding programmes. Plant J 90(5):898–917

Gilroy S, Bialasek M, Suzuki N, Gorecka M, Devireddy AR, Karpinski S, Mittler R (2016) ROS, calcium, and electric signals: key mediators of rapid systemic signaling in plants. Plant Physiol 171(3):1606–1615

Gondim FA, Gomes-Filho E, Costa JH, Alencar NLM, Prisco JT (2012) Catalase plays a key role in salt stress acclimation induced by hydrogen peroxide pretreatment in maize. Plant Physiol Biochem 56:62–71

Gondim FA, de Souza Miranda R, Gomes-Filho E, Prisco JT (2013) Enhanced salt tolerance in maize plants induced by H2O2 leaf spraying is associated with improved gas exchange rather than with non-enzymatic antioxidant system. Theor Exp Plant Physiol 25(4):251–260

Gong M, Chen B, Li Z-G, Guo L-H (2001) Heat-shock-induced cross adaptation to heat, chilling, drought and salt stress in maize seedlings and involvement of H_2O_2. J Plant Physiol 158(9):1125–1130

Hasanuzzaman M, Fujita M (2011) Selenium pretreatment upregulates the antioxidant defense and methylglyoxal detoxification system and confers enhanced tolerance to drought stress in rapeseed seedlings. Biol Trace Elem Res 143(3):1758–1776

Hasanuzzaman M, Hossain MA, Fujita M (2011) Selenium-induced up-regulation of the antioxidant defense and methylglyoxal detoxification system reduces salinity-induced damage in rapeseed seedlings. Biol Trace Elem Res 143(3):1704–1721

He L, Gao Z (2009) Pretreatment of seed with H_2O_2 enhances drought tolerance of wheat (*Triticum aestivum* L.) seedlings. Afr J Biotechnol 8(22):6151–6157

Hemavathi, Upadhyaya CP, Akula N, Young KE, Chun SC, Kim DH, Park SW (2010) Enhanced ascorbic acid accumulation in transgenic potato confers tolerance to various abiotic stresses. Biotechnol Lett 32(2):321–330

Hossain MA, Fujita M (2013) Hydrogen Peroxide priming stimulates drought tolerance in mustard (*Brassica juncea* L.) seedlings. Plant Gene Trait 4(1):109–123

Hossain MA, Hasanuzzaman M, Fujita M (2010) Up-regulation of antioxidant and glyoxalase systems by exogenous glycinebetaine and proline in mung bean confer tolerance to cadmium stress. Physiol Mol Biol Plants 16(3):259–272

Hossain MA, Hasanuzzaman M, Fujita M (2011) Coordinate induction of antioxidant defense and glyoxalase system by exogenous proline and glycinebetaine is correlated with salt tolerance in mung bean. Front Agric China 5(1):1–14

Hossain MA, Mostofa MG, Fujita M (2013a) Cross protection by cold-shock to salinity and drought stress-induced oxidative stress in mustard (*Brassica campestris* L.) seedlings. Mol Plant Breed 4(7):50–70

Hossain MA, Mostofa MG, Fujita M (2013b) Heat-shock positively modulates oxidative protection of salt and drought-stressed mustard (*Brassica campestris* L.) seedlings. J Plant Sci Mol Breed 2(1):2

Hossain MA, Bhattacharjee S, Armin SM, Qian P, Xin W, Li HY, Burritt DJ, Fujita M, Tran LS (2015) Hydrogen peroxide priming modulates abiotic oxidative stress tolerance: insights from ROS detoxification and scavenging. Front Plant Sci 6:420

Huang S, Van Aken O, Schwarzlander M, Belt K, Millar AH (2016) The roles of mitochondrial reactive oxygen species in cellular signaling and stress response in plants. Plant Physiol 171(3):1551–1559

İşeri ÖD, Körpe DA, Sahin FI, Haberal M (2013) Hydrogen peroxide pretreatment of roots enhanced oxidative stress response of tomato under cold stress. Acta Physiol Plant 35(6):1905–1913

Ishibashi Y, Yamaguchi H, Yuasa T, Iwaya-Inoue M, Arima S, Zheng S-H (2011) Hydrogen peroxide spraying alleviates drought stress in soybean plants. J Plant Physiol 168(13):1562–1567

Islam T, Manna M, Reddy MK (2015) Glutathione Peroxidase of *Pennisetum glaucum* (PgGPx) is a functional Cd 2+ dependent peroxiredoxin that enhances tolerance against salinity and drought stress. PLoS One 10(11):e0143344

Jha B, Sharma A, Mishra A (2011) Expression of SbGSTU (tau class glutathione S-transferase) gene isolated from *Salicornia brachiata* in tobacco for salt tolerance. Mol Biol Rep 38(7):4823–4832

Ji W, Zhu Y, Li Y, Yang L, Zhao X, Cai H, Bai X (2010) Over-expression of a glutathione S-transferase gene, *GsGST*, from wild soybean (*Glycine soja*) enhances drought and salt tolerance in transgenic tobacco. Biotechnol Lett 32(8):1173–1179

Kaushal N, Gupta K, Bhandhari K, Kumar S, Thakur P, Nayyar H (2011) Proline induces heat tolerance in chickpea (Cicer arietinum L.) plants by protecting vital enzymes of carbon and antioxidative metabolism. Physiol Mol Biol Plants 17(3):203

Kavitha K, George S, Venkataraman G, Parida A (2010) A salt-inducible chloroplastic monodehydroascorbate reductase from halophyte *Avicennia marina* confers salt stress tolerance on transgenic plants. Biochimie 92(10):1321–1329

Kerchev P, Waszczak C, Lewandowska A, Willems P, Shapiguzov A, Li Z, Alseekh S, Muhlenbock P, Hoeberichts FA, Huang J, Van Der Kelen K, Kangasjarvi J, Fernie AR, De Smet R, Van de Peer Y, Messens J, Van Breusegem F (2016) Lack of GLYCOLATE OXIDASE1, but not GLYCOLATE OXIDASE2, attenuates the Photorespiratory phenotype of CATALASE2-deficient Arabidopsis. Plant Physiol 171(3):1704–1719

Kerdnaimongkol K, Woodson WR (1999) Inhibition of catalase by antisense RNA increases susceptibility to oxidative stress and chilling injury in transgenic tomato plants. J Am Soc Hortic Sci 124(4):330–336

Kim K, Portis AR (2004) Oxygen-dependent H_2O_2 production by Rubisco. FEBS Lett 571(1–3):124–128

Kim MD, Kim YH, Kwon SY, Yun DJ, Kwak SS, Lee HS (2010) Enhanced tolerance to methyl viologen-induced oxidative stress and high temperature in transgenic potato plants overexpressing the *CuZnSOD*, *APX* and *NDPK2* genes. Physiol Plant 140(2):153–162

Kissoudis C, van de Wiel C, Visser RG, van der Linden G (2014) Enhancing crop resilience to combined abiotic and biotic stress through the dissection of physiological and molecular crosstalk. Front Plant Sci 5:207

Kouřil R, Lazár D, Lee H, Jo J, Nauš J (2003) Moderately elevated temperature eliminates resistance of Rice plants with enhanced expression of glutathione reductase to intensive photooxidative stress. Photosynthetica 41(4):571–578

Koussevitzky S, Suzuki N, Huntington S, Armijo L, Sha W, Cortes D, Shulaev V, Mittler R (2008) Ascorbate peroxidase 1 plays a key role in the response of *Arabidopsis thaliana* to stress combination. J Biol Chem 283(49):34197–34203

Krishnamurthy A, Rathinasabapathi B (2013) Oxidative stress tolerance in plants: novel interplay between auxin and reactive oxygen species signaling. Plant Signal Behav 8(10):e25761

Kumari R, Singh S, Agrawal S (2010) Response of ultraviolet-B induced antioxidant defense system in a medicinal plant, *Acorus calamus*. J Environ Biol 31(6):907–911

Lawlor DW (2009) Musings about the effects of environment on photosynthesis. Ann Bot 103(4):543–549

Lawlor DW, Tezara W (2009) Causes of decreased photosynthetic rate and metabolic capacity in water-deficient leaf cells: a critical evaluation of mechanisms and integration of processes. Ann Bot 103(4):561–579

Le Martret B, Poage M, Shiel K, Nugent GD, Dix PJ (2011) Tobacco chloroplast transformants expressing genes encoding dehydroascorbate reductase, glutathione reductase, and glutathione-S-transferase, exhibit altered anti-oxidant metabolism and improved abiotic stress tolerance. Plant Biotechnol J 9(6):661–673

Lee Y-P, Kim S-H, Bang J-W, Lee H-S, Kwak S-S, Kwon S-Y (2007) Enhanced tolerance to oxidative stress in transgenic tobacco plants expressing three antioxidant enzymes in chloroplasts. Plant Cell Rep 26(5):591–598

Li Z, Wakao S, Fischer BB, Niyogi KK (2009) Sensing and responding to excess light. Annu Rev Plant Biol 60:239–260

Li F, Wu QY, Sun YL, Wang LY, Yang XH, Meng QW (2010) Overexpression of chloroplastic monodehydroascorbate reductase enhanced tolerance to temperature and methyl viologen-mediated oxidative stresses. Physiol Plant 139(4):421–434

Li C, Jiang D, Wollenweber B, Li Y, Dai T, Cao W (2011) Waterlogging pretreatment during vegetative growth improves tolerance to waterlogging after anthesis in wheat. Plant Sci 180(5):672–678

Li Q, Li Y, Li C, Yu X (2012) Enhanced ascorbic acid accumulation through overexpression of dehydroascorbate reductase confers tolerance to methyl viologen and salt stresses in tomato. Czech J Genet Plant Breed 48:74–86

Lim S, Kim Y-H, Kim S-H, Kwon S-Y, Lee H-S, Kim J-S, Cho K-Y, Paek K-Y, Kwak S-S (2007) Enhanced tolerance of transgenic sweetpotato plants that express both CuZnSOD and APX in chloroplasts to methyl viologen-mediated oxidative stress and chilling. Mol Breed 19(3):227–239

Lim MY, Pulla RK, Park JM, Harn CH, Jeong BR (2012) Over-expression of L-gulono-γ-lactone oxidase (GLOase) gene leads to ascorbate accumulation with enhanced abiotic stress tolerance in tomato. In Vitro Cell Dev Biol Plant 48(5):453–461

Lisko KA, Torres R, Harris RS, Belisle M, Vaughan MM, Jullian B, Chevone BI, Mendes P, Nessler CL, Lorence A (2013) Elevating vitamin C content via overexpression of myo-inositol oxygenase and L-gulono-1, 4-lactone oxidase in Arabidopsis leads to enhanced biomass and tolerance to abiotic stresses. In Vitro Cell Dev Biol Plant 49(6):643–655

Liu Z-J, Guo Y-K, Bai J-G (2010) Exogenous hydrogen peroxide changes antioxidant enzyme activity and protects ultrastructure in leaves of two cucumber ecotypes under osmotic stress. J Plant Growth Regul 29(2):171–183

Liu X-F, Sun W-M, Li Z-Q, Bai R-X, Li J-X, Shi Z-H, Hongwei G, Zheng Y, Zhang J, Zhang G-F (2013) Over-expression of *ScMnSOD*, a SOD gene derived from Jojoba, improve drought tolerance in Arabidopsis. J Integr Agric 12(10):1722–1730

Matsumura T, Tabayashi N, Kamagata Y, Souma C, Saruyama H (2002) Wheat catalase expressed in transgenic rice can improve tolerance against low temperature stress. Physiol Plant 116(3):317–327

McKersie BD, Chen Y, de Beus M, Bowley SR, Bowler C, Inze D, D'Halluin K, Botterman J (1993) Superoxide Dismutase Enhances Tolerance of Freezing Stress in Transgenic Alfalfa (*Medicago sativa* L.). Plant Physiol 103(4):1155–1163

Melchiorre M, Robert G, Trippi V, Racca R, Lascano HR (2008) Superoxide dismutase and glutathione reductase overexpression in wheat protoplast: photooxidative stress tolerance and changes in cellular redox state. Plant Growth Regul 57(1):57–68

Miller G, Suzuki N, CIFTCI-YILMAZ S, Mittler R (2010) Reactive oxygen species homeostasis and signalling during drought and salinity stresses. Plant Cell Environ 33(4):453–467

Mittler R (2002) Oxidative stress, antioxidants and stress tolerance. Trends Plant Sci 7(9):405–410

Mittler R (2006) Abiotic stress, the field environment and stress combination. Trends Plant Sci 11(1):15–19

Mittler R, Blumwald E (2010) Genetic engineering for modern agriculture: challenges and perspectives. Annu Rev Plant Biol 61:443–462

Mittler R, Vanderauwera S, Gollery M, Van Breusegem F (2004) Reactive oxygen gene network of plants. Trends Plant Sci 9(10):490–498

Mohamed EA, Iwaki T, Munir I, Tamoi M, Shigeoka S, Wadano A (2003) Overexpression of bacterial catalase in tomato leaf chloroplasts enhances photo-oxidative stress tolerance. Plant Cell Environ 26(12):2037–2046

Mostofa MG, Seraj ZI, Fujita M (2014) Exogenous sodium nitroprusside and glutathione alleviate copper toxicity by reducing copper uptake and oxidative damage in rice (*Oryza sativa* L.) seedlings. Protoplasma 251(6):1373–1386

Moller IM (2001) PLANT MITOCHONDRIA AND OXIDATIVE STRESS: electron transport, NADPH turnover, and metabolism of reactive oxygen species. Annu Rev Plant Physiol Plant Mol Biol 52:561–591

Moller IM, Jensen PE, Hansson A (2007) Oxidative modifications to cellular components in plants. Annu Rev Plant Biol 58:459–481

Mostofa MG, Fujita M (2013) Salicylic acid alleviates copper toxicity in rice (Oryza sativa L.) seedlings by up-regulating antioxidative and glyoxalase systems. Ecotoxicology 22(6):959–973

Mostofa MG, Hossain MA, Fujita M (2015) Trehalose pretreatment induces salt tolerance in rice (Oryza sativa L.) seedlings: oxidative damage and co-induction of antioxidant defense and glyoxalase systems. Protoplasma 252(2):461–475

Murata Y, Pei Z.-M, Mori IC, Schroeder J (2001) Abscisic acid activation of plasma membrane Ca^{2+} channels in guard cells requires cytosolic NAD(P)H and is differentially disrupted upstream and downstream of reactive oxygen species production in abi1-1 and abi2-1 protein phosphatase 2C Mutants. The Plant Cell 13(11):2513

Nahar K, Hasanuzzaman M, Alam MM, Fujita M (2015) Exogenous glutathione confers high temperature stress tolerance in mung bean (Vigna radiata L.) by modulating antioxidant defense and methylglyoxal detoxification system. Environ Exp Bot 112:44–54

Nishiyama Y, Murata N (2014) Revised scheme for the mechanism of photoinhibition and its application to enhance the abiotic stress tolerance of the photosynthetic machinery. Appl Microbiol Biotechnol 98(21):8777–8796

Noctor G, Foyer CH (1998) Simultaneous measurement of foliar glutathione, gamma-glutamylcysteine, and amino acids by high-performance liquid chromatography: comparison with two other assay methods for glutathione. Anal Biochem 264(1):98–110

Noctor G, Mhamdi A, Foyer CH (2014) The roles of reactive oxygen metabolism in drought: not so cut and dried. Plant Physiol 164(4):1636–1648

Ohnishi N, Murata N (2006) Glycinebetaine counteracts the inhibitory effects of salt stress on the degradation and synthesis of D1 protein during photoinhibition in Synechococcus sp. PCC 7942. Plant Physiol 141(2):758–765

Padmanabhan MS, Dinesh-Kumar S (2010) All hands on deck—the role of chloroplasts, endoplasmic reticulum, and the nucleus in driving plant innate immunity. Mol Plant-Microbe Interact 23(11):1368–1380

Pandey P, Irulappan V, Bagavathiannan MV, Senthil-Kumar M (2017) Impact of combined abiotic and biotic stresses on plant growth and avenues for crop improvement by exploiting physio-morphological traits. Front Plant Sci 8(537)

Pastori GM, Foyer CH (2002) Common components, networks, and pathways of cross-tolerance to stress. The central role of "redox" and abscisic acid-mediated controls. Plant Physiol 129(2):460–468

Perl A, Perl-Treves R, Galili S, Aviv D, Shalgi E, Malkin S, Galun E (1993) Enhanced oxidative-stress defense in transgenic potato expressing tomato Cu,Zn superoxide dismutases. Theor Appl Genet 85(5):568–576

Pfannschmidt T, Bräutigam K, Wagner R, Dietzel L, Schröter Y, Steiner S, Nykytenko A (2009) Potential regulation of gene expression in photosynthetic cells by redox and energy state: approaches towards better understanding. Ann Bot 103(4):599–607

Pitzschke A, Forzani C, Hirt H (2006) Reactive oxygen species signaling in plants. Antioxid Redox Signal 8(9–10):1757–1764

Pogson BJ, Woo NS, Förster B, Small ID (2008) Plastid signaling to the nucleus and beyond. Trends Plant Sci 13(11):602–609

Prashanth S, Sadhasivam V, Parida A (2008) Over expression of cytosolic copper/zinc superoxide dismutase from a mangrove plant *Avicennia marina* in indica rice var Pusa Basmati-1 confers abiotic stress tolerance. Transgenic Res 17(2):281–291

Qi YC, Liu WQ, Qiu LY, Zhang SM, Ma L, Zhang H (2010) Overexpression of glutathione S-transferase gene increases salt tolerance of arabidopsis. Russ J Plant Physiol 57(2):233–240

Radhakrishnan R, Lee I-J (2013) Spermine promotes acclimation to osmotic stress by modifying antioxidant, abscisic acid, and jasmonic acid signals in soybean. J Plant Growth Regul 32(1):22–30

Rani B, Dhawan K, Jain V, Chhabra ML, Singh D (2013) High temperature induced changes in antioxidative enzymes in Brassica juncea (L) Czern & Coss. Recuperado el, 12.

Ramegowda V, Senthil-Kumar M (2015) The interactive effects of simultaneous biotic and abiotic stresses on plants: mechanistic understanding from drought and pathogen combination. J Plant Physiol 176:47–54

Ravindran K, Indrajith A, Pratheesh P, Sanjiviraja K, Balakrishnan V (2010) Effect of ultraviolet-B radiation on biochemical and antioxidant defence system in *Indigofera tinctoria* L. seedlings. Int J Eng Sci Technol 2(5):226–232

Rhoads DM, Subbaiah CC (2007) Mitochondrial retrograde regulation in plants. Mitochondrion 7(3):177–194

Rhoads DM, Umbach AL, Subbaiah CC, Siedow JN (2006) Mitochondrial reactive oxygen species. Contribution to oxidative stress and inter-organellar signaling. Plant Physiol 141(2):357–366

Rizhsky L, Liang H, Shuman J, Shulaev V, Davletova S, Mittler R (2004) When defense pathways collide. The response of Arabidopsis to a combination of drought and heat stress. Plant Physiol 134(4):1683–1696

Robson CA, Vanlerberghe GC (2002) Transgenic plant cells lacking mitochondrial alternative oxidase have increased susceptibility to mitochondria-dependent and-independent pathways of programmed cell death. Plant Physiol 129(4):1908–1920

Robinson JM, Bunce JA (2000) Influence of drought-induced water stress on soybean and spinach leaf ascorbate-dehydroascorbate level and redox status. Int J Plant Sci 161(2):271–279

Rodriguez-Serrano M, Romero-Puertas MC, Sanz-Fernandez M, Hu J, Sandalio LM (2016) Peroxisomes extend peroxules in a fast response to stress via a reactive oxygen species-mediated induction of the peroxin PEX11a. Plant Physiol 171(3):1665–1674

Rubio MC, Gonzalez EM, Minchin FR, Webb KJ, Arrese-Igor C, Ramos J, Becana M (2002) Effects of water stress on antioxidant enzymes of leaves and nodules of transgenic alfalfa overexpressing superoxide dismutases. Physiol Plant 115(4):531–540

Sage RF, Kubien DS (2007) The temperature response of C3 and C4 photosynthesis. Plant Cell Environ 30(9):1086–1106

Sagor G, Berberich T, Takahashi Y, Niitsu M, Kusano T (2013) The polyamine spermine protects Arabidopsis from heat stress-induced damage by increasing expression of heat shock-related genes. Transgenic Res 22(3):595–605

Sanda S, Yoshida K, Kuwano M, Kawamura T, Munekage YN, Akashi K, Yokota A (2011) Responses of the photosynthetic electron transport system to excess light energy caused by water deficit in wild watermelon. Physiol Plant 142(3):247–264

Sathiyaraj G, Srinivasan S, Kim Y-J, Lee OR, Parvin S, Balusamy SRD, Khorolragchaa A, Yang DC (2014) Acclimation of hydrogen peroxide enhances salt tolerance by activating defense-related proteins in Panax ginseng CA Meyer. Mol Biol Rep 41(6):3761–3771

Sato Y, Masuta Y, Saito K, Murayama S, Ozawa K (2011) Enhanced chilling tolerance at the booting stage in rice by transgenic overexpression of the ascorbate peroxidase gene, OsAPXa. Plant Cell Rep 30(3):399–406

Savvides A, Ali S, Tester M, Fotopoulos V (2016) Chemical priming of plants against multiple abiotic stresses: mission possible? Trends Plant Sci 21(4):329–340

Scheibe R, Dietz K-J (2012) Reduction–oxidation network for flexible adjustment of cellular metabolism in photoautotrophic cells. Plant Cell Environ 35(2):202–216

Selote DS, Khanna-Chopra R (2010) Antioxidant response of wheat roots to drought acclimation. Protoplasma 245(1–4):153–163

Shapiguzov A, Vainonen J, Wrzaczek M, Kangasjärvi J (2012) ROS-talk–how the apoplast, the chloroplast, and the nucleus get the message through. Front Plant Sci 3:292

Singh N, Mishra A, Jha B (2014) Over-expression of the peroxisomal ascorbate peroxidase (SbpAPX) gene cloned from halophyte *Salicornia brachiata* confers salt and drought stress tolerance in transgenic tobacco. Mar Biotechnol 16(3):321–332

Smirnoff N (2005) Ascorbate, tocopherol and carotenoids: metabolism, pathway engineering and functions. In: Antioxidants and Reactive Oxygen Species in Plants. Blackwell Publishing Ltd., Oxford, pp 53–86

Spoel SH, Loake GJ (2011) Redox-based protein modifications: the missing link in plant immune signalling. Curr Opin Plant Biol 14(4):358–364

Sreenivasulu N, Sopory S, Kishor PK (2007) Deciphering the regulatory mechanisms of abiotic stress tolerance in plants by genomic approaches. Gene 388(1):1–13

Sreenivasulu N, Harshavardhan VT, Govind G, Seiler C, Kohli A (2012) Contrapuntal role of ABA: does it mediate stress tolerance or plant growth retardation under long-term drought stress? Gene 506(2):265–273

Sultana S, Khew C-Y, Morshed MM, Namasivayam P, Napis S, Ho C-L (2012) Overexpression of monodehydroascorbate reductase from a mangrove plant (AeMDHAR) confers salt tolerance on rice. J Plant Physiol 169(3):311–318

Sun WH, Duan M, Shu DF, Yang S, Meng QW (2010) Over-expression of *StAPX* in tobacco improves seed germination and increases early seedling tolerance to salinity and osmotic stresses. Plant Cell Rep 29(8):917–926

Suzuki N, Koussevitzky S, Mittler R, Miller G (2012) ROS and redox signaling in the response of plants to abiotic stress. Plant Cell Environ 35(2):259–270

Suzuki N, Rivero RM, Shulaev V, Blumwald E, Mittler R (2014) Abiotic and biotic stress combinations. New Phytol 203(1):32–43

Szalai G, Kellős T, Galiba G, Kocsy G (2009) Glutathione as an antioxidant and regulatory molecule in plants under abiotic stress conditions. J Plant Growth Regul 28(1):66–80

Szechyńska-Hebda M, Karpiński S (2013) Light intensity-dependent retrograde signaling in higher plants. J Plant Physiol 170(17):1501–1516

Taiz L (2013) Agriculture, plant physiology, and human population growth: past, present, and future. Theoretical and experimental. Plant Physiol 25(3):167–181

Takagi D, Takumi S, Hashiguchi M, Sejima T, Miyake C (2016) Superoxide and singlet oxygen produced within the thylakoid membranes both cause photosystem I photoinhibition. Plant Physiol 171(3):1626–1634

Takahashi S, Murata N (2008) How do environmental stresses accelerate photoinhibition? Trends Plant Sci 13(4):178–182

Takahashi S, Milward SE, Fan D-Y, Chow WS, Badger MR (2009) How does cyclic electron flow alleviate photoinhibition in Arabidopsis? Plant Physiol 149(3):1560–1567

Tanaka Y, Hibino T, Hayashi Y, Tanaka A, Kishitani S, Takabe T, Yokota S (1999) Salt tolerance of transgenic rice overexpressing yeast mitochondrial Mn-SOD in chloroplasts. Plant Sci 148(2):131–138

Tang L, Kwon SY, Kim SH, Kim JS, Choi JS, Sung CK, Kwak SS, Lee HS (2006) Enhanced tolerance of transgenic potato plants expressing both superoxide dismutase and ascorbate peroxidase in chloroplasts against oxidative stress and high temperature. Plant Cell Rep 25(12):1380–1386

Tausz M, Šircelj H, Grill D (2004) The glutathione system as a stress marker in plant ecophysiology: is a stress-response concept valid? J Exp Bot 55(404):1955–1962

Teng K, Li J, Liu L, Han Y, Du Y, Zhang J, Sun H, Zhao Q (2014) Exogenous ABA induces drought tolerance in upland rice: the role of chloroplast and ABA biosynthesis-related gene expression on photosystem II during PEG stress. Acta Physiologiae Plantarum 36(8):2219–2227

Tertivanidis K, Goudoula C, Vasilikiotis C, Hassiotou E, Perl-Treves R, Tsaftaris A (2004) Superoxide dismutase transgenes in sugarbeets confer resistance to oxidative agents and the fungus *C. beticola*. Transgenic Res 13(3):225–233

Tseng MJ, Liu C-W, Yiu J-C (2007) Enhanced tolerance to sulfur dioxide and salt stress of transgenic Chinese cabbage plants expressing both superoxide dismutase and catalase in chloroplasts. Plant Physiol Biochem 45(10):822–833

Uchida A, Jagendorf AT, Hibino T, Takabe T, Takabe T (2002) Effects of hydrogen peroxide and nitric oxide on both salt and heat stress tolerance in rice. Plant Sci 163(3):515–523

Upadhyaya CP, Hossain MA (2016) Transgenic plants for higher antioxidant content and drought stress tolerance. In: Drought stress tolerance in plants, vol 2. Springer, Cham, pp 473–511

Upadhyaya CP, Venkatesh J, Gururani MA, Asnin L, Sharma K, Ajappala H, Park SW (2011) Transgenic potato overproducing L-ascorbic acid resisted an increase in methylglyoxal under salinity stress via maintaining higher reduced glutathione level and glyoxalase enzyme activity. Biotechnol Lett 33(11):2297

Van Breusegem F, Slooten L, Stassart JM, Moens T, Botterman J, Van Montagu M, Inze D (1999a) Overproduction of *Arabidopsis thaliana FeSOD* confers oxidative stress tolerance to transgenic maize. Plant Cell Physiol 40(5):515–523

Van Breusegem F, Slooten L, Stassart J-M, Botterman J, Moens T, Van Montagu M, Inzé D (1999b) Effects of overproduction of tobacco *MnSOD* in maize chloroplasts on foliar tolerance to cold and oxidative stress. J Exp Bot:71–78

Vieira Dos Santos C, Rey P (2006) Plant thioredoxins are key actors in the oxidative stress response. Trends Plant Sci 11(7):329–334

Wahid A, Perveen M, Gelani S, Basra SM (2007) Pretreatment of seed with H 2 O 2 improves salt tolerance of wheat seedlings by alleviation of oxidative damage and expression of stress proteins. J Plant Physiol 164(3):283–294

Wang FZ, Wang QB, Kwon SY, Kwak SS, Su WA (2005) Enhanced drought tolerance of transgenic rice plants expressing a pea manganese superoxide dismutase. J Plant Physiol 162(4):465–472

Wang-Yueju, Wisniewski M, Meilan R, Cui M, Webb R, Fuchigami L (2005) Overexpression of cytosolic ascorbate peroxidase in tomato confers tolerance to chilling and salt stress. J Am Soc Hortic Sci 130(2):167–173

Wang Y, Wisniewski M, Meilan R, Cui M, Fuchigami L (2006) Transgenic tomato (*Lycopersicon esculentum*) overexpressing cAPX exhibits enhanced tolerance to UV-B and heat stress. J Appl Hortic 8:87–90

Wang Y, Wisniewski M, Meilan R, Uratsu SL, Cui M, Dandekar A, Fuchigami L (2007) Ectopic expression of Mn-SOD in *Lycopersicon esculentum* leads to enhanced tolerance to salt and oxidative stress. J Appl Hortic 9:3–8

Wang Y, Li J, Wang J, Li Z (2010) Exogenous H_2O_2 improves the chilling tolerance of manila grass and mascarene grass by activating the antioxidative system. Plant Growth Regul 61(2):195–204

Wang X, Guo X, Li Q, Tang Z, Kwak S, Ma D (2011) Studies on salt tolerance of transgenic sweetpotato which harbors two genes expressing CuZn superoxide dismutase and ascorbate peroxidase with the stress-inducible SWPA2 promoter. Plant Gene Trait 3(2):6–12

Wang Y, Zhang J, Li J-L, Ma X-R (2014) Exogenous hydrogen peroxide enhanced the thermotolerance of *Festuca arundinacea* and *Lolium perenne* by increasing the antioxidative capacity. Acta Physiol Plant 36(11):2915–2924

Wani SH, Sah SK, Hossain MA, Kumar V, Balachandran SM (2016) Transgenic approaches for abiotic stress tolerance in crop plants. In: Al-Khayri JM, Jain SM, Johnson DV (eds) Advances in plant breeding strategies: agronomic, abiotic and biotic stress traits. Springer, Cham, pp 345–396

Wilson KE, Ivanov AG, Öquist G, Grodzinski B, Sarhan F, Huner NP (2006) Energy balance, organellar redox status, and acclimation to environmental stress. Botany 84(9):1355–1370

Wise RR (1995) Chilling-enhanced photooxidation: the production, action and study of reactive oxygen species produced during chilling in the light. Photosynth Res 45(2):79–97

Woodson JD, Chory J (2008) Coordination of gene expression between organellar and nuclear genomes. Nat Rev Genet 9(5):383–395

Xu J, Yang J, Duan X, Jiang Y, Zhang P (2014) Increased expression of native cytosolic Cu/Zn superoxide dismutase and ascorbate peroxidase improves tolerance to oxidative and chilling stresses in cassava (*Manihot esculenta* Crantz). BMC Plant Biol 14(1):208

Yamamoto Y, Aminaka R, Yoshioka M, Khatoon M, Komayama K, Takenaka D, Yamashita A, Nijo N, Inagawa K, Morita N (2008) Quality control of photosystem II: impact of light and heat stresses. Photosynth Res 98(1–3):589–608

Yan K, Chen W, He X, Zhang G, Xu S, Wang L (2010) Responses of photosynthesis, lipid peroxidation and antioxidant system in leaves of *Quercus mongolica* to elevated O_3. Environ Exp Bot 69(2):198–204

Yang H, Wu F, Cheng J (2011) Reduced chilling injury in cucumber by nitric oxide and the antioxidant response. Food Chem 127(3):1237–1242

Yoshimura K, Miyao K, Gaber A, Takeda T, Kanaboshi H, Miyasaka H, Shigeoka S (2004) Enhancement of stress tolerance in transgenic tobacco plants overexpressing Chlamydomonas glutathione peroxidase in chloroplasts or cytosol. Plant J 37(1):21–33

Zagorchev L, Seal CE, Kranner I, Odjakova M (2013) A central role for thiols in plant tolerance to abiotic stress. Int J Mol Sci 14(4):7405–7432

Zandalinas SI, Mittler R, Balfagón D, Arbona V, Gómez-Cadenas A (2017) Plant adaptations to the combination of drought and high temperatures. Physiologia Plantarum. https://doi.org/10.1111/ppl.12540

Zhai CZ, Zhao L, Yin LJ, Chen M, Wang QY, Li LC, Xu ZS, Ma YZ (2013) Two wheat glutathione peroxidase genes whose products are located in chloroplasts improve salt and H_2O_2 tolerances in Arabidopsis. PLoS One 8(10):e73989

Zhang C, Liu J, Zhang Y, Cai X, Gong P, Zhang J, Wang T, Li H, Ye Z (2011) Overexpression of SlGMEs leads to ascorbate accumulation with enhanced oxidative stress, cold, and salt tolerance in tomato. Plant Cell Rep 30(3):389–398

Zhang D-Y, Yang H L, Li X-S, Li H-Y, Wang Y-C (2014) Overexpression of *Tamarix albiflonum* TaMnSOD increases drought tolerance in transgenic cotton. Mol Breed 34(1):1–11

Zhao F, Zhang H (2006) Salt and paraquat stress tolerance results from co-expression of the *Suaeda salsa* glutathione S-transferase and catalase in transgenic rice. Plant Cell Tissue Organ Cult 86(3):349–358

Zhao D-Y, Shen L, Yu M-M, Zheng Y, Sheng J-P (2009) Relationship between activities of antioxidant enzymes and cold tolerance of postharvest tomato fruits. Food Sci 14:309–313

Zhao F.-Y, Liu W, Zhang S-Y (2009) Different Responses of Plant Growth and Antioxidant System to the Combination of Cadmium and Heat Stress in Transgenic and Nontransgenic Rice. J Int Plant Biol 51(10):942–950

Chapter 12
Strategies to Alleviate Salinity Stress in Plants

Sara Francisco Costa, Davide Martins, Monika Agacka-Mołdoch, Anna Czubacka, and Susana de Sousa Araújo

Abstract Soil salinization is a major threat to agriculture in arid and semiarid regions. Besides the identification and use of salt-adapted species or cultivars in saline areas, the use of treatments to alleviate the effects of salinity stress is a promising solution to ensure crop production in such adverse conditions. Chemical, biological, and physical treatments are being successfully applied to seeds, seedlings, or plants before exposure to salinity stress. These treatments activate physiological and molecular pathways enabling the seed or plant to respond more quickly and/or more vigorously after exposure to salinity. Coupled to this, agricultural management practices have also contributed to mitigation of the effects of excessive salt accumulation in the soil. The acquired fundamental knowledge about how a plant reacts to high salt concentrations has been essential for the development of educated and applied strategies for salinity alleviation. In this chapter, we provide a general overview of the main strategies applied to alleviate salinity effects in plants, with a critical discussion of the main achievements described in this field.

Authors Sara Francisco Costa and Davide Martins have contributed equally to this work.

S. F. Costa
Plant Cell Biotechnology Laboratory, Instituto de Tecnologia Química e Biológica António Xavier (ITQB NOVA), Oeiras, Portugal

D. Martins
Genetics and Genomics of Plant Complex Traits Laboratory, Instituto de Tecnologia Química e Biológica António Xavier (ITQB NOVA), Oeiras, Portugal

M. Agacka-Mołdoch · A. Czubacka
Department of Plant Breeding and Biotechnology, Institute of Soil Science and Plant Cultivation—State Research Institute, Puławy, Poland

S. d. S. Araújo (✉)
Plant Cell Biotechnology Laboratory, Instituto de Tecnologia Química e Biológica António Xavier (ITQB NOVA), Oeiras, Portugal

Plant Biotechnology Laboratory, Department of Biology and Biotechnology 'L. Spallanzani', Università degli Studi di Pavia, Pavia, Italy
e-mail: saraujo@itqb.unl.pt

V. Kumar et al. (eds.), *Salinity Responses and Tolerance in Plants, Volume 1*,
https://doi.org/10.1007/978-3-319-75671-4_12

Keywords Abiotic stresses · Alleviation strategies · Biological treatments · Chemical treatments · Crops · Field management practices · Halopriming · Ionizing radiation · Magnetic field · Osmopriming · Physical treatments · Plant growth–promoting rhizobacteria · Rhizospheric fungi · Salinity

Abbreviations

ABA	Abscisic acid
ACC	1-Aminocyclopropane-1-carboxylic acid
AMF	Arbuscular mycorrhizal fungi
APX	Ascorbate peroxidase
BABA	β-Aminobutyric acid
$CaCl_2$	Calcium chloride
CAT	Catalase
$CuSO_4$	Copper sulfate
EC	Electrical conductivity
EMF	Electromagnetic field
ET	Ethylene
FAO	Food and Agriculture Organization
GA_3	Gibberellic acid
H_2O_2	Hydrogen peroxide
H_2S	Hydrogen sulfide
IAA	Indoleacetic acid
JA	Jasmonate
K^+	Potassium
K_3PO_4	Tripotassium phosphate
KCl	Potassium chloride
KH_2PO_4	Monopotassium phosphate
KNO_3	Potassium nitrate
KOH	Potassium hydroxide
MDA	Malondialdehyde
MF	Magnetic field
$MgSO_4$	Magnesium sulfate
Na^+	Sodium
NaCl	Sodium chloride
NCBI	National Center for Biotechnology Information
NO	Nitric oxide
NO_3	Nitrate
n-Si	Nanosilicon particles
n-SiO_2	Nanosilicon dioxide particles
O_2^-	Superoxide radical

OH^-	Hydroxyl radical
P5CR	Pyrroline-5-carboxylate reductase
P5CS	Pyrroline-5-carboxylate synthetase
PEG	Polyethylene glycol
PGPB	Plant growth–promoting bacteria
PGPR	Plant growth–promoting rhizobacteria
POX	Peroxidase
Put	Putrescine
QTL	Quantitative trait locus
RONSS	Reactive oxygen–nitrogen–sulfur species
ROS	Reactive oxygen species
SA	Salicylic acid
SMF	Static magnetic field
SNP	Sodium nitroprusside
SOD	Superoxide dismutase
Spd	Spermidine
Spm	Spermine
UV	Ultraviolet radiation
$ZnSO_4$	Zinc sulfate

12.1 Introduction

Soil salinization is a major threat to agriculture in arid and semiarid regions, where water scarcity and inadequate drainage of irrigated lands severely reduce crop yield (Hanin et al. 2016). According to information available at the Food and Agriculture Organization (FAO) Soils Portal (http://www.fao.org/soils-portal/soil-management), more than 6% of the world's total land area is affected by salt accumulation. The same source indicates that the total area of saline soils has been estimated at 397 million hectares, while sodic soils represent 434 million hectares. This threatening scenario could be exacerbated presently by the emerging climate changes observed in different areas of the world.

Salinization can broadly refer to the accumulation of different salts, including potassium, magnesium, calcium and sodium carbonates, bicarbonates chlorides, and sulfates (Bockheim and Gennadiyev 2000). Consequently, this diverse ionic composition will result in a wide range of physiochemical properties. The literature discussing salt-affected soils often uses two different concepts: soil salinization and soil sodicity. According to the FAO Soils Portal, salt-affected soils can be divided into saline, saline–sodic, and sodic, depending on salt amounts, types of salt, the amount of sodium present, and soil alkalinity. Saline soils are those that have a saturation soil paste extract electrical conductivity (EC) of more than 4 dS m^{-1} at 25 °C (which corresponds to approximately 40 mM sodium chloride (NaCl)), generating an osmotic pressure of approximately 0.2 MPa (Grieve et al. 2008;

Munns and Tester 2008). Soil sodicity is a term more restricted to the amount of Na^+ held in the soil. High sodicity (more than 5% of Na^+ in the overall cation content) causes clay to swell excessively when wet, therefore severely limiting air and water movements and resulting in poor drainage (Munns 2005; Hanin et al. 2016). Because of the nature of salts present in sodic soils, they are usually alkaline—an aspect also limiting crop cultivation. Saline–sodic soils have, as the name indicates, intermediate properties.

A growing body of research has discussed the effects of soil salinity on plant responses (for recent reviews, see Munns and Gilliham (2015); Hanin et al. (2016); Negrão et al. (2017)). The response of plants to salinity can be described in two main phases, reflecting a time frame of biological processes occurring: shoot ion-independent (early) responses and ion-dependent (late) responses (Negrão et al. 2017). An early response (a couple of minutes or a day after stress imposition) is characterized as a rapid response by the plant, in which the osmotic effect of the salt is sensed in the soil. Early responses are also known as the osmotic phase (Roy et al. 2014). Decreased soil water potential leads to a decline in water uptake by the plant, causing stomatal closure, reducing photosynthesis, and inhibiting leaf expansion, which results in a decrease in the shoot growth rate (Munns and Tester 2008; Das et al. 2015; Nongpiur et al. 2016; Hanin et al. 2016). The ionic phase happens when salts' toxic effects are sensed within the plant because of their accumulation (Munns 2005). The second phase is a slower response occurring after several days to weeks and takes place after toxic accumulation of Na^+ in photosynthetic tissues (Roy et al. 2014). As a result of the salt accumulation over time, there is a slower inhibition of growth, occurring especially in older leaves, causing their senescence (Munns and Tester 2008). In contrast to young leaves, old leaves are not capable of diluting the salt that enters their cells, and this leads to their death. If the leaves' death rate is higher than the leaves' production rate, then the photosynthetic capacity will be impaired, leading to a reduced growth rate (Munns and Tester 2008; Nongpiur et al. 2016).

According to Negrão et al. (2017), salinity tolerance mechanisms in plants can be classified into three different categories: ion exclusion, which refers to net exclusion of toxic ions from the shoot; tissue tolerance, which refers to compartmentalization of toxic ions in specific tissues, cells, and subcellular organelles; and shoot ion-independent tolerance, which focuses on maintenance of growth and water uptake independent of the extent of Na^+ accumulation in the shoot. For a detailed description of the main physiological, cellular, and metabolic responses activated in each case, see Munns and Tester (2008); Roy et al. (2014). Nevertheless, it is worth highlighting that ion transport via cell membranes is the basic factor determining salinity tolerance (Ismail and Horie 2017). Moreover, salinity leads to oxidative stress in plants because of production of reactive oxygen species (ROS) such as superoxide radicals (O_2^-), hydrogen peroxide (H_2O_2), and hydroxyl radicals (OH^-), which can trigger DNA and cellular damage (El-Mashad and Mohamed 2012). To cope with this, antioxidative stress defenses are triggered through enzymatic antioxidant mechanisms including catalase (CAT), superoxide dismutase (SOD), and peroxidase (POX), among others (Gharsallah et al. 2016).

Under salt stress, concomitant accumulation of compatible solutes occurs with the accumulation of solutes in the cytosol (Munns and Tester 2008). Low-weight solutes such as proline, polyols, amino acids, proteins, and betaine—commonly referred to as compatible solutes—play a role in both osmoprotection and osmotic adjustment under high salt concentrations (El-Mashad and Mohamed 2012; Gharsallah et al. 2016). Indeed, overproduction of proline in plants imparts stress tolerance by maintaining cell turgor or osmotic balance; stabilizing membranes, thereby preventing electrolyte leakage; and bringing concentrations of ROS within normal ranges, thus preventing oxidative bursts in plant (Hayat et al. 2012).

Plant morphology, biochemistry, and physiology also play a major role in shaping the different degrees of salinity tolerance (Nongpiur et al. 2016). Among plants there is a gradient of NaCl tolerance in the environment, from the very tolerant ones to the very sensitive ones. Species of the genera *Tecticornia* or *Atriplex* fall into the salt-tolerant group and are often termed halophytes (Munns and Tester 2008; Flowers and Colmer 2015). Rice (*Oryza sativa*) has been described as one of the species most sensitive to salinity (Das et al. 2015). In the field, where the salinity can rise to 100 mM NaCl (about 10 dS m^{-1}), rice will die before maturity, while wheat (*Triticum aestivum*) will still be able to produce a reduced yield (Munns et al. 2006). Thus, wheat is considered a moderately salt-tolerant crop. On the other hand, barley (*Hordeum vulgare*) is considered one of the most salt-tolerant cereals, but it dies after extended periods at salt concentrations higher than 250 mM NaCl (equivalent to 50% seawater) (Munns et al. 2006). These examples allow us to conclude that understanding of the mechanism by which crop production could be maintained in a saline growing scenario is needed to ensure our food security (Landi et al. 2017). Differences in salinity responses can occur within the same genus. As an example, in the *Brassica* genus, *B. napus* is the most tolerant species, followed by *B. juncea* and then by *B. oleracea* (Chakraborty et al. 2016).

There have been significant breakthroughs in the understanding of the mechanisms, control, and modulation of Na^+ accumulation in plants (Munns and Tester 2008). Numerous genes have been identified that could be used in molecular breeding programs. As an example, *Saltol*, a major quantitative trait locus (QTL) for salt tolerance, is being transferred into seven popular locally adapted rice varieties: ADT45, CR1009, Gayatri, MTU1010, PR114, Pusa 44, and Sarjoo 52 (Singh et al. 2016). In another study, transgenic tomato plants overexpressing a vacuolar Na^+/H^+ antiporter were able to grow, flower, and produce fruit in the presence of 200 mM NaCl (Zhang and Blumwald 2001). Despite the success of these approaches, they are time consuming and likely not accessible to all salinity-devoted researchers or farmers, because of their cost. In this context, the development of easy and cost-effective approaches to mitigate the impacts of salinity stress in plants and crop production is highly desirable.

One strategy to mitigate the effects of salt stress in plants could be the application of chemical, biological, and physical treatments in seeds, seedlings, or plants before exposure to salinity stress (see Fig. 12.1). Those treatments, mainly applied before seed sowing, are expected to activate physiological and molecular pathways

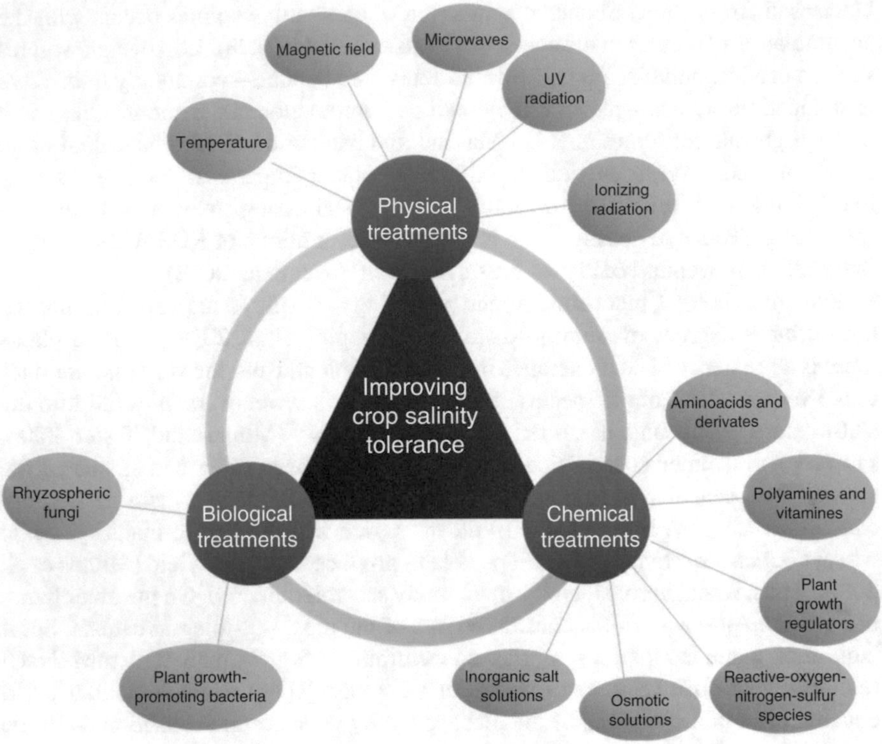

Fig. 12.1 Schematic representation of the multiple strategies available to alleviate salinity stress in plants. *UV* ultraviolet

enabling the seed to respond more quickly and/or more vigorously after exposure to an abiotic stress factor. This chapter aims to provide a general overview of the main treatments applied to plants to alleviate salinity effects, with a critical discussion of the main achievements described in this field.

12.2 Chemical Strategies to Alleviate Salinity Stress

Seed and plant priming constitutes a technology to enhance plant tolerance of various abiotic stresses including salinity. Seed priming is generally defined as a presowing treatment, applied before germination, to improve the speed and uniformity of germination (Paparella et al. 2015). The efficiency of the priming technology depends on the plant species, the method of treatment and its duration, and the dose of the protectant/priming agent used (Nawaz et al. 2013; Jisha and

Puthur 2016). Nevertheless, the concept of priming can be further extended to seedlings and whole plants. In this context, plant priming is a mechanism leading to a physiological state that enables plants to respond more rapidly and/or more robustly after exposure to biotic or abiotic stresses (Aranega-Bou et al. 2014). Plant priming can be initiated naturally in response to an environmental stress event, which acts as a cue indicating an increased probability of facing that specific stress factor in the future (Savvides et al. 2016).

Salinity stress can be alleviated by the use of many chemical agents. Some of them are natural metabolites produced by plants in small amounts but, if applied exogenously, they alter gene expression and regulation of metabolism, leading to enhanced salt tolerance. Those substances include sugars (e.g., trehalose), amino acids and their derivatives (e.g., proline, glycine betaine, melatonin, β-aminobutyric acid (BABA), and glutathione), plant growth regulators (e.g., jasmonate (JA) and salicylic acid (SA)), polyamines (e.g., spermine (Spm), spermidine (Spd), and putrescine (Put)), and vitamins, which are involved in the integration of stress signals (Vaishnav et al. 2016). Under salt stress conditions, they could also play a role as osmoprotectants or antioxidants. Another group of chemical agents effective in inducing plant tolerance of abiotic stresses are reactive oxygen–nitrogen–sulfur species (RONSS), which are known also for their damaging activity if they occur in a high concentration. This group of compounds includes reactive molecules containing oxygen (e.g., H_2O_2), sulfur (e.g., hydrogen sulfide (H_2S)), and nitrogen (e.g., nitric oxide (NO)). Numerous plant stress physiology studies have been devoted to studying the role of sodium nitroprusside (SNP), an inorganic compound that is a source of NO (Vaishnav et al. 2016; Savvides et al. 2016).

Chemical pretreatment of plants before stress occurrence increases their tolerance of adverse environmental conditions, since it contributes to plants responding faster and more strongly to unfavorable conditions (Savvides et al. 2016). A nonexhaustive list of chemical priming treatments targeting the improvement of seed, seedling, and plant salt stress tolerance is presented in Table 12.1. Seed germination is a critical step in plant development. In many crops, germination may be delayed and growth of seedlings inhibited under saline conditions. Consequently, it is not surprising that a large body of research is focused on understanding seed priming, as well as developing new priming protocols. The application of chemical agents for plant and seed pretreatment seems to be an inexpensive method of crop protection from salinity stress because the compounds have been shown to be effective in low concentrations (Savvides et al. 2016). Another benefit is that the tolerance of salinity stress induced by the use of chemical agents becomes systemic although they are usually applied only to the seeds, roots, or leaves. However, most studies have determined plant tolerance of the stress soon after chemical priming. Therefore, the durability of crop protection provided by the aforementioned chemical compounds remains an open question. In Sects. 12.2.1–12.2.4 we discuss some chemical priming approaches described in the literature.

Table 12.1 Priming treatments adopted for developing salinity stress tolerance in plants

Species	Plant growth stage	Priming agent	Method of application	Reference
Eggplant (*Solanum melongena*)	Plant	Proline	Foliar sprayed	Shahbaz et al. (2013)
Rice (*Oryza sativa*)	Seedling	Proline, trehalose	Added to medium	Nounjan et al. (2012)
		Silicon	Added to medium	Kim et al. (2014)
		Silicon	Added to soil	Farooq et al. (2015)
	Seed	NaCl, KCl, $CaCl_2$, KNO_3, ascorbic acid, mannitol, PEG, sorbitol, wood vinegar	Soaked in solution	Theerakulpisut et al. (2016)
		Spd, GA_3	Soaked in solution	Chunthaburee et al. (2014)
Soybean (*Glycine max*)	Seed	Proline, glycine betaine	Soaked in solution	Vaishnav et al. (2016)
		KNO_3	Soaked in solution	Miladinov et al. (2015)
Wheat (*Triticum aestivum*)	Seed	Cysteine	Soaked in solution	Nasibi et al. (2016)
		Choline	Soaked in solution	Salama and Mansour (2015)
		$CaCl_2$	Soaked in solution	Tamini (2016)
	Seedling	Proline	Foliar sprayed	Talat et al. (2013)
Faba bean (*Vicia faba*)	Seed	Melatonin	Soaked in solution	Dawood and El-Awadi (2015)
		Ascorbic acid, nicotinamide	Soaked in solution	Azooz et al. (2013)
		Salicylic acid	Soaked in solution	Anaya et al. (2018)
	Seed/seedling	Silicon	Added to medium/foliar sprayed	Abdul Qados and Moftah (2015)
Mung bean (*Vigna radiata*)	Seed	Chitosan	Soaked in solution	Sen and Mandal (2016)
		BABA	Soaked in solution	Jisha and Puthur (2016)
Broccoli (*Brassica oleracea*)	Seedling	Methyl jasmonate, urea	Foliar sprayed	del Amor and Cuadra-Crespo (2011)

(continued)

Table 12.1 (continued)

Species	Plant growth stage	Priming agent	Method of application	Reference
Tomato (*Solanum lycopersicum*)	Seedling	Spd	Foliar sprayed	Zhang et al. (2016)
		Silicon	Added to medium	Li et al. (2015)
	Seed	Silicon	Treated with solution	Almutaiori (2016)
		PEG	Soaked in solution	Pradhan et al. (2015)
Kentucky bluegrass (*Poa pratensis*)	Seedling	Spd	Foliar sprayed	Puyang et al. (2016)
Lentil (*Lens culinaris*)	Seed	Silicon	Soaked in solution	Sabaghnia and Janmohammadi (2015)
Zinnia (*Zinnia elegans*)	Seedling	Silicon	Added to medium	Manivannan et al. (2015)
Winter cherry (*Physalis angulata*)	Seed	PEG	Soaked in solution	de Souza et al. (2016)
Rape (*Brassica napus*)	Seed	PEG	Soaked in solution	Kubala et al. (2015)
Cotton (*Gossypium spp.*)	Seed	KNO_3	Soaked in solution	Nazir et al. (2014)
Pepper (*Capsicum annuum*)	Seed	KCl, NaCl, $CaCl_2$	Soaked in solution	Aloui et al. (2014)
Maize (*Zea mays*)	Seed	NaCl, $CaCl_2$	Soaked in solution	Gebreegziabher and Qufa (2017)
Black seed (*Nigella sativa*)	Seed	KNO_3, $CaCl_2$, NaCl, $ZnSO_4$, $CuSO_4$	Soaked in solution	Gholami et al. (2015)
Pea (*Pisum sativum*)	Seed	KCl, KOH	Soaked in solution	Naz et al. (2014)
Okra (*Abelmoschus esculentus*)	Seed	KCl, mannitol, $CaCl_2$	Soaked in solution	Dkhil et al. (2014)
Common bean (*Phaseolus vulgaris*)	Seed	Vitamin B_{12}	Soaked in solution	Keshavarz and Modares Sanavy (2015)
Tobacco (*Nicotiana rustica*)	Plant	Ethanolamine	Added to medium	Rajaeian and Ehsanpour (2015)

BABA β-aminobutyric acid, *GA_3* gibberellic acid, *PEG* polyethylene glycol, *Spd* spermidine

12.2.1 Amino Acids and Their Derivatives

Proline is an amino acid acting as an osmoprotectant, which plays an important role in reducing oxidative stress when plants are coping with abiotic stresses (Kavi Kishor and Sreenivasulu 2014; Per et al. 2017). Under stress, plants usually increase the production of endogenous proline. Therefore, exogenous application of this amino acid has been considered a promising way to alleviate salinity effects. Shahbaz et al. (2013) reported that foliar-sprayed proline ameliorated the adverse effect of salinity on the shoot fresh weights of two tested eggplant cultivars and water use efficiency in one of them. Addition of proline to the media in which rice seedlings were grown caused upregulation of proline synthesis genes encoding pyrroline-5-carboxylate synthetase (P5CS) and pyrroline-5-carboxylate reductase (P5CR) and a further increase in endogenous proline (Nounjan et al. 2012). However, exogenous proline also reduced the activity of four antioxidant enzymes (SOD, POX, ascorbate peroxidase (APX), and CAT) and uptake of sodium ions, resulting in a low Na^+/K^+ ratio. In the same experiment, trehalose was also tested. This nonreducing sugar was shown to be less effective in limiting Na^+ uptake and simultaneously decreased production of endogenous proline. Although proline and trehalose did not overcome growth limitation under salt stress, they showed a beneficial impact during the stress recovery period (Nounjan et al. 2012).

Pretreatment of wheat (*T. aestivum*) seeds with cysteine enhanced tolerance of salinity stress, increasing the activity of antioxidant enzymes and decreasing ion toxicity (Nasibi et al. 2016). The beneficial effect of cysteine on wheat plants can be attributed to production of either glutathione or H_2S—molecules that are well known for their antioxidant roles. Mung bean (*Vigna radiata*) seeds treated with BABA showed positive effects against salt stress via accumulation of proline, total protein, and carbohydrate in seedlings (Jisha and Puthur 2016). Moreover, the activities of antioxidant enzymes increased in BABA-primed seeds. The biomass of seedlings raised from primed seeds was increased in comparison with nonprimed ones, under stressed and unstressed conditions. On the basis of these findings, the authors concluded that seed priming with BABA is a cost-effective technique and a promising strategy to overcome negative effects of salinity stress and, likely, other abiotic stresses in crops.

Dawood and El-Awadi (2015) studied the suitability of using melatonin as a priming agent. In this work, seeds of faba bean (*Vicia faba*) were soaked in a melatonin solution and the seedlings were irrigated with diluted seawater. Interestingly, the application of melatonin showed beneficial features, since it improved growth parameters, relative water content and the contents of photosynthetic pigments, total carbohydrates, total phenolics, and indoleacetic acid (IAA). Melatonin treatment also reduced the contents sodium and chloride ions.

12.2.2 Polyamines and Vitamins

The effect of salt stress can be alleviated in crops through prior treatment with compounds belonging to the polyamine family (Zhang et al. 2016). Spd, Spm, and Put are polyamines that are naturally present in plant tissues. Their content usually increases under salinity stress because they are able to alleviate its effects. They can interact with membrane phospholipids, thus stabilizing cellular structures. Polyamines also stabilize the structures of many enzymes and maintain K^+/Na^+ homeostasis by limiting Na^+ uptake by roots and limiting loss of K^+ by shoots (Zhao et al. 2007). Furthermore, polyamines neutralize ROS induced under salt stress and thus protect cells from oxidative damage (Puyang et al. 2016). Exogenous Spd was used for foliar spraying of young tomato plants, which were then exposed to salinity–alkalinity stress 7 days later (Zhang et al. 2016). The results showed that Spd caused a decreases in ROS and malondialdehyde (MDA) content. Also, a simultaneous increase in antioxidant enzyme activities and nonenzymatic components of the antioxidant system was noticed, which resulted in chloroplast protection from damage because of the stress. Likewise, an improvement in the salinity tolerance of Kentucky bluegrass by exogenous Spd treatment was reported by Puyang et al. (2016). The grass leaves were sprayed with Spd twice before being exposed to NaCl solution for 28 days. Spd treatment mitigated decreases in the K^+/Na^+ ratio and in the content of chlorophyll, potassium, calcium, and magnesium ions. Moreover, the treatment also resulted in a reduction in electrolyte leakage and Na^+ content and caused an increase in the content of endogenous osmoprotectants such as proline, Spm, and Spd. Khan et al. (2012) reported that seed priming with polyamines (Spd, Spm, and Put) enhanced the germination and early growth of hot pepper seedlings (*Capsicum annuum*) in comparison with untreated seeds.

Other interesting findings were described by Chunthaburee et al. (2014), who treated rice seeds with Spd and gibberellic acid (GA_3). This combined treatment enhanced the antioxidant system in the plants and reduced the production of H_2O_2. Moreover, Spd and GA_3 priming improved ion homeostasis and delayed the loss of pigments. Although the application of these substances improved the growth of seedlings, the beneficial effect was quantitatively small, which might limit their use in a seed technology context.

Seed priming with vitamins has been revealed to be an approach that could stimulate the growth of legume and cereal seedlings under unfavorable conditions. Faba bean seeds treated with ascorbic acid and nicotinamide showed increased content of photosynthetic pigments, soluble carbohydrates and proteins, proline, and other amino acids, whereas transpiration and ion leakage decreased (Azooz et al. 2013). The application of nicotinamide was found to be more effective in the alleviation of salt stress effects than ascorbic acid. However, the combination of these two vitamins resulted in an enhanced synergistic effect. Priming seeds of common bean (*Phaseolus vulgaris*) with vitamin B_{12} also increased the survival capacity of bean plants. This positive effect was due to concomitant stimulation of antioxidant enzyme activity and reduction of ROS (Keshavarz and Modares Sanavy 2015). Application of choline enhanced salt tolerance in wheat through stimulation of membrane biosynthesis to maintain plasma membrane stability, fluidity, and therefore ion homeostasis (Salama and Mansour 2015).

12.2.3 Plant Growth Regulators and Organic or Inorganic Compounds

Another essential group of chemical agents that play a major role in adaptation to stresses, including salinity, are plant growth regulators. They affect plant growth and development, and induce defense against a variety of abiotic stresses. Application of SA on faba bean seeds resulted in a positive effect on germination parameters and improved establishment of seedlings (Anaya et al. 2018). This work also highlighted that this beneficial response is dose dependent, with the lowest applied concentration of SA being the one with the most positive effect, even under strong salinity.

Salinity stress may also be alleviated by application of nitrogen fertilizer such as urea, which is commonly used in agriculture. Foliar application of this compound on broccoli plants grown at 40 mM NaCl allowed maintenance of growth, gas exchange, and leaf $N\text{–}NO_3$ concentrations at levels similar to those in nonsalinized plants (del Amor and Cuadra-Crespo 2011). Similar results were obtained when methyl jasmonate was used in the same way. Foliar application of both urea and methyl jasmonate became ineffective when a threefold increase in salinity was applied.

In soybean, priming of seeds with NO, applied as SNP, played a protective role in germination and seedling development under salinity stress (Vaishnav et al. 2016). The germination rate of SNP-primed seeds under salinity conditions was 82% higher than that of untreated seeds. Interestingly, SNP treatment seemed to have an impact on plant architecture, resulting in longer seedlings with more lateral roots. Treated seedlings also accumulated more chlorophyll and showed a higher content of endogenous proline than untreated seedlings or those treated with other tested substances (glycine betaine, mannitol, or proline). They also exhibited a reduced MDA content, which indicated lower lipid peroxidation. The overall results indicated that a myriad of plant protective responses are activated in response to SNP priming, which may be of relevance to development of new seed treatments.

Much attention has been devoted to investigating the usefulness of silicon (Si) in the alleviation of salinity stress in crops. It is a very common element in nature, and its role in plant stress tolerance is complex and multifaceted (Yin et al. 2016). First of all, it is a component of the double-layer cuticle, being the first physical barrier limiting transpiration during salinity stress. Other roles of silicon in response to salinity have been postulated, including maintenance/improvement of water status and a decrease in oxidative damage (Yin et al. 2016). Last, but not least, silicon reduces concentrations of Na^+ ions and limits their transportation, resulting in alleviation of salinity (Manivannan et al. 2015; Kim et al. 2014). Priming treatments with nanosilicon (n-Si) and nanosilicon dioxide (n-SiO_2) particles resulted in improved germination of lentil (*Lens culinaris*) (Sabaghnia and Janmohammadi 2015) and tomato seeds (Almutaiori 2016) under salt stress. Seedling traits such as weight, root length, and shoot length were also positively affected. The authors concluded that application of nanomaterials on seeds may stimulate defense mechanisms of plants and increase salinity tolerance in other crops. The role of silicon and nanosilicon in the mitigation of salt stress in faba beans was investigated by Abdul Qados and Moftah (2015). Both forms of silicon were effective in improving the salt tolerance of plants by increasing plant height, fresh and dry weight, and total yield under salt conditions. In a hydroponic system, silicon supplementation of *Zinnia elegans* resulted in alleviation of salinity stress induced by NaCl treatment (Manivannan et al. 2015). Silicon-treated plants displayed enhanced growth and improved photosynthetic parameters. Moreover, silicon improved membrane integrity, resulting in reduced electrolyte leakage potential and lipid peroxidation levels (Manivannan et al. 2015). Silicon was also shown to increase the activity of antioxidant enzymes in rice, which contributed to reduced levels of oxidative damage (Kim et al. 2014). In tomato, exogenous silicon application affected seedling tolerance of salinity stress (Li et al. 2015). Silicon treatment improved tomato growth, photosynthetic pigment and soluble protein content, the net photosynthetic rate, and root morphological traits under salt stress. One interesting aspect of this study was the fact that the leaf transpiration rate and stomatal conductance were not decreased but increased by application of silicon under salt stress (Li et al. 2015). Silicon has been described as modulating the level of stress-related plant growth regulators such as JA, SA, and abscisic acid (ABA). Kim et al. (2014) monitored the content of these molecules in rice roots under salinity stress after application of silicon and noted that the levels of ABA were higher 6 and 12 hours after treatment but became insignificant after 24 hours. The authors concluded that the uptake of greater silicon amounts limits the activation of ABA metabolism, which regulates stomatal conductance during stress conditions, contributing to the improved performance of silicon-treated plants under salinity.

Some natural polysaccharides have shown beneficial effects in plant protection against salt stress. As an example, Sen and Mandal (2016) tested the response of mung bean seeds primed with chitosan under NaCl stress. Chitosan promoted seed germination and seedling growth, particularly in the early phase of growth.

12.2.4 Exploring the Potentialities of Osmopriming and Halopriming

Osmopriming has unexploited potential to improve crop response to salinity. This seed-priming method involves treatment of seeds with solutions with a low water potential (e.g., an osmotic solution). Numerous chemicals are used in the seed osmopriming technique, including polyethylene glycol (PEG), mannitol, and sorbitol (Nawaz et al. 2013; Paparella et al. 2015). Soaking of seeds in such a low-water-potential solution enables a slow imbibition that is sufficient to trigger seed germination metabolic processes; however, germination remains inhibited. The impacts of osmopriming on the activation of many enzymes, DNA damage, and repair responses, as well as on the modulation of ROS-mediated damage of cellular components, has been reviewed extensively (for an example, see Paparella et al. 2015).

Primed seeds usually germinate faster and more uniformly (Nawaz et al. 2013). The most commonly used osmotic agent in osmopriming is PEG, which is often used in combination with ionic salts. Osmoprimed seeds have showed improved vigor and germination under salt stress. As an example, Pradhan et al. (2015) primed tomato (*Solanum lycopersicum*) seeds with PEG 6000 and observed resulting beneficial effects on the germination percentage, vigor index, and seedling dry weight under salt stress. Kubala et al. (2015) also used PEG 6000 to prime seeds of *B. napus* and noted that PEG treatment improved germination and seedling growth under salinity stress. Additionally, these authors demonstrated that this improved phenotype was connected with a significant increase in proline content. PEG 8000 was used as a seed-priming agent in *Physalis angulata* under saline conditions (de Souza et al. 2016). The authors observed that this treatment resulted in higher germination percentages and uniformity under saline conditions. Moreover, seedlings showed relatively normal growth and had slightly greater biomass. The authors concluded that priming induced the activation of physiological mechanisms responsible for osmotic adjustment and synthesis of proteins involved in defense against free radicals. The authors also suggested that such responses reflect reprogramming of the transcriptome under such conditions, resulting in upregulation of genes related to ion transport and genes encoding antioxidant enzymes (de Souza et al. 2016).

Another promising priming technique is halopriming, which involves soaking seeds in solutions of inorganic salts such as NaCl, potassium chloride (KCl), potassium nitrate (KNO_3), tripotassium phosphate (K_3PO_4), monopotassium phosphate (KH_2PO_4), magnesium sulfate ($MgSO_4$), or calcium chloride ($CaCl_2$) (Nawaz et al. 2013). Priming seeds of cotton and soybean with KNO_3 had a positive effect on germination and seed vigor under salinity stress (Miladinov et al. 2015; Nazir et al. 2014). Similarly, seed priming with different salts—KCl, NaCl, and $CaCl_2$—showed a positive effect on germination in pepper under salt stress by speeding up imbibition, which enabled faster metabolic activity in the seeds (Aloui et al. 2014). In maize (*Zea mays*), seed priming with NaCl and $CaCl_2$ increased germination and seedling growth parameters in comparison with nonprimed seeds under salt stress (Gebreegziabher and Qufa 2017). Interestingly, those two priming treatments showed different impacts on maize physiology. While NaCl priming improved crop maturity and yield, $CaCl_2$ priming accelerated the germination process.

The efficiency of salt priming for seedling growth has been extensively examined in the literature. Priming agents such as KNO_3, $CaCl_2$, NaCl, zinc sulfate ($ZnSO_4$), and copper sulfate ($CuSO_4$) were tested to overcome salinity effects in black seed (*Nigella sativa*) (Gholami et al. 2015). Among the tested compounds, NaCl was the most effective in the alleviation of an adverse effect of salt stress, promoting efficient germination. Despite the primacy of NaCl, KNO_3, $CaCl_2$, and $ZnSO_4$ were also effective in promoting germination performance and seedling development but to a lesser extent (Gholami et al. 2015). *Pisum sativum* seeds primed with KCl and potassium hydroxide (KOH) showed improvements in the germination percentage and seedling growth (Naz et al. 2014). When wheat seeds were primed with $CaCl_2$, salinity tolerance of the seedlings was improved, as seen by their enhanced growth (Tamini 2016). A deeper investigation of mechanisms triggered by priming showed that cell membranes were stabilized, chlorophyll content and activity of nitrate reductase were enhanced, and accumulation of proline, total soluble sugars, and proteins occurred. On the basis of this, the author suggested that priming of seeds with $CaCl_2$ successfully triggers physiological and metabolic processes that could support wheat cultivation in salt-affected soils. In okra (*Abelmoschus esculentus*), seed priming with KCl, $CaCl_2$, and mannitol improved the germination percentage, seedling dry weight, and final emergence percentage (Dkhil et al. 2014). Theerakulpisut et al. (2016) compared the effectiveness of ten priming agents—NaCl, KCl, $CaCl_2$, KNO_3, ascorbic acid, mannitol, PEG 6000, sorbitol, wood vinegar (made from eucalyptus), and distilled water—in protection of rice seedlings against salinity stress. Of all tested agents, KNO_3, mannitol, and wood vinegar were most effective in enhancing growth parameters of seedlings. The application of these compounds allowed maintenance of ion homeostasis while preventing chlorophyll degradation and membrane damage in plant tissues (Theerakulpisut et al. 2016).

Besides seeds, salt treatments have been also successfully applied in plant priming. As an example, Yan et al. (2015) pretreated sweet sorghum plants with NaCl for 10 days, then the plants were stressed with twice as high a concentration of that compound for an additional 7 days. The authors found a higher photosynthetic rate during stress and a smaller reduction of dry matter in pretreated plants than in nonpretreated ones. According to these authors, pretreatment of the plants with the lower salt concentration enhanced their osmotic resistance and reduced root uptake of sodium ions, which highlighted a concerted mechanism to deal with the constraint (Yan et al. 2015).

12.3 Biological Treatments to Improve Crop Salinity Tolerance

The identification of physiological responses, as well as biochemical networks involved in halophyte responses to salinity, has provided new salt stress–related genes available for plant improvement. This finding has supported the development of new approaches to alleviate salinity in plants and possibly stimulate crop cultivation in soils with enhanced salt accumulation (Shrivastava and Kumar 2015). Some of these genes have been used in molecular breeding approaches, as genetic engineering, but the effects of introduced genes need to be evaluated in the field to determine their effects on salinity tolerance and yield improvement (for a review, see Roy et al. 2014). Unfortunately, the ecological interaction of microorganisms inhabiting the roots (the rhizosphere) and leaves (the phyllosphere) of halophytes has been neglected in these studies, but it is important to keep in mind that they may contribute significantly to the halophytes' well-being and salinity tolerance (Ruppel et al. 2013).

Plants are capable of recruiting and forming mutualistic associations with a number of soil microorganisms, with beneficial effects on the plants' productivity and resilience in harsh conditions (De-La-Peña and Loyola-Vargas 2014; Munns and Gilliham 2015; Kasim et al. 2016). The contribution of a specialized microbiome in assisting plants to withstand salinity is often overlooked; however, it can be used as an alternative to the development of salt-tolerant crops through genetic modification (Yuan et al. 2016). Mounting evidence highlights the fact that rhizospheric fungi and plant growth–promoting rhizobacteria (PGPR) found in association with plants help them to acquire some degree of salinity stress tolerance (Campanelli et al. 2012; Habib et al. 2016; Kang et al. 2014; Bal et al. 2013; Gururani et al. 2013; Hajiboland 2013).

12.3.1 Rhizospheric Fungi

Campanelli et al. (2012) investigated how alfalfa (*Medicago sativa*) plants inoculated with *Glomus viscosum* dealt with NaCl stress. Inoculated alfalfa plants showed diminished progression of tissue wilting and increased plant height, root density, and leaf area in comparison with noninoculated plants. Additionally, mycorrhizal dependency increased at higher NaCl concentrations but, on the other hand, mycorrizhal colonization was negatively affected (Campanelli et al. 2012). Such findings highlight the need to carefully choose microbial species to be used in specific crops. Also, the source of the inoculum—either derived from a culture collection or isolated from a naturally saline environment—must be considered when assessing the potential of these microorganisms to assist plant growth under stressful conditions (Estrada et al. 2013). Isolation of microbial inoculants from plants' natural habitat has been shown to be more efficient in assisting plant acclimation to harsh environments (Bal et al. 2013; Querejeta et al. 2006; Liu et al. 2016). Estrada et al. (2013) concluded that use of native arbuscular mycorrhizal fungi (AMF) from a saline environment was more effective than use of AMF from a culture collection, providing greater protection against oxidative damage and more efficient photosynthetic apparatus and stomatal conductance. Bearing that in mind, some studies have demonstrated that native microbial inoculants isolated from saline soils can perform better and display a greater capacity to promote plant growth (Bharti et al. 2013; Campagnac and Khasa 2013; Damodaran et al. 2014; Liu et al. 2016; Azad and Kaminskyj 2016).

12.3.2 *Plant Growth–Promoting Bacteria*

Plant growth–promoting bacteria (PGPB) or rhizobacteria (PGPR) have been studied for their potential to promote plant growth, serving as an important tool to increase worldwide crop productivity even under salinity. For example, the interaction of *Burkholderia cepacia* SE4 and *Promicromonospora* sp. SE188 with *Cucumis sativus* seedlings (Kang et al. 2014) has been studied. Other studies have included assessment of the effects of *Bacillus pumilus* strain DH-11 in *Solanum tuberosum* plants (Gururani et al. 2013) and *Exiguobacterium oxidotolerans* in *Bacopa monnieri* (Bharti et al. 2013). Common findings in all of these studies were that the microorganism–plant interactions enhanced shoot and root growth, nutrient uptake, chlorophyll content, photosynthetic efficiency, and the K^+/Na^+ ratio in saline environments.

Plant growth regulators have a crucial role in controlling a number of physiological and metabolic processes during plant development and interaction with environmental conditions. Ethylene (ET) is thought to be a coordinator between plant development and their response to stress conditions, including high salt levels (Ma et al. 2012; Ellouzi et al. 2014; Peng et al. 2014; Tao et al. 2015). Concentrations of ET and its precursor 1-aminocyclopropane-1-carboxylic acid (ACC) can increase in plants exposed to such stressful conditions (Habib et al. 2016). Whether the accumulation of ET or ACC has a beneficial or detrimental effect on plant responses to salinity still remains unclear (Tao et al. 2015; Habib et al. 2016). Some studies have pointed out that ET may play a negative role in regulating plant adaptation to salinity, with potential inhibitory effects on root and shoot length and overall plant growth (Barnawal et al. 2014; Nadeem et al. 2014). One way to prevent an inhibitory effect of ET is through accumulation of compounds, such as ACC deaminase, that can interfere with its synthesis, reducing the excess of ET in plants. ACC deaminase–containing PGPR can be used as a way to reduce the deleterious effects that ET accumulation may exert in plants tissues (Bal et al. 2013; Ali et al. 2014; Barnawal et al. 2014; Tao et al. 2015; Habib et al. 2016). Inoculation of rice seed with ACC-utilizing bacteria (*Alcaligenes* spp.) improved plant performance under NaCl-imposed stress (Bal et al. 2013). Association with two bacterial strains containing ACC deaminase—*Bacillus megaterium* UPKR2 and *Enterobacter* sp. UPMR18—led to higher germination rates and biomass accumulation in okra seedlings under NaCl stress, with improved chlorophyll content and ROS-scavenging enzyme activity (Habib et al. 2016). Similar results were obtained with *P. sativum* inoculated with the ACC deaminase–containing rhizobacterium *Arthrobacter protophormiae* SA3, in which a 60% decrease in ACC content was noted (Barnawal et al. 2014). Besides overcoming the toxic effects of ET-induced damage, inoculation with this rhizobacterium also proved beneficial for nodulation and mycorrhization (Barnawal et al. 2014). The use of other PGPB endophytes from the genus *Pseudomonas* has also been a successful approach to alleviate the toxic effects of salinity stress. Ali et al. (2014) used the ACC deaminase–containing PGPB endophytes *P. fluorescens* YsS6, *P. migulae* 8R6, and ACC deaminase–deficient mutants of them to promote tomato plant growth in the absence or presence of salt stress. The results showed that tomato plants inoculated with *P. fluorescens* YsS6 and *P. migulae* 8R6 were healthier and accumulated more biomass than noninoculated control plants (Ali et al. 2014). Interestingly, when treated with the corresponding bacterial ACC deaminase–knockout mutants, plants were shown to be more salt sensitive, with a performance similar to that of noninoculated plants, which may lead one to conclude that this enzyme has a direct effect on plant adaptation in response to salt stress (Ali et al. 2014).

As a result of successful plant–microbe interactions, PGPR can further multiply into microcolonies or biofilms, providing additional protection against environmental stresses. In a study carried out by Kasim et al. (2016), *Bacillus amyloliquefaciens* showed the highest biofilm formation while efficiently enhancing growth and recovery of salt-sensitive barley cultivar Giza 123 seedlings under NaCl stress. Similar results were obtained by Chen et al. (2016), who found that *B. amyloliquifaciens* SQR9 conferred salt tolerance in maize plants, improving chlorophyll content and accumulation of total soluble sugar, while alleviating oxidative damage with increasing POX/CAT activity.

12.4 Physical Treatments to Alleviate Salinity Stress in Crops

The use of physical methods to increase plant production offers eco-friendly advantages and the possibility of use on a high-throughput scale. Physical invigoration methods (also known as "physical priming") are an alternative approach to current chemically or biologically based ones. The use of physical invigoration methods has been established as a promising approach to develop new biotech-based solutions for the growing seed market (Araújo et al. 2016). Several reports have described successful use of agents such as temperature; magnetic fields (MFs); or microwave, ultraviolet (UV), or ionizing radiation as promising presowing seed treatments (for reviews, see Paparella et al. 2015; Araújo et al. 2016). In general, presowing exposure to nonlethal doses of these physical agents has a positive impact not only in stimulating germination but also in improving the final quality and physiological performance of the plants, in terms of growth and yield. Presowing temperature treatments are a frequent approach to break seed dormancy, enhancing the overall quality of seed lots (Liu and El-Kassaby 2015). However, in this section we will not focus on thermopriming, because usually it is combined with other priming approaches (hydropriming or osmopriming; see Paparella et al. 2015). To the best of our knowledge, there have been only a few reports describing the use of physical invigorations methods to stimulate salinity tolerance. Nevertheless, we have collected some examples from literature databases, and we discuss their potential application as treatments to alleviate salinity stress in plants.

12.4.1 Ionizing Radiation

Gamma rays are known to have several potential applications in agriculture, including crop improvement, depending on the dose of exposure (Ahuja et al. 2014). The characterization of the effects of the gamma rays on seeds is a topic receiving growing attention. Several studies have demonstrated the suitability of low-/high-dose gamma radiation as an efficient seed invigoration treatment (Araújo et al. 2016). Moreover, some recent reports have supported its use as an effective treatment to alleviate salinity effects in plants.

Rejili et al. (2008) examined the effects of the interaction of salinity and gamma radiation on the growth and K^+/Na^+ ionic balance of Gannouch and Mareth populations of *M. sativa* cultivar Gabès. This work analyzed the morphological changes and physiological responses of the two alfalfa populations irradiated with a 350 Gy dose and cultivated in the presence of different salt concentrations for 50 days. The results showed that exposure to gamma radiation, alone or in combination with salt stress, significantly increased shoot numbers, stem height and chlorophyll *b* content in the Gannouch population, while no change occurred in the Mareth population. The authors concluded that the growth of gamma-irradiated plants was stimulated and that these plants were more tolerant of salt stress (Rejili et al. 2008). In a similar approach, Kumar et al. (2017) investigated the effects of presowing gamma radiation seed treatments on pigeon pea (*Cajanus cajan*) plant growth, seed yield, and seed quality under salt stress (80 and 100 mM NaCl) and control (0 mM NaCl) conditions. The study was conducted in two genetically diverse varieties, Pusa-991 and Pusa-992. A positive effect of presowing exposure of seed to low-dose gamma radiation (<0.01 kGy) under salt stress was evident in pigeon pea. Pigeon pea variety Pusa-992 showed a better salt tolerance response than Pusa-991 and, importantly, irradiated plants performed better than nonirradiated plants even at increasing salinity levels. Gamma radiation caused a favorable alteration in the source–sink (shoot–root) partitioning of recently fixed carbon (^{14}C), enhanced glycine betaine content, reduced protease activity, reduced the partitioning of Na^+, and promoted accumulation of K^+ under salt stress. Changes in these traits were important to differentiate between salt-tolerant and salt-susceptible varieties of pigeon pea and impacted on the seed yield and quality under salinity (Kumar et al. 2017). Importantly, these works revealed that the metabolic pathways activated in response to irradiation under salt stress are also genotype dependent.

In another study, Ahmed et al. (2011) investigated the effects of gamma radiation doses (0, 20, 40, and 80 Gy) on damsissa (*Ambrosia maritima*) plants under salt stress. In this work, salt stress was imposed in sandy loam soil (sand-to-loam ratio 3:2) with a mixture of salts (NaCl, $CaCl_2$, and $MgSO_4$ in a 2:2:1 ratio). The authors observed that irradiation of damsissa seeds with 40 or 80 Gy increased plant tolerance of salinity, in comparison with control, with regard to plant height, fresh/dry weights, and photosynthetic pigment. Importantly, they reported that irradiation alleviated the adverse effects of salinity by increasing total sugar and total soluble phenols in damsissa plant shoots. These findings provide new clues about metabolic pathways triggered by irradiation and might contribute to improving the salinity response of these plants.

12.4.2 Magneto-Priming

The study of the effects of MFs on biological organisms, including plants, dates back to the second half of the nineteenth century (Pietruszewski and Martínez 2015). Several recent reviews have summarized the impacts of MFs on many biological processes in plants, such as growth, development, and metabolism (Maffei 2014; Wolff et al. 2014). Static magnetic fields (SMFs) and electromagnetic fields (EMFs) are used in agriculture for seed priming, also known as “magneto-priming,” with proven beneficial effects on seed germination, vigor, and crop yield (Baby et al. 2011). Recently, some researchers have investigated the suitability of using MFs to alleviate the impacts of salinity in crops and postulated some molecular mechanisms behind this improved response.

Rathod and Anand (2016) reported that magneto-priming (50 mT for 2 h) caused a significant increase in the height, leaf area, and dry weight of wheat plants under nonsaline and saline conditions. In this work, the authors described a decrease in the Na^+/K^+ ratio in plants from primed seeds compared with those from unprimed seeds under salinity. Interestingly, magneto-priming seemed to be more beneficial to the salt-sensitive variety (HD 2967) that was studied than to the salt-tolerant one (Kharchia) under saline conditions. The authors concluded that magneto-priming induced tolerance, allowing primed plants to yield similarly under saline conditions as unprimed plants did under normally sown conditions.

The effects of MFs in seed germination and growth of sweet corn under NaCl stress (0, 50, and 100 mM) were investigated (Karimi et al. 2017). Seeds were exposed to weak (15 mT) or strong (150 mT) MFs for 6, 12, and 24 hours. While salinity reduced germination and seedling growth in nontreated seeds, MF-treated seeds showed better performance in terms of the aforementioned traits, regardless of the NaCl concentration tested. The authors also found that the improved response of MF-treated seeds was related to the maintenance of the water content and suggested that MF treatment primed the plant for salinity by H_2O_2 signaling. Magnetic field priming for 6 hours was suggested for enhancing germination and growth of sweet corn under salt stress (Karimi et al. 2017). In another study, Kataria et al. (2017) investigated the effects of presowing seed treatment with MFs (200 mT for 1 h) in alleviating the adverse effects of salt stress on germination in soybean and maize. In both species, magneto-primed seeds had enhanced percentage germination and early seedling growth parameter values (root and shoot length, and vigor indexes) under different salinity levels (0–100 mM NaCl) in comparison with untreated seeds. Interestingly, these authors also noticed that the levels of O_2^- and

H_2O_2 in germinating magneto-primed seeds of maize and soybean were also increased. They suggested that increased water uptake, greater activity of hydrolytic enzymes (α-amylase and protease), and increased free radical content in MF-treated seeds as compared with untreated seeds under both nonsaline and saline conditions enhanced the rate of germination and seedling vigor (Kataria et al. 2017).

12.4.3 *Ultraviolet Radiation*

UV radiation is a part of the nonionizing region of the electromagnetic spectrum and constitutes approximately 8–9% of total solar radiation (Coohill 1989). As a consequence of the stratospheric ozone layer being depleted, the amounts of solar UV radiation reaching the earth's surface are increasing, as is interest in understanding the mechanisms by which plants may protect themselves from this threat (Hollósy 2002). UV exposure triggers specific protective responses in plants, whose molecular signature could be used to design new treatments to improve the response to abiotic constraints as salinity.

The information available regarding the potential use of UV radiation as a salinity alleviation treatment is very limited. To the best of our knowledge, only the work of Ouhibi et al. (2014) is available in the PubMed bibliographic repository of the National Center for Biotechnology Information (NCBI; https://www.ncbi.nlm.nih.gov, accessed on the 10th August 2017). In this work, lettuce (*Lactuca sativa*) seeds were treated with UV-C radiation in an attempt to improve the resulting seedlings' tolerance of salt stress (Ouhibi et al. 2014). Two levels of UV-C radiation (0.85 or 3.42 kJ m^{-2}) were tested, and UV-primed seeds and nonprimed seeds were grown under either nonsaline conditions or 100 mM NaCl. In nonprimed seeds, salt stress resulted in a smaller increase in the fresh weights of roots and leaves, accompanied by a restriction in tissue hydration and K^+ ion uptake, as well as an increase in Na^+ ion concentrations in all organs. These effects were mitigated in plants grown from the UV-C primed seeds, with the salt-mitigating effect of UV-C being more pronounced at 0.85 kJ m^{-2} than at 3.42 kJ m^{-2}, which suggested a dose-dependent effect of the treatment. The authors suggested that UV-C priming could be used as a simple and cost-effective strategy to alleviate NaCl-induced stress in lettuce (Ouhibi et al. 2014).

UV-C light has been described as a powerful tool for stimulating the synthesis and accumulation of health-promoting phytochemicals (before and after harvest), extending the shelf life of fresh plant products or stimulating plant defenses against biotic attacks (Urban et al. 2016). Consequently, the ability of UV-C to stimulate plant adaptation to abiotic constraints is postulated, and it needs to be further investigated.

12.5 Field and Water Management Practices

We have previously described some chemical, physical, and biological treatments that can be exogenously applied to plants. Nevertheless, added value may be obtained if these treatments are coupled to adequate field and water management practices, constituting a useful approach to improvement of soil quality (Li et al. 2016). Such efforts should aim to minimize unproductive water losses, maintain soil salinity at tolerable levels, and enhance organic matter and nutrient availability (Bezborodov et al. 2010). Several technologies have been developed and used to control the aforementioned constraints, including mulching of the soil surface with plastic or crop residues (Pang et al. 2010; Maomao et al. 2014; Zhao et al. 2014; Li et al. 2016), a buried straw layer (Zhao et al. 2014), repellent salinity agents (Shahein et al. 2015), or conservation agricultural practices (Devkota et al. 2015). Mulching of the soil surface, using materials such as crop residues or plastic, is considered to be one of the main options to mitigate the toxic effects of soil salinization, promoting soil quality and consequently increasing crop yield (Bezborodov et al. 2010; Zhang et al. 2014; Xie et al. 2017). As residues are placed on the soil surface, they will shade the topsoil and prevent water loss by evaporation, increase soil moisture accumulation, and help reduce salt accumulation (Li et al. 2013; Li et al. 2016). When the material used derives from crop residues, it additionally assists in building soil organic carbon (Bezborodov et al. 2010; Maomao et al. 2014). As these substrates of natural origin decompose, organic carbon is released into the saline soil and mitigates the nutrient limitation of the microbial community, increasing soil microbial activity (Kamble et al. 2014; Li et al. 2016; Xie et al. 2017). Xie et al. (2017) concluded that maize straw application was successful in alleviating topsoil salinity. As straw decomposes, organic carbon fractions are released into the soil, promoting soil aggregation, increasing soil porosity, and reducing salt accumulation in the surface soil (Zhao et al. 2016a, Xie et al. 2017). Moreover, incorporation of a straw layer into the soil interrupts the continuity of capillary movement of salt from deeper soil layers and ultimately reduces salt levels in topsoil (Zhao et al. 2016b; Xie et al. 2017). Practices such as burying a straw layer also have a significant effect on soil water distribution. Zhao et al. (2016a) observed that use of a buried maize straw layer at a depth of 40 cm, combined with a plastic mulch cover, led to greater soil water content and significantly reduced salt content in the upper 40 cm depth, throughout the growth season of sunflower. Buried straw layer treatment alone induced a decrease in salt content in the early growth period (Zhao et al. 2016a). A buried straw layer combined with plastic mulching also had a positive effect on the microbial community, leading to significantly greater populations of bacteria, actinomyces, and fungi (Li et al. 2016). Recent findings have demonstrated that adoption of conservation agriculture practices—involving reduced tillage, residue retention, and crop rotation—have great potential to positively affect salt accumulation by reducing evaporation and upward salt transport (Devkota et al. 2015; Murphy et al. 2016). As soil water evaporates, it leads to salt accumulation on the surface. This process can explain the more pronounced soil salinization occurring near the soil surface (10 cm) than in the top 90 cm soil fraction (Devkota et al. 2015). Therefore, the effect of crop residues on salt accumulation is more efficient up to 30 cm from the soil surface; below this depth, its effect is insignificant (Davkota et al. 2015).

Soil remediation can also be achieved through the addition of antisalinity agents such as Dinamic, Uni-sal, or humic acid applied through drip irrigation. Shahein et al. (2015) demonstrated that a combination of Dinamic and Uni-sal led to successful increases in the leaf nutrient profile, yield, fruit quality, shoot weight, and plant length in tomato plants irrigated with saline water (Shahein et al. 2015). In a similar study, the results obtained by El-Khawaga (2013) corroborated the positive influence of Uni-sal and other antisalinity agents such as Cal-Mor and citric acid on palm trees under salinity stress. Soil supplementation with sulfur also had a positive effect on date palm offshoots' tolerance of salinity, increasing accumulation of compatible solutes and chlorophyll content, which resulted in enhanced dry weight accumulation and water status (Abbas et al. 2015). The benefits of sulfur and sulfur-containing compounds could be related to their involvement in improving nutrient assimilation and stimulating ROS-scavenging activity through the sulfur metabolite glutathione (Khan et al. 2014; Abbas et al. 2015).

12.6 Conclusions

The development of new strategies to alleviate salinity in plants and stimulate crop cultivation under salinity stress is a relevant research purpose to face the needs of a world with growing soil salinity issues. Chemical, biological, and physical treatments are now available to address this challenge, being able to induce salinity tolerance at a systemic level, even if they are only applied on the seeds, roots, or leaves. Importantly, new agricultural practices focused on field and water management are also available to mitigate the harmful effects of excessive saline soils on growth and productivity of crops.

As described in this chapter, salinity can affect plant metabolic processes in different ways, depending not only on the time frame of the response to the salt constraint but also on the organ/tissue in which the response took place, as well as the plant genotype or species studied. A considerable number of studies have been aimed at understanding the molecular mechanism behind salinity tolerance in plants, with the ultimate goal being to reduce the effects of salinity on growth and yield (Munns and Tester 2008; Roy et al. 2014, Negrão et al. 2017). Physiological responses and biochemical networks involved in salt-tolerant plant responses to salinity have been revealed, and new salt stress–related genes have been made available for plant improvement.

The majority of the strategies described herein are seed treatments—also known as priming treatments—which trigger a physiological state that enables the growing plant to respond faster and in a more robust way when exposed to environmental constraints (Aranega-Bou et al. 2014; Savvides et al. 2016). Consequently, the previous fundamental knowledge about how a plant reacts to the effects of a high salt concentration has been essential for the development of strategies for salinity alleviation. As we have described, treatments applied to alleviate the effects of salinity, aimed at accumulation of compatible solutes (e.g., proline and glycine betaine) or photosynthetic pigments, ROS scavenging (e.g., polyamines and antioxidant enzymes), or stimulation of ROS-mediated signaling cascades (e.g., H_2O_2 accumulation) have been designed. Those approaches have been revealed to enhance the plant's ability to ensure water and nutrient uptake, photosynthesis, biomass accumulation and, ultimately, yield. An important future application of these "biomarkers" is that they can be used to monitor the suitability of new formulations or to screen for plant accessions that are better adapted to coping with saline conditions.

Omics technologies - global molecular profiling - applied to target plant species and crops can provide new insights into the molecular mechanism underlying plant adaptation to salinity. Transcriptomics, proteomics, and metabolomics are extending our knowledge of the molecular mechanisms associated with plant response to salinity. In time, they will reveal new target regulatory points that might possibly be modulated by the application of treatments "by design," addressing the particularities of the soil and the crop's requirements. These approaches also need to be combined with an investment in new and sustainable agricultural practices that can ensure food security for local populations under the threat of the growing salinity scenario.

Acknowledgements Financial support from Fundação para a Ciência e a Tecnologia (Lisbon, Portugal) is acknowledged through the research unit "GREEN-it: Bioresources for Sustainability" (UID/Multi/04551/2013), a SSA postdoctoral grant (SFRH/BPD/108032/2015), a DM Plants for Life PhD grant (PD/BD/128498/2017), and a SFC research fellowship in the scope of the FCT project (PTDC/AGR-PRO/4261/2014).

References

Abbas MF, Abbas MJ, Shareef HJ (2015) Role of sulphur in salinity tolerance of date palm (*Phoenix dactylifera* L.) offshoots cvs. Berhi and Sayer. Int J Agric Food Sci 5(3):92–97

Abdul Qados AMS, Moftah AE (2015) Influence of silicon and nano-silicon on germination, growth and yield of faba bean (*Vicia faba* L.) under salt stress conditions. AJEA 5(6):509–524. https://doi.org/10.9734/AJEA/2015/14109

Ahmed AHH, Ghalab ARM, Hussein OS, Hefny AM (2011) Effect of gamma rays and salinity on growth and chemical composition of *Ambrosia maritima* L. Plants J Rad Res Appl Sci 4:1139–1162

Ahuja S, Kumar M, Kumar P et al (2014) Metabolic and biochemical changes caused by gamma irradiation in plants. J Radioanal Nucl Chem 300:199–212. https://doi.org/10.1007/s10967-014-2969-5

Ali S, Charles TC, Glick BR (2014) Amelioration of high salinity stress damage by plant growth–promoting bacterial endophytes that contain ACC deaminase. Plant Physiol Biochem 80:160–167. https://doi.org/10.1016/j.plaphy.2014.04.003

Almutaiori ZM (2016) Effect of nano-silicon application on the expression of salt tolerance genes in germinating tomato (*Solanum lycopersicum* L.) seedlings under salt stress. Plant Omics J 9(1):106–114

Aloui H, Souguir M, Latique S, Hannachi C (2014) Germination and growth in control and primed seeds of pepper as affected by salt stress. Cercetari Agronomice in Moldova 47(3). https://doi.org/10.2478/cerce-2014-0029

Anaya F, Fghire R, Wahbi S, Loutfi K (2018) Influence of salicylic acid on seed germination of *Vicia faba* L. under salt stress. Journal of the Saudi Society of Agricultural Sciences. https://doi.org/10.1016/j.jssas.2015.10.002

Aranega-Bou P, de la O Leyva M, Finiti I et al (2014) Priming of plant resistance by natural compounds: hexanoic acid as a model. Front Plant Sci 5:488. https://doi.org/10.3389/fpls.2014.00488

Araújo SS, Paparella S, Dondi D et al (2016) Physical methods for seed invigoration: advantages and challenges in seed technology. Front Plant Sci 7:646. https://doi.org/10.3389/fpls.2016.00646

Azad K, Kaminskyj S (2016) A fungal endophyte strategy for mitigating the effect of salt and drought stress on plant growth. Symbiosis 68:73–78. https://doi.org/10.1007/s13199-015-0370-y

Azooz MM, El-Zahrani AM, Youssef MM (2013) The potential role of seed priming with ascorbic acid and nicotinamide and their interactions to enhance salt tolerance of *Vicia faba* L. Aust J Crop Sci 7(13):2091–2100

Baby SM, Narayanaswamy GK, Anand A (2011) Superoxide radical production and performance index of photosystem II in leaves from magnetoprimed soybean seeds. Plant Signal Behav 6:1635–1637. https://doi.org/10.4161/psb.6.11.17720

Bal HB, Nayak L, Das S, Adhya TK (2013) Isolation of ACC deaminase producing PGPR from rice rhizosphere and evaluating their plant growth promoting activity under salt stress. Plant Soil 366:93–105. https://doi.org/10.1007/s11104-012-1402-5

Barnawal D, Bharti N, Maji D et al (2014) ACC deaminase–containing *Arthrobacter protophormiae* induces NaCl stress tolerance through reduced ACC oxidase activity and ethylene production resulting in improved nodulation and mycorrhization in *Pisum sativum*. J Plant Physiol 171:884–894. https://doi.org/10.1016/j.jplph.2014.03.007

Bezborodov GA, Shadmanov DK, Mirhashimov RT et al (2010) Mulching and water quality effects on soil salinity and sodicity dynamics and cotton productivity in Central Asia. Agric Ecosyst Environ 138:95–102. https://doi.org/10.1016/j.agee.2010.04.005

Bharti N, Yadav D, Barnawal D et al (2013) *Exiguobacterium oxidotolerans*, a halotolerant plant growth promoting rhizobacteria, improves yield and content of secondary metabolites in *Bacopa monnieri* (L.) Pennell under primary and secondary salt stress. World J Microbiol Biotechnol 29:379–387. https://doi.org/10.1007/s11274-012-1192-1

Bockheim JG, Gennadiyev AN (2000) The role of soil-forming processes in the definition of taxa in Soil Taxonomy and the World Soil Reference Base. Geoderma 95:53–72. https://doi.org/10.1016/S0016-7061(99)00083-X

Campagnac E, Khasa DP (2013) Relationship between genetic variability in *Rhizophagus irregularis* and tolerance to saline conditions. Mycorrhiza 24:121–129. https://doi.org/10.1007/s00572-013-0517-8

Campanelli A, Ruta C, De MG, Morone-Fortunato I (2012) The role of arbuscular mycorrhizal fungi in alleviating salt stress in *Medicago sativa* L. var. icon. Symbiosis 59:65–76. https://doi.org/10.1007/s13199-012-0191-1

Chakraborty K, Bose J, Shabala L, Shabala S (2016) Difference in root K^+ retention ability and reduced sensitivity of K^+-permeable channels to reactive oxygen species confer differential salt tolerance in three *Brassica* species. J Exp Bot 67:4611–4625. https://doi.org/10.1093/jxb/erw236

Chen L, Liu Y, Wu G et al (2016) Induced maize salt tolerance by rhizosphere inoculation of *Bacillus amyloliquefaciens* SQR9. Physiol Plant 158:34–44. https://doi.org/10.1111/ppl.12441

Chunthaburee S, Pattanagul W, Theerakulpisut P, Sanitchon J (2014) Alleviation of salt stress in seedlings of black glutinous rice by seed priming with spermidine and gibberellic acid. Not Bot Horti Agrobo 42(2):405–413. https://doi.org/10.1583/nbha4229688

Coohill TP (1989) Ultraviolet action spectra (280 to 380 nm) and solar effectiveness spectra for higher plants. Photochem Photobiol 50:451–457. https://doi.org/10.1111/j.1751-1097.1989.tb05549.x

Damodaran T, Rai RB, Jha SK et al (2014) Rhizosphere and endophytic bacteria for induction of salt tolerance in gladiolus grown in sodic soils. J Plant Interact 9:577–584. https://doi.org/10.1080/17429145.2013.873958

Das P, Nutan KK, Singla-Pareek SL, Pareek A (2015) Understanding salinity responses and adopting "omics-based" approaches to generate salinity tolerant cultivars of rice. Front Plant Sci 6:712. https://doi.org/10.3389/fpls.2015.00712

Dawood MG, El-Awadi ME (2015) Alleviation of salinity stress on *Vicia faba* plants via seed priming with melatonin. Acta Biol Colomb 20(2):223–235. https://doi.org/10.15446/abc.v20n2.43291

de Souza MO, Pelacani CR, Willems LA, Castro RD, Hilhorst HW, Ligterink W (2016) Effect of osmopriming on germination and initial growth of *Physalis angulata* L. under salt stress and on expression of associated genes. An Acad Bras Cienc 88(Suppl 1):503–516. https://doi.org/10.1590/0001-3765201620150043

del Amor FM, Cuadra-Crespo P (2011) Alleviation of salinity stress in broccoli using foliar urea or methyl-jasmonate: analysis of growth, gas exchange, and isotope composition. Plant Growth Regul 63(1):55–62. https://doi.org/10.1007/s10725-010-9511-8

De-la-Peña C, Loyola-Vargas VM (2014) Biotic interactions in the rhizosphere: a diverse cooperative enterprise for plant productivity. Plant Physiol 166:701–719. https://doi.org/10.1104/pp.114.241810

Devkota M, Martius C, Gupta RK et al (2015) Managing soil salinity with permanent bed planting in irrigated production systems in Central Asia. Agric Ecosyst Environ 202:90–97. https://doi.org/10.1016/j.agee.2014.12.006

Dkhil BB, Issa A, Denden M (2014) Germination and seedling emergence of primed okra (*Abelmoschuis esculentus* L.) seeds under salt stress and low temperature. Am. J Plant Physiol 9(2):38–45. https://doi.org/10.3923/ajpp.2014.38.45

El-Khawaga AS (2013) Effect of antisalinity agents on growth and fruiting of different date palm cultivars. Asian J Crop Sci 5:65–80. https://doi.org/10.3923/ajcs.2013.65.80

Ellouzi H, Ben Hamed K, Hernández I et al (2014) A comparative study of the early osmotic, ionic, redox and hormonal signaling response in leaves and roots of two halophytes and a glycophyte to salinity. Planta 240:1299–1317. https://doi.org/10.1007/s00425-014-2154-7

El-Mashad AAA, Mohamed HI (2012) Brassinolide alleviates salt stress and increases antioxidant activity of cowpea plants (*Vigna sinensis*). Protoplasma 249:625–635. https://doi.org/10.1007/s00709-011-0300-7

Estrada B, Aroca R, Barea JM, Ruiz-Lozano JM (2013) Native arbuscular mycorrhizal fungi isolated from a saline habitat improved maize antioxidant systems and plant tolerance to salinity. Plant Sci 201–202:42–51. https://doi.org/10.1016/j.plantsci.2012.11.009

Farooq MA, Saqib ZA, Akhtar J et al (2015) Protective role of silicon (Si) against combined stress of salinity and boron (B) toxicity by improving antioxidant enzymes activity in rice. Silicon:1–5. https://doi.org/10.1007/s12633-015-9346-z

Flowers TJ, Colmer TD (2015) Plant salt tolerance: adaptations in halophytes. Ann Bot 115:327–331. https://doi.org/10.1093/aob/mcu267

Gebreegziabher BG, Qufa CA (2017) Plant physiological stimulation by seeds salt priming in maize (*Zea mays*): prospect for salt tolerance. Afr J Biotechnol 16(5):209–223. https://doi.org/10.5897/AJB2016.15819

Gharsallah C, Fakhfakh H, Grubb D, Gorsane F (2016) Effect of salt stress on ion concentration, proline content, antioxidant enzyme activities and gene expression in tomato cultivars. AoB Plants. https://doi.org/10.1093/aobpla/plw055

Gholami M, Mokhtarian F, Baninasab B (2015) Seed halopriming improves the germination performance of black seed (*Nigella sativa*) under salinity stress conditions. J Crop Sci Biotechnol 18(1):21–26. https://doi.org/10.1007/s12892-014-0078-1

Grieve CM, Grattan SR, Maas E V. (2008) Plant salt tolerance. In: WW Wallender and KK Tanji (eds.) Agricultural salinity assessment and management. 2nd edition. Reston: American Society of Civil Engineers, pp. 405–459.

Gururani MA, Upadhyaya CP, Baskar V et al (2013) Plant growth–promoting rhizobacteria enhance abiotic stress tolerance in *Solanum tuberosum* through inducing changes in the expression of ROS-scavenging enzymes and improved photosynthetic performance. J Plant Growth Regul 32:245–258. https://doi.org/10.1007/s00344-012-9292-6

Habib SH, Kausar H, Saud HM (2016) Plant growth–promoting rhizobacteria enhance salinity stress tolerance in okra through ROS-scavenging enzymes. Biomed Res Int. https://doi.org/10.1155/2016/6284547

Hajiboland R (2013) Role of arbuscular mycorrhiza in amelioration of salinity. In: Ahmad P, Azooz MA, Prasad MNV (eds) Salt stress in plants: signalling, omics and adaptations. Springer Science Business Media, New York, pp 301–354

Hanin M, Ebel C, Ngom M et al (2016) New insights on plant salt tolerance mechanisms and their potential use for breeding. Front Plant Sci 7:1787. https://doi.org/10.3389/fpls.2016.01787

Hayat S, Hayat Q, Alyemeni MN et al (2012) Role of proline under changing environments: a review. Plant Signal Behav 7:1456–1466. https://doi.org/10.4161/psb.21949

Hollósy F (2002) Effects of ultraviolet radiation on plant cells. Micron 33:179–197. https://doi.org/10.1016/S0968-4328(01)00011-7

Ismail AM, Horie T (2017) Genomics, physiology, and molecular breeding approaches for improving salt tolerance. Annu Rev Plant Biol 68:annurev-arplant-042916-040936. https://doi.org/10.1146/annurev-arplant-042916-040936

Jisha KC, Puthur JT (2016) Seed priming with BABA (beta-amino butyric acid): a cost-effective method of abiotic stress tolerance in *Vigna radiata* (L.) Wilczek. Protoplasma 253(2):277–289. https://doi.org/10.1007/s00709-015-0804-7

Kamble PN, Gaikwad VB, Kuchekar SR, Bååth E (2014) Microbial growth, biomass, community structure and nutrient limitation in high pH and salinity soils from Pravaranagar (India). Eur J Soil Biol 65:87–95. https://doi.org/10.1016/j.ejsobi.2014.10.005

Kang SM et al (2014) Plant growth promoting rhizobacteria reduce adverse effects of salinity and osmotic stress by regulating phytohormones and antioxidants in *Cucumis sativus*. J Plant Interact 9(1):673–682. https://doi.org/10.1080/17429145.2014.894587

Karimi S, Eshghi S, Karimi S, Hasan-Nezhadian S (2017) Inducing salt tolerance in sweet corn by magnetic priming. Acta Agriculturae Slovenica 109:89. https://doi.org/10.14720/aas.2017.109.1.09

Kasim WA, Gaafar RM, Abou-Ali RM et al (2016) Effect of biofilm forming plant growth promoting rhizobacteria on salinity tolerance in barley. Ann Agric Sci 61:217–227. https://doi.org/10.1016/j.aoas.2016.07.003

Kataria S, Baghel L, Guruprasad KN (2017) Pre-treatment of seeds with static magnetic field improves germination and early growth characteristics under salt stress in maize and soybean. Biocatal Agric Biotechnol 10:83–90. https://doi.org/10.1016/j.bcab.2017.02.010

Kavi Kishor PB, Sreenivasulu N (2014) Is proline accumulation per se correlated with stress tolerance or is proline homeostasis a more critical issue? Plant Cell Environ 37:300–311. https://doi.org/10.1111/pce.12157

Keshavarz H, Modares Sanavy SAM (2015) Biochemical and morphological response of common bean (*Phaseolus vulgaris* L.) to salinity stress and vitamin B_{12}. Intl J Farm Alli Sci 4(7):585–593

Khan HA, Ziaf K, Amjad M, Iqbal Q (2012) Exogenous application of polyamines improves germination and early seedling growth of hot pepper. Chil J Agric Res 72(3):429–433. https://doi.org/10.4067/S0718-58392012000300018

Khan NA, Khan MIR, Asgher M et al (2014) Plant biochemistry & physiology salinity tolerance in plants: revisiting the role of sulfur metabolites. J Plant Biochem Phsiol 2:1–8. https://doi.org/10.4172/2329-9029.1000120

Kim YH, Khan AL, Waqas M, Shim JK, Kim DH, Lee KY, Lee IJ (2014) Silicon application to rice root zone influenced the phytohormonal and antioxidant responses under salinity stress. J Plant Growth Regul 33(2):137–149. https://doi.org/10.1007/s00344-013-9356-2

Kubala S, Wojtyla L, Quinet M, Lechowska K, Lutts S, Garnczarska M (2015) Enhanced expression of the proline synthesis gene P5CSA in relation to seed osmopriming improvement of *Brassica napus* germination under salinity stress. J Plant Physiol 183:1–12. https://doi.org/10.1016/j.jplph.2015.04.009

Kumar P, Sharma V, Atmaram CK, Singh B (2017) Regulated partitioning of fixed carbon (^{14}C), sodium (Na^+), potassium (K^+) and glycine betaine determined salinity stress tolerance of gamma irradiated pigeonpea [*Cajanus cajan* (L.) Millsp]. Environ Sci Pollut Res 24:7285–7297. https://doi.org/10.1007/s11356-017-8406-x

Landi S, Hausman J-F, Guerriero G, Esposito S (2017) Poaceae vs. abiotic stress: focus on drought and salt stress, recent insights and perspectives. Front Plant Sci 8:1214. https://doi.org/10.3389/fpls.2017.01214

Li SX, Wang ZH, Li SQ et al (2013) Effect of plastic sheet mulch, wheat straw mulch, and maize growth on water loss by evaporation in dryland areas of China. Agric Water Manag 116:39–49. https://doi.org/10.1016/j.agwat.2012.10.004

Li H, Zhu Y, Hu Y, Han W, Gong H (2015) Beneficial effects of silicon in alleviating salinity stress of tomato seedlings grown under sand culture. Acta Physiol Plant 37(4). https://doi.org/10.1007/s11738-015-1818-7

Li Y-Y, Pang H-C, Han X-F et al (2016) Buried straw layer and plastic mulching increase microflora diversity in salinized soil. J Integr Agric 15:1602–1611. https://doi.org/10.1016/S2095-3119(15)61242-4

Liu S, Guo X, Feng G et al (2016) Indigenous arbuscular mycorrhizal fungi can alleviate salt stress and promote growth of cotton and maize in saline fields. Plant Soil 398:195–206. https://doi.org/10.1007/s11104-015-2656-5

Liu Y, El-Kassaby YA (2015) Timing of seed germination correlated with temperature-based environmental conditions during seed development in conifers. Seed Sci Res 25:29–45. https://doi.org/10.1017/S0960258514000361

Ma H, Song L, Shu Y et al (2012) Comparative proteomic analysis of seedling leaves of different salt tolerant soybean genotypes. J Proteome 75:1529–1546. https://doi.org/10.1016/j.jprot.2011.11.026

Maffei ME (2014) Magnetic field effects on plant growth, development, and evolution. Front Plant Sci 5:445. https://doi.org/10.3389/fpls.2014.00445

Manivannan A, Soundararajan P, Arum LS, Ko CH, Muneer S, Jeong BR (2015) Silicon-mediated enhancement of physiological and biochemical characteristics of *Zinnia elegans* 'Dreamland Yellow' grown under salinity stress. Hortic Environ Biotechnol 56(6):721–731. https://doi.org/10.1007/s13580-015-1081-2

Maomao H, Xiaohou S, Yaming Z (2014) Effects of different regulatory methods on improvement of greenhouse saline soils, tomato quality, and yield. Sci World J. https://doi.org/10.1155/2014/953675

Miladinov ZJ, Balešević-Tubić SN, ĐorĐević VB, Đukić VH, Ilić AD, Čobanović LM (2015) Optimal time of soybean seed priming and primer effect under salt stress conditions. JAS 60(2):109–117. https://doi.org/10.2298/JAS1502109M

Munns R (2005) Genes and salt tolerance: bringing them together. New Phytol 167:645–663. https://doi.org/10.1111/j.1469-8137.2005.01487.x

Munns R, Gilliham M (2015) Salinity tolerance of crops—what is the cost? New Phytol 208:668–673. https://doi.org/10.1111/nph.13519

Munns R, James RA, Läuchli A (2006) Approaches to increasing the salt tolerance of wheat and other cereals. J Exp Bot 57:1025–1043. https://doi.org/10.1093/jxb/erj100

Munns R, Tester M (2008) Mechanisms of salinity tolerance. Annu Rev Plant Biol 59:651–681. https://doi.org/10.1146/annurev.arplant.59.032607.092911

Murphy RP, Montes-Molina JA, Govaerts B et al (2016) Crop residue retention enhances soil properties and nitrogen cycling in smallholder maize systems of Chiapas, Mexico. Appl Soil Ecol 103:110–116. https://doi.org/10.1016/j.apsoil.2016.03.014

Nadeem SM, Ahmad M, Zahir ZA et al (2014) The role of mycorrhizae and plant growth promoting rhizobacteria (PGPR) in improving crop productivity under stressful environments. Biotechnol Adv 32:429–448. https://doi.org/10.1016/j.biotechadv.2013.12.005

Nasibi F, Kalantari KM, Zanganeh R, Mohammadinejad G, Oloumi H (2016) Seed priming with cysteine modulates the growth and metabolic activity of wheat plants under salinity and osmotic stresses at early stages of growth. Indian J Plant Physiol 21(3):279–286. https://doi.org/10.1007/s40502-016-0233-4

Nawaz J, Hussain M, Jabbar A, Nadeem GA, Sajid M, Subtain M, Shabbir I (2013) Seed priming: a technique. Int J Agri Crop Sci 6(20):1373–1381

Naz F, Gul H, Hamayun M, Sayyed A, Husna SK, Sherwani (2014) Effect of NaCl stress on *Pisum sativum* germination and seedling growth with the influence of seed priming with potassium (KCl and KOH). American-Eurasian J Agric Environ Sci 14 (11):1304–1311. doi:https://doi.org/10.5829/idosi.aejaes.2014.14.11.748

Nazir MS, Saad A, Anjum Y, Ahmad W (2014) Possibility of seed priming for good germination of cotton seed under salinity stress. J Biol Agric Health 4(8):66–68

Negrão S, Schmöckel SM, Tester M (2017) Evaluating physiological responses of plants to salinity stress. Ann Bot 119:1–11. https://doi.org/10.1093/aob/mcw191

Nongpiur RC, Singla-Pareek SL, Pareek A (2016) Genomics approaches for improving salinity stress tolerance in crop plants. Curr Genomics 17:343–357. https://doi.org/10.2174/1389202917666160331202517

Nounjan N, Nghia PT, Theerakulpisut P (2012) Exogenous proline and trehalose promote recovery of rice seedlings from salt-stress and differentially modulate antioxidant enzymes and expression of related genes. J Plant Physiol 169(6):596–604. https://doi.org/10.1016/j.jplph.2012.01.004

Ouhibi C, Attia H, Rebah F et al (2014) Salt stress mitigation by seed priming with UV-C in lettuce plants: growth, antioxidant activity and phenolic compounds. Plant Physiol Biochem 83:126–133. https://doi.org/10.1016/j.plaphy.2014.07.019

Pang HC, Li YY, Yang JS, Sen LY (2010) Effect of brackish water irrigation and straw mulching on soil salinity and crop yields under monsoonal climatic conditions. Agric Water Manag 97:1971–1977. https://doi.org/10.1016/j.agwat.2009.08.020

Paparella S, Araujo SS, Rossi G, Wijayasinghe M, Carbonera D, Balestrazzi A (2015) Seed priming: state of the art and new perspectives. Plant Cell Rep 34(8):1281–1293. https://doi.org/10.1007/s00299-015-1784-y

Peng J, Li Z, Wen X et al (2014) Salt-induced stabilization of EIN3/EIL1 confers salinity tolerance by deterring ROS accumulation in *Arabidopsis*. PLoS Genet 10(10):e1004664. https://doi.org/10.1371/journal.pgen.1004664

Per TS, Khan NA, Reddy PS et al (2017) Approaches in modulating proline metabolism in plants for salt and drought stress tolerance: phytohormones, mineral nutrients and transgenics. Plant Physiol Biochem 115:126–140. https://doi.org/10.1016/j.plaphy.2017.03.018

Pietruszewski S, Martínez E (2015) Magnetic field as a method of improving the quality of sowing material: a review. Int Agrophy 29:377–389. https://doi.org/10.1515/intag-2015-0044

Pradhan M, Prakash P, Manimurugan C, Tiwari SK, Sharma RP, Singh PM (2015) Screening of tomato genotypes using osmopriming with PEG 6000 under salinity stress. Res Environ Life Sci 8(2):245–250

Puyang X, An M, Xu L, Han L, Zhang X (2016) Protective effect of exogenous spermidine on ion and polyamine metabolism in Kentucky bluegrass under salinity stress. Hortic Environ Biotechnol 57(1):11–19. https://doi.org/10.1007/s13580-016-0113-x

Querejeta JI, Allen MF, Caravaca F, Roldán A (2006) Differential modulation of host plant delta^{13}C and delta^{18}O by native and nonnative arbuscular mycorrhizal fungi in a semiarid environment. New Phytol 169:379–387. https://doi.org/10.1111/j.1469-8137.2005.01599.x

Rajaeian SO, Ehsanpour AA (2015) Physiological responses of tobacco plants (*Nicotiana rustica)* pretreated with ethanolamine to salt stress. Russ J Plant Physiol 62:246–252. https://doi.org/10.1134/S1021443715020156

Rathod GR, Anand A (2016) Effect of seed magneto-priming on growth, yield and Na/K ratio in wheat (*Triticum aestivum* L.) under salt stress. Indian J Plant Physiol 21:15–22. https://doi.org/10.1007/s40502-015-0189-9

Rejili M, Telahigue D, Lachiheb B et al (2008) Impact of gamma radiation and salinity on growth and K^+/Na^+ balance in two populations of *Medicago sativa* (L.) cultivar Gabès. Prog Nat Sci 18:1095–1105. https://doi.org/10.1016/j.pnsc.2008.04.004

Roy SJ, Negrão S, Tester M et al (2014) Salt resistant crop plants. Curr Opin Biotechnol 26:115–124. https://doi.org/10.1016/j.copbio.2013.12.004

Ruppel S, Franken P, Witzel K (2013) Properties of the halophyte microbiome and their implications for plant salt tolerance. Funct Plant Biol 40:940–951. https://doi.org/10.1071/FP12355

Sabaghnia N, Janmohammadi JM (2015) Effect of nano-silicon particles application on salinity tolerance in early growth of some lentil genotypes. Annales UMCS Sectio C 69(2):39–55. https://doi.org/10.1515/umcsbio-2015-0004

Salama KHA, Mansour MMF (2015) Choline priming–induced plasma membrane lipid alterations contributed to improved wheat salt tolerance. Acta Physiol Plantarum 37(7):170. https://doi.org/10.1007/s11738-015-1934-4

Savvides A, Ali S, Tester M, Fotopoulos V (2016) Chemical priming of plants against multiple abiotic stresses: mission possible? Trends Plant Sci 21(4):329–340. https://doi.org/10.1016/j.tplants.2015.11.003

Sen S, Mandal P (2016) Solid matrix priming with chitosan enhances seed germination and seedling invigoration in mung bean under salinity stress. JCEA 17(3):749–762. https://doi.org/10.5513/jcea01/17.3.1773

Shahbaz M, Mushtaq Z, Andaz F, Masood A (2013) Does proline application ameliorate adverse effects of salt stress on growth, ions and photosynthetic ability of eggplant (*Solanum melongena* L.)? Scientia Hortic 164:507–511. https://doi.org/10.1016/j.scienta.2013.10.001

Shahein MM, Husein ME, Abou-El-Hassan S (2015) Alleviating adverse effect of saline irrigation water on growth and productivity of tomato plants via some repellant salinity agents. Int J Biosci 8(5):659–665

Shrivastava P, Kumar R (2015) Soil salinity: a serious environmental issue and plant growth promoting bacteria as one of the tools for its alleviation. Saudi J Biol Sci 22:123–131. https://doi.org/10.1016/j.sjbs.2014.12.001

Singh R, Singh Y, Xalaxo S et al (2016) From QTL to variety-harnessing the benefits of QTLs for drought, flood and salt tolerance in mega rice varieties of India through a multi-institutional network. Plant Sci 242:278–287. https://doi.org/10.1016/j.plantsci.2015.08.008

Talat A, Nawaz K, Hussian K, Bhatti KH, Siddiqi EH, Khalid A, Anwer S, Sharif MU (2013) Foliar application of proline for salt tolerance of two wheat (*Triticum aestivum* L.) cultivars. World Appl Sci J 22(4):547–554. https://doi.org/10.5829/idosi.wasj.2013.22.04.19570

Tamini SM (2016) Effect of seed priming on growth and physiological traits of five Jordanian wheat (*Triticum aestivum* L.) landraces under salt stress. J Biosci Agric Res 11(1):906–922. https://doi.org/10.18801/jbar.110116.111

Tao J-J, Chen H-W, Ma B et al (2015) The role of ethylene in plants under salinity stress. Front Plant Sci 6:1059. https://doi.org/10.3389/fpls.2015.01059

Theerakulpisut P, Kanawapee N, Panwong B (2016) Seed priming alleviated salt stress effects on rice seedlings by improving Na^+/K^+ and maintaining membrane integrity. Int. J Plant Biol 7(6402):53–58. https://doi.org/10.4081/pb.2016.6402

Urban L, Charles F, de Miranda MRA, Aarrouf J (2016) Understanding the physiological effects of UV-C light and exploiting its agronomic potential before and after harvest. Plant Physiol Biochem 105:1–11. https://doi.org/10.1016/j.plaphy.2016.04.004

Vaishnav A, Kumari S, Jain S, Choudhary DK, Sharma KP (2016) Exogenous chemical mediated induction of salt tolerance in soybean plants. IJALS 2(3):43–47. https://doi.org/10.9379/sf.ijals.122064-007-0081-x

Wolff SA, Coelho LH, Karoliussen I, Jost A-IK (2014) Effects of the extraterrestrial environment on plants: recommendations for future space experiments for the MELiSSA higher plant compartment. Life (Basel, Switzerland) 4:189–204. https://doi.org/10.3390/life4020189

Xie W, Wu L, Zhang Y et al (2017) Effects of straw application on coastal saline topsoil salinity and wheat yield trend. Soil Tillage Res 169:1–6. https://doi.org/10.1016/j.still.2017.01.007

Yan K, Xu H, Cao W, Chen X (2015) Salt priming improved salt tolerance in sweet sorghum by enhancing osmotic resistance and reducing root Na^+ uptake. Acta Physiol Plant 37:203. https://doi.org/10.1007/s11738-015-1957-x

Yin L, Wang S, Tanaka K, Fujihara S, Itai A, Den X, Zhang S (2016) Silicon-mediated changes in polyamines participate in silicon-induced salt tolerance in *Sorghum bicolor* L. Plant Cell Environ 39:245–258. https://doi.org/10.1111/pce.12521

Yuan Z, Druzhinina IS, Labbé J et al (2016) Specialized microbiome of a halophyte and its role in helping non-host plants to withstand salinity. Sci Rep 6:1–13. https://doi.org/10.1038/srep32467

Zhang H-X, Blumwald E (2001) Transgenic salt-tolerant tomato plants accumulate salt in foliage but not in fruit. Nat Biotechnol 19:765–768. https://doi.org/10.1038/90824

Zhang P, Wei T, Jia Z et al (2014) Effects of straw incorporation on soil organic matter and soil water-stable aggregates content in semiarid regions of Northwest China. PLoS One 9:1–11. https://doi.org/10.1371/journal.pone.0092839

Zhang Z, Chang XX, Zhang L, Li JM, Hu XH (2016) Spermidine application enhances tomato seedling tolerance to salinity–alkalinity stress by modifying chloroplast antioxidant systems. Russ J Plant Physiol 63(4):461–468. https://doi.org/10.1134/s102144371604018x

Zhao S, Li K, Zhou W et al (2016a) Changes in soil microbial community, enzyme activities and organic matter fractions under long-term straw return in North-Central China. Agric Ecosyst Environ 216:82–88. https://doi.org/10.1016/j.agee.2015.09.028

Zhao Y, Li Y, Wang J et al (2016b) Buried straw layer plus plastic mulching reduces soil salinity and increases sunflower yield in saline soils. Soil Tillage Res 155:363–370. https://doi.org/10.1016/j.still.2015.08.019

Zhao Y, Pang H, Wang J et al (2014) Effects of straw mulch and buried straw on soil moisture and salinity in relation to sunflower growth and yield. F. Crop Res 161:16–25. https://doi.org/10.1016/j.fcr.2014.02.006

Zhao FG, Song CP, He JQ, Zhu H (2007) Polyamines improve K^+/Na^+ homeostasis in barley seedlings by regulating root ion channel activities. Plant Physiol 145:1061–1072. https://doi.org/10.1104/pp.107.105882

Chapter 13
Polyamines and Their Metabolic Engineering for Plant Salinity Stress Tolerance

Tushar Khare, Amrita Srivastav, Samrin Shaikh, and Vinay Kumar

Abstract Polyamines (PAs) are small polycationic aliphatic amines and are ubiquitous in the plant kingdom. They play important roles in plant growth, development, and stress responses. Several research reports have established a correlation between their accumulation and salt stress tolerance in different plant species. Creditable research in the recent past has proved their vital roles in stress responses and adaptation strategies employed by plants, including scavenging of free radicals, neutralization of acids, and stabilization of cell membranes. They are able to bind several charged molecules including DNA, proteins, membrane phospholipids, and pectic polysaccharides, and have been credited with roles in protein phosphorylation and post-transcriptional modifications. They also play important roles in plant growth regulation, as well as acting as signaling molecules. Owing to their diverse functions in plant growth, development, and stress responses, they have emerged as potent targets for metabolic engineering to confer salt stress tolerance on manipulated plants. This chapter highlights their biosynthesis and transport, their exogenous applications to alleviate salt stress, and their metabolic engineering for developing salt-tolerant plants.

Keywords Salinity stress · Polyamines · Metabolic engineering · Spermine · Spermidine · Putrescine · Transgenics

T. Khare · A. Srivastav · S. Shaikh
Department of Biotechnology, Modern College of Arts, Science and Commerce (Savitribai Phule Pune University), Pune, India

V. Kumar (✉)
Department of Biotechnology, Modern College of Arts, Science and Commerce (Savitribai Phule Pune University), Pune, India

Department of Environmental Science, Savitribai Phule Pune University, Pune, India

V. Kumar et al. (eds.), *Salinity Responses and Tolerance in Plants, Volume 1*,
https://doi.org/10.1007/978-3-319-75671-4_13

Abbreviations

ABA	Abscisic acid
ACC	Aminocyclopropane carboxylate
ADC	Arginine decarboxylase
APOX	Ascorbate peroxidase
cDNA	Complementary DNA
DAO	Diamine oxidase
dcSAM	Decarboxylated S-adenosylmethionine
EMS	Ethyl methanesulfonate
FAD	Flavin adenine dinucleotide
Fv/Fm	Variable fluorescence/maximum fluorescence
GABA	Gamma aminobutyric acid
GOT	Glutamate oxaloacetate transaminase
GPT	Glutamate pyruvate transaminase
GR	Glutathione reductase
GS	Glutamine synthetase
LAT	L-type amino acid transporter
NAD	Nicotinamide adenine dinucleotide
NADH	Reduced nicotinamide adenine dinucleotide
NADH-GDH	NADH-dependent glutamate dehydrogenase
NADH-GOGAT	NADH–glutamine oxoglutarate aminotransferase
NADPH	Reduced nicotinamide adenine dinucleotide phosphate
NDPK	Nucleoside diphosphate kinase
NO	Nitric oxide
ODC	Ornithine decarboxylic acid
ORF	Open reading frame
PA	Polyamine
PSII	Photosystem II
Put	Putrescine
PUT	Polyamine uptake transporter
ROS	Reactive oxygen species
SAM	S-adenosylmethionine
SAMDC	S-adenosylmethionine decarboxylase
SAMS	S-adenosylmethionine synthetase
SOD	Superoxide dismutase
Spd	Spermidine
SPDS	Spermidine synthase
Spm	Spermine
SPMS	Spermine synthase
tSpm	Thermospermine

13.1 Introduction

Polyamines (PAs) are small polycationic aliphatic amines. They are abundantly found in plants and are involved in several plant growth and development processes. The common PAs spermidine (Spd), spermine (Spm), and putrescine (Put) are apparently involved in plant responses to microbial symbionts that are critical for plant nutrition (El Ghachtoul et al. 1996), and are involved in molecular signaling events in interactions between plants and pathogens (Martin-Tanguy 1987). Several biological processes ranging from growth to development, as well as responses to environmental stimuli, are included in the functions of PAs (Evans and Malmberg 1989; Galston and Kaur-Sawhney 1990).

PAs are present in free, soluble conjugated, and insoluble bound forms. They are synthesized from the amino acids ornithine or arginine. Their capacities to neutralize acids and act as antioxidant agents—as well as their roles in membrane/cell wall stabilization—make them indispensable for normal functioning of cells. They are, indeed, able to bind several negatively charged molecules, such as DNA (Basu et al. 1990; Pohjanpelto and Höltta 1996), proteins, membrane phospholipids (Beigbeder 1995; Tassoni et al. 1996), and pectic polysaccharides (D'Oraci and Bagni 1987). They have been shown to be involved in protein phosphorylation (Ye et al. 1994), post-transcriptional modifications (Mehta et al. 1994), and conformational transition of DNA (Basu et al. 1990). There is direct evidence that they are essential for growth and development in prokaryotes and eukaryotes (Tabor and Tabor 1984; Heby and Persson 1990; Slocum 1991; Tiburcio et al. 1990). PAs are considered to be plant growth regulators, which mediate hormone effects or act as independent signaling molecules, especially in stress situations (Evans and Malmberg 1989; Galston and Kaur-Sawhney 1995).

13.2 Biosynthesis and Metabolism of Polyamines

Putrescine [$NH_2(CH_2)_4NH_2$], spermidine [$NH_2(CH_2)_3NH(CH_2)_4NH_2$], and spermine [$NH_2(CH_2)_3NH(CH_2)_4NH(CH_2)_3NH_2$] are the most commonly found PAs in higher plants. These three PAs differ in the number of amino groups, each of which carries a positive charge (NH_3^+) at physiological pH. In addition, uncommon PAs such as homo-spermidine, 1,3-diaminopropane, cadaverine, and canavalmine have also been detected in plants, animals, algae and bacteria.

The precursors for PA biosynthesis are the amines ornithine and arginine. Arginine is converted to Put by sequential action of the enzymes arginine decarboxylase (ADC), N-carbamoyl putrescine amidohydrolase, and agmatine deiminase. N-carbamoyl putrescine amidohydrolase is allosterically regulated by its substrate. Ornithine is directly converted to Put by the enzyme ornithine decarboxylase. Put acts as a precursor for the triamine Spd. The third amino group of Spd is donated from decarboxylated S-adenosylmethionine (dcSAM), formed by decarboxylation of S-adenosylmethionine (SAM), catalyzed by Spd synthase (SPDS). Another amino group is added to Spd, again by dcSAM, to form the tetramine Spm, and this reaction is catalyzed by Spm synthase (SPMS).

Homeostasis of PAs is achieved by regulation of the activities of PA biosynthesis enzymes, particularly SAM decarboxylase (SAMDC). This enzyme is synthesized as a proenzyme, which is then processed to form the active enzyme. Processing of SAMDC and an increase in its activity are promoted when the Put concentration is high in cells, thus leading to synthesis of Spd and Spm. Accumulation of the product of SAMDC—dcSAM—is known to inactivate SAMDC by irreversible modification at the cysteine residue in one of its subunits, which leads to its inactivation. SAMDC is also known to be regulated at the translational level. SAMDC has two upstream open reading frames (ORFs) in the 5' untranslated region of its transcript. When PA levels in the cell are high, the peptide coded by one of these upstream ORFs is known to suppress translation of the main SAMDC ORF, thus stopping synthesis of dcSAM and hence that of Spd and Spm. When PA levels are low, the translation product of the other upstream ORF induces translation of SAMDC and hence promotes PA biosynthesis. Figure 13.1 postulates the processes involved in homeostasis of PAs in cells.

13.2.1 Polyamine Transport and Conjugation

PAs are synthesized in cells, but preformed PAs may be transported intracellularly, as well as across the plasma membrane, according to the PA requirement of cells. Transport of PAs is mediated by PA uptake transporters (PUTs), which belong to the

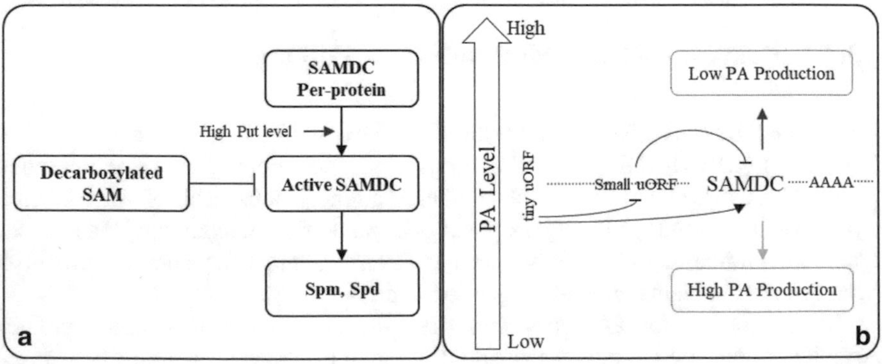

Fig. 13.1 Processes involved in homeostasis of polyamine in cells. (**a**) Regulation of S-adenosylmethionine decarboxylase (SAMDC) activity by putrescine and decarboxylated S-adenosylmethionine (dcSAM). (**b**) Role of upstream open reading frames in translational regulation of the SAMDC protein under low and high polyamine availability

L-type amino acid transporter (LAT) family, located in the endoplasmic reticulum, Golgi membranes, and plasma membrane. Long-distance transport of PAs is thought to occur through the xylem and depends on the transpiration pull.

Storage forms of PAs occur as conjugates with phenolic acids such as cinnamic acid, coumaric acid, or sinapic acid. These conjugates may exist as basic conjugates if only one of the amines is used in conjugate formation or as neutral conjugates if all of the amines are bound to phenolic acid. Phenolic acid–PA conjugates are formed by amide linkages between the PA and the coenzyme A (coA) esters of phenolic acids in reactions catalyzed by transferases such as Spd dicoumaroyltransferase, Spd disinapoyltransferase, or Spd hydroxycinnamoyl transferase. The genes coding for these transferases are expressed in actively growing tissues such as root tips, developing embryos, or actively metabolizing tissues such as the tapetum, and they may represent sources of PAs required for growth and proliferation.

PAs are also covalently linked to proteins through the activity of transglutaminases. Transglutaminases are a widely distributed enzyme family, which brings about post-translational modification of proteins with PAs. The PAs may attach to either a glutamine or a lysine residue of a protein to make them positively charged or cationic, thus altering the solubility and properties of the protein. PAs may also form cross-links between two proteins, thus altering their stability and conformation.

13.2.2 Physiological Roles of Polyamines

PAs are present in actively growing tissues, where they play a role in regulating cell division and differentiation. Put and Spd are seen to be essential for survival of cells, while Spm may not be required, as evidenced from silencing of genes in the PA biosynthetic pathway. PAs are known to regulate fundamental processes, including cell proliferation, enzyme activities, ion transport, and membrane stabilization. PAs are also known to regulate plant growth and developmental processes such as embryogenesis, germination, morphogenesis, reproductive development, and responses to biotic and abiotic stresses. However, given the multiple effects of PAs, it is difficult to ascribe a definite mode of action to them at the molecular level.

13.2.3 Cross Talk Between Polyamines and Other Biosynthetic Pathways

Spd and Spm are known to inhibit aminocyclopropane carboxylate (ACC) synthase activity, which converts SAM into ACC, a precursor of ethylene. On the other hand, ethylene inhibits SAM decarboxylase activity and thus the biosynthesis of Spd and Spm. Hence, ethylene and PA biosynthesis pathways act antagonistically to each other.

Breakdown of Put is mediated by the enzyme diamine oxidase (DAO), which converts Put to Δ^1-pyrroline, NH_3, and H_2O_2. PA oxidase metabolizes Spd and Spm to form pyrroline and aminopyrroline, respectively, diaminopropane, and H_2O_2. Pyrroline is further converted to the nonprotein amino acid gamma aminobutyric acid (GABA) by a nicotinamide adenine dinucleotide (NAD)–dependent dehydrogenase, which feeds into the Kreb's cycle. Methyl-Put provides the pyrrolidine ring in the synthesis of nicotine and tropane alkaloids. The H_2O_2 formed by PA oxidation is thought to play a role in lignin polymerization. Cross talk between PA metabolism and a number of other biosynthetic pathways (ethylene, GABA, alkaloids, lignin) suggests a regulatory mechanism for stress signaling in plants.

13.3 Polyamines in Stress Responses

There is creditable evidence confirming the role of PAs in stress tolerance. Transcription of PA biosynthesis genes, their translation, and enzyme activities are known to be induced under abiotic and biotic stress. Scavenging of excess reactive oxygen species (ROS) is a direct strategy by which plants adapt to adverse environments. However, plants can also alter their metabolism and accumulate beneficial metabolites, including PAs. Often changes in PA metabolism and expression levels of their pathway genes are positively linked with enhanced tolerance of abiotic stresses and pathogenic attack in plants (Shi and Chan 2014). For example, overexpression of *FcWRKY70* increased Put accumulation to confer drought tolerance on *Fortunella crassifolia* (Gong et al. 2015). Spd promotes biomass accumulation and upregulates proteins involved in cell rescue. Spd was involved in inducing antioxidant enzymes in tomato (*Lycopersicon esculentum*) seedlings subjected to high-temperature conditions (Sang et al. 2017). Exogenous Spm treatment induced defense responses and resistance against *Phytophthora capsici,* a root rot pathogen, in *Capsicum annuum* (Koc et al. 2017). Furthermore, Spd was found to be vital for adjustment of intracellular PA pathways and endogenous PA homeostasis, which enhanced salt tolerance in rice (Saha and Giri 2017).

13.3.1 Mode of Action of Polyamines in Stress Responses

PAs are reported to protect plants against stress through mechanisms including the following:

1. Polycationic PAs are known to bind anionic molecules such as nucleic acids and proteins, thus stabilizing them. This property may be important in preventing stress-induced damage to these macromolecules. Spd and Spm have been shown to prevent radiation and oxidative stress–induced strand breaks in DNA.

2. PAs are important for membrane transport regulation in plants. They have been shown to block two slow and fast vacuolar cation channels. The effect of PAs is direct, and the channels open when PAs are withdrawn. PAs also affect vacuolar and plasma membrane H^+ and Ca^{2+} pumps, and have been reported to bring about stomatal closure in response to drought stress by blocking the activity of a KAT1-like inward K^+ channel in the guard cell membrane. This is an indirect effect caused by low-affinity PA binding to the channel protein. PA-induced stomatal closure provides a link between stress conditions and PA levels.
3. PAs are known to modulate ROS homeostasis. They are known for inhibiting auto-oxidation of metals, which eventually reduces the electron supply required for ROS generation. Experimental evidence of the role of PAs in directly acting as scavengers of alkyl, hydroxyl, peroxyl, and superoxide radicals at physiological concentrations is not available. PAs are known to induce antioxidant enzymes in stress situations, as evidenced by use of inhibitors of PA biosynthesis or transgenics overexpressing enzymes involved in PA biosynthesis. For example, application of D-arginine, which is an inhibitor of PA biosynthesis enzymes, resulted in reduced levels of PAs and increased levels of ROS. In addition, overexpression of *ADC* resulted in enhanced drought stress tolerance, attributed to reduced ROS generation in transgenic plants. However, better and more concrete evidence is needed to prove a direct link between increased PA levels and antioxidant enzymes. PAs—especially Spm—seemingly act as signaling molecules to activate the antioxidant machinery. Generation of H_2O_2 by PA catabolism, which may be promoted when PA levels are above a specific threshold, is known to play a vital role in regulation of signaling cascades under abiotic and biotic stress conditions. For example, generation of H_2O_2 by PA oxidase in the apoplast is thought to play a role in initiating a hypersensitive response in plants, which leads to the induction of defense response genes. Hence, the ratio of PA biosynthesis and catabolism may be important in inducing stress tolerance either by prevention of oxidative damage or by induction of programmed cell death.

Additionally, abiotic stresses often induce accumulation of abscisic acid (ABA) and nitric oxide (NO), and the interactions between them. PAs trigger protective responses, including regulation of the channels for ion homeostasis and stomatal responses to improve and maintain water status, which induces the antioxidant machinery to check excess ROS generation, with synthesis and accumulation of compatible osmolytes. All of these phenomena help to enhance the abiotic stress tolerance of plants (Shi and Chan 2014).

13.4 Role of Polyamines in Salt Stress Tolerance

PAs are present in all compartments of the plant cell, including the nucleus, which indicates their participation in diverse fundamental processes in the cell, ranging from replication, transcription, translation, and preservation of membrane integrity to acting as signaling molecules, modulation of enzyme activities, plant growth and development, and plant adaptations to challenging environments (Galston et al. 1997; Walden et al. 1997; Kaur-Sawhney et al. 2003).

PAs are potential biochemical indicators of salinity tolerance in plants (Ashraf and Harris 2004). They play crucial roles in tolerance of abiotic stresses, including salinity, and increases in the levels of PAs are correlated with stress tolerance in plants (Bouchereau et al. 1999; Shi and Chan 2014; Gupta et al. 2014; Liu et al. 2005). PA overaccumulation has been reported in different plant species in response to salt stress, and has been credited with stress-alleviating or protective effects, mainly via scavenging of free radicals (Groppa and Benavides 2008), stabilizing cell membranes and intracellular ionic balances (Ruiz et al. 2007). Because of their cationic nature, PAs interact with anionic macromolecules, and these reversible ionic interactions lead to stabilization of DNA, membranes, and selective proteins (Bachrach 2005), and also act as modulators of stress-regulated gene expression. In addition, PAs play an important role in maintaining membrane integrity (Legocka and Kluk 2005).

13.4.1 Endogenous Content of Polyamines and Salinity Tolerance

Since the early reports of active Put accumulation in plants growing under salt stress, one of the key issues has been deciphering the mechanisms underlying stress-induced changes in PA metabolism and coinciding abiotic stress tolerance (Kaur-Sawhney et al. 2003). In recent years, an abundance of data concerning stress-dependent PA accumulation have appeared. Cellular PA levels were changed under most abiotic stress conditions (Alcázar et al. 2011; Alet et al. 2012; Hussain et al. 2011). PAs have been shown to increase the survival of various plants under salt stress (Kuznetsov and Shevyakova 2007).

Increases in endogenous PA levels were reported in *Arabidopsis* and sunflower plants when the plants were exposed to salinity stress (Kasinathan and Wingler 2004; Mutlu and Bozcuk 2005; Quinet et al. 2010). In rice roots and *Lupinus luteus* seedlings, very early and temporary increases in Put and Spm accumulation under salt stress were observed (Legocka and Kluk 2005). The intracellular PA level is regulated by PA catabolism. PAs are oxidatively catabolized by amine oxidases, which include copper-binding DAOs and flavin adenine dinucleotide (FAD)–binding PA oxidases. These enzymes play a significant role in stress tolerance (Drolet et al. 1986; Gupta et al. 2014). PAs block both inward and outward currents through nonselective cationic channels in pea mesophyll protoplasts, thus assisting the process of adaptation to salt stress by reducing Na^+ uptake and K^+ leakage from mesophyll cells. In *Vicia faba*, PAs targeted inward K^+ channels in guard cells and modulated stomatal movements, providing a link between stress conditions, PA levels, ion channels, and stomata regulation (Alcázar et al. 2010). Higher endogenous levels of PAs, particularly Spd, are positively correlated with greater increases in antioxidant potentials. Put has also been observed to activate the antioxidant machinery to curb ROS generation and consequently prevent lipid peroxidation and denaturation of biomolecules, which ultimately resulted in better plant growth under salt stress (Gill and Tuteja 2010; Amri et al. 2011).

13.4.2 Exogenous Application of Polyamines and Salinity Tolerance

The effects of exogenous PAs in improving salt tolerance have been observed in various plant species such as rice, belladonna, barley, tobacco, cucumber, and rapeseed (Tavladoraki et al. 2012; Li et al. 2013; Sagor et al. 2013). Application of exogenous PAs has been found to increase the level of endogenous PAs during stress. PA treatments have been associated with maintenance of membrane integrity, accumulation of compatible osmolytes, reduced ROS generation, lesser organelle levels of Na^+ and Cl^-, and less negative impacts on photosynthetic machinery, in comparison with nontreated plants (Gupta et al. 2014). It has been suggested that binding of PAs to membranes prevents lipid peroxidation and quenching of free radicals. PAs binding have protected the composition of thylakoid membranes from degradation in oat leaves, and also reduced membrane fluidity in microsomal membranes in *Phaseolus*. Binding of PAs to microsomal membranes in *L. luteus* most likely rigidified microsomal membrane surfaces, stabilizing them against NaCl (Tonon et al. 2004; Legocka and Kluk 2005). Further, the production of superoxide-dependent radicals by senescing microsomal membranes was inhibited by PAs (Legocka and Kluk 2005). Elevated levels of cellular PA may modulate the activity of plasma membrane ion channels, improving ionic relations by preventing NaCl-induced K^+ efflux from the leaf mesophyll and assisting in the plant's adaptation to salinity (Shabala et al. 2007). Application of exogenous PAs (Spd and Put) partially prevented reductions in phospholipid and PA content in tonoplast vesicles, attenuating salt injury in barley seedlings (Zhao and Qin 2004), and also increased the length of internodes and the numbers of internodes and leaves in a commercial genotype of pomegranate, *Rabab*, increasing the growth of seedlings. The exogenous treatment have been considered to be responsible to protect the plasma membrane against stress damage by maintaining membrane integrity, thereby preventing superoxide-generating reduced NAD phosphate (NADPH) oxidase activation or inhibiting protease and RNase activity (Amri et al. 2011). A comparative proteomic analysis was performed to investigate PA-mediated responses; 36 commonly regulated proteins were successfully identified in Bermuda grass; among them, proteins involved in electron transport and energy pathways were largely enriched, and nucleoside diphosphate kinase (NDPK) and three antioxidant enzymes were extensively regulated by PAs. Taking these findings together, it has been proposed that

PAs could activate multiple pathways that enhance Bermuda grass adaptation to salt and drought stresses (Shi et al. 2013).

13.4.2.1 Application of Exogenous Putrescine and Salinity Tolerance

Exogenously applied Put (1 mM) has been shown to alleviate salt stress (200 mM NaCl) in terms of fresh weight increase and electrolyte leakage in in vitro callus from the *Malus sylvestris* Mill., cv. 'Domestica' (Kuznetsov and Shevyakova 2007) apple. In the green alga *Scenedesmus obliquus* it conferred some kind of tolerance of enhanced NaCl salinity on the photosynthetic apparatus and permitted cell growth (Demetriou et al. 2007). In seedlings of Indian mustard and Virginia pine, exogenous addition of Put improved seedling growth by preventing lipid peroxidation and denaturation of macromolecules through induction of the antioxidative enzymes ascorbate peroxidase (APOX), glutathione reductase (GR), and superoxide dismutase (SOD), increasing glutathione and carotenoid levels and reducing the activities of acid phosphatase and V-type H^+-ATPase (Verma and Mishra 2005; Tang and Newton 2005). Exogenous Put reduced Na^+ accumulation and increased alkaloid and endogenous Put levels in the roots of a salt-sensitive rice cultivar and in tissues from *Atropa belladonna* plants grown under saline conditions (Gill and Tuteja 2010; Dupont-gillain et al. 2017; Ali 2000). In soybean (*Glycine max* L.) seedlings, exogenous Put increased endogenous Put levels but decreased ADC activity. However, Put increased SAMDC and DAO levels, resulting in increased Spd and Spm levels (Verma and Mishra 2005; Gu-wen et al. 2014).

13.4.2.2 Application of Exogenous Spermidine and Salinity Tolerance

Investigations have established that accumulation of Na^+, loss of K^+ ions, and inhibition of plasma membrane–bound H^+-ATPase activity in salt-stressed plants can be overcome by exogenous Spd treatment (Roy et al. 2005). Spm and Spd treatments significantly prevented electrolyte and amino acid leakage from roots and shoots of rice subjected to salinity; they also prevented chlorophyll loss, inhibition of photochemical reactions of photosynthesis, and downregulation of chloroplast-encoded genes such as psbA, psbB, psbE, and rbcL, indicating a positive correlation between salt tolerance and accumulation of PAs in rice. Exogenous treatment increased plant growth by preventing chlorophyll degradation and increasing PA levels, as well as antioxidant enzyme levels in cucumber and *Panax ginseng* seedlings (Parvin et al. 2014; Duan et al. 2008). These PAs are known to interact with phospholipids or other anionic groups of membranes, and in this process they stabilize the membranes. There have been several recent reports affirming the participation of PAs—especially thylakoid-bound PAs—in regulating the structure and/or function of the photosynthetic apparatus in challenging environments (Amri et al. 2011). Exogenous Spd application promoted assimilation of excess NH_4^+ by coordinating and strengthening the synergistic action of reduced NAD (NADH)–dependent glutamate dehydrogenase (NADH-GDH), glutamine synthetase (GS)/NADH–glutamine oxoglutarate aminotransferase (NADH-GOGAT), and transamination pathways. Subsequently, NH_4^+ and its related enzymes GDH, GS, glutamate oxaloacetate transaminase (GOT), and glutamate pyruvate transaminase (GPT), in vivo, are maintained in a proper and balanced state to enable mitigation of stress-resulted damage in salt-sensitive cultivars of tomato (*Solanum lycopersicum*) (Zhang et al. 2013). Under salinity–alkalinity stress conditions, exogenous Spd enhanced the activities of SAMDC and DAO in zoysia grass cultivars (Li et al. 2016) and reduced the activities of ADC. These results suggest that exogenous Spd treatment can regulate the metabolic status of PAs caused by salinity–alkalinity stress, and eventually enhance tomato (*S. lycopersicum*) plant tolerance of that stress (Hu et al. 2012).

PAs (Spd or Spm) have also been shown to inhibit the extent of salt-induced protein carbonylation. While the salt-induced increase in anthocyanin or reducing sugar levels was further prompted by Spd or Spm, the proline content was elevated by Spd, particularly in rice (Roychoudhury et al. 2011). This resulted in reduced Na^+ content and higher chlorophyll values, higher variable fluorescence/maximum fluorescence (Fv/Fm) values, and a higher net photosynthetic rate in wheat seedlings (Gill and Tuteja 2010). There were slight improvements in leaf Ca^{2+} and Mg^{2+} content. Increased Ca^{2+} and Mg^{2+} uptake may be beneficial in stressful and recovery phases, as these ions are vital for maintaining membrane integrity, as well as reducing Na^+ uptake. Thus, reduced uptake of toxic ions such as Na^+ and higher uptake of beneficial ions such as Ca^{2+} and Mg^{2+} under Spd treatment may be attributed to better salt tolerance in plants (Anjum 2011). Observable changes in level of proteins involved in protein-metabolism and stress defense were seen on exogenous Spd application in cucumber seedlings (Li et al. 2013). It has been postulated that Spd treatment might play an important role in salt tolerance via regulation of the Calvin cycle, protein-folding assembly, and inhibition of protein proteolysis (Li et al. 2013; Gupta et al. 2014).

13.4.2.3 Application of Exogenous Spermine and Salinity Tolerance

When seedlings of *Sorghum bicolor* treated with 0.25 mM Spm were subjected to salt stress, they showed improvement in growth and partial increases in the activities of peroxidase and GR, with a concomitant decrease in the level of membrane lipid peroxidation (Chai et al. 2010). Shu et al. (2013) reported that Spm treatment induced antioxidant systems in chloroplasts and concomitant improved the efficiency of photosystem II (PSII) in cucumber. During salinity, Spm may directly regulate plasma and tonoplast membrane channels and/or regulatory proteins involved in signaling. It would also be interesting to consider that Spm, as a substrate of flavin-containing PA oxidases, produces H_2O_2 in the apoplast and accordingly seems to be implicated in oxidative stress signaling. This may be relevant to composition of apoplastic barriers, having pivotal roles during salt stress responses and adaptation (Alet et al. 2012). Under salinity conditions, *Mesembryanthemum crystallinum* could accumulate both free and bound Spm in both leaves and roots, which supports the involvement of PAs in the development of salt resistance (Shevyakova et al. 2006a, b).

13.5 Transgenic Approaches to Obtaining Polyamine-Overaccumulating Salt-Tolerant Plants

Several studies have demonstrated positive effects of increments in PA levels on plant growth and survival in one or more biotic/abiotic stress conditions. PA accumulation is correlated with enhanced stress tolerance (Groppa and Benavides 2008; Marco et al. 2015). Differential regulation of PA biosynthesis pathway genes during stress is evidenced by their different expression levels, which show the involvement of stress-induced pathways in PA regulation (Alcázar et al. 2010). The importance of PA regulation has been confirmed by demonstrations of poor performance of transgenic plants carrying loss-of-function mutations in PA synthesis genes (Urano et al. 2004; Marco et al. 2015). For example, *Arabidopsis thaliana spe1-1* and *spe 2-1* ethyl methanesulfonate (EMS) mutants showed low ADC activity and were unable to accumulate PAs after acclimation to a higher saline environment. The mutants showed decreased salt tolerance in comparison with their wild-type counterparts (Kasinathan and Wingler 2004). Similarly, *Arabidopsis acl5/spms* double mutants, compromised in Spm and thermospermine (tSpm) production, displayed hypersensitivity to salt and drought stress (Yamaguchi et al. 2006, 2007). Equivalent approaches of gene silencing/knockout were used by Alet et al. (2012), Sagor et al. (2013), and Cuevas et al. (2008), which indicated the vital role of different genes and their expression levels in PA biosynthesis, not only under salinity stress but also in different stress environments. A strategy for production of plants with improved salt tolerance includes transgenic plant production with altered PA levels, overexpressing PA biosynthesis genes. Though the extent of amendment of endogenous levels of one or more specific PAs overexpressed in transgenic lines varies, all of the transgenic lines share a mutual augmentation in tolerance of many abiotic stresses, including salinity. The genes procured for such approaches include *ODC*, *ADC*, *SAMDC*, *SPDS*, and *SPMS* (Marco et al. 2015).

One of the key enzymes involved in biosynthesis of SAM is S-adenosyl-L-methionine synthetase, a common precursor of PA as well as ethylene. Gong et al. (2014) introduced complementary DNA (cDNA) coding for SAM synthetase (SAMS) (*SlSAMS1*) in the tomato genome via *Agrobacterium tumefaciens*–mediated transformation. The raised transgenic line exhibited a substantial improvement in tolerance of alkali stress, with a sustained nutrient balance, high photosynthetic ability, and low oxidative damage in comparison with nontransformed lines. The overexpression of *SlSAMS1* markedly improved biomass production under high salt (sodic–alkali) stress, as well as prolonging plant survival, showing the importance of SAMS in improving tolerance of alkali–salt stress in tomato. The full-length SAMS gene from *Suadea salsa* (*SsSAMS2*) was introduced into *Nicotiana tabacum* via *Agrobacterium*-mediated genetic transformation (Qi et al. 2010). The transgenic line showed more tolerance of salt stress, with improvements in seed numbers, weight, and PA accumulation in comparison with wild-type plants.

The putative effects of Put on salinity-mediated responses in *Lotus tenuis* was analyzed by Espasandin et al. (2017). Micropropagated plantlets from wild-type and transgenic lines harboring a construct for ADC (*pRD29A::oat arginine decarboxylase*) were cultivated under gradually increasing saline conditions. Healthier transgenic lines were observed in comparison with wild-type plants with reduced growth. The overexpression of *ADC* resulted in better osmotic adjustment (improved by 5.8-fold) via proline release. The transgenic plants also showed balanced Na^+/K^+ ratios, correlating with their healthier growth.

SPDS cDNA from *Cucurbita ficifolia* was cloned and the gene was introduced into *A. thaliana* under the controlled CaMV 35S promoter (Kasukabe et al. 2004). The stably integrated and actively transcribed transgene in T_2 and T_3 was proved to be responsible for improved SPDS activity and Spd content. The transgenic line also showed improved tolerance of multiple abiotic stresses (chilling, hyperosmosis, drought, paraquat toxicity), including salinity.

Transgenic *Oryza sativa* harboring tritordeum *SAMDC* was induced with 50 µmol of ABA (Roy and Wu 2002). The transgenic lines showed 3- to 4-fold increments in Spm and Spd content in comparison with nontransformed plants under salt stress. In the regrowth analysis, transgenic plants showed reasonable growth at 150 mM salt concentration. Constitutive overexpression of *SAMDC* (of human origin) in tobacco gave rise to elevated Put and Spm levels, as well as salt and osmotic stress tolerance (Waie and Rajam 2003). On the other hand, modulation of genes other than those involved in PA biosynthesis also influenced PA levels in the plants and hence their tolerance of abiotic stresses (Moschou et al. 2012; Tavladoraki et al. 2012). *ACC synthase* and *ACC oxidase* genes are important for ethylene biosynthesis. Antisense silencing of these genes in transgenic tobacco resulted in improved tolerance of multiple stress factors, including salinity (Tavladoraki et al. 2012). Similarly, carnation antisense *ACC synthase*– or *ACC oxidase*–transformed tobacco plants displayed higher *SAMDC* activity as well as higher Put and Spd levels, contributing to increased tolerance of oxidative stress, high salinity, and low pH. Examples of transgenic approaches for improved salt tolerance are listed in Table 13.1. A schematic representation of implementation of genes involved in PA synthesis to improve salt tolerance in plants is shown in Fig. 13.2.

13.6 Summary and Conclusions

Polyamines are an interesting group of polycationic amines that affect a number of processes in plant growth, metabolism, and development. They are present in almost all cellular compartments, and their biosynthesis is fairly well understood. The biosynthetic pathway of PAs shares components with other metabolic pathways in the

Table 13.1 Transgenic plants overexpressing genes related to polyamine biosynthesis to achieve better salt tolerance

Gene	Source	Transgenic plant	Enhanced characters	Reference
ADC	*Avena sativa* L.	*Oryza sativa* L.	Increased Put levels and biomass under saline conditions	Roy and Wu (2001)
ODC	Mouse	*Nicotiana tabacum*	Increased Put levels; improved salt tolerance	Kumria and Rajam (2002)
SAMDC	Tritordium	*Oryza sativa* L.	Increased Spd and Spm levels; increased seedling growth under salt stress	Roy and Wu (2002)
SAMDC	Human	*Nicotiana tabacum* var. *Xanthi*	Increased Put and Spd levels; improved salt tolerance up to 250 mM level	Wie and Rajam (2003)
SPDS	*Cucurbita ficifolia*	*Arabidopsis thaliana*	Improved SPDS activity and Spd content; improved tolerance of multiple abiotic stresses, including salinity	Kasukabe et al. (2004)
SAMS1	*Solanum lycopersicum*	*Solanum lycopersicum*	Substantial growth in tolerance of alkali stress; sustained nutrient balance; high photosynthetic ability; low oxidative damage	Gong et al. (2014)
SAMS2	*Suadea salsa*	*Nicotiana tabacum*	Greater tolerance of salt stress; improved seed numbers and weight; high polyamine accumulation	Qi et al. (2010)
ADC	*Avena sativa* L.	*Lotus tenuis*	Better osmotic adjustment (improved by 5.8-fold) via proline release; balanced Na^+/K^+ ratio	Espasandin et al. (2017)

ADC arginine decarboxylase, *ODC* ornithine decarboxylic acid, *Put* putrescine, *SAMDC* S-adenosylmethionine decarboxylase, *SAMS* S-adenosylmethionine synthetase, *Spd* spermidine, *SPDS* spermidine synthase, *Spm* spermine

cell, and this cross talk may play an important role in their homeostasis. PAs exist in free or conjugated forms and are known to bind to macromolecules such as proteins and nucleic acids. Though the mechanisms underlying regulation of plant growth and development by PAs are far from understood, their contribution to stress tolerance has been well established by using various approaches. Their role in stress tolerance is thought to be mediated primarily through stabilization of macromolecules, preventing oxidative damage and regulating ion transport. H_2O_2 generated as a result of PA catabolism is also thought to play a role in stress signaling. However, a causal role of PAs in stress tolerance has not yet been demonstrated. PAs differ from plant hormones in that they are generally present in abundance in cells—in millimolar concentrations versus the micromolar concentrations reported for hormones. PAs therefore qualify as growth regulators or metabolites required for growth and development of plants, as well as for stress tolerance.

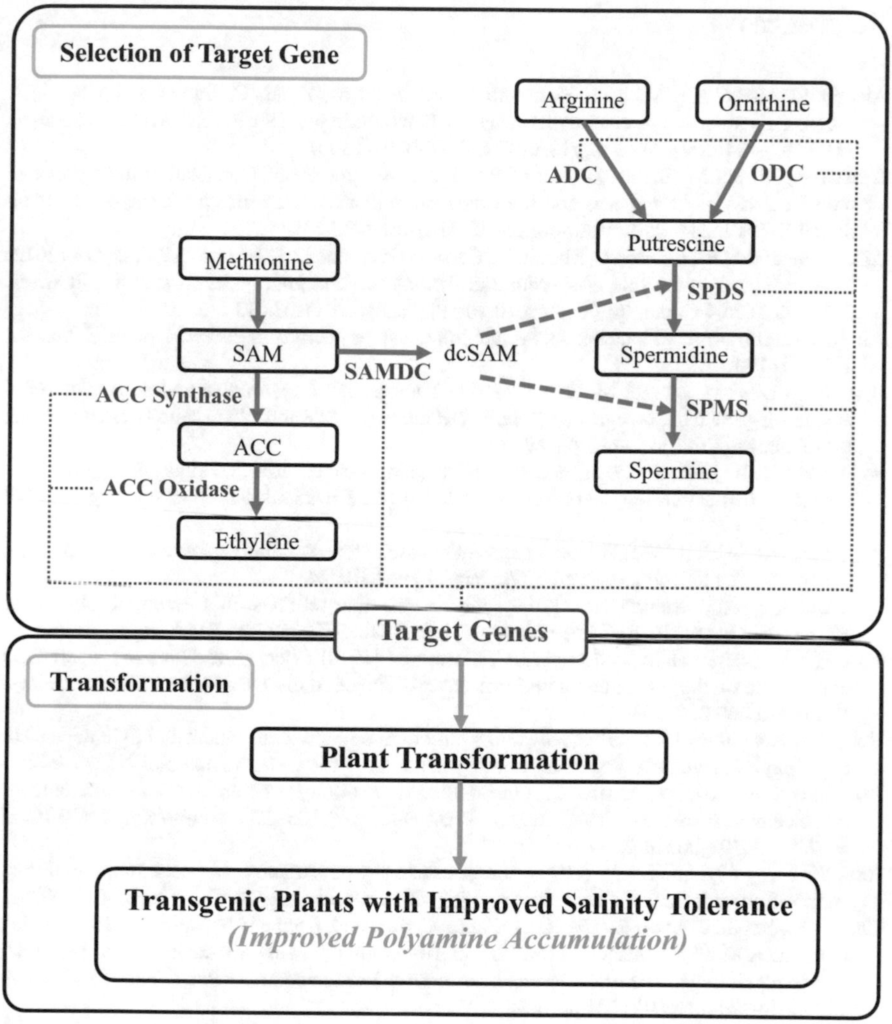

Fig. 13.2 Implementation of genes involved in polyamine synthesis to improve salt tolerance in plants. ACC: Aminocyclopropane-1-carboxylic acid synthase, ADC: Arginine decarboxylase, ODC: ornithine decarboxylase, SAMDC: S-adenosylmethionine decarboxylase, SPDS: spermidine synthase, SPMS: spermine synthase

Acknowledgements The research work in the corresponding author's laboratory is supported by Science and Engineering Research Board (SERB), Department of Science and Technology (DST), Government of India funds (grant number EMR/2016/003896). The authors acknowledge the use of facilities created under DST-FIST and DBT Star College Schemes implemented at Modern College, Ganeshkhind, Pune, India.

References

Alcázar R, Altabella T, Marco F, Bortolotti C, Reymond M, Koncz C, Carrasco P, Tiburcio A (2010) Polyamines: molecules with regulatory functions in plant abiotic stress tolerance. Planta 231:1237–1249. https://doi.org/10.1007/s00425-010-1130-0

Alcázar R, Bitrián M, Bartels D et al (2011) Polyamine metabolic canalization in response to drought stress in *Arabidopsis* and the resurrection plant *Craterostigma plantagineum*. Plant Signal Behav 6:243–250. https://doi.org/10.4161/psb.6.2.14317

Alet AI, Sánchez DH, Cuevas JC, Marina M, Carrasco P, Altabella T, Tiburcio AF, Ruiz OA (2012) New insights into the role of spermine in *Arabidopsis thaliana* under long-term salt stress. Plant Sci 182:94–100. https://doi.org/10.1016/j.plantsci.2011.03.013

Ali RM (2000) Role of putrescine in salt tolerance of *Atropa belladonna* plant. Plant Sci 152:173–179

Amri E, Mirzaei M, Moradi M, Zare K (2011) The effects of spermidine and putrescine polyamines on growth of pomegranate (*Punica granatum* L. cv "Rabbab") in salinity circumstance. Int J Plant Physiol Biochem 3:43–49

Anjum MA (2011) Effect of exogenously applied spermidine on growth and physiology of citrus rootstock *Troyer citrange* under saline. Turk J Agric For 35:43–53. https://doi.org/10.3906/tar-0912-563

Ashraf M, Harris PJC (2004) Potential biochemical indicators of salinity tolerance in plants. Plant Sci 166:3–16. https://doi.org/10.1016/j.plantsci.2003.10.024

Bachrach U (2005) Naturally occurring polyamines: interaction with macromolecules. Curr Protein Pept Sci 6:559–566. https://doi.org/10.2174/138920305774933240

Basu HS, Schwietert HCA, Feuerstein BG, Marton LJ (1990) Effect of variation in the structure of spermine on the association with DNA and the induction of DNA conformational changes. Biochem J 269:329–334

Beigbeder AR (1995) Influence of polyamine inhibitors on light independent and light dependent chlorophyll biosynthesis and on the photosyntetic rate. J Photochem Photobiol 28:235–242

Bouchereau A, Aziz A, Larher F, Martin-Tanguy J (1999) Polyamines and environmental challenges: recent development. Plant Sci 140:103–125. https://doi.org/10.1016/S0168-9452(98)00218-0

Chai YY, Jiang CD, Shi L et al (2010) Effects of exogenous spermine on sweet sorghum during germination under salinity. Biol Plant 54:145–148. https://doi.org/10.1007/s10535-010-0023-1

Cuevas JC, Lopez-Cobollo R, Alcazar R, Zarza X, Koncz C, Altabella T, Salinas J, Tiburcio AF, Ferrando A (2008) Putrescine is involved in *Arabidopsis* freezing tolerance and cold acclimation by regulating abscisic acid levels in response to low temperature. Plant Physiol 148:1094–1105. https://doi.org/10.1104/pp.108.122945

D'Oraci D, Bagni N (1987) In vitro interactions between polyamines and pectic substances. Biochem Biophys Res Commun 148:1159–1163

Demetriou G, Neonaki C, Navakoudis E, Kotzabasis K (2007) Salt stress impact on the molecular structure and function of the photosynthetic apparatus—the protective role of polyamines. 1767:272–280. https://doi.org/10.1016/j.bbabio.2007.02.020

Drolet G, Dumbroff EB, Legge RL, Thompson JE (1986) Radical scavenging properties of polyamines. Phytochemistry 25:367–371. https://doi.org/10.1016/S0031-9422(00)85482-5

Duan JJ, Li J, Guo S, Kang Y (2008) Exogenous spermidine affects polyamine metabolism in salinity-stressed *Cucumis sativus* roots and enhances short-term salinity tolerance. J Plant Physiol 165:1620–1635. https://doi.org/10.1016/j.jplph.2007.11.006

Dupont-gillain CC, Quinet M, Ndayiragije A et al (2017) Putrescine differently influences the effect of salt stress on polyamine metabolism and ethylene synthesis in rice cultivars differing in salt resistance. 61:2719–2733. https://doi.org/10.1093/jxb/erq118

El Ghachtoul N, Martin-Tangu J, Paynot M, Gianinazz S (1996) First report of the inhibition of arbuscular mycorrhizal infection of *Pisum sativum* by specific and irreversible inhibition of polyamine biosynthesis or by gibberellic treatment. FEBS Lett 385:189–192

Espasandin FD, Calzadilla PI, Maiale SJ, Ruiz OA, Sansberro PA (2017) Overexpression of the arginine decarboxylase gene improves tolerance to salt stress in *Lotus tenuis* plants. J Plant Growth Regul:1–10. https://doi.org/10.1007/s00344-017-9713-7

Evans PT, Malmberg RL (1989) Do polyamines have roles in plant development? Annu Rev Plant Physiol Plant Mol Biol 40:235–269

Galston AW, Kaur-Sawhney RK (1990) Polyamines in plant physiology. Plant Physiol 94:406–410

Galston AW, Kaur-Sawhney R (1995) Polyamines as endogenous growth regulators. In: Davies PJ (ed) lant hormones: physiology, biochemistry and molecular biology, 2nd edn. Kluwer Academic Publishers, Dordrecht, pp 158–178

Galston AW, Kaur-Sawhney R, Altabella T, Tiburcio AF (1997) Plant polyamines in reproductive activity and response to abiotic stress. Bot Acta 110:197–207. https://doi.org/10.1111/j.1438-8677.1997.tb00629.x

Gill SS, Tuteja N (2010) Polyamines and abiotic stress tolerance in plants. Plant Signal Behav 5:26–33. https://doi.org/10.4161/psb.5.1.10291

Gong B, Li X, VandenLangenberg KM, Wen D, Sun S, Wei M, Li Y, Yang F, Shi Q, Wang X (2014) Overexpression of S-adenosyl-l-methionine synthetase increased tomato tolerance to alkali stress through polyamine metabolism. Plant Biotechnol J 12:694–708. https://doi.org/10.1111/pbi.12173

Gong XQ, Zhang JY, Hu JB, Wang W, Wu H, Zhang QH, Liu JH (2015) FcWRKY70, a WRKY protein of *Fortunella crassifolia*, functions in drought tolerance and modulates putrescine synthesis by regulating arginine decarboxylase gene. Plant Cell Environ 38:2248–2262

Groppa MD, Benavides MP (2008) Polyamines and abiotic stress: recent advances. Amino Acids 34:35–45. https://doi.org/10.1007/s00726 007-0501-8

Gupta B, Huang B, Gupta B, Huang B (2014) Mechanism of salinity tolerance in plants: physiological, biochemical, and molecular characterization. Int J Genomics 2014:1–18. https://doi.org/10.1155/2014/701596

Gu-wen Z, Sheng-chun XU, Qi-zan HU et al (2014) Putrescine plays a positive role in salt-tolerance mechanisms by reducing oxidative damage in roots of vegetable soybean. J Integr Agric 13:349–357. https://doi.org/10.1016/S2095-3119(13)60405-0

Heby O, Persson L (1990) Molecular genetics of polyamine synthesis in eukaryotic cells. Trends Biochem Sci 15:153–158

Hu X, Zhang Y, Shi Y et al (2012) Effect of exogenous spermidine on polyamine content and metabolism in tomato exposed to salinity–alkalinity mixed stress. Plant Physiol Biochem 57:200–209. https://doi.org/10.1016/j.plaphy.2012.05.015

Hussain SS, Ali M, Ahmad M, Siddique KHM (2011) Polyamines: natural and engineered abiotic and biotic stress tolerance in plants. Biotechnol Adv 29:300–311. https://doi.org/10.1016/j.biotechadv.2011.01.003

Kasinathan V, Wingler A (2004) Effect of reduced arginine decarboxylase activity on salt tolerance and on polyamine formation during salt stress in *Arabidopsis thaliana*. Physiol Plant 121:101–107. https://doi.org/10.1111/j.0031-9317.2004.00309.x

Kasukabe Y, He L, Nada K, Misawa S, Ihara I, Tachibana S (2004) Overexpression of spermidine synthase enhances tolerance to multiple environmental stresses and up-regulates the expression of various stress regulated genes in transgenic *Arabidopsis thaliana*. Plant Cell Physiol 45:712–722. https://doi.org/10.1093/pcp/pch083

Kaur-Sawhney R, Tiburcio AF, Altabella T, Galston AW (2003) Polyamines in plants: an overview. J Cell Mol Biol 2:1–12

Koc E, Islek C, Kasko Arici Y (2017) Spermine and its interaction with proline induce resistance to the root rot pathogen *Phytophthora capsici* in pepper (*Capsicum annuum*). Hortic Environ Biotechnol 58:254–267

Kumria R, Rajam MV (2002) Ornithine decarboxylase transgene in tobacco affects polyamines, in vitro-morphogenesis and response to salt stress. J Plant Physiol159:983–90

Kuznetsov VV, Shevyakova NI (2007) Polyamines and stress tolerance of plants. Plant Stress 1:50–71

Legocka J, Kluk A (2005) Effect of salt and osmotic stress on changes in polyamine content and arginine decarboxylase activity in *Lupinus luteus* seedlings. J Plant Physiol 162:662–668. https://doi.org/10.1016/j.jplph.2004.08.009

Li B, He L, Guo S et al (2013) Proteomics reveal cucumber Spd-responses under normal condition and salt stress. Plant Physiol Biochem 67:7–14. https://doi.org/10.1016/j.plaphy.2013.02.016

Li S, Jin H, Zhang Q (2016) The effect of exogenous spermidine concentration on polyamine metabolism and salt tolerance in zoysia grass (*Zoysia japonica* Steud) subjected to short-term salinity stress. Front Plant Sci 7:1221. https://doi.org/10.3389/fpls.2016.01221

Liu J, Jiang MY, Zhou YF, Liu Y-L (2005) Production of polyamines is enhanced by endogenous abscisic acid in maize seedlings subjected to salt stress. J Integr Plant Biol 47:1326–1334. https://doi.org/10.1111/j.1744-7909.2005.00183.x

Marco F, Bitrián M, Carrasco P, Alcázar R, Tiburcio AF (2015) Polyamine biosynthesis engineering as a tool to improve plant resistance to abiotic stress. In: Genetic manipulation in plants for mitigation of climate change. Springer India, pp 103–116. https://doi.org/10.1007/978-81-322-2662-8_5

Martin-Tanguy J (1987) Hydroxycinnamic acid amides, hypersensitivity, flowering and sexual organogenesis in plants. In: Von Wettstein D, Chua DN (eds) Plant molecular biology. Plenum Publishing Corporation, New York, pp 253–263

Mehta HS, Saftner RA, Mehta RA, Davies PJ (1994) Identification of posttranscriptionally modified 18-kilodalton protein from rice as eukaryotic translocation initiation factor 5A. Plant Physiol 106:1413–1419

Moschou PN, Wu J, Tavladoraki P, Angelini R, Roubelakis-Angelakis KA (2012) The polyamines and their catabolic products are significant players in the turnover of nitrogenous molecules in plants. J Exp Bot 63:5003–5015. https://doi.org/10.1093/jxb/ers202

Mutlu F, Bozcuk S (2005) Effects of salinity on the contents of polyamines and some other compounds in sunflower plants differing in salt tolerance. Russ J Plant Physiol 52:29–34. https://doi.org/10.1007/s11183-005-0005-x

Parvin S, Ran O, Sathiyaraj G et al (2014) Spermidine alleviates the growth of saline-stressed ginseng seedlings through antioxidative defense system. Gene 537:70–78

Pohjanpelto P, Höltta E (1996) Phosphorylation of Okazaki-like DNA fragments in mammalian cells and role of polyamines in the processing of this DNA. EMBO J 15:1193–1200

Qi YC, Wang FF, Zhang H, Liu WQ (2010) Overexpression of *Suadea salsa* S-adenosylmethionine synthetase gene promotes salt tolerance in transgenic tobacco. Acta Physiol Plant 32:263–269. https://doi.org/10.1007/s11738-009-0403-3

Quinet M, Ndayiragije A, Lefèvre I et al (2010) Putrescine differently influences the effect of salt stress on polyamine metabolism and ethylene synthesis in rice cultivars differing in salt resistance. J Exp Bot 61:2719–2733. https://doi.org/10.1093/jxb/erq118

Roy M, Wu R (2001) Arginine decarboxylase transgene expression and analysis of environmental stress tolerance in transgenic rice. Plant Sci 160:869–875. https://doi.org/10.1016/S0168-9452(01)00337-5

Roy M, Wu R (2002) Overexpression of S-adenosylmethionine decarboxylase gene in rice increases polyamine level and enhances sodium chloride stress tolerance. Plant Sci 163:987–992. https://doi.org/10.1016/S0168-9452(02)00272-8

Roy P, Niyogi K, Sengupta DN, Ghosh B (2005) Spermidine treatment to rice seedlings recovers salinity stress-induced damage of plasma membrane and PM-bound H^+-ATPase in salt-tolerant and salt-sensitive rice cultivars. Plant Sci 168:583–591. https://doi.org/10.1016/j.plantsci.2004.08.014

Roychoudhury A, Basu S, Sengupta DN (2011) Amelioration of salinity stress by exogenously applied spermidine or spermine in three varieties of Indica rice differing in their level of salt tolerance. J Plant Physiol 168:317–328. https://doi.org/10.1016/j.jplph.2010.07.009

Ruiz OA, Rodrı M, Jime JF (2007) Modulation of spermidine and spermine levels in maize seedlings subjected to long-term salt stress. Plant Physiol Biochem 45:812–821. https://doi.org/10.1016/j.plaphy.2007.08.001

Sagor GHM, Berberich T, Takahashi Y, Niitsu M, Kusano T (2013) The polyamine spermine protects *Arabidopsis* from heat stress-induced damage by increasing expression of heat shock-related genes. Transgenic Res 22:595–605. https://doi.org/10.1007/s11248-012-9666-3

Saha J, Giri K (2017) Molecular phylogenomic study and the role of exogenous spermidine in the metabolic adjustment of endogenous polyamine in two rice cultivars under salt stress. Gene 609:88–103

Sang QQ, Shan X, An YH, Shu S, Sun J, Guo SR (2017) Proteomic analysis reveals the positive effect of exogenous spermidine in tomato seedlings' response to high-temperature stress. Front Plant Sci 8:120

Shabala S, Cuin TA, Pottosin I (2007) Polyamines prevent NaCl-induced K^+ efflux from pea mesophyll by blocking non-selective cation channels. FEBS Lett 581:1993–1999. https://doi.org/10.1016/j.febslet.2007.04.032

Shevyakova NI, Rakitin VY, Stetsenko LA et al (2006a) Oxidative stress and fluctuations of free and conjugated polyamines in the halophyte *Mesembryanthemum crystallinum* L. under NaCl salinity. Plant Growth Regul 50:69–78. https://doi.org/10.1007/s10725-006-9127-1

Shevyakova NI, Shorina MV, Rakitin VY, Kuznetsov VV (2006b) Stress-dependent accumulation of spermidine and spermine in the halophyte *Mesembryanthemum crystallinum* under salinity conditions. Russ J Plant Physiol 53:739–745. https://doi.org/10.1134/S1021443706060021

Shi H, Chan Z (2014) Improvement of plant abiotic stress tolerance through modulation of the polyamine pathway. J Integr Plant Biol 56:114–121

Shi H, Ye T, Chan Z (2013) Comparative proteomic and physiological analyses reveal the protective effect of exogenous polyamines in the bermudagrass (*Cynodon dactylon*) response to salt and drought stresses. J Proteome Res 12:4951–4964. https://doi.org/10.1021/pr400479k

Shu S, Yuan L, Guo S et al (2013) Effects of exogenous spermine on chlorophyll fluorescence, antioxidant system and ultrastructure of chloroplasts in *Cucumis sativus* L. under salt stress. Plant Physiol Biochem 63:209–216

Slocum RD (1991) Tissue and subcellular localisation of polyamines and enzymes of polyamine metabolism. In: Slocum RD, Flores HE (eds) Biochemistry and physiology of polyamines in plants. CRC Press, Boca Raton, pp 93–105

Tabor CW, Tabor H (1984) Polyamines. Annu Rev Biochem 53:749–790

Tang W, Newton RJ (2005) Polyamines reduce salt-induced oxidative damage by increasing the activities of antioxidant enzymes and decreasing lipid peroxidation in Virginia pine. Plant Growth Regul 46:31–43. https://doi.org/10.1007/s10725-005-6395-0

Tassoni A, Antognoni F, Bagni N (1996) Polyamine binding to plasma membrane vesicles from zucchini hypocotyls. Plant Physiol 110:817–824

Tavladoraki P, Cona A, Federico R, Tempera G, Viceconte N, Saccoccio S, Battaglia V, Toninello A, Agostinelli E (2012) Polyamine catabolism: target for antiproliferative therapies in animals and stress tolerance strategies in plants. Amino Acids 42:411–426. https://doi.org/10.1007/s00726-011-1012-1

Tiburcio AF, Kaur-Sawhney R, Galston AW (1990) Polyamine metabolism. In: Miflin BJ, Lea PJ (eds) The biochemistry of plants, intermediary nitrogen fixation. Academic Press, New York, pp 283–325

Tonon G, Kevers C, Faivre-Rampant O, Graziani M, Gaspar T (2004) Effect of NaCl and mannitol iso-osmotic stresses on proline and free polyamine levels in embryogenic *Fraxinus angustifolia* callus. J Plant Physiol 161:701–708

Urano K, Yoshiba Y, Nanjo T, Ito T, Yamaguchi-Shinozaki K, Shinozaki K (2004) *Arabidopsis* stress-inducible gene for arginine decarboxylase *AtADC2* is required for accumulation of putrescine in salt tolerance. Biochem Biophys Res Commun 313:369–375. https://doi.org/10.1016/j.bbrc.2003.11.119

Verma S, Mishra SN (2005) Putrescine alleviation of growth in salt stressed *Brassica juncea* by inducing antioxidative defense system. J Plant Physiol 162:669–677. https://doi.org/10.1016/j.jplph.2004.08.008

Waie B, Rajam MV (2003) Effect of increased polyamine biosynthesis on stress response in transgenic tobacco by introduction of human S-adenosylmethionine gene. Plant Sci 164:727–734. https://doi.org/10.1016/S0168-9452(03)00030-X

Walden R, Cordeiro A, Tiburcio AF (1997) Polyamines: small molecules triggering pathways in plant growth and development. Plant Physiol 113:1009–1013

Yamaguchi K, Takahashi Y, Berberich T, Imai A, Miyazaki A, Takahashi T, Michael AJ, Kusano T (2006) The polyamine spermine protects against high salt stress in *Arabidopsis thaliana*. FEBS Lett 580:6783–6788. https://doi.org/10.1016/j.febslet.2006.10.078

Yamaguchi K, Takahashi Y, Berberich T, Imai A, Takahashi T, Michael AJ, Kusano T (2007) A protective role for the polyamine spermine against drought stress in *Arabidopsis*. Biochem Biophys Res Commun 352:486–490. https://doi.org/10.1016/j.bbrc.2006.11.041

Ye XS, Avdiushko SA, Kuc J (1994) Effect of polyamines on in vitro phosphorylation of soluble and plasma membrane proteins in tobacco, cucumber and *Arabidopsis thaliana*. Plant Sci 97: 109–118

Zhang Y, Hu X, Shi Y et al (2013) Beneficial role of exogenous spermidine on nitrogen metabolism in tomato seedlings exposed to saline–alkaline stress. J Am Soc Hortic Sci 138:38–49

Zhao FG, Qin P (2004) Protective effect of exogenous polyamines on root tonoplast function against salt stress in barley seedlings. Plant Growth Regul 42:97–103. https://doi.org/10.1023/B:GROW.0000017478.40445.bc

Chapter 14
Single-Gene Versus Multigene Transfer Approaches for Crop Salt Tolerance

Satpal Turan

Abstract Plants face many challenges during biotic and abiotic stresses during their lifetime. Salinity stress is the most typical abiotic stress and combines water stress and ionic stress. It affects plants in many aspects at the molecular, cellular, and morphological levels. In response and adaptation to salt stress, plant gene regulation is modulated at the transcriptional or post-transcriptional level. Efforts have been made to overcome salinity by traditional approaches such as breeding, priming, and modern techniques of genetic engineering. However, because salt tolerance depends on multigenic properties, it is hard to control this problem simply by a single gene transfer. Although the response and signaling mechanisms of plants under salt stress have not been completely elucidated, thorough understanding of the salt stress response in plants has enabled scientists to make transgenic plants showing salt tolerance, mostly by a single transfer but also by multigene transfer. In addition to a purely gene-based approach, epigenetics and noncoding RNA have been found to play roles in salt stress/tolerance in plants. This chapter provides a brief introduction to salt stress responses and strategies for salt tolerance in plants. Moreover, single-gene versus multigene transfer and/or regulation of salt tolerance in plants are described.

Keywords Salinity stress · Antioxidative enzymes · Proline · Heat shock proteins · Ion homeostasis · Transcription factor · Salinity tolerance · Transcriptional regulation · Gene transfer · Plant breeding · Water use efficiency

Abbreviations

ABA Abscisic acid
APX Ascorbate peroxidase

S. Turan (✉)
National Research Centre on Plant Biotechnology, IARI PUSA, New Delhi, India

V. Kumar et al. (eds.), *Salinity Responses and Tolerance in Plants, Volume 1*,
https://doi.org/10.1007/978-3-319-75671-4_14

ATP	Adenosine triphosphate
BADH	Betaine aldehyde dehydrogenase
CDPK	Calcium-dependent protein kinase
CIPK	Calcineurin B–like protein interacting kinase
CRISPER	Clustered regularly interspaced short tandem repeats
ER	Endoplasmic reticulum
GORK	Guard cell outward-rectifying potassium channel
GST	Glutathione-S-transferase
Hsp31	Heat shock protein 31
IAA	Indole-3-acetic acid
MAPK	Mitogen-activated protein kinase
miRNA	MicroRNA
NSSC	Nonselective cation channel
NUE	Nitrogen use efficiency
PDH45	Pea DNA helicase 45
QTL	Quantitative trait locus
ROS	Reactive oxygen species
SAHA	Suberoylanilide hydroxamic acid
SALT	Salt stress-induced plant protein
sHSP	Small heat shock protein
siRNA	Small interfering RNA
SOD	Superoxide dismutase
SOS	Salt Overly Sensitive
TALEN	Transcription activator–like nuclease
WUE	Water use efficiency
ZFN	Zinc finger nuclease

14.1 Introduction

Salinity stress is the most typical stress affecting crop production, and its effects are expected to become more severe with global warming (Beebe et al. 2011). Nearly 20% of the world's irrigated land is salt affected (UNESCO Water Portal 2007). This is a severe problem in hot and temperate areas, where excessive irrigation and evaporation occur. Salinity stress involves changes in various physiological and metabolic processes in plants. The effects of salt stress on plants depend on its severity and duration (Munns 2005; Rozema and Flowers 2008). Salt stress consists of two kinds of stress: one is osmotic and the other is ionic. Initially, plant growth is affected by osmotic stress; later, severe injury is caused by an ionic imbalance (Rahnama et al. 2010; James et al. 2011). During the initial phases, because of the high salt concentration around plant roots, the water absorption capacity of the root system decreases and water loss from leaves is accelerated by osmotic stress; therefore, salinity stress is also considered to be hyperosmotic stress (Munns 2005).

Osmotic stress at the initial stage causes various physiological changes, such as disruption of membranes, nutrient imbalance, impaired ability to detoxify reactive oxygen species (ROS), differences in antioxidant enzymes, and decreased photosynthetic activity (Munns and Tester 2008; Pang et al. 2010). The next phase of salt stress is ionic stress; one of the most damaging effects of ionic stress is accumulation of Na^+ and Cl^- ions in tissues of plants exposed to soils with high NaCl concentrations. Higher Na^+ blocks K^+ uptake, resulting in inhibition of various enzymatic reactions, production of more ROS and, finally, lower productivity, which may even lead to cell or plant death (James et al. 2011; Wang et al. 2013).

Plant adaptation to or tolerance of salinity stress involves a complex signaling network, physiological changes, metabolic pathways, and molecular or gene networks (Zhu 2001). Comprehensive knowledge of how plants respond to salinity stress at different levels and an integrated approach combining molecular, physiological, and biochemical techniques are imperative for developing salt-tolerant varieties in salt-affected areas (Roy et al. 2014; Cabello et al. 2014; Zhang et al. 2016a). The response to salt stress involves sensing of stress and further passage of signals to other molecules in a cascade for an effect. The sensors may be membrane proteins, transporters, or cytosolic proteins, which may sense Na^+ or Cl^- ions in the rhizosphere, osmotic potential, or altered Ca^{2+}, K^+, and ROS concentrations in cells. Also, signaling depends upon the severity of stress; for example, at low salt concentrations, cell signaling is different and the cell defense is able to overcome harmful effects, but at high salt concentrations, cells opt for apoptosis. Different types of signaling can occur in the same cells, and eventually there is involvement of all tissues and the whole plant in mounting a final response. Many hormones such as abscisic acid (ABA) (Vishwakarma et al. 2017), ethylene (Kissoudis et al. 2017), auxins (Sereflioglu et al. 2017), jasmonate (Ahmad et al. 2016; Ding et al. 2016), salicylic acid (Yadu et al. 2013), and brassinosteroids (Kaur et al. 2016); enzymes such as kinases and phosphatases; and various transcription factors and effecter genes/proteins are involved in the signaling cascade (Park et al. 2016). There can be cross talk in different signaling cascades (Gupta and Huang 2014; Shabala et al. 2015). Among the mechanisms of salinity tolerance are ion transport, uptake, and compartmentalization; biosynthesis of osmoprotectants and compatible solutes; activation and synthesis of antioxidant enzymes/compounds and polyamines; and hormonal modulation (Reddy et al. 1992; Roy et al. 2014). Salt stress tolerance is a quantitative trait, as salt stress in plants changes an array of gene expressions involved in various metabolic regulations. Breeding has been used to produce a lot of salt-tolerant lines of crop plants (Singh et al. 2010), but because it is a time consuming and nonspecific approach, and because of the limitations of crossing different genotypes, there is a need for new technology. Molecular breeding has been much more successful and is faster than traditional breeding, and many desirable traits are transferred with molecular breeding in crop plants. Saltol, a quantitative trait locus (QTL) for salt tolerance, mapped in the short arm of chromosome 1 in rice, was introgressed with molecular breeding in two commonly grown rice varieties, BR11 and BR28. Pokkali, a salt-tolerant rice line, was the donor for Saltol (Thomson et al. 2010; Linh et al. 2012). Traditional breeding and molecular breeding involve exchange of many genes other than those that are desired, so a genetic engineering approach has been adopted successfully for engineering of target-specific genes. From molecular markers and mutational approaches to more recent omics approaches such as transcriptomics, proteomics, subtractive hybridization, microarrays, next-generation sequencing, and bioinformatic tools, the salinity response in plants has been studied in detail, and many genes directly or indirectly involved in salinity response and tolerance have been found. A lot of transgenic plants have been produced, showing tolerance of salt stress at the greenhouse level or field level. Largely in response to salt stress, crop varieties/genotypes vary in their inherent ability to adjust several physiological and biochemical processes (Grewal 2010; Turan et al. 2012; Turan and Tripathy 2013; Singh et al. 2014).

14.2 Plant Responses and Mechanism of Salinity Stress Tolerance

Tolerance of stress is related to the responses of plants to stress, as stress causes turgor loss, a decrease in cell expansion, yield loss, and degradation of cellular structures, so to protect the plants from stress, osmolytes are produced. Similarly, to detoxify ROS, antioxidative enzymes are produced in adaptation to stress. Salt stress also affects the general health of plants and photosynthesis, so many genes in some pathways that enhance photosynthesis also cause salinity stress tolerance. The primary tolerance mechanism is a blockade against the entry of salts or Na^+, or its exclusion or compartmentalization. Other mechanisms of salinity tolerance are signaling networks and transcriptions factors affecting target genes involved directly in salinity tolerance. Some epigenetic mechanisms have also been found to be involved in salinity tolerance. Here, some examples are given of different kinds of tolerance mechanisms prevailing in plants. Plant hormones such as transcription factors also have a big role in salinity stress tolerance, as hormones are endogenous molecules that, in small concentrations, control regulation of many genes. A few examples of salinity stress tolerance at the cellular level in plants are shown in Fig. 14.1) at cellular level.

14.2.1 Plant Hormones

Photohormones such as auxins, cytokinins, gibberellic acid, ABA, ethylene, salicylic acid, methyl jasmonate, and brassinosteroids control plant growth and development by regulating various physiological processes such as flowering, ripening, germination, and elongation. Apart from these actions, they also protect plants in adverse conditions such as salinity stress.

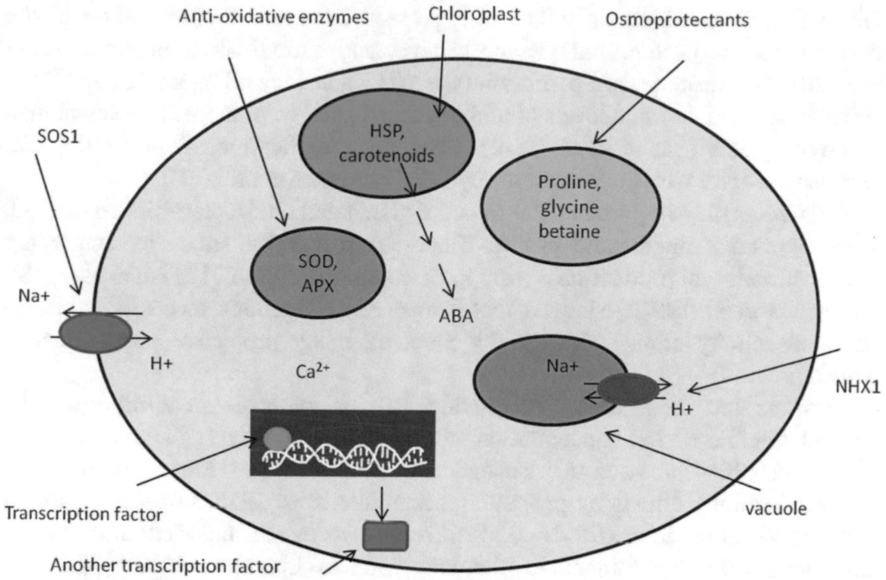

Fig. 14.1 A few types of mechanism involved in salinity stress tolerance in plants, including ion homeostasis with Na^+/H^+ antiporters, antioxidative enzymes, osmoprotectants, transcriptional control, heat shock proteins, and plant hormones. ABA abscisic acid, APX ascorbate peroxidase, NHX1 vacuolar Na^+/H^+ antiporter, SOD superoxide dismutase, SOS1 plasma membrane Na^+/H^+ antiporter

ABA has a role in leaf abscission and seed dormancy in plants. ABA levels have been found to be enhanced in salinity stress. The binding of ABA to receptors inactivates protein phosphatases such as ABI1 and ABI2, whose inactivation in turn activates SNF-like kinases by inducing ABA-dependent genes and other ion channels, and finally the stomata close to prevent water loss. *Bacillus licheniformis* enhanced saline tolerance in *Chrysanthemum* plants but was unable to do so when ABA biosynthesis was stopped (Zhou et al. 2017). NaCl and ABA induced levels of the transcription factor MtMYBS1 in *Medicago truncatula* which, when overexpressed in *Arabidopsis*, resulted in salinity tolerance (Dong et al. 2017). When *Arabidopsis ATAF1*—which is inducible by salt and ABA—was overexpressed in rice, it provided salinity tolerance (Liu et al. 2016).

Auxins play a role in root formation and meristem formation in plants. Indole-3-acetic acid (IAA) has been found to play a role in seed germination under salt stress through the transcription factor NTM2. Sereflioglu et al. (2017) showed that alpha tocopherol–induced salinity tolerance was provided by new auxin synthesis in soybean roots. Sahoo et al. (2014) showed enhanced levels of IAA in transgenic rice plants with the *SUV3* gene, which were comparatively salt tolerant.

Gibberellic acid plays the main role in elongation. Gibberellin reduced NaCl growth inhibition in a concentration-dependent manner in rice by regulating enolase, glutamyl reductase, and salt stress–induced plant protein (SALT) (Wen et al. 2010).

Gibberellin in combination with $CaCl_2$ protected plants by upregulation of the osmoprotectants proline and glycine betaine, and antioxidative enzymes (Khan et al. 2010). Transgenic rice plants with the *SUV3* gene, encoding a DNA and RNA helicase, showed elevated levels of gibberellic acid and were more salt tolerant than wild-type plants (Sahoo et al. 2014). Exogenous application of gibberellic acid enhanced salinity stress tolerance in soybean (Hamayun et al. 2010).

Cytokinins play important roles in cell division and chloroplast biogenesis and, when applied exogenously, protect plants from salinity stress by improving photosynthesis and production of osmoprotectants (Ha et al. 2012; Wu et al. 2013). Nishiyama et al. (2012) showed that decreased endogenous levels of cytokinin imparted salinity stress tolerance by inducing many protective metabolites in *Arabidopsis*.

Ethylene has been found to provide salinity tolerance in *Arabidopsis* by upregulating the K^+ ion concentration with respect to Na^+ ions (Yang et al. 2013). Peng et al. (2014) showed that salt-induced EIN3 and EIL1 transcription factors alleviated salinity effects by preventing accumulation of ROS. Overexpression of ethylene response factor *HbERF-IXc5* in *Hevea brasiliensis* imparted salinity tolerance along with improvements in other latex features (Lestari et al. 2018).

Leymus chinensis showed enhanced salt tolerance when a gene for S-adenosylmethionine decarboxylase, *LcSAMDC1*, was overexpressed (Liu et al. 2017). SOS2 provides salinity tolerance in *Arabidopsis* and is known to upregulate the ethylene pathway genes *EIN3* and ethylene response factor 1; also, SOS2 phosphorylates EIN3 (Quan et al. 2017).

Brassinosteroids protect plants from salinity stress by improving physiological processes (Ashraf et al. 2010). Exogenous application of brassinosteroid provides salinity tolerance in rice (Sharma et al. 2013). It was shown by Wu et al. (2017) that when the brassinosteroid 24-epibrassinolide was applied to perennial ryegrass, it enhanced its salinity tolerance by induction of selected hormones and antioxidative enzymes. Brassinosteroid application in *Arabidopsis* caused plant protection from salt stress by mitochondrial heat shock protein DnaJ induction (Bekh-Ochir et al. 2013).

14.2.2 Ion Homeostasis

Ion homeostasis in cells is maintained by various water and ion channels and ion transporters regulated by plant hormones. Normally plants maintain concentrations of all elements as required by the plants for optimal growth. During salinity, sodium and chloride ion accumulation, especially in old leaves, causes necrosis, leaf senescence, and production of ROS (Niu et al. 1995; Assaha et al. 2017). Further ion homeostasis is impacted as the Na^+/K^+ ratio is disturbed by increased salt stress, inhibiting many enzymatic reactions in which K^+ is an important cofactor (Silva et al. 2015). High Na^+ reduces K^+ uptake by inhibiting potassium transporters such as AKT1, a hyperpolarization-activated inward-rectifying potassium channel. A high cytosolic Na^+ increases K^+ efflux through transporters such as the guard cell outward-rectifying potassium channel (GORK) (Adams and Shin 2014; Gariga et al. 2017) and nonselective cation channels (NSSCs) (Sun et al. 2009). A high K^+/Na^+ ratio in shoots is considered a feature of salt tolerance. High-affinity transporters in the HKT1 family enhance salt tolerance by reabsorption of Na^+ from the xylem into the roots and prevention of Na^+ entry into the shoots. Various P- and V-type ATPases help with preparation of a proton gradient in vacuoles or the cytoplasm, which in turn is utilized for energy generation required for transport. Apart from these, there are various transporters/antiporters that help with ion homeostasis and salt tolerance—for example, H^+/Ca^{2+} antiporter help in calcium homeostasis (Cheng et al. 2004). Ion homeostasis is achieved under salt stress by many transporters such as SOS1, a plasma membrane Na^+/H^+ antiporter which extrudes Na^+ into the soil, and many sodium–hydrogen exchangers in the tonoplast, which help with sodium compartmentalization into vacuoles (Almeida et al. 2017). GhSOS1, an Na^+/H^+ antiporter gene from *Gossypium hirsutum*, causes salinity tolerance when overexpressed in *Arabidopsis thaliana* (Chen et al. 2017). Apart from plant hormones, there are various signaling molecules that are known to play roles in ion homeostasis in plants; for example, nitic oxide balances the K^+/Na^+ ratio in *Kandelia obovata* by AKT1-type K^+ channels and an Na^+/H^+ antiporter (Chen et al. 2013).

14.2.3 Osmolytes

In response to a water deficit created by salt stress, plants synthesize compounds called osmolytes. These help in osmotic adjustment by lowering cytosolic osmotic potential. The mechanism of protection by osmolytes is replacement of water at the protein surface and stabilization of molecular structures under stress by the osmolytes acting as molecular chaperones. They are not inhibitory even at high concentrations (Rontein et al. 2002; Cheong and Yun 2007; Park et al. 2016). They comprise various sugars such as sucrose, glucose, fructose, trehalose, mannitol, sorbitol, and ononitol; amino acids such as proline, aspartate, and glycine; ammonium compounds such as glycine betaine; charged metabolites such as betaine, choline, oxalate, malate, putrescine, and spermidine; and ions such as K^+ (Rontein et al. 2002, Sakamoto and Murata 2000). Puniran-Hartley et al. (2014) showed that in wheat and barley, production of osmolytes was associated with increased oxidative stress tolerance. Reddy et al. (2015) suggested that overproduction of proline provided salinity tolerance and enhanced photosynthetic and antioxidant enzyme activities in transgenic *Sorghum bicolor* L. Glycine betaine produced as a result of overexpression of the betaine aldehyde dehydrogenase (BADH) gene from *Suaeda corniculata* in *Arabidopsis* provided salinity tolerance. The osmotin-like protein gene *SindOLP*, from *Solanum nigrum*, imparted salinity stress tolerance in sesame when was overexpressed (Chowdhury et al. 2017).

14.2.4 Antioxidative Enzymes

ROS are produced under both normal and stressed conditions, but under normal conditions the rates of their production and detoxification are balanced (Mittler 2002; Turan and Tripathy 2013). Under stressed conditions the rate of production is greater, so ROS such as H_2O_2, singlet oxygen, and hydroxyl radicals produce damaged cell molecules and membranes. Plants have devised multiple strategies to control ROS in the forms of nonenzymatic and enzymatic antioxidative molecules (Hossain et al. 2017). Nonenzymatic molecules such as ascorbate, glutathione, melatonin, and carotenoids play important roles in plant defense under salt stress (Turan and Tripathy 2013). Among the enzymatic molecules, superoxide reductase, ascorbate peroxidase, glutathione reductase, dehydroascorbate reductase, and monodehydroascorbate reductase help in detoxification of harmful ROS (Zhang et al. 2016b). Overexpression of two genes, *Gly1* and *Gly2*, in the methylglyoxal detoxification pathway in rice resulted in multiple stress tolerance, including tolerance of salinity stress (Gupta et al. 2017). Ectopic expression of tomato glutathione-S-transferase (GST) in *Arabidopsis* caused salinity and drought tolerance (Xu et al. 2015a). Upregulation of the antioxidative enzymes superoxide dismutase and catalase in transgenic tobacco after overexpression of the JcWRKY transcription factor gene from *Jatropha curcas* was responsible for enhanced salinity stress tolerance (Agarwal et al. 2016).

14.2.5 Heat Shock Proteins or Molecular Chaperones

These molecules help in stabilizing molecular structures of proteins and preventing their aggregation under salt stress. Expression of endoplasmic reticulum small heat shock proteins (ER-sHSPs) in tomato resulted in enhanced salinity tolerance by improving photosynthesis and root architecture, increasing relative water content, and reducing accumulation of Na^+ (Fu et al. 2016). Similarly, Jiang et al. (2009) showed that when the cytosolic class I small heat shock protein coding gene *RcHSP17.8*, from *Rosa chinensis*, was overexpressed in *Arabidopsis*, it conferred salt and heat tolerance. When overexpressed in *Arabidopsis*, the 70 kDa protein coding gene *MuHSP70*, from *Macrotyloma uniflorum*, caused salinity tolerance (Masand and Yadav 2016).

14.2.6 Transcription Factors

Transcription factors are the proteins that bind to *cis*-elements and regulate the expression of various genes. A number of transcription factors such as AP2/EREBP, MYB, NAC, WRKY, bZIP, bHLH, Dof and heat shock transcription factors. Zhang et al. (2015) showed that when an SsDREBa AP2/EREBP family transcription factor from *Suaeda salsa* was overexpressed in tobacco, it conferred salinity and drought tolerance. Overexpression of *Solanum lycopersicum SlWRKY3* imparted salinity stress tolerance in tomato (Hichri et al. 2017). Yu et al. (2017) showed that the R2R3Myb transcription factor TaSIM caused salinity stress tolerance in *Arabidopsis*. When *OsMYB91* was overexpressed in rice, it caused salinity tolerance (Zhu et al. 2015). Ectopic expression of *JcWRKY* from *Jatropha curcas* conferred salinity tolerance in tobacco by involving salicylic acid signaling (Agarwal et al. 2016). Similarly, when *TaNAC29* from *Triticum aestivum* was transferred into *Arabidopsis*, it conferred enhanced salinity and drought tolerance (Huang et al. 2015). When *TabZIP60* from *T. aestivum* was overexpressed in *Arabidopsis*, it caused tolerance of multiple stresses such as freezing, salinity, and drought. A transcription factor can regulate many genes by binding to *cis*-elements of various genes; also, a gene can have many types of *cis*-elements, so it can be regulated by many transcription factors, meaning that transcription factors not only work independently but also cross talk during various abiotic stresses.

14.2.7 Signaling Molecules

A salt stress signal is perceived by some receptors and a signal cascade is set up, causing a stress response in plants. The signal is mediated by many messengers and intermediates until it reaches the final effector. Some signaling pathways are ABA dependent and others are not. Ca^{2+} plays a role as the most common secondary messenger, as in the most common pathway (Salt Overly Sensitive [SOS]), a change in cytosolic Ca^{2+} is sensed by SOS3 (a calcium-binding protein), which in turn activates SOS2 (a serine–threonine protein kinase), which activates an Na^+/H^+ antiporter by phosphorylation. Another common pathway in the plant stress response is the mitogen-activated protein kinase (MAPK) cascade, which involves a relay of phosphorylation. Overexpression of the MAPK *OsMKK6* caused salinity tolerance in rice. Both positive and negative signaling can occur; for example, when *GhRaf19* from cotton was overexpressed in *Nicotiana benthamiana*, it resulted in decreased tolerance of salt and drought but increased tolerance of cold (Jia et al. 2016). Other signaling cascades involve calcium-dependent protein kinase (CDPK) and calcineurin B–like protein interacting kinase (CIPK). *ZoCDPK1* from ginger provided salinity tolerance when overexpressed in tobacco, by improving photosynthetic capacity and seed germination (Vivek et al. 2013). *TaCIPK29* transgenic tobacco plants showed salinity tolerance through upregulation of antioxidative enzymes (Deng et al. 2013).

14.2.8 *Other Mechanisms of Salinity Tolerance*

Late embryogenic abundant proteins, which are produced during the late phase of embryonic development, aid salinity tolerance in many ways, such as acting as hydrating buffers, protecting and stabilizing, renaturing cellular proteins, acting as chaperones, and sequestrating ions (Turan et al. 2012). For example, ectopic expression of the LEA protein *Ad*LEA (from *Arachis diogoi*) in tobacco provided salinity and drought tolerance by upregulating oxidative defense and improving photosynthesis under stressed conditions (Sharma et al. 2016). Some helicases also provide salinity tolerance by assisting basic processes of replication and transcription involving adenosine triphosphate (ATP)–dependent DNA or RNA unwinding. Transgenic tobacco expressing pea DNA helicase 45 (*PDH45*) showed salinity tolerance (Garg et al. 2017). Recently, microRNAs have also been shown to play role in salt stress tolerance. For example, when the microRNA (miRNA) miR528 from rice was constitutively expressed in wild creeping bent grass, it provided salinity tolerance associated with increased water retention, chlorophyll, and proline content (Yuan et al. 2015). Small interfering RNAs (siRNAs) are also effective in salinity tolerance, as their targets are transcription factors, so they can control expression of many genes together. For example, *AtMYB74*, which is upregulated in salt stress, was found to be regulated by RNA-directed DNA methylation (Xu et al. 2015b). The role of epigenetics in salinity response and tolerance has also been demonstrated (Pandey et al. 2017; Kumar et al. 2017; Patanun et al. 2017). According to Patanun et al. (2017), the histone deacetylase inhibitor suberoylanilide hydroxamic acid (SAHA) caused salinity tolerance in cassava by inducing acetylation in H3 and H4 histones.

14.3 Single-Gene Versus Multigene Transfer Approaches for Salinity Stress Tolerance

Most salinity-tolerant transgenic plants are produced with single-gene transfer encoding for a product directly or indirectly involved in salinity stress tolerance (reviewed by Turan et al. 2012; Hanin et al. 2016; Mishra and Tanna 2017). However, in some cases, multiple genes are transferred, and it has been found that a combination of genes gives more salinity stress tolerance (Table 14.1). Basically, any response or biosynthesis of any osmolyte is catalyzed by many enzymatic steps. So, for an effective response or better production, a multigene strategy may be more helpful for salinity stress tolerance (Fig. 14.2). Similarly, regulation of any ion transport involves multiple proteins or cascade of proteins. Success has been achieved in many cases where a whole metabolic pathway has been transferred to achieve a high concentration of a particular product, although such approaches have not yet been tried for salinity tolerance. Paddon et al. (2013) published an article in *Nature* that reported a high level of artemisinin by overexpression of many genes

Table 14.1 Genes that cause enhanced salinity stress tolerance when transferred together into plants

Gene sources	Genes transferred	Host	Effect on salinity tolerance	Reference
Brassica juncea, Oryza sativa	*GlyI, GlyII*	*Tobacco*	Increased	Singla-Pareel et al. (2003)
Erianthus arundinaceus, pea	*EaDREB2, PDH45*	Sugarcane (*Saccharum* spp.)	Increased	Ghosh et al. (2014)
Hordeum vulgare, bacteria	*Hva1, mtlD*	*Maize*	Increased	Nguyen et al. (2013)
Arabidopsis	*SOS2, SOS3*	*Arabidopsis*	Increased	Yang et al. (2009)
Rice	*OsP5CS1, OsP5CS2*	Tobacco	Increased	Zhang et al. (2014)
Salicornia europaea, tobacco	*BADH, SeNHX1*	Tobacco	Increased	Zhou et al. (2008)
Arabidopsis	*AtDREB2A, AtHB7, AtABF3*	*Arachis hypogaea*	Increased	Pruthvi et al. (2014)
Gossypium hirsutum	*GhSOD1, GhCAT1*	*Gossypium hirsutum*	Increased	Luo et al. (2013)
Aphanothece halophytica	*ApGSMT, ApDMT*	*Arabidopsis*	Increased	Waditee et al. (2005)

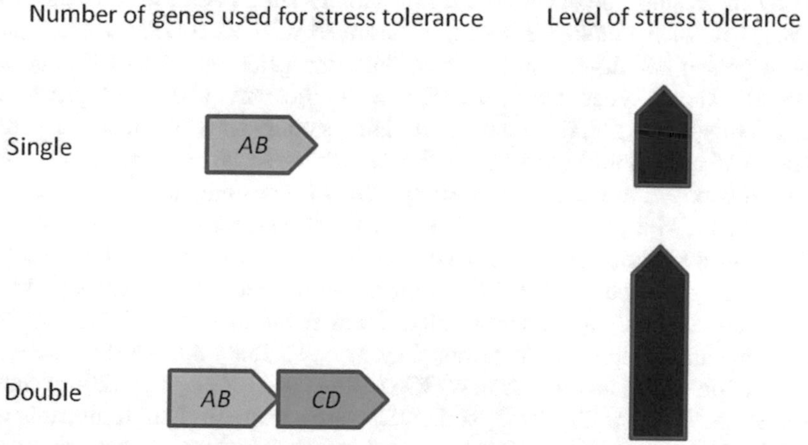

Fig. 14.2 The level of stress tolerance in plants is generally greater when two genes involved in salt tolerance are transferred to a plant instead of a single gene, although this is not true in every case

and downregulation of some other genes. Traditional breeding and priming also involve multiple genes for the final response of plants to salinity stress. Regulatory genes such as transcription factors, microRNAs, and signaling molecules such as MAPKs, which target and control gene expression of many downstream genes, are better candidates for salinity tolerance. Salinity, drought, temperature, and metal stresses share common transcription factors, which cross talk, so transcription factors are good candidates for multiple stress tolerance. For example, Yang et al. (2017) showed that *Th*DREB, a dehydration response element binding factor from *Tamarix hispida*, enhanced salinity and drought tolerance when overexpressed in tobacco.

14.3.1 More Than One Gene and Salinity Tolerance

Methylglyoxal is an oxidant produced in salt stress, which harms plants because of its toxic nature. Plants have adapted to this stress by producing glyoxalase enzymes, which detoxify methylglyoxal. There have been a lot of studies on transgenics expressing methyl glyoxalase enzyme genes singly and in combination for detoxification of methylglyoxal. Veena and Sopory (1999) showed that transcripts of *Gly*1 were upregulated in salt stress, and tobacco that was transgenic for *Gly*1 showed enhanced tolerance of salt stress, which was correlated with the level of *Gly*1 expression. Singla-Pareek et al. (2003) showed that when *Gly*I and *Gly*II together were expressed in tobacco, double transgenic plants showed improved salinity tolerance and a higher yield under salinity stress than either single transgenic. Subsequently, Ghosh et al. (2014) showed that *Gly*II enhances salinity tolerance by increasing photosynthesis under salinity stress. Augustine et al. (2015) showed that when *EaDREB2* was cotransformed with *PDH45*, transgenic plants showed greater salinity stress tolerance than transgenic plants transformed with EaDREB2 alone. When *SacB*, *JERF36*, *vgb*, *BtCry3A*, and *OC-I* genes were cotransformed in poplar, this resulted in higher salinity stress tolerance, as well as tolerance of other abiotic and biotic stresses (Su et al. 2011). Garg et al. (2017) produced double transgenics, transferring *PDH45* for salinity and ESPS for herbicide tolerance. Nguyen et al. (2013) showed that cotransformation of *HVA*1 and *mtl*D enhanced salinity stress tolerance in maize more greatly than either one alone. Earlier studies showed that *SOS1* encoding an Na^+/H^+ antiporter, when expressed in plants, caused salinity tolerance; further, it was found that when *SOS2* and *SOS3* were individually expressed in plants, they caused salinity tolerance because they are part of the upstream regulation of SOS1 (Zhu 2000). Yang et al. (2009) showed that when *AtNHX1*, *SOS1*, *SOS2*, and *SOS3* were coexpressed, all transcripts were found but there was no significant increase in salt tolerance in comparison with expression of *SO3* alone. This is because SOS3 and SOS2 have different forms, which are also tissue specific, indicating that not only is the expression of genes essential but also their temporal and spatial expressions are important. When two genes (*OsP5CS1* and *OsP5CS2*) involved in osmoprotection were coexpressed in tobacco, transgenic plants were more salt tolerant (Zhang et al. 2014). Similarly Zhou et al. (2008) showed that when two genes—one for an Na^+/H^+ antiporter (*SeNHX1*) and one for betaine synthesis (*BADH*)—were cotransformed and individually transformed, coexpressed transgenic plants showed greater salinity tolerance than plants with either individual gene transformation. SOS2 and EIN3 interact with each other and regulate salinity stress tolerance synergistically (Quan et al. 2017). Coexpression of the stress-responsive transcription factor genes *AtDREB2A*, *AtHB7*, and *AtABF3* improved salinity stress tolerance in peanut (*Arachis hypogaea* L.) (Pruthvi et al. 2014). Glycine betaine is a very good osmoprotectant, which causes salinity tolerance. The homozygous Mpgsmt (G), Mpsdmt (S), and their cross, Mpgsmt and Mpsdmt (G × S), showed increased accumulation of betaine (Lai et al. 2014). Coexpression of two genes, *GhSOD1* and *GhCAT1*, in chloroplasts resulted in greater salinity and methylglyoxal tolerance than individual expression of either gene in chloroplasts or the cytosol (Luo et al. 2013). Also, engineering of different metabolic pathways for the same product can have differential effects on salinity tolerance. For example, when the genes for N-methyltransferases (*ApGSMT* and *ApDMT*) were transferred to *Arabidopsis*, transgenic plants were more salt tolerant than when the transgenic plants were engineered through the choline oxidizing pathway (Waditee et al. 2005).

14.3.2 Multistress Tolerance Genes

Apart from the aforementioned examples where many genes were involved in tolerance of a single stress, there are genes that provide multistress tolerance (Table 14.2). Kong et al. (2011) isolated a maize gene encoding a novel group C MAPK, *ZmMKK4* which, when overexpressed in *Arabidopsis*, increased salinity and cold tolerance by upregulating production of osmolytes such as proline, soluble sugar content, antioxidative enzymes, and lateral root growth. Overexpression of the yeast heat shock protein 31 (Hsp31) in tobacco caused both biotic and abiotic stress tolerance by playing a role in detoxification of methylglyoxal (Melvin et al. 2017). Similarly, Ben Romdhane et al. (2017) showed that ectopic expression of the

Table 14.2 Single genes that cause multistress tolerance when transferred into plants

Gene name	Gene source	Host	Provides resistance against
ZmMKK4	*Zea mays*	*Arabidopsis*	Salinity, cold
Hsp31	*Saccharomyces cerevisiae*	Tobacco	Biotic and abiotic stress
AITMP1	*Aeluropus littoralis*	Tobacco	Salt, cold, heat, osmotic stress, drought, H_2O_2
GmNEK1	*Glycine max*	Tobacco	Salinity, cold
SindOLP	*Solanum nigrum*	*Sesamum indicum*	Salt, drought, oxidative, and biotic stress
AtNDPK2	*Arabidopsis*	*Arabidopsis*	Salinity, cold, oxidative stress
BrCIPK1	*Brassica rapa*	Rice	Salinity, drought
OsMSR2	Rice	*Arabidopsis*	Salinity, drought
OsMAK5	Rice	Rice	Salinity, drought, cold
MdCIPK6L	Apple	Tomato	Salt, drought, chilling

Aeluropus littoralis gene *AlTMP1* in tobacco caused tolerance of multiple stresses such as salt, freezing, H_2O_2, osmotic stress, heat, and drought by ionic homeostasis and improvement of water status under stressed conditions. It is not always that there is positive regulation of stress by transcription factors; for example, overexpression of *OsNAC2* caused sensitivity to saline and drought stress, whereas *RNAi* plants with *OsNAC2* were more salinity and drought tolerant, with higher yield (Shen et al. 2017). Moon et al. (2003) showed that when *AtNDPK2*, coding for NDP kinase 2, was overexpressed in *Arabidopsis*, it increased salinity, cold, and oxidative tolerance. Overexpression of BrCIPK1 in rice enhanced salinity and drought tolerance by increasing osmoprotectant proline biosynthesis (Abdula et al. 2016). Xu et al. (2011) showed that expression of the calmodulin-like gene OsMSR2 from rice provided enhanced salinity and drought tolerance in *Arabidopsis*. Overexpression of the *OsMAK5* gene caused enhanced activity of OsMAPK kinase and provided tolerance of salt, drought, and cold, while its RNAi lines had enhanced levels of the pathogenesis-related proteins PR1 and PR10, with tolerance of fungal and bacterial biotic stresses (Xiong and Yang 2003). Overexpression of *MdCIPK6L* from apple conferred tolerance of multiple abiotic stresses—including salt, drought, and chilling—in tomato (Wang et al. 2012). Overexpression of *GmNEK1*—a kinase from *Glycine max*—caused salinity and cold tolerance in tobacco (Pan et al. 2017). According to Chowdhury et al. (2017) when an OLP gene (*SindOLP*) was overexpressed in sesame, it resulted in tolerance of salt, drought, oxidative stress, and the charcoal rot pathogen. This clearly indicates cross talk in different abiotic and biotic stresses.

14.4 Perspectives

Scientists worldwide are working on salinity tolerance, and much success has been achieved in understanding salinity responses in plants. Initially through traditional breeding, and then through molecular breeding with molecular markers and marker-assisted selection, many salt-tolerant lines/cultivars have been produced (Varshney et al. 2007; Turan et al. 2012). However, because of reproductive barriers in breeding, and because those processes are time consuming, there has been a need for newer techniques. Genetic engineering has proved more successful in targeting specific transfer of genes responsible for salt tolerance into desired genotypes, even from far distant lines or completely different genotypes. Many transgenic plants for salinity tolerance have been produced (Turan et al. 2012; Hanin et al. 2016). Broadly, two kinds of gene have been used for creating transgenic plants for salt tolerance: ones that are directly involved in salinity stress tolerance (genes for osmolytes, antioxidative enzymes, ion transporters, etc.) and ones that are indirectly involved in salinity tolerance (genes for transcription factors and signaling molecules such as kinases and hormones). The latter approach is more successful, as it involves many genes indirectly; for example, transcription factors can control expression of many genes. But still there is a greater need and more scope for improvement of approaches for salinity stress tolerance in plants. One of the most fundamental requirements for this is to study plant responses to salt stress in more detail. Many recent omics techniques and next-generation sequencing are being used to reveal the roles of many unknown genes in various species. Day by day, advanced techniques such as zinc finger nuclease (ZFN), transcription activator–like nuclease (TALEN), and clustered regularly interspaced short tandem repeats (CRISPER) can be used for a precise cut in the genome and for introducing desired changes at one or many loci. Most transgenic plants have been produced by single gene transfers, whereas more than one gene or a set of genes for metabolic pathways can be more effective in salinity tolerance. Sometimes use of constitutive promotors also affects plant growth under normal conditions. There is a need for identification of the roles of more *cis*-regulatory elements and use of cell-specific and tissue-specific inducible promotors for finely controlling gene expression, instead of constitutive promotors. Salinity tolerance has been tested mostly at the laboratory level; there is a need for more field testing or more restrictive conditions. A mixture of different stresses representing a climate similar to field conditions should be used for plant salt tolerance testing. Along with genes for salinity tolerance, other genes that help plants grow healthier—such as those for water use efficiency (WUE), nitrogen use efficiency (NUE), or any other nutrient uptake—can be of additive advantage. However, there is a limit to foreign DNA transfer through genetic engineering, as the gene load affects plant growth height, yield, and other parameters, so a combined approach of genetic engineering and molecular breeding—gene pyramiding—should prove more successful in the future.

References

Abdula SE, Lee HJ, Ryu H, Kang KK, Nou I, Sorrells ME, Cho YG (2016) Over expression of BrCIPK1 gene enhances abiotic stress tolerance by increasing proline biosynthesis in rice. Plant MolBiol Rep 34:501

Adams E, Shin R (2014) Transport, signaling, and homeostasis of potassium and sodium in plants. J Integr Plant Biol 56:231–249

Agarwal P, Dabi M, Sapara KK, Joshi PS, Agarwal PK (2016) Ectopic expression of JcWRKY transcription factor confers salinity tolerance via salicylic acid signaling. Front Plant Sci 7:1541

Ahmad P, Rasool S, Gul A, Sheikh SA, Akram NA, Ashraf M, Kazi AM, Gucel S (2016) Jasmonates: multifunctional roles in stress tolerance. Front Plant Sci 7:813

Almeida DM, Oliveira MM, Saibo NJ (2017) Regulation of Na^+ and K^+ homeostasis in plants: towards improved salt stress tolerance in crop plants. Genet Mol Biol 40:326–345

Ashraf M, Akram NA, Arteca RN, Foolad MR (2010) The physiological, biochemical and molecular roles of brassinosteroids and salicylic acid in plant processes and salt tolerance. Crit Rev Plant Sci 29:162–190

Assaha DVM, Ueda A, Saneoka H, Al-Yahyai R, Yaish MW (2017) The role of Na^+ and K^+ transporters in salt stress adaptation in glycophytes. Front Physiol 8:509

Augustine SM, Ashwin Narayan J, Syamaladevi DP, Appunu C, Chakravarthi M, Ravichandran V, Tuteja N, Subramonian N (2015) Overexpression of *EaDREB2* and pyramiding of *EaDREB2* with the pea DNA helicase gene (*PDH45*) enhance drought and salinity tolerance in sugarcane (*Saccharum spp*. hybrid). Plant Cell Rep 34:247–263

Beebe S, Ramirez J, Jarvis A, Rao I, Mosquera G, Bueno G et al (2011) Genetic improvement of common beans and the challenges of climate change. In: Yadav SS, Redden RJ, Hatfield JL, Lotze-Campen H, Hall AE (eds) Crop adaptation of climate change., 1st edn. Wiley, New York, pp 356–369

Bekh-Ochir D, Shimada S, Yamagami A, Kanda S, Ogawa K, Nakazawa M, Matsui M, Sakuta M, Osada H, Asami T, Nakano T (2013) A novel mitochondrial DnaJ/Hsp40 family protein BIL2 promotes plant growth and resistance against environmental stress in brassinosteroid signaling. Planta 237:1509–1525

Ben Romdhane W, Ben-Saad R, Meynard D, Verdeil JL, Azaza J, Zouari N, Fki L, Guiderdoni E, Al-Doss A, Hassairi A (2017) Ectopic expression of *Aeluropus littoralis* plasma membrane protein gene AlTMP1 confers abiotic stress tolerance in transgenic tobacco by improving water status and cation homeostasis. Int J Mol Sci 18:pii: E692

Cabello JV, Lodeyro AF, Zurbriggen MD (2014) Novel perspectives for the engineering of abiotic stress tolerance in plants. Curr Opin Biotechnol 26:62–70

Chen J, Xiong DY, Wang WH, Hu WJ, Simon M, Xiao Q, Chen J, Liu TW, Liu X, Zheng HL (2013) Nitric oxide mediates root K^+/Na^+ balance in a mangrove plant, Kandelia obovata, by enhancing the expression of AKT1-type K^+ channel and Na^+/H^+ antiporter under high salinity. PLoS One 8:e71543

Chen X, Lu X, Shu N, Wang D, Wang S, Wang J, Guo L, Guo X, Fan W, Lin Z, Ye W (2017) *GhSOS1*, a plasma membrane Na^+/H^+ antiporter gene from upland cotton, enhances salt tolerance in transgenic Arabidopsis thaliana. PLoS One 12:e0181450

Cheng NH, Pittman JK, Zhu JK, Hirschi KD (2004) The protein kinase SOS2 activates the *Arabidopsis* H^+/Ca^{2+} antiporter CAX1 to integrate calcium transport and salt tolerance. J Biol Chem 279:2922–2926

Cheong M, Yun D-J (2007) Salt-stress signaling. J Plant Biol 50:148–155

Chowdhury S, Basu A, Kundu S (2017) Overexpression of a new osmotin-like protein gene (SindOLP) confers tolerance against biotic and abiotic stresses in sesame. Front Plant Sci 8:410

Deng X, Hu W, Wei S, Zhou S, Zhang F, Han J, Chen L, Li Y, Feng J, Fang B, Luo Q, Li S, Liu Y, Yang G, He G (2013) TaCIPK29, a CBL-interacting protein kinase gene from wheat, confers salt stress tolerance in transgenic tobacco. PLoS One 8:e69881

Ding H, Lai J, Wu Q, Zhang S, Chen L, Dai YS, Wang C, Du J, Xiao S, Yang C (2016) Jasmonate complements the function of *Arabidopsis* lipoxygenase 3 in salinity stress response. Plant Sci 244:1–7

Dong W, Song Y, Zhao Z, Qiu NW, Liu X, Guo W (2017) The *Medicago truncatula* R2R3-MYB transcription factor gene *MtMYBS1* enhances salinity tolerance when constitutively expressed in *Arabidopsis thaliana*. Biochem Biophys Res Commun 490:225–230

Fu C, Liu XX, Yang WW, Zhao CM, Liu J (2016) Enhanced salt tolerance in tomato plants constitutively expressing heat-shock protein in the endoplasmic reticulum. Genet Mol Res 15 https://doi.org/10.4238/gmr.15028301

Garg B, Gill SS, Biswas DK, Sahoo RK, Kunchge NS, Tuteja R, Tuteja N (2017) Simultaneous expression of *PDH45* with *EPSPS* gene improves salinity and herbicide tolerance in transgenic tobacco plants. Front Plant Sci 8:364

Gariga M, Raddatz N, Véry AA, Sentenac H, Rubio-Meléndez ME, González W, Dreyer I (2017) Cloning and functional characterization of HKT1 and AKT1 genes of *Fragaria* spp. Relationship to plant response to salt stress. J Plant Physiol 210:9–17

Ghosh A, Pareek A, Sopory SK, Singla-Pareek SL (2014) A glutathione responsive rice glyoxalase II, OsGLYII-2, functions in salinity adaptation by maintaining better photosynthesis efficiency and anti-oxidant pool. Plant J 80:93–105

Grewal HS (2010) Water uptake, water use efficiency, plant growth and ionic balance of wheat, barley, canola and chickpea plants on a sodic vertosol with variable subsoil NaCl salinity. Agric Water Manag 97:148–156

Gupta B, Haung B (2014) Mechanism of salinity tolerance in plants: physiological, biochemical, and molecular characterization. Int J Genomics. https://doi.org/10.1155/2014/701596

Gupta BK, Sahoo KK, Ghosh A, Tripathi AK, Anwar K, Das P, Singh AK, Pareek A, Sopory SK, Singla-Pareek SL (2017) Manipulation of glyoxalase pathway confers tolerance to multiple stresses in rice. Plant Cell Environ. https://doi.org/10.1111/pce.12968. [Epub ahead of print]

Ha S, Vankova R, Yamaguchi-Shinozaki K, Shinozaki K, Tran L-SP (2012) Cytokinins: metabolism and function in plant adaptation to environmental stresses. Trends Plant Sci 17:172–179

Hamayun M, Khan SA, Khan AL, Shin JH, Ahmad B, Shin DH, Lee IJ (2010) Exogenous gibberellic acid reprograms soybean to higher growth and salt stress tolerance. J Agric Food Chem 58:7226–7232

Hanin M, Ebel C, Ngom M, Laplaze L, Masmoudi K (2016) New insights on plant salt tolerance mechanisms and their potential use for breeding. Front Plant Sci 7:1787

Hichri I, Muhovski Y, Žižková E, Dobrev PI, Gharbi E, Franco-Zorrilla JM, Lopez-Vidriero I, Solano R, Clippe A, Errachid A, Motyka V, Lutts S (2017) The *Solanum lycopersicum* WRKY3 transcription factor SlWRKY3 is involved in salt stress tolerance in tomato. Front Plant Sci 8:1343

Hossain MS, El Sayed AI, Moore M, Dietz KJ (2017) Redox and reactive oxygen species network in acclimation for salinity tolerance in sugar beet. J Exp Bot 68:1283–1298

Huang Q, Wang Y, Li B, Chang J, Chen M, Li K, Yang G, He G (2015) TaNAC29, a NAC transcription factor from wheat, enhances salt and drought tolerance in transgenic *Arabidopsis*. BMC Plant Biol 15:268

James RA, Blake C, Byrt CS, Munns R (2011) Major genes for Na^+ exclusion, Nax1 and Nax2 (wheat HKT1;4 and HKT1;5), decrease Na^+ accumulation in bread wheat leaves under saline and waterlogged conditions. J Exp Bot 62:2939–2947

Jia H, Hao L, Guo X, Liu S, Yan Y, Guo X (2016) A Raf-like MAPKKK gene, GhRaf19, negatively regulates tolerance to drought and salt and positively regulates resistance to cold stress by modulating reactive oxygen species in cotton. Plant Sci 252:267–281

Jiang C, Xu J, Zhang H, Zhang X, Shi J, Li M, Ming F (2009) A cytosolic class I small heat shock protein, RcHSP17.8, of *Rosa chinensis* confers resistance to a variety of stresses to *Escherichia coli*, yeast and *Arabidopsis thaliana*. Plant Cell Environ 32:1046–1059

Kaur N, Kirat K, Saini S, Sharma I, Gantet P, Pati PK (2016) Reactive oxygen species generating system and brassinosteroids are linked to salt stress adaptation mechanisms in rice. Plant Signal Behav 11:e1247136

Khan MN, Siddiqui MH, Mohammad F, Naeem M, Masroor M, Khan K (2010) Calcium chloride and gibberellic acid protect linseed (*Linum usitatissimum* L.) from NaCl stress by inducing antioxidative defence system and osmoprotectant accumulation. ActaPhysiol Plant 32:121–132

Kissoudis C, Seifi A, Yan Z, Islam AT, van der Schoot H, van de Wiel CC, Visser RG, van der Linden CG, Bai Y (2017) Ethylene and abscisic acid signaling pathways differentially influence tomato resistance to combined powdery mildew and salt stress. Front Plant Sci 7:2009

Kong X, Pan J, Zhang M, Xing X, Zhou Y, Liu Y, Li D, Li D (2011) ZmMKK4, a novel group C mitogen-activated protein kinase kinase in maize (*Zea mays*), confers salt and cold tolerance in transgenic *Arabidopsis*. Plant Cell Environ 34:1291–1303

Kumar S, Beena AS, Awana M, Singh A (2017) Salt-induced tissue-specific cytosine methylation downregulates expression of HKT genes in contrasting wheat (*Triticum aestivum* L.) genotypes. DNA Cell Biol 36:283–294

Lai SJ, Lai MC, Lee RJ, Chen YH, Yen HE (2014) Transgenic Arabidopsis expressing osmolyte glycine betaine synthesizing enzymes from halophilic methanogen promote tolerance to drought and salt stress. Plant Mol Biol 85:429–441

Lestari R, Rio M, Martin F, Leclercq J, Woraathasin N, Roques S, Dessailly F, Clément-Vidal A, Sanier C, Fabre D, Melliti S, Suharsono S, Montoro P (2018) Overexpression of *Hevea brasiliensis* ethylene response factor HbERF-IXc5 enhances growth, tolerance to abiotic stress and affects laticifer differentiation. Plant Biotechnol J 16:322–336

Linh LH, Linh TH, Xuan TD, Ham LH, Ismail AM, Khanh TD (2012) Molecular breeding to improve salt tolerance of rice (*Oryza sativa* L.) in the Red River Delta of Vietnam. Int J Plant Genom 2012:949038

Liu Y, Sun J, Wu Y (2016) *Arabidopsis* ATAF1 enhances the tolerance to salt stress and ABA in transgenic rice. J Plant Res 129:955–962

Liu Z, Liu P, Qi D, Peng X, Liu G (2017) Enhancement of cold and salt tolerance of *Arabidopsis* by transgenic expression of the S-adenosylmethionine decarboxylase gene from *Leymus chinensis*. J Plant Physiol 211:90–99

Luo X, Wu J, Li Y, Nan Z, Guo X, Wang Y, Zhang A, Wang Z, Xia G, Tian Y (2013) Synergistic effects of GhSOD1 and GhCAT1 overexpression in cotton chloroplasts on enhancing tolerance to methyl viologen and salt stresses. PLoS One 8:e54002

Masand S, Yadav SK (2016) Overexpression of MuHSP70 gene from *Macrotyloma uniflorum* confers multiple abiotic stress tolerance in transgenic *Arabidopsis thaliana*. Mol Biol Rep 43:53–64

Melvin P, Bankapalli K, D'Silva P, Shivaprasad PV (2017) Methylglyoxal detoxification by a DJ-1 family protein provides dual abiotic and biotic stress tolerance in transgenic plants. Plant Mol Biol 94:381–397

Mishra A, Tanna B (2017) Halophytes: potential resources for salt stress tolerance genes and promoters. Front Plant Sci 8:829

Mittler R (2002) Oxidative stress, antioxidants and stress tolerance. Trends Plant Sci 7:405–410

Moon H, Lee B, Choi G, Shin D, Prasad DT, Lee O et al (2003) NDP kinase 2 interacts with two oxidative stress activated MAPKs to regulate cellular redox state and enhances multiple stress tolerance in transgenic plants. Proc Natl Acad Sci U S A 100:358–363

Munns R (2005) Genes and salt tolerance: bringing them together. New Phytol 167:645–663

Munns R, Tester M (2008) Mechanisms of salinity tolerance. Annu Rev Plant Biol 59:651–681

Nguyen TX, Nguyen T, Alameldin H, Goheen B, Loescher W, Sticklen M (2013) Transgene pyramiding of the HVA1 and mtlD in T3 maize (*Zea mays* L.) plants confers drought and salt tolerance, along with an increase in crop biomass. Int J Agron. https://doi.org/10.1155/2013/598163

Nishiyama R, Le DT, Watanabe Y, Matsui A, Tanaka M, Seki M, Yamaguchi-Shinozaki K, Shinozaki K, Tran LS (2012) Transcriptome analyses of a salt-tolerant cytokinin-deficient mutant reveal differential regulation of salt stress response by cytokinin deficiency. PLoS One 7:e32124

Niu X, Bressan RA, Hasegawa PM, Pardo JM (1995) Ion homeostasis in NaCl stress environments. Plant Physiol 109:735–742

Paddon CJ, Westfall PJ, Pitera DJ, Benjamin K, Fisher K, McPhee D, Leavell MD, Tai A, Main A, Eng D, Polichuk DR, Teoh KH, Reed DW, Treynor T, Lenihan J, Fleck M, Bajad S, Dang G, Dengrove D, Diola D, Dorin G, Ellens KW, Fickes S, Galazzo J, Gaucher SP, Geistlinger T, Henry R, Hepp M, Horning T, Iqbal T, Jiang H, Kizer L, Lieu B, Melis D, Moss N, Regentin R, Secrest S, Tsuruta H, Vazquez R, Westblade LF, Xu L, Yu M, Zhang Y, Zhao L, Lievense J, Covello PS, Keasling JD, Reiling KK, Renninger NS, Newman JD (2013) High-level semi-synthetic production of the potent antimalarial artemisinin. Nature 496:528–532

Pan WJ, Tao JJ, Cheng T, Shen M, Ma JB, Zhang WK, Lin Q, Ma B, Chen SY, Zhang JS (2017) Soybean NIMA-related kinase 1 promotes plant growth and improves salt and cold tolerance. Plant Cell Physiol 58:1268–1278

Pandey G, Yadav CB, Sahu PP, Muthamilarasan M, Prasad M (2017) Salinity induced differential methylation patterns in contrasting cultivars of foxtail millet (*Setaria italica* L.) Plant Cell Rep 36:759–772

Pang Q, Chen S, Dai S, Chen Y, Wang Y, Yan X (2010) Comparative proteomics of salt tolerance in *Arabidopsis thaliana* and *Thellungiella halophila*. J Proteome Res 9:2584–2599

Park HJ, Kim WY, Yun DJ (2016) A new insight of salt stress signaling in plant. Mol Cells 39:447–459

Patanun O, Ueda M, Itouga M, Kato Y, Utsumi Y, Matsui A, Tanaka M, Utsumi C, Sakakibara H, Yoshida M, Narangajavana J, Seki M (2017) The histone deacetylase inhibitor suberoylanilide hydroxamic acid alleviates salinity stress in cassava. Front Plant Sci 7:2039

Peng J, Li Z, Wen X, Li W, Shi H, Yang L, Zhu H, Guo H (2014) Salt-induced stabilization of EIN3/EIL1 confers salinity tolerance by deterring ROS accumulation in *Arabidopsis*. PLoS Genet 10:e1004664

Pruthvi V, Narasimhan R, Nataraja KN (2014) Simultaneous expression of abiotic stress responsive transcription factors, *AtDREB2A*, *AtHB7* and *AtABF3* improves salinity and drought tolerance in peanut (*Arachis hypogaea* L.) PLoS One 9:e111152

Puniran-Hartley N, Hartley J, Shabala L, Shabala S (2014) Salinity-induced accumulation of organic osmolytes in barley and wheat leaves correlates with increased oxidative stress tolerance: in planta evidence for cross-tolerance. Plant Physiol Biochem 83:32–39

Quan R, Wang J, Yang D, Zhang H, Zhang Z, Huang R (2017) EIN3 and SOS2 synergistically modulate plant salt tolerance. Sci Rep 7:44637

Rahnama A, Poustini K, Munns R, James RA (2010) Stomatal conductance as a screen for osmotic stress tolerance in durum wheat growing in saline soil. Funct Plant Biol 37:255–263

Reddy MP, Sanish S, Iyengar ERR (1992) Photosynthetic studies and compartmentation of ions in different tissues of *Salicornia brachiata* Roxb. under saline conditions. Photosynthetica 26:173–179

Reddy PS, Jogeswar G, Rasineni GK, Maheswari M, Reddy AR, Varshney RK et al (2015) Proline over-accumulation alleviates salt stress and protects photosynthetic and antioxidant enzyme activities in transgenic sorghum [*Sorghum bicolor* (L.) Moench]. Plant Physiol Biochem 94:104–113

Rontein D, Basset G, Hanson AD (2002) Metabolic engineering of osmoprotectant accumulation in plants. Metab Eng 4:49–56

Roy SJ, Negrão S, Tester M (2014) Salt resistant crop plants. Curr Opin Biotechnol 26:115–124

Rozema J, Flowers T (2008) Ecology. Crops for salinized world. Science 322:1470–1480

Sahoo RK, Ansari MW, Tuteja R, Tuteja N (2014) OsSUV3 transgenic rice maintains higher endogenous levels of plant hormones that mitigates adverse effects of salinity and sustains crop productivity. Rice (NY) 7:17

Sakamoto A, Murata N (2000) Genetic engineering of glycine betaine synthesis in plants: current status and implications for enhancement of stress tolerance. J Exp Bot 51:81–88

Sereflioglu S, Dinler BS, Tasci E (2017) Alpha-tocopherol-dependent salt tolerance is more related with auxin synthesis rather than enhancement antioxidant defense in soybean roots. Acta Biol Hung 68:115–125

Shabala S, Wu H, Bose J (2015) Salt stress sensing and early signalling events in plant roots: current knowledge and hypothesis. Plant Sci 241:109–119

Sharma I, Ching E, Saini S, Bhardwaj R, Pati PK (2013) Exogenous application of brassinosteroid offers tolerance to salinity by altering stress responses in rice variety Pusa Basmati-1. Plant Physiol Biochem 69:17–26

Sharma A, Kumar D, Kumar S, Rampuria S, Reddy AR, Kirti PB (2016) Ectopic expression of an atypical hydrophobic group 5 LEA protein from wild peanut, *Arachis diogoi*, confers abiotic stress tolerance in tobacco. PLoS One 11:e0150609

Shen J, Lv B, Luo L, He J, Mao C, Xi D, Ming F (2017) The NAC-type transcription factor OsNAC2 regulates ABA-dependent genes and abiotic stress tolerance in rice. Sci Rep 7:40641

Silva EN, Silveira JA, Rodrigues CR, Viégas RA (2015) Physiological adjustment to salt stress in *Jatropha curcas* is associated with accumulation of salt ions, transport and selectivity of K^+, osmotic adjustment and K^+/Na^+ homeostasis. Plant Biol (Stuttg) 17:1023–1029

Singh AK, Sopory SK, Wu K, Singla-Pareek SK (2010) Transgenic approaches. In: Pareek A, Sopory SK, Bohnert HJ, Govindjee (eds) Abiotic stress adaptation in plants: physiological, molecular and genomic foundation. Springer, New York, pp 417–450

Singh M, Kumar J, Singh VP, Prasad SM (2014) Plant tolerance mechanism against salt stress: the nutrient management approach. Biochem Pharmacol 3:e165

Singla-Pareek SL, Reddy MK, Sopory SK (2003) Genetic engineering of the glyoxalase pathway in tobacco leads to enhanced salinity tolerance. Proc Natl Acad Sci U S A 100:14672–14677

Su X, Chu Y, Li H, Hou Y, Zhang B, Huang Q, Hu Z, Huang R, Tian Y (2011) Expression of multiple resistance genes enhances tolerance to environmental stressors in transgenic poplar (Populus × euramericana 'Guariento'). PLoS One 6:e24614

Sun J, Dai S, Wang R, Chen S, Li N, Zhou X, Lu C, Shen X, Zheng X, Hu Z et al (2009) Calcium mediates root K^+/Na^+ homeostasis in poplar species differing in salt tolerance. Tree Physiol 9:1175–1186

Thomson MJ, de Ocampo M, Egdane J, Rahman MA, Sajise AG, Adorada DL et al (2010) Characterizing the saltol quantitative trait locus for salinity tolerance in rice. Rice 3:148–160

Turan, S, Cornish K, Kumar S (2012) Salinity tolerance in plants: Breeding and genetic engineering. AJCS 6:1337–1348

Turan S, Tripathy BC (2013) Salt and genotype impact on antioxidative enzymes and lipid peroxidation in two rice cultivars during de-etiolation. Protoplasma 250:209–222

Unesco Water Portal (2007) http://www.unesco.org/water

Varshney RK, Langridge P, Graner A (2007) Application of genomics to molecular breeding of wheat and barley. Adv Genet 58:121–155

Veena Reddy VS, Sopory SK (1999) Glyoxalase I from *Brassica juncea*: molecular cloning, regulation and its over-expression confer tolerance in transgenic tobacco under stress. Plant J 17:385–395

Vishwakarma K, Upadhyay N, Kumar N, Yadav G, Singh J, Mishra RK, Kumar V, Verma R, Upadhyay RG, Pandey M, Sharma S (2017) Abscisic acid signaling and abiotic stress tolerance in plants: a review on current knowledge and future prospects. Front Plant Sci 8:161

Vivek PJ, Tuteja N, Soniya EV (2013) CDPK1 from ginger promotes salinity and drought stress tolerance without yield penalty by improving growth and photosynthesis in *Nicotiana tabacum*. PLoS One 8:e76392

Waditee R, Bhuiyan MN, Rai V, Aoki K, Tanaka Y, Hibino T, Suzuki S, Takano J, Jagendorf AT, Takabe T, Takabe T (2005) Genes for direct methylation of glycine provide high levels of glycine betaine and abiotic-stress tolerance in *Synechococcus* and *Arabidopsis*. Proc Natl Acad Sci U S A 102:1318–1323

Wang RK, Li LL, Cao ZH, Zhao Q, Li M, Zhang LY, Hao YJ (2012) Molecular cloning and functional characterization of a novel apple MdCIPK6L gene reveals its involvement in multiple abiotic stress tolerance in transgenic plants. Plant Mol Biol 79:123–135

Wang M, Zheng Q, Shen Q, Guo S (2013) The critical role of potassium in plant stress response. Int J Mol Sci 14:7370–7390

Wen F-P, Zhang Z-H, Bai T, Xu Q, Pan Y-H (2010) Proteomics reveals the effects of gibberellic acid (GA3) on salt stressed rice (*Oryza sativa* L.) shoots. Plant Sci 178:170–175

Wu X, He J, Chen J, Yang S, Zha D (2013) Alleviation of exogenous 6-benzyladenine on two genotypes of eggplant (*Solanum melongena* Mill.) growth under salt stress. Protoplasma 51:169–176

Wu W, Zhang Q, Ervin EH, Yang Z, Zhang X (2017) Physiological mechanism of enhancing salt stress tolerance of perennial ryegrass by 24-epibrassinolide. Front Plant Sci 8:1017

Xiong L, Yang Y (2003) Disease resistance and abiotic stress tolerance in rice are inversely modulated by an abscisic acid–inducible mitogen-activated protein kinase. Plant Cell 15:745–759

Xu GY, Pedro SCFR, Wang ML, Xu ML, Cui YC, Li LY, Zhu YX, Xia X (2011) A novel rice calmodulin-like gene, OsMSR2, enhances drought and salt tolerance and increases ABA sensitivity in *Arabidopsis*. Planta 234(1):47–59

Xu J, Xing XJ, Tian YS, Peng RH, Xue Y, Zhao W, Yao QH (2015a) Transgenic *Arabidopsis* plants expressing tomato glutathione S-transferase showed enhanced resistance to salt and drought stress. PLoS One 10:e0136960

Xu R, Wang Y, Zheng H, Lu W, Wu C, Huang J, Yan K, Yang G, Zheng C (2015b) Salt-induced transcription factor MYB74 is regulated by the RNA-directed DNA methylation pathway in *Arabidopsis*. J Exp Bot 66:5997–6008

Yadu S, Dewangan TL, Chandrakar V, Keshavkant S (2013) Imperative roles of salicylic acid and nitric oxide in improving salinity tolerance in *Pisum sativum* L. Physiol Mol Biol Plant 23:43–58

Yang Q, Chen ZZ, Zhou XF, Yin HB, Li X, Xin XF, Hong XH, Zhu JK, Gong Z (2009) Overexpression of SOS (Salt Overly Sensitive) genes increases salt tolerance in transgenic *Arabidopsis*. Mol Plant 2:22–31

Yang L, Zu Y-G, Tang Z-H (2013) Ethylene improves *Arabidopsis* salt tolerance via K^+ in shoots and roots rather than decreasing tissue Na^+ content. Environ Exp Bot 86:60–69

Yang G, Yu L, Zhang K, Zhao Y, Guo Y, Gao C (2017) A *ThDREB* gene from *Tamarix hispida* improved the salt and drought tolerance of transgenic tobacco and *T. hispida*. Plant Physiol Biochem 113:187–197

Yu Y, Ni Z, Chen Q, Qu Y (2017) The wheat salinity-induced R2R3-MYB transcription factor TaSIM confers salt stress tolerance in *Arabidopsis thaliana*. Biochem Biophys Res Commun 491:642–648

Yuan S, Li Z, Li D, Yuan N, Hu Q, Luo H (2015) Constitutive expression of rice microRNA 528 alters plant development and enhances tolerance to salinity stress and nitrogen starvation in creeping bentgrass. Plant Physiol 169:576–593

Zhang X, Tang W, Liu J, Liu Y (2014) Co-expression of rice *OsP5CS1* and *OsP5CS2* genes in transgenic tobacco resulted in elevated proline biosynthesis and enhanced abiotic stress tolerance. Chin J Appl Environ Biol 20:717–722

Zhang X, Liu X, Wu L, Yu G, Wang X, Ma H (2015) The SsDREB transcription factor from the succulent halophyte *Suaeda salsa* enhances abiotic stress tolerance in transgenic tobacco. Int J Genomics 2015:875497

Zhang M, Smith JA, Harberd NP, Jiang C (2016a) The regulatory roles of ethylene and reactive oxygen species (ROS) in plant salt stress responses. Plant Mol Biol 91:651–659

Zhang S, Gan Y, Xu B (2016b) Application of plant-growth-promoting fungi *Trichoderma longibrachiatum* T6 enhances tolerance of wheat to salt stress through improvement of antioxidative defense system and gene expression. Front Plant Sci 7:1405

Zhou S, Chen X, Zhang X, Li Y (2008) Improved salt tolerance in tobacco plants by co-transformation of a betaine synthesis gene BADH and a vacuolar Na^+/H^+ antiporter gene SeNHX1. Biotechnol Lett 30:369–376

Zhou C, Zhu L, Xie Y, Li F, Xiao X, Ma Z, Wang J (2017) *Bacillus licheniformis* SA03 Confers Increased Saline-Alakaline Tolerance in Chrysanthemum Plants by Induction of Abscisic Acid Accumulation. Front Plant Sci 8:1143

Zhu JK (2000) Genetic analysis of plant salt tolerance using *Arabidopsis*. Plant Physiol 124:941–948

Zhu JK (2001) Plant salt tolerance. Trends Plant Sci 6:66–71

Zhu N, Cheng S, Liu X, Du H, Dai M, Zhou DX, Yang W, Zhao Y (2015) The R2R3-type MYB gene OsMYB91 has a function in coordinating plant growth and salt stress tolerance in rice. Plant Sci 236:146–156

Chapter 15
Molecular Markers and Their Role in Producing Salt-Tolerant Crop Plants

Sagar Satish Datir

Abstract Besides drought and temperature, soil salinity is a severe abiotic environmental constraint to world agriculture, widely affecting yield and quality of crops. Increasing world population and simultaneous depletion of agricultural land created an alarming situation to meet the global food requirement. To meet this demand, it is imperative to utilize salt-affected land for agricultural produce. Researchers are actively engaged in developing salt-tolerant crop varieties using both conventional and molecular breeding technologies. Identification of salt-tolerant crop varieties using physiological and biochemical indices has become a routine technique in many laboratories. However, screening for salinity tolerance based on these indices relative to unstressed controls is labor-intensive and time-consuming. Moreover, results produced using these screening techniques are subjected to fluctuation in environmental conditions. In recent years, DNA-based molecular marker technique has been developed for screening salinity tolerance in crops. Molecular markers such as RFLP (restriction fragment length polymorphism), RAPD (random amplified polymorphic DNA), AFLP (amplified fragment length polymorphism), SSRs (simple sequence repeats), SRAP (sequence related amplified polymorphism), ILPs (intron length polymorphism), and EST-SSRs (expressed sequence tags and simple sequence repeats) have been proven useful for rapid and sensitive screening of germplasm for salinity tolerance. However, raid advances in high-throughput sequencing technology make single-nucleotide polymorphism (SNP) become the marker of choice for salinity tolerance studies. Identification of genomic regions associated with salinity stress tolerance is of great significance which offers new avenues in marker-assisted selection breeding program. Molecular marker-assisted breeding may help in advancing fundamental understanding of salinity tolerance in crops. This book chapter provides a comprehensive review on the role of molecular markers as selection criteria for salinity stress tolerance in plants.

Keywords Amplified fragment length polymorphism · Marker-assisted selection · Molecular marker · Quantitative trait loci · Random amplified polymorphic DNA ·

S. S. Datir (✉)
Department of Biotechnology, Savitribai Phule Pune University, Pune, India

V. Kumar et al. (eds.), *Salinity Responses and Tolerance in Plants, Volume 1*,
https://doi.org/10.1007/978-3-319-75671-4_15

Restriction fragment length polymorphism · Salinity tolerance · Simple sequence repeats · Single-nucleotide polymorphism

Abbreviations

AFLP	Amplified length polymorphism
ESTs	Expressed sequence tags
ISSRs	Inter-simple sequence repeats
MAS	Marker-assisted selection
QTL	Quantitative trait loci
RAPD	Rapid amplified polymorphic DNA
RFLP	Restriction fragment length polymorphism
SNP	Single-nucleotide polymorphism
SRAP	Sequence-related amplified polymorphism
SSRs	Simple sequence repeats
ST	Salinity tolerance

15.1 Introduction

Plants provide essential nutritional components to human diet. The global concern about rise inhuman population and concomitant increase in food demand created an alarming situation about crop production. Furthermore, this condition is aggravated by simultaneous depletion of agricultural area, and hence, it is imperative to utilize salt-affected land for agricultural produce. Besides drought and temperature, soil salinity is considered as a major environmental limitation to world agriculture which causes adverse impact on yield and affects quality of several commercially important crops. Salinity not only causes severe reduction in yield but also diminishes nutritional quality of agricultural crops (Yokoi et al. 2002; Machado and Serralheiro 2017). It has been reported that at least 20% of all irrigated lands are salt-affected, and recent statistics showed that the total global area of salt-affected soils is approximately 45 million hectares irrigated land (Pitman and Läuchli 2002; Martinez-Beltran and Manzur 2005; Shrivastava and Kumar 2015). Determination of salinity tolerance (ST) is a complex process, and therefore, direct selection in natural field conditions is intricate because uncontrollable climatic agents negatively influence the repeatability and precision of such trials. However, it has been suggested that varieties developed at laboratory level tend to perform differently in controlled conditions rather than on field; hence, there is a need for field evaluation of crop varieties, especially in different salt concentrations and different regions (Richards 1993; Reddy et al. 2017). In view of this, researchers are actively engaged in developing salinity tolerance indices using morphological, physiological, and biochemical parameters (Chen et al. 2007; Smethurst et al. 2009; Rana et al. 2015; Krishnamurthy et al. 2016). However, screening for salinity tolerance based on these indices relative to unstressed controls is labor-intensive and time-consuming. Moreover, results produced using these screening techniques are subjected to fluctuation in environmental conditions (Heiskanen 2006; Masuka et al. 2012). Hence, it is crucial to create and identify salinity tolerance genotypes using molecular genetic techniques. Breeding approach accompanied with molecular marker-assisted selection is the most promising approach in terms of efficiency to increase the productivity under salt-affected soils (Bizimana et al. 2017). Development of sustainable high-yielding crop varieties that persist under abiotic stresses is a prerequisite for meeting the food requirements of a growing world population (Leonforte et al. 2013). Progress in developing salt-tolerant crops has been slowed mainly because the physiological, biochemical, and molecular mechanisms of salt tolerance in plants are not yet sufficiently understood (Gregorio et al. 2002; Läuchli and Grattan 2007; Gupta and Huang 2014). As traditional crop improvement techniques reach their limits, modern agriculture has to implement new techniques to meet the food demands of an ever-growing world population. Therefore, to combat the crop production losses, there is a need to develop molecular marker system for the identification of stress-responsive genes (Telem et al. 2016). It has been documented that salinity tolerance in plants is a complex trait both genetically and physiologically, found to be controlled by polygenes (Baby et al. 2010). Hence, to develop new varieties with a high level of salinity tolerance, it requires an understanding of the genetic control underlying salt tolerance mechanisms (Bizimana et al. 2017). Traits such as salt tolerance are quantitatively inherited. Hence, mapping quantitative trait loci (QTL) with molecular markers can be very helpful to plant breeders (Ghomi et al. 2013). Identification of marker(s) associated with QTLs contributing to ST is a promising approach of crop improvement and to use them as indirect selection criteria (Greenway and Munns 1980; Foolad et al. 1998; Bizimana et al. 2017) for selecting ST in a process known as marker-assisted selection (MAS) (Ashraf et al. 2012; Ashraf and Foolad 2013). Molecular markers have allowed the identification of QTLs associated with ST traits at various developmental stages in several plant species like *Arabidopsis* (*Arabidopsis thaliana* L.) (Quesada et al. 2002; DeRose-Wilson and Gaut 2011), rice (*Oryza sativa* L.) (Takehisa et al. 2004; Singh et al. 2007; Haq et al. 2010; De Leon et al. 2017), wheat (*Triticum aestivum* L.) (Lindsay et al. 2004; James et al. 2006; Turki et al. 2015), barley (*Hordeum vulgare* L.) (Mano and Takeda 1997; Liu et al. 2017), and tomato (*Solanum* sp.) (Villalta et al. 2008; Li et al. 2011). Theoretically, the use of DNA markers associated with genes or QTLs associated with ST at the various developmental stages can facilitate development of plants with improved ST throughout the ontogeny of the plant. Although progress has been done in identification of molecular markers associated with QTLs in several crop species, there has been a limited use of identified markers and MAS for breeding for ST in any crop species. Numerous factors have been suggested for slow progress in MAS selection for ST such as developed markers. For instance, they are population-specific, as markers developed based on ST evaluation under controlled greenhouse or growth chamber conditions that may not reflect ST under field conditions, markers were identified based on populations derived from wide crosses (e.g., crosses with wild species) that may not be directly useful in breeding populations (i.e., lack of correspondence between QTLs identified in interspecific populations and those existing in breeding populations), lack of repeated experimentation for marker validation, and lack of marker polymorphism in breeding populations. Thus, such issues can be alleviated to facilitate the use of markers for ST breeding in different crop species (Ashraf et al. 2012).

Using molecular marker technology gives the physiologist and the breeders the opportunity to enhance the efficiency of conventional plant breeding by tightness between DNA markers to the trait of interest (Amin and Diab 2013). Understanding the molecular mechanisms underlying salinity stress and subsequently developing ST crops can be a solution for increasing food production (Reddy et al. 2017). Molecular markers such as RFLP (restriction fragment length polymorphism) (Botstein et al. 1980), RAPD (random amplified polymorphic DNA), AFLP (amplified fragment length polymorphism), SSRs (simple sequence repeats), ISSRs (intersimple sequence repeats), SRAP (sequence-related amplified polymorphism), and SNPs (single-nucleotide polymorphism) are extensively utilized and have been proven to be extremely useful in plant genetic research (Khlestkina 2014; Lateef 2015; Robarts and Wolfe 2014). It has been suggested that high-throughput screening methods are required to exploit novel sources of genetic variation in crops like rice and further improve salinity tolerance in breeding programs (Sónia et al. 2013). However, the application of molecular markers in MAS for physiologically complex traits like salinity tolerance has barely been accomplished (Ashraf and Foolad 2013). Implementation of various DNA markers found to speed up the plant selection process through MAS with special reference to particular traits or through the selection of chromosomal segments flanked by the markers at the genomic level (Collard and Mackill 2008).This can be done by speeding up discovery of gene and allele, and delivery of marker-assisted selection (Reddy et al. 2017).This book chapter aims to provide a comprehensive review on the role of various molecular markers used in discovery for salinity stress tolerance in plants.

15.2 Use of Hybridization and PCR-Based Markers (RFLP, RAPD, AFLP, and SSRs) in Salinity Stress Tolerance Studies

Molecular markers such as RFLP, RAPD, AFLP, and SSRs have been proven useful to detect their association with salt tolerance in several crops like wheat, barley, rice, potato, tomato, sorghum, eggplant, cotton, sunflower, etc. (Rao et al. 2007; Aghaei et al. 2008; Barchi et al. 2011; Amin and Diab 2013; Khatab and Samah 2013; Mondal et al. 2013; Shehata et al. 2014; Gharsallah et al. 2016; Saleh 2016). Figure 15.1 depicts molecular markers used in salinity stress tolerance studies. Table 15.1 explains some QTLs mapped for ST in various crop plants.

The genetic basis of salinity responses was investigated in Egyptian bread wheat (*Triticum aestivum* L.) using a doubled haploid (DH) population of 139 individuals

Fig. 15.1 Various molecular markers used in salinity stress tolerance studies

derived from the cross between salt-tolerant and salt-susceptible Egyptian breeding cultivars. The progeny was genotyped with RFLP, SSRs, and AFLP markers. Furthermore, interval and composite mapping used to identify the genomic regions controlling traits related to ST. Markers significantly associated with 12 traits related to ST traits were identified. This work provides a molecular tool for breeders and physiologists to facilitate the selection of wheat varieties under salt stress in a strategic improvement program in wheat using MAS (Amin and Diab 2013). A number of QTLs/genes associated with ST have been identified and mapped in wheat using association analysis. Genes related to Na^+ and K^+ homoeostasis were detected in different genomic regions of wheat. For example, the gene *Kna1* for sodium exclusion was detected on chromosome 4D in bread wheat (Dubcovsky et al. 1996; Byrt et al. 2007), while another two Na^+ exclusion genes were identified in durum: *Nax1* on chromosome 2AL (Lindsay et al. 2004; Hussain et al. 2017) and *Nax2* on chromosome 5AL (James et al. 2006; Byrt et al. 2007). These QTLs have been iden-

Table 15.1 Some QTLs mapped for salinity stress in various plants

Crop	Molecular marker	QTL	Remark	Reference
Cicer arietinum L.	SSR	*DF*	Days to flowering	Vadez et al. (2012)
		SDW	Shoot dry weight	
		SYLD	Seed yield	
		PN	Pod number	
		SN	Seed number	
		YLD_R	Yield ratio	
		YLD_D	Yield difference	
		100SW	100-seed weight	
Hordeum vulgare L.	SSR and DArT	*QSl.YyFr.1H, QSl.YyFr.2H, QSl.YyFr.5H, QSl.YyFr.6H,QSl.YyFr.7H*	Salinity tolerance at the vegetative stage	Zhou et al. (2012)
*Gossypium tomentosum*X *Gossypium hirsutum*	SSR	*qRL*	Longest root length	Oluoch et al. (2016)
Oryza sativa L.	SSR and SNP	*qNa11.5*	Na^+ concentration	De Leon et al. (2016)
		qK1.3863	K^+ concentration	
		qNaK3.32	NaK	
	RFLP, SSL, and SSR	*Saltol*	Controlling Na^+/K^+ homeostasis	Bonilla et al. (2002), Thomson et al. (2010)
Pisum sativum L.	SNP	q_SALT	Salt index symptom score	Leonforte et al. (2013)
Solanum sp.	RFLP	*Stpq* *Stlq*	Salt tolerance from *S. pennellii* QTL Salt tolerance from *S. lycopersicoides* QTL	Li et al. (2011)
Triticum aestivum L.	SSR	NT NL LL TFW SL RL SDW RDW	Number of tillers per plant, number of leaves per tiller, leaf length, total fresh weight of shoot and root, shoot length, root length, shoot dry weight and root dry weight	Turki et al. (2015)
Triticum aestivum L.	RFLP	*Kna1*	K^+/Na^+	Dubcovsky et al. (1996), Byrt et al. (2007)

(continued)

Table 15.1 (continued)

Crop	Molecular marker	QTL	Remark	Reference
Triticum turgidum L. Ssp. Durum (Desf.)	AFLP, RFLP and microsatellite	*Nax1*	Low Na^+ concentration	Lindsay et al. (2004), Hussain et al. (2017)
Zea mays L.	SNP	*SPH*	Plant height under salt stress	Luo et al. (2017)
		PHI	Plant height-based salt tolerance index	

tified using various molecular markers. For instance, AFLP, RFLP, microsatellite markers, and SNPs identified a locus, named *Nax1* (Na exclusion), in several wheat populations (Lindsay et al. 2004; Hussain et al. 2017).

Like wheat, salinity is considered as the most widespread soil problem in rice-growing countries which limits the rice production worldwide (Jain et al. 2014). Mardani et al. (2014) generated an $F_{2:4}$ population derived from the cross between salt-tolerant and salt-sensitive variety to determine the germination traits. AFLP (105) and SSR (131) markers used to construct a linkage map. A total of 17 QTLs were detected related to germination traits under salt stress condition, and some of them were reported for the first time. These studies suggested that the identification of genomic regions associated with salt tolerance and its components under salt stress will be useful for marker-based approaches to improve ST for farmers in salt-prone rice environments. A number of mapping studies have identified QTLs associated with salinity stress tolerance in rice such as *Saltol*, *QNa*, and *SKC1/OsHKT8*, along with QNa:K (Singh et al. 2007; Haq et al. 2010). For example, a study employing an $F_{2:3}$ population between the tolerant *indica* landrace Nona Bokra and the susceptible *japonica* Koshihikari using RFLP marker system identified QTLs controlling tolerance traits, including major QTLs for shoot K^+ concentration on chromosome 1 (*qSKC-1*) and shoot Na^+ concentration on chromosome 7 (*qSNC-7*; Lin et al. 2004). In another approach, 33 rice landrace genotypes were assessed at molecular level using 11 SSR markers coupled with morphological markers linked with ST QTL at germination stage. Out of 11, three markers were discovered as the most competent descriptors to screen the salt-tolerant genotypes with higher polymorphic information content coupled with higher marker index value, significantly distinguished the salt-tolerant genotypes. It has been suggested that combining morphological and molecular assessment, four rice landraces were considered as true salt-tolerant genotypes which may further contribute in greater way in the development of salt-tolerant genotypes in rice (Ali et al. 2014).

Mung bean is one of the most important pulse crops. However, its production is decreasing mainly due to increasing soil salinity in irrigated land agriculture (Saha et al. 2010). Research on identification and development of breed cultivars that are adapted to salt stress condition is not sufficiently undertaken (Singh and Singh 2011). Thirty eight novel SSRs specific to candidate genes involved in salt tolerance were developed for detection of genetic variations in 12 mung bean genotypes variably adapted to salt stress. Several putative polymorphic alleles were detected. These markers may be coupled with specific loci linked with ST and will help to identify the QTLs or other important genes in mung bean (Sehrawat et al. 2014). Du et al. (2016) evaluated 43 advanced salt-tolerant and 31 highly salt-sensitive cultivars for ten ST-related traits in 304 upland cotton cultivars. An association analysis of ten ST-related traits and 145 SSRs showed a total of 95 significant associations with germination index and seedling stage physiological index and four seedling stage biochemical indexes. These studies laid solid foundations for further improvements in cotton ST by referencing elite germplasms, alleles associated with ST traits, and optimal crosses. Likewise, use of ISSR, RFLP, and RAPD has successfully helped to identify salt-tolerant and salt-sensitive genotypes in sorghum (Rao et al. 2007; Khalil 2013). Similarly, SSR-based marker-assisted screening of commercial tomato genotypes under salt stress may guide strategies for the introgression of valuable traits in target tomato varieties to overcome salinity (Gharsallah et al. 2016). Though barley is highly tolerant to salinity compared with wheat, rice, and oat, it still suffers from salt toxicity in many areas of the world, thus offering a means for efficient utilization of saline soil and improvement of productivity in these environments (Li et al. 2007). Khatab and Samah (2013) developed molecular markers associated with ST using ISSR and SSR markers and described their usefulness in the future breeding program for ST in barley. In studies conducted by Zhou et al. (2012), a high-density genetic linkage map was constructed using SSRs, and DArT markers identified five QTLs for salinity tolerance in barley. These markers have potential application for marker-assisted selection in breeding for enhanced salt tolerance in barley.

Newly developed sequence-related amplified polymorphism (SRAP) PCR-based molecular marker (Li and Quiros 2001; Robarts and Wolfe 2014) used to genotype F2 population derived from salt-tolerant and salt-sensitive maize inbred lines. Six SRAP molecular markers closely linked to salt tolerance were determined. These studies claimed that SRAP markers identified in this research will accelerate maize marker-assisted selection breeding and lay the foundation for salt-tolerant gene cloning (Xiang et al. 2015).

15.3 Expressed Sequence Tags-Simple Sequence Repeats (EST-SSRs)

In silico development of expressed sequence tag (EST) derived SSRs from transcribed regions of the DNA through mining of publicly available databases is relatively simpler and cost effective than genomic SSRs from the un-transcribed regions (Cuadrado and Schwarzacher 1998; Tang et al. 2008; Kalia et al. 2011). Additionally, EST-SSRs have higher cross-species transferability compared to genomic SSRs, and the functions of EST-SSRs can be presumed, and marker development is easy to perform at low cost (Wang et al. 2014). Such EST-SSRs find potential use in genetic research for molecular breeding for ST in cotton, chickpea, and maize (Varshney et al. 2009; Wang et al. 2014; Yumurtaci et al. 2017).

Salt stress considerably hinders the growth and productivity of maize. Therefore, there is a need to identify and develop salt-tolerant genotypes using molecular markers that would enhance the further breeding processes in commercially important crops like maize. Efforts were taken to assemble 3308 ESTs from salt stress-related libraries to mine repetitive sequences for development of applicable markers. EST data revealed 208 simple and 18 non-simple repetitive regions. Further, 59 EST-SSR markers were validated using two contrasting parental genotypes and their recombinant inbred lines F35 salt sensitive and F63 salt tolerant to understand the genetic basis of ST. Identified EST-SSRs might be used as new functional molecular markers in the diversity analysis, identification of quantitative trait loci (QTLs), and comparative genomic studies in maize in the future. With the improvement of feasible molecular markers, these candidate EST-SSRs might have the ability to broaden the genetic base of maize. Despite an increasing number of DNA markers, the identification of comprehensive markers for screening of salinity-specific regions in maize is lagging behind and needs to be further explored (Yumurtaci et al. 2017). QTLs responsible for the ST of field-grown maize plants are still unknown. Recently, Luo et al. (2017) produced a genetic map to identify main genetic factors contributing to ST in mature maize using a double haploid population and 1317 SNP markers. Studies identified a major QTL for *SPH* (plant height under stress) which explained 31.2% of the phenotypic variation in addition to *PHI* (plant height-based salt tolerance index) which accounted for 12% of the phenotypic variation. QTLs detected in adult maize in this study established a foundation for the map-based cloning of genes associated with ST and provided a potential target for MAS in developing maize varieties with ST.

Cotton is moderately salt tolerant, grown commercially worldwide as a leading fiber crop. There is promising potential to improve ST in cultivated cotton using molecular breeding program aiming to develop salt-tolerant molecular markers. In total, 132 pairs of EST-SSR primers were developed on the basis of cotton ST ESTs obtained from BLAST with *Arabidopsis* ST genes, so as to serve cotton ST molecular breeding efforts. These primer pairs were tested on 38 cotton accessions differencing in their ST capacities. Out of 132, 106 effective markers showed different amplification efficiencies, and cross-species transferability, within different cotton species. Therefore, these salt-tolerant markers exhibit polymorphism among 38 cotton accessions and help reveal their utilization in marker-assisted breeding in developing salt-tolerant cotton. However, studies on the identification of quantitative trait loci (QTLs) for traits related to ST in cotton are meager. Recently, 11 QTLs were mapped for ST at seedling stage in $F_{2:3}$ population derived from the interspecific cross of *Gossypium tomentosum* and *Gossypium hirsutum*. A total 1295 SSRs were used for molecular genotyping. Mapping of QTLs related to ST was carried out for several traits. *qRL* for longest root length was discovered as the a major QTL explaining the phenotypic variance of 11.97% and 18.44% in two environments.

Therefore, molecular mapping studies could facilitate the development of cotton cultivars with ST (Oluoch et al. 2016).

A very limited number of molecular markers and candidate genes are available for undertaking molecular breeding in chickpea to tackle salinity stress. A total of 8258 ESTs (2595 unigenes) from salinity-challenged libraries were generated which further resulted into 2029 sequences containing 3728 simple sequence repeats, and 177 new EST-SSR markers were developed. Experimental validation of a set of 77 SSR markers on 24 genotypes revealed 230 alleles. Generated set of chickpea ESTs, therefore, serves as a resource of high-quality transcripts for gene discovery and development of functional markers associated with salinity stress tolerance that will be helpful to facilitate chickpea breeding (Varshney et al. 2009). Though chickpea exhibits sensitivity to salinity stress, it has been reported that tolerant and sensitive lines exist that can be used to better understand tolerance mechanisms and assist in breeding lines with improved tolerance (Munns and Tester 2008; Pushpavalli et al. 2015). Vadez et al. (2012) performed genotyping of ICCV 2 × JG 62 chickpea progenies that showed sensitivity of reproduction to salt stress using 216 SSRs and identified several QTLs for seed yield and yield components. These results confirm that ST in chickpea is closely related to the success of reproduction under salinity stress and shoot biomass development in early-flowering genotypes. These QTLs can be further utilized in chickpea breeding program.

15.4 Intron Length Polymorphism (ILP) Markers

Besides RFLP, RAPD, AFLP, SSRs, and SNP, intron length polymorphism (ILP) (Choi et al. 2004; Jaikishan et al. 2015) is a new type of functional molecular marker, which has not been reported extensively in salinity stress tolerance in crops. The availability of large expressed sequence tag (EST) databases for many crop species provides a valuable resource for the development of molecular markers like ILP which specifically explores the variation in the intron sequences (Wang et al. 2005; Gupta et al. 2012). The advantages of ILPs over the other so far reported DNA markers including gene-specific, codominant nature, hypervariable, neutral behavior, convenient, reliability, and high transferability rates among the plant species (Braglia et al. 2010; Gupta et al. 2011). So far the studies on the use of ILPs in salinity ST studies are meager. Functional markers (FMs) are a type of gene-based marker that was developed from sequence polymorphisms present in allelic variants of a functional gene at a given locus and so far used in association studies salinity stress tolerance in cotton and foxtail millet (Gupta et al. 2011; Cai et al. 2017). FMs accurately discriminate between traits associated with alleles of a target gene and are ideal molecular markers for MAS in plant breeding (Liu et al. 2012).

A large number of genome-wide cotton ILP markers have been developed. Differences in the intron lengths of orthologous A- and D-genome genes in cotton were screened, by comparing genome sequences and annotation information from *G. raimondii* and *G. arboreum*. A total of 10,180 putative ILP markers have been identified from 5021 orthologous genes. Among these, 535 ILP markers from 9 gene families related to stress were selected for experimental verification. Studies found a total of 25 marker-trait associations involved in 9 ILPs for 10 salt stress traits. The nine genes showed the various expressions in different organs and tissues, and five genes were significantly upregulated after salt treatment. The five genes found to play an important role in ST. Particularly, silencing of WRKY DNA-binding protein and mitogen-activated protein kinase can significantly enhance cotton susceptibility to salt stress. It has been concluded that five genes verified to be related to salt stress tolerance and have potential to improve ST in cotton abiotic-resistance breeding (Cai et al. 2017). Therefore, ILP markers would be useful in identification of genes with ST for developing abiotic-resistance cultivars in plant breeding programs.

15.5 Single-Nucleotide Polymorphisms

A number of molecular markers such as RFLP, RAPD, AFLP, SSRs, and diversity array technology (DArT) are labor-intensive and time-consuming compared with SNP (Kruglyak 1997; Mammadov et al. 2012) markers. Advances in high-throughput sequencing technologies such as next-generation sequencing (NGS) and the availability of sequence information have paved the way for the identification and development of SNP markers for several crop species (Rafalski 2002; Mammadov et al. 2012). However, the role of SNPs in salinity stress is relatively unexplored in several crop plants especially compared to conventional molecular markers such as RFLP, RAPD, SSRs, and AFLPs. SNPs have become the most widely used preferable molecular marker system for genetic analysis of plants because of their frequent occurrence in the genome and low rate of mutations (Ganal et al. 2012; Telem et al. 2016). Identification of SNPs marker in crops like eggplant, pea, rice, and cotton in close linkage with the relevant genomic regions can be implemented for marker-assisted selection and in varietal improvement programs (Barchi et al. 2011; Leonforte et al. 2013; De Leon et al. 2016; Wang et al. 2016).

Recently, a comprehensive linkage maps were constructed using SNP markers associated with ESTs in recombinant inbred line population in field pea. A total of 705 SNPs (91.7%) successfully detected segregating polymorphisms. Studies identified QTLs for ST on linkage groups Ps III and VII, with flanking SNP markers which were suitable for selection of salinity-resistant cultivars in *Pisum sativum*. Furthermore, the sequences comparison underpinning these SNP markers to the *Medicago truncatula* genome defined genomic regions containing candidate genes associated with saline ST. The SNP assays and associated genetic linkage maps developed not only permitted the identification of ST QTLs and candidate genes, but this piece of information also constitutes an important set of tools for MAS programs in field pea cultivars (Leonforte et al. 2013).

Significant abiotic/environmental factors influence the direct selection of superior salt-tolerant genotypes under field conditions (Richards 1993; Hanin et al. 2016).

Therefore, several QTL mapping studies using molecular marker approaches have been conducted for the seedling stage ST using different kinds of mapping populations from early segregating to the fixed lines in rice (Masood et al. 2004; Mingzhe et al. 2005; Lee et al. 2006; Thomson et al. 2010; Islam et al. 2011; Bizimana et al. 2017). For instance, being one of the most sensitive crops to salt stress, a number of studies identified that SNPs are robust molecular markers for QTLs responsible for seedling stage ST in rice (Bimpong et al. 2014a; Bimpong et al. 2014b; Kumar et al. 2015; Bizimana et al. 2017). For instance, a total of 194 polymorphic SNP markers were used to construct a genetic linkage map involving 142 selected recombinant-inbred lines (RILs) derived from a cross between salt sensitive, IR29, and salt tolerant, Hasawi. Furthermore, 20 quantitative trait loci (QTLs) were identified for salinity tolerance in rice, and it has been concluded the level of ST could be further enhanced by pyramiding of the different QTLs in one genetic background through marker-assisted selection (Bizimana et al. 2017). New allelic variants of key salt-tolerant genes found to affect ST suggested as a fundamental tool for developing a salt-tolerant rice variety (Sónia et al. 2013). In this study, an EcoTILLING approach was used to search for genotypic differences related to salt stress. The allelic variations in five key salt-related genes, namely, OsCPK17, OsRMC, OsNHX1, OsHKT1;5, and SalT involved in Na^+/K^+ ratio equilibrium, signalling cascade, and stress protection, were explored to associate with salt-tolerant phenotype. A total of 40 new allelic variants were discovered in coding region, and association analyses identified 11 significant SNPs related to salinity. Further evaluation based on bioinformatics analysis revealed that among the five nonsynonymous SNPs significantly associated with salt stress traits, a T67 K mutation which is supposing to diminish one transmembrane domain in OsHKT1 and a P140A transition that highly elevate the possibility of OsHKT1, five phosphorylations were identified. Moreover, the K24E mutation could be to affecting SalT interaction with other proteins, thus disturbing its function. Therefore, developing functional markers based on candidate gene alleles developed from SNPs that determine the genetic basis of trait is an important aspect of using genetic information in practical plant breeding (Andersen and Lubberstedt 2003; Yu et al. 2009; Mammadov et al. 2012). Although significant SNPs were identified and some candidate genes were suggested with close association with *Saltol* locus (Kumar et al. 2015), tight association of candidate genes in or around a single variant still needs enrichment with more markers at a locus to avoid false association (De Leon et al. 2016).

The genetic variability for salinity traits in cultivated tomato is limited, and it has been reported that it is moderately sensitive to salinity. Limited progress has been made in developing salt-tolerant tomato which can be attributed to complexity of the trait, insufficient genetic knowledge of tolerance components, lack of efficient selection criteria, difficulties in the identification and transfer of tolerance-related genes from unadapted germplasm into the cultivated background, and limited breeding efforts (Foolad et al. 1998; Foolad 2006). However, the sources of tolerance have been reported among several wild *Solanum* species (Tal and Shannon 1983; Scholberg and Locascio 1999; Rao et al. 2013). An association analysis was conducted in a core collection of 94 genotypes of wild tomato (*Solanum pimpinel-*

lifolium) to identify variations using SNP/indels in four candidate genes, viz., DREB1A, VP1.1, NHX1, and TIP, linked to ST traits (physiological and yield traits under salt stress). The candidate gene analysis identified five SNP/indels in DREB1A and VP1.1 genes explaining 17.0% to 25.8% phenotypic variation for various salt tolerance traits. Two alleles SpDREB1A_297_6 and SpDREB1A_297_12 exhibited in-frame deletion of 6 bp and 12 bp individually or as haplotypes accounted for maximum phenotypic variance (about 25%) for various ST traits. Studies suggested that design of SNP-based molecular markers for selection of the favorable alleles/haplotypes will hasten marker-assisted introgression of salt tolerance from *S. pimpinellifolium* into cultivated tomato (Rao et al. 2015). The use of functional markers derived from SNPs present in candidate gene in molecular plant breeding is more advantageous than linked markers, and hence, the molecular information can be used confidently across breeding programs to select promising alleles for a trait of interest (Bagge and Lubberstedt 2008; Kujur et al. 2016).

Patil et al. (2015) conducted genomic-assisted haplotype analysis and the developed high-throughput SNP markers for ST in soybean. *GmCHX1* gene has been identified as salt-tolerant determinant in soybean (Qi et al. 2014). High-quality whole-genome re-sequencing on 106 diverse soybean lines identified three major structural variants and allelic variation in the promoter and genic regions of the *GmCHX1* gene. When identified SNP markers were validated using a panel of 104 soybean lines and an interspecific biparental population (F8) from PI483463 x Hutcheson, a strong correlation was observed between the genotype and salt treatment phenotype. SNP observed in *GmCHX1* gene precisely identified salt-tolerant/salt-sensitive soybean genotypes. The newly developed SNP markers and genotype information will accelerate marker-assisted selection programs and greatly benefit soybean breeders in the development of salt-tolerant varieties (Patil et al. 2015).

15.6 Conclusion and Future Perspectives

Increasing soil salinity problems causes major concern in crop production. Successful exploitation of QTL dissection and its applications to enhance crop productivity under salinity conditions will largely depend on their successful integration with conventional breeding methodologies and a thorough understanding of the biochemical and physiological processes limiting yield under such adverse conditions (Tuberosa and Salvi 2007). Therefore, improving salinity tolerance using molecular genetic aspects such as identification of molecular markers associated with salinity tolerance QTLs in crop plants has been considered as an important but largely unfulfilled aim of modern agriculture. Although conventional markers such as RFLP, RAPD, AFLP, and SSRs have been employed in MAS for conferring salinity tolerance in crops, the usefulness of gene-based functional and SNP markers in identification of candidates for salt-tolerant QTLs in crop plants offer new approaches to understand genomic control of salt tolerance. However, SNP-based identification of QTLs in ST studies is still at their first phase. Therefore, high-throughput technologies and complete genome sequencing projects may offer new approaches to understand genomic control of salt tolerance and solve this persistent problem. This further may entail efficient marker-assisted breeding for salinity stress-tolerant crops with economically acceptable yield under saline conditions which further help in selection and quicker variety release.

Acknowledgment The author is thankful to Prof. Ameeta Ravikumar, HoD, Department of Biotechnology, Savitribai Phule Pune University, for her kind support during writing this book chapter.

References

Aghaei K, Ehsanpour AA, Balali G, Mostajeran A (2008) *In vitro* screening of potato (*Solanum tuberosum* L.) cultivars for salt tolerance using physiological parameters and RAPD analysis. American-Eurasian J Agric Environ Sci 3:159–164

Ali Md N, Yeasmin K, Gantait S, Goswami R, Chakraborty S (2014) Screening of rice landraces for salinity tolerance at seedling stage through morphological and molecular markers. Physiol Mol Biol Plants 20:411–423

Amin AY, Diab AA (2013) QTL mapping of wheat (*Triticum aestivum* L.) in response to salt stress. Int J Biotechnol Res 3:47–60

Andersen J, Lubberstedt T (2003) Functional markers in plants. Trends Plant Sci 18:554–560

Ashraf M, Foolad MR (2013) Crop breeding for salt tolerance in the era of molecular markers and marker-assisted selection. Plant Breed 132:10–20

Ashraf M, Akram M, Foolad MR (2012) Marker-assisted selection in plant breeding for salinity tolerance. In: Shabala S, Cuin TA (eds) Plant salt tolerance methods and protocols, methods in molecular biology. Springer/Humana press, Hatfield, Hertfordshire, UK, pp 305–334

Baby J, Jini D, Sujatha S (2010) Biological and physiological perspectives of specificity in abiotic salt stress response from various rice plants. Asian J Agri Sci 2:99–105

Bagge M, Lubberstedt T (2008) Functional markers in wheat: technical and economic aspects. Mol Breed 22:319–328

Barchi L, Lanteri S, Portis E, Acquadro A, Valè G, Toppino L, Gl R (2011) Identification of SNP and SSR markers in eggplant using RAD tag sequencing. BMC Genomics 12:304

Bimpong IK, Manneh B, Diop B, Ghislain K, Sow A, Amoah NKA et al (2014a) New quantitative trait loci for enhancing adaptation to salinity in rice from Hasawi, a Saudi landrace into three African cultivars at the reproductive stage. Euphytica 200:45–60

Bimpong IK, Manneh B, Namaky R, Diaw F, Amoah NKA, Sanneh B et al (2014b) Mapping QTLs related to salt tolerance in rice at the young seedling stage using 384-plex single nucleotide polymorphism SNP, marker sets. Mol Plant Breed 5:47–63

Bizimana JB, Luzi-Kihupi A, Murori RW, Singh RK (2017) Identification of quantitative trait loci for salinity tolerance in rice (*Oryza sativa* L.) using IR29/Hasawi mapping population. J Genet 96:571–582

Bonilla P, Dvorak J, Mackill D, Deal K, Gregorio G (2002) RFLP and SSLP mapping of salinity tolerance genes in chromosome 1 of rice (*Oryza sativa* L.) using recombinant inbred lines. Philipp Agric Sci 85:68–76

Botstein D, White RL, Skolnick M, Davis RW (1980) Construction of a genetic linkage map in man using restriction fragment length polymorphisms. Am J Hum Genet 32:314–331

Braglia L, Manca A, Mastromauro F, Breviario D (2010) cTBP: a successful intron length polymorphism (ILP)-based genotyping method targeted to well-defined experimental needs. Diversity 2:572–585

Byrt CS, Platten JD, Spielmeyer W, James RA, Lagudah ES, Dennis ES, Tester M, Munns R (2007) HKT1;5-like cation transporters linked to Na+ exclusion loci in wheat, *Nax2* and *Kna1*. Plant Physiol 143:1918–1928

Cai C, Wu S, Niu E, Cheng C, Guo W (2017) Identification of genes related to salt stress tolerance using intron-length polymorphic markers, association mapping and virus-induced gene silencing in cotton. Sci Rep 7:528

Chen ZH, Zhou MX, Newman IA, Mendham NJ, Zhang GP, Shabala S (2007) Potassium and sodium relations in salinised barley tissues as a basis of differential salt tolerance. Funct Plant Biol 34:150–162

Choi HK, Kim DJ, Uhm T, Limpens E, Lim H, Mun J-H, Kalo P, Penmetsa RV, Seres A, Kulikova O, Roe BA, Bisseling T, Kiss GB, Cook DR (2004) A sequence-based genetic map of *Medicago truncatula* and comparison of marker colinearity with *M. sativa*. Genetics 166:1463–1502

Collard BC, Mackill DJ (2008) Marker-assisted selection: an approach for precision plant breeding in the twenty first century. Philos Trans R Soc Lond Ser B Biol Sci 363:557–572

Cuadrado A, Schwarzacher T (1998) The chromosomal organization of simple sequence repeats in wheat and rye genomes. Chromosoma 107:587–594

De Leon TB, Linscombe S, Subudhi PK (2016) Identification and validation of QTLs for seedling salinity tolerance in introgression lines of a salt tolerant rice landrace 'Pokkali'. PLoS One. https://doi.org/10.1371/journal.pone.0175361

DeRose-Wilson L, Gaut BS (2011) Mapping salinity tolerance during *Arabidopsis thaliana* germination and seedling growth. PLoS One 6. https://doi.org/10.1371/journal.pone.0022832

Du L, Cai C, Wu S, Zhang F, Hou S, Gua W (2016) Evaluation and exploration of favorable QTL alleles for salt stress related traits in cotton cultivars (*G. hirsutum* L.) PLoS One 11:e0151076

Dubcovsky J, Santa Maŕia G, Epstein E, Luo MC, Dvorak J (1996) Mapping of the K^+/Na^+ discrimination locus *Kna1*in wheat. Theor Appl Genet 92:448–454

Foolad MR (2006) Tolerance to abiotic stresses. In: Razdan MK, Mattoo AK (eds) Genetic improvement of solanaceous crops, vol Volume 2, Tomato. CRC Press, Taylor and Francis, pp 521–592

Foolad MR, Chen FQ, Lin GY (1998) RFLP mapping of QTLs conferring salt tolerance during germination in an interspecific cross of tomato. Theor Appl Genet 97:1133–1144

Ganal MW, Polley A, Graner EM, Plieske J, Wieseke R, Luerssen H, Durstewitz G (2012) Large SNP arrays for genotyping in crop plants. J Bioscience 37:821–828

Gharsallah C, Abdelkrim AB, Fakhfakh H, Salhi-Hannachi A, Gorsane F (2016) SSR marker-assisted screening of commercial tomato genotypes under salt stress. Breed Sci 66:823–830

Ghomi K, Rabiei B, Sabouri H, Sabouri A (2013) Mapping QTLs for traits related to salinity tolerance at seedling stage of Rice (*Oryza sativa* L.): an Agrigenomics study of an Iranian Rice population. OMICS Int J Integr Biol 17:242–251

Greenway H, Munns R (1980) Mechanism of salt tolerance in nonhalophytes. Annu Rev Plant Physiol 31:149–190

Gregorio GB, Senadhira D, Mendoza RD, Manigbas NL, Roxas JP, Guerta CQ (2002) Progress in breeding for salinity tolerance and associated abiotic stresses in rice. Field Crop Res 76:91–101

Gupta B, Huang B (2014) Mechanism of salinity tolerance in plants: physiological, biochemical, and molecular characterization. Int J Genomics 2014. https://doi.org/10.1155/2014/701596

Gupta S, Kumari K, Das J, Lata C, Puranik S, Prasad M (2011) Development and utilization of novel intron length polymorphic markers in foxtail millet (*Setaria italica* (L.) P. Beauv.) Genome 54:586–602

Gupta SK, Bansal R, Gopalakrishna T (2012) Development of intron length polymorphism markers in cowpea [*Vigna unguiculata* (L.) Walp.] and their transferability to other Vigna species. Mol Breed 30:1363–1370

Hanin M, Ebel C, Ngom M, Laplaze L, Masmoudi K (2016) New insights on plant salt tolerance mechanisms and their potential use for breeding. Frontiers in. Plant Sci 6. https://doi.org/10.3389/fpls.2016.01787

Haq TU, Gorham J, Akhtar J, Akhtar N, Steele KA (2010) Dynamic quantitative trait loci for salt stress components on chromosome 1 of rice. Funcl. Plant Biol 37:634–645

Heiskanen J (2006) Estimated aboveground tree biomass and leaf area index in a mountain birch forest using ASTER satellite data. Int J Remote Sens 27:1135–1158

Hussain B, Lucas SJ, Ozturk L, Budak H (2017) Mapping QTLs conferring salt tolerance and micronutrient concentrations at seedling stage in wheat. Sci Report 7:15662. https://doi.org/10.1038/s41598-017-15726-6

Islam MR, Salam MA, Hassan L, Collard BCY, Singh RK, Gregorio GB (2011) QTL mapping for salinity tolerance at seedling stage in rice. Emir J Food Agric 23:137–146

Jaikishan I, Rajendrakumar P, Madhusudhana R, Elangovan M, Patil JV (2015) Development and utility of PCR-based intron polymorphism markers in sorghum [*Sorghum bicolor* (L.) Moench]. J Crop Sci Biotechnol 18:309–318

Jain M, Moharana KC, Shankar R, Kumari R, Garg R (2014) Genome wide discovery of DNA polymorphisms in rice cultivars with contrasting drought and salinity stress response and their functional relevance. Plant Biotechnol J 12:253–264

James RA, Davenport RJ, Munns R (2006) Physiological characterization of two genes for Na+ exclusion in durum wheat: *Nax1* and *Nax2*. Plant Physiol 142:1537–1547

Kalia RK, Rai MK, Kalia S, Singh R, Dhawan AK (2011) Microsatellite markers: an overview of the recent progress in plants. Euphytica 177:309–334

Khalil RMA (2013) Molecular and biochemical markers associated with salt tolerance in some *Sorghum* genotypes. World Appl Sci J 22:459–469

Khatab IA, Samah MA (2013) Development of agronomical and molecular genetic markers associated with salt stress tolerance in some barley genotypes. Curr Res J Biol Sci 5:198–204

Khlestkina EK (2014) Molecular markers in genetic studies and breeding. App Res 2014, 4: 236–244

Krishnamurthy SL, Sharma PC, Sharma SK, Batra V, Kumar V, Rao LVS, (2016) Effect of salinity and use of stress indices of morphological and physiological traits at the seedling stage in rice. Indian J Exper Biol 54:843–850

Kruglyak L (1997) The use of a genetic map of biallelic markers in linkage studies. Nat Genet 17:21–24

Kujur A, Upadhyaya HD, Bajaj D, Gowda CLL, Sharma S, Tyagi AK, Parida SK (2016) Identification of candidate genes and natural allelic variants for QTLs governing plant height in chickpea. Sci Rep 6:27968. https://doi.org/10.1038/srep27968

Kumar V, Singh A, Amitha Mithra SV, Krishnamurthy SL, Parida SK, Jain S et al (2015) Genome-wide association mapping of salinity tolerance in rice (*Oryza sativa*). DNA Res 22:133–145

Lateef DD (2015) DNA marker technologies in plants and applications for crop improvements. J Biosci Med 3:7–18

Läuchli A, Grattan SR (2007) The plant growth and development under salinity stress. In: Jenks MA et al (eds) Advances in molecular breeding toward drought and salt tolerant crops. The Netherlands, Springer, pp 1–32

Lee SY, Ahn JH, Cha YS, Yun DW, Lee MC, Ko JC et al (2006) Mapping of quantitative trait loci for salt tolerance at the seedling stage in rice. Mol Cells 21:192–196

Leonforte A, Sudheesh S, Cogan NOI, Salisbury PA, Nicolas ME, Materne M, Forster JW, Kaur S (2013) SNP marker discovery, linkage map construction and identification of QTLs for enhanced salinity tolerance in field pea (*Pisum sativum* L.) BMC Plant Biol 13:161

Li G, Quiros CF (2001) Sequence-related amplified polymorphism (SRAP). A new marker system based on a simple PCR reaction: its application to mapping and gene tagging in *Brassica*. Theor Appl Genet 103:455–461

Li C, Zhang G, Lance R (2007) Recent advances in breeding barley for drought and saline stress tolerance. In: Jenks MA et al (eds) Advances in molecular breeding toward drought and salt tolerant crops. Springer, The Netherlands, pp 603–626

Li J, Liu L, Bai Y, Zhang P, Finkers R, Du Y et al (2011) Seedling salt tolerance in tomato. Euphytica 178:403–414

Lin HX, Zhu MZ, Yano M, Gao JP, Liang ZW, Su WA et al (2004) QTLs for Na^+ and K^+ uptake of the shoots and roots controlling rice salt tolerance. Theor Appl Genet 108:253–260

Lindsay MP, Lagudah ES, Hare RA, Munns R (2004) A locus for sodium exclusion (*Nax1*), a trait for salt tolerance, mapped in durum wheat. Funct Plant Biol 31:1105–1114

Liu Y, He Z, Appels R, Xia X (2012) Functional markers in wheat: current status and future prospects. Theor Appl Genet 125:1–10

Liu X, Fan Y, Mak M, Babla M, Holford P et al (2017) QTLs for stomatal and photosynthetic traits related to salinity tolerance in barley. BMC Genomics 18:9. https://doi.org/10.1186/s12864-016-3380-0

Luo M, Zhao Y, Zhang R, Xing J, Duan X, Li J et al (2017) Mapping of a major QTL for salt tolerance of mature field-grown maize plants based on SNP markers. BMC Plant Biol 17:140

Machado RMA, Serralheiro RP (2017) Soil salinity: effect on vegetable crop growth. Management practices top and mitigate soil salinization. Horticulturae 3:30. https://doi.org/10.3390/horticulturae3020030

Mammadov J, Aggarwal R, Buyyarapu R, Kumpatla S (2012) SNP markers and their impact on plant breeding. Int J Plant Genomics 2012. https://doi.org/10.1155/2012/728398

Mano Y, Takeda K (1997) Mapping quantitative trait loci for salt tolerance at germination and the seedling stage in barley (*Hordeum vulgare* L.) Euphytica 94:263–272

Mardani Z, Rabiei B, Sabouri H, Sabouri A (2014) Identification of molecular markers linked to salt-tolerant genes at germination stage of rice. Plant Breed 133:196–202

Martinez-Beltran J,Manzur CL(2005) Overview of salinity problems in the world and FAO strategies to address the problem. In Proceedings of International Salinity Forum, Riverside, California, pp 311–313

Masood MS, Sciji Y, Shinwari ZK, Anwar R (2004) Mapping quantitative trait loci (QTLs) for salt tolerance in rice (*Oryza sativa*) using RFLPs. Pak J Bot 36:825–834

Masuka B, Araus JL, Das B, Sonder K, Cairns JE (2012) Phenotyping for abiotic stress tolerance in maize. J Integr Plant Biol 54:238–249

Ming-zhe Y, Jian-fei W, Hong-you C, Hu-qu Z, Hongsheng Z (2005) Inheritance and QTL mapping of salt tolerance in rice. Rice Sci 2:25–32

Mondal U, Khanom MSR, Hassan L, Begum SN (2013) Foreground selection through SSRs markers for the development of salt tolerant rice variety. J Bangladesh Agri Univ 11:67–72

Munns R, Tester M (2008) Mechanisms of salinity tolerance. Ann Rev Plant Biol 59:651–681

Oluoch G, Zheng J, Wang X, Khan MKR, Zhou Z et al (2016) QTL mapping for salt tolerance at seedling stage in the interspecific cross of *Gossypium tomentosum* with *Gossypium hirsutum*. Euphytica 209:223–235

Patil G, Do T, Vuong TD, Valliyodan B, Lee J-D, Chaudhary J, Shannon JG, Nguyen T (2015) Genomic-assisted haplotype analysis and the development of high-throughput SNP markers for salinity tolerance in soybean. Sci Rep 6:19199

Pitman MG, Läuchli A (2002) Global impact of salinity and agricultural ecosystems. In: Läuchli A, Lüttge U (eds) Salinity: environment – plants – molecules. Kluwer Academic Publishers, Dordrecht, pp 3–20

Pushpavalli R, Krishnamurthy L, Thudi M, Gaur PM, Rao MV, Siddique KHM et al (2015) Two key genomic regions harbour QTLs for salinity tolerance in ICCV 2 × JG 11 derived chickpea (*Cicer arietinum* L.) recombinant inbred lines. BMC Plant Biol 15:124. https://doi.org/10.1186/s12870-015-0491-8

Qi X et al (2014) Identification of a novel salt tolerance gene in wild soybean by whole-genome sequencing. Nat Commun 5:4340

Quesada V, García-Martínez S, Piqueras P, Ponce MR, Micol JL (2002) Genetic architecture of NaCl tolerance in *Arabidopsis thaliana*. Plant Physiol 130:951–963

Rafalski A (2002) Applications of single nucleotide polymorphisms in crop genetics. Curr Opin Plant Biol 5:94–100

Rana V, Ram S, Sendhil R, Nehra K, Sharma I (2015) Physiological, biochemical and morphological study in wheat (*Triticum aestivum* L.) RILs population for salinity tolerance. J Agric Sci 7:119–128

Rao MVS, Kumar PK, Manga V, Mani NS (2007) Molecular markers for screening salinity response in *Sorghum*. IJBT 6:271–273

Rao ES, Kadirvel P, Symonds RC, Ebert AW (2013) Relationship between survival and yield related traits in *Solanum pimpinellifolium* under salt stress. Euphytica 190:215–228

Rao ES, Kadirvel P, Symonds RC, Geethanjali S, Thontadarya RN, Ebert AW (2015) Variations in DREB1A and VP1.1 genes show association with salt tolerance traits in wild tomato (*Solanum pimpinellifolium*). PLoS One. https://doi.org/10.1371/journal.pone.0132535

Reddy INBL, Kim B-K, Yoon I-N, Kim K-H, Kwon T-R (2017) Salt tolerance in Rice: focus on mechanisms and approaches. Rice Sci 24:123–144

Richards RA (1993) Should selection for yield in saline regions be made on saline or non saline soils. Euphytica 32:431–438

Robarts DWH, Wolfe AD (2014) Sequence-related amplified polymorphism (SRAP) markers: a potential resource for studies in plant molecular biology. Appl Plant Sci 2. https://doi.org/10.3732/apps.1400017

Saha P, Chatterjee P, Biswas AK (2010) NaCl pretreatment alleviates salt stress by enhancement of antioxidant defense and osmolyte accumulation in mungbean (*Vigna radiata* L. Wilczek). Indian J Exp Biol 48:593–600

Saleh B (2016) DNA changes in cotton (*Gossypium hirsutum* L.) under salt stress as revealed by RAPD marker. Adv Hort Sci 30:13–21

Scholberg JMS, Locascio SJ (1999) Growth response of snap bean and tomato as affected by salinity and irrigation method. Hortscience 34:259–264

Sehrawat N, Bhat KV, Kaga A, Tomooka N, Yadav M, Jaiwal PK (2014) Development of new gene-specific markers associated with salt tolerance for mungbean (*Vigna radiata* L. Wilczek). Spanish J Agric Res 12:732–741

Shehata AI, Messaitfa ZH, Quraini FE, Rizwana H, Hazzani AAA, Wahabi MSE (2014) Genotypic of salt stressed sunflower *(Helianthus annuus*). Int J Pure App Biosci 2:40–61

Shrivastava P, Kumar R (2015) Soil salinity: a serious environmental issue and plant growth promoting bacteria as one of the tools for its alleviation. Saudi J Biol Sci 22:123–131

Singh DP, Singh BB (2011) Breeding for tolerance to abiotic stresses in mungbean. J Food Legumes 24:83–90

Singh RK, Gregorio GB, Jain RK (2007) QTL mapping for salinity tolerance in rice. Physiol Mol Biol Plants 13:87–99

Smethurst CF, Gill WM, Shabala S (2009) Using excised leaves to screen lucerne for salt tolerance: physiological and cytological evidence. Plant Signal Behav 4:39–41

Sónia N, Cecília MA, Ines SP et al (2013) New allelic variants found in key rice salt-tolerance genes: an association study. Plant Biotechnol J 11:87–100

Takehisa H, Shimodate T, Fukuta Y et al (2004) Identification of quantitative trait loci for plant growth of rice in paddy field flooded with salt water. Field Crop Res 89:85–95

Tal M, Shannon MC (1983) Salt tolerance in the wild relatives of the cultivated tomato: responses of *Lycopersicon esculentum, L. cheesmaniae, L. peruvianum, Solanum pennellii* and F_1 hybrids to high salinity. Aust J Plant Physiol 10:109–117

Tang J, Baldwin SJ, Jacobs JME, Linden CG, Voorrips RV, Leunissen JAM et al (2008) Large-scale identification of polymorphic microsatellites using an *in silico* approach. BMC Bioinformatics 9:374

Telem RS, Wani SH, Singh NB, Sadhukhan R, Mandal N (2016) Single nucleotide polymorphism (SNP) marker for abiotic stress tolerance in crop plants. In: Al-Khayri JM et al (eds) Advances in plant breeding strategies: agronomic, abiotic and biotic stress traits. Springer, Switzerland, pp 327–343

Thomson MJ, De Ocampo M, Egdane J, Rahman MA, Sajise AG, Adorada DL et al (2010) Characterizing the *Saltol* quantitative trait locus for salinity tolerance in rice. Rice 3:148–160

Tuberosa R, Salvi S (2007) Dissecting QTLs for tolerance to drought and salinity. In: Jenks MA et al (eds) Advances in molecular breeding toward drought and salt tolerant crops, Springer, The Netherlands, pp 381–411

Turki N, Shehzad T, Harrabi M, Okuno K (2015) Detection of QTLs associated with salinity tolerance in durum wheat based on association analysis. Euphytica 201:29–41

Vadez V, Krishnamurthy L, Thudi M, Chetukuri A, Timothy DC, Turner NC, Kadambot HMS, Pooran MG, Varshney RK (2012) Assessment of ICCV 2 × JG 62 chickpea progenies shows sensitivity of reproduction to salt stress and reveals QTL for seed yield and yield components. Mol Breed 30:9–21

Varshney RK, Hiremath PJ, Lekha P, Kashiwagi J, Balaji J, Deokar AA, Vadez V, Xiao Y, Srinivasan R, Gaur PM, Siddique KH, Town CD, Hoisington DA (2009) A comprehensive resource of drought- and salinity- responsive ESTs for gene discovery and marker development in chickpea (*Cicer arietinum* L). BMC Genomics 2009 10:523. https://doi.org/10.1186/1471-2164-10-523

Villalta I, Reina-Sánchez A, Bolarín MC et al (2008) Genetic analysis of Na^+ and K^+concentrations in leaf and stem as physiological components of salt tolerance in tomato. Theor Appl Genet 116:869–880

Wang X, Zhao X, Zhu J, Wu W (2005) Genome-wide investigation of intron length polymorphisms and their potential as molecular markers in rice (*Oryza sativa* L.) DNA Res 12:417–427

Wang X, Lu X, Wang J, Wang D, Yin Z, Fan W, Wang S, Ye W (2016) Mining and analysis of SNP in response to salinity stress in upland cotton (*Gossypium hirsutum* L.) PLoS One. https://doi.org/10.1371/journal.pone.0158142

Wang BH, Zhu P, Yuan YL, Wang CB, Yu CM, Zhang HH, Zhu XY, Wang W, Yao CB, Zhuang ZM, Li P (2014) Development of EST-SSR markers related to salt tolerance and their application in genetic diversity and evolution analysis in Gossypium. Genet Mol Res 13:3732–3746

Xiang C, Du J, Zhang P, Cao G, Wang D (2015) Preliminary study on salt resistance seedling trait in maize by SRAP molecular markers. In: Zhang T C, Nakajima M (eds) Advances in applied biotechnology, lecture notes in electrical engineering, vol 332. Springer-Verlag, Berlin, pp 11–18

Yokoi S, Bressan RA, Hassegawa PM (2002) Salt Stress tolerance of Plant. JIRCAS working report pp. 25–33

Yu JM, Zhang ZW, Zhu CS, Tabanao DA, Pressoir G et al (2009) Simulation appraisal of the adequacy of number of background markers for relationship estimation in association mapping. Plant Genome 2:63–77

Yumurtaci A, Sipahi H, Zhao L (2017) Genetic analysis of microsatellite markers for salt stress in two contrasting maize parental lines and their RIL population. Acta Bot Croat 76:55–63

Zhou G, Johnson P, Ryan PR, Delhaize E, Zhou M (2012) Quantitative trait loci for salinity tolerance in barley (*Hordeum vulgare* L.) Mol Breed 29:427–436